To My father, Dick Parker, who taught me to love agriculture, and to My mother, Louise Parker, who taught me to love life

CONTENTS

Preface..xi
Acknowledgments...xiii
About the Author..xiv

CHAPTER 1 *Aquaculture Basics and History*..2
    Introduction and Definitions...3
    Historical Patterns and Practices..6
    Aquaculture Activities...18
    The Future of Aquaculture...20

CHAPTER 2 *Aquatic Plants and Animals*..28
    U.S. Aquatic Plant Species...29
    U.S. Aquatic Animal Species..29
    Common Characteristics of Aquatic Species...35
    Structure and Functions of Aquatic Animals and Plants..............................40

CHAPTER 3 *Marketing Aquaculture*..56
    International Production..57
    Consumption..57
    Marketing Basics..57
    Processing..70
    Inspection..74

CHAPTER 4 *Management Practices for Finfish*......................................90
    Spawning...91
    Sex Determination..95
    Finfish..96
    Controls..155
    Genetics...159

CHAPTER 5 Recreational Fishing Industry 168
- Brief History of Recreational Fishing 169
- Methods of Fishing 172
- Types of Fish 174
- Impact of Recreational Fishing 175
- U. S. Fish & Wildlife Service 177
- Pacific Salmon 182

CHAPTER 6 Raising Ornamental Fish 186
- Sources of Species 187
- Specific Ornamental Fish 191
- Florida and Tropical Ornamental Fish Culture 197

CHAPTER 7 Management Practices for Crustaceans and Mollusks 200
- Culture Status 201
- Other Commercial Species 229

CHAPTER 8 Management Practices for Alligators, Frogs, and Plants 232
- Alligators 233
- Frogs 239
- Aquatic Plants 242

CHAPTER 9 Fundamentals of Nutrition in Aquaculture 252
- Nutrition of Fish 253
- Energy Requirements 256
- Protein Requirements 261
- Vitamin Requirements 263
- Mineral Requirements 265
- Other Dietary Components 269

CHAPTER 10 Feeds and Feeding 278
- Diet Formulation and Processing 279
- Feeding Aquatic Animals 283
- Other Warmwater Fish 297
- Time of First Feeding 298
- Feed Calculations 299
- Aquatic Plants 304

CHAPTER 11 *Health of Aquatic Animals* 310

    Health Management 311

    Stress and Disease.. 311

    Disease Resistance 314

    Protective Barriers Against Infection 314

    Disease Types 316

    Parasitic Diseases... 318

    Fungus.. 326

    Bacterial Diseases 326

    Viral Diseases... 331

    Noninfectious Diseases 333

    Determining the Presence of Disease 337

    Disease Treatment... 339

    Calculating Treatments 341

    Immunization 342

CHAPTER 12 *Water Requirements for Aquaculture* 348

    Water Qualities, Measurements, and Alterations 349

    Other Factors 366

    Obtaining Water... 368

    Managing Water.. 373

    Calculating Treatments 378

    Disposing of Water 381

CHAPTER 13 *Aquatic Structures and Equipment* 390

    Ponds.. 391

    Construction of Levee-Type Ponds 399

    Raceways and Tanks 404

    Cages and Pens 408

    Other Major Equipment 415

CHAPTER 14 *Aquariums* 430

    Background.. 431

    Fish for the Aquarium 431

    Choosing and Establishing an Aquarium 433

    Managing the Aquarium.. 437

    Using a Beginner Aquarium... 439

    Water Quality 441

    Aquarium Checklist.. 442

CHAPTER 15 Recirculating Systems ... 446
- Background ... 447
- System Design ... 447
- System Management ... 456

CHAPTER 16 Sustainable Aquaculture and Aquaponics ... 464
- Attempts to Define "Sustainable" ... 465
- Standards of Sustainable Aquaculture ... 466
- Aquaponics ... 479
- Sustainable Standards Scorecard ... 481

CHAPTER 17 Aquaculture Business ... 486
- Counting the Cost ... 487
- Managing the Business ... 489
- Planning—The Secret of Business Success ... 496
- Setting Goals for Business Management Decisions ... 503
- Business and Risky Decisions ... 504
- Business Structures ... 506
- Records Improve Profitability ... 508
- Using an Accounting System for Analysis ... 517
- Computers and Management Decisions ... 519
- Obtaining Credit ... 520
- Human Resources ... 523
- Business Managers of Tomorrow ... 525

CHAPTER 18 Career Opportunities in Aquaculture ... 536
- General Skills and Knowledge ... 537
- Intangible Skills ... 542
- Entrepreneurship ... 545
- Jobs in Aquaculture ... 546
- Supervised Agricultural Experience ... 554
- Education and Experience ... 555
- Identifying a Job ... 556
- Getting a Job ... 558

Appendix ... 569
Glossary ... 589
Index ... 623

PREFACE

WRITING A TEXTBOOK ON AQUACULTURE is a lot like writing a textbook on all of agriculture. If this textbook included all of aquaculture, it would be several volumes. Deciding what to include and how much was a challenge. With the explosion of information, those who survive and thrive in the future must learn how to find information and how to use it for their circumstances. The author hopes that this book will help train the reader to find, evaluate, and use information—to learn to learn.

CONTENT AND ORGANIZATION

The 18 chapters of this third edition include five new chapters: Chapter 5, Recreational Fishing Industry; Chapter 6, Raising Ornamental Fish; Chapter 14, Aquariums; Chapter 15, Recirculating Systems; and Chapter 16, Sustainable Aquaculture and Aquaponics.

Chapters 1 and 2 introduce the role of aquaculture in the past, present, and future. Aquatic plants and animals are introduced early on in Chapter 2, but before getting into specifics about each, the book discusses marketing in Chapter 3. Without markets, aquaculture could not continue to grow and improve. Management practices for different groups of species and industry components are described in Chapters 4, 5, 6, 7, and 8. The management practices are not meant to be all inclusive or absolute—only a starting point for knowledge to grow. Chapters 9 and 10 cover nutrition principles, feeds, and feeding practices of finfish. Because feeding is a part of management, it was also included in Chapters 4, 6, 7 and 8.

Health and water quality are interrelated and important to the success of aquaculture. These are covered in Chapters 11 and 12, respectively. Aquatic structures such as ponds, raceways, and pen including some of the unique equipment of aquaculture are discussed in Chapter 13. Chapters 14 and 15 cover specific types of aquatic structures—aquariums and recirculating systems.

These days no book would be complete without some discussion of sustainability. Chapter 16 introduces the concept of sustainability through 12 standards of sustainable aquaculture. The chapter provides some detail on aquaponics—a system that could exemplify sustainable concepts.

Getting an aquaculture business going requires some business savvy that is introduced in Chapter 17. Finally, anyone with a knowledge of and a love of aquaculture will want to get a job in or make a career of aquaculture. Chapter 18 steers the reader toward finding a job or making a career of aquaculture.

NEW IN THE THIRD EDITION

The third edition is in full color. New and updated information in the third edition includes charts, graphs, and various tables in many of the chapters, and specific chapters on recreational fishing, ornamental fish, aquariums, recirculating systems, and sustainability. Many chapters provide more URLs for websites on the Internet.

SCIENCE IN AQUACULTURE

Anyone who attempts to learn of aquaculture soon realizes how much science is involved. Aquaculture demands a reasonable understanding of chemistry to deal with water quality. A species cannot be cultured until its biology is known. Being able to produce significant numbers of an aquatic species for culture requires a thorough understanding of reproductive life cycles. To recognize healthy animals and prevent diseases, an understanding of anatomy and physiology is necessary. Feed costs represent a significant share of the cost of production, so an understanding of the science of nutrition is essential.

FEATURES

Each chapter and each feature must be used as a whole. Each part complements the others.

An education prepares students for a productive life. Preparation is difficult without knowing what is required. Each chapter in this book starts with a list of learning objectives. These help the student identify what concepts are really important from all the information in the chapter.

The beginning of each chapter also features a list of key words. Knowing the meaning of these key words is essential to reading and understanding the chapter. Many of the words are defined within the text and all are defined in the glossary.

Throughout the book, tables, charts, graphs, and illustrations provide quick and understandable access to information without wading through excess words. Students will quickly learn how to read these and grasp the information they contain.

Knowledge and information alone are useless unless they can be applied. In the Knowledge Applied section at the end of each chapter, students and instructors will find opportunities for learning by doing. For more information the student can go to the list of Learning/Teaching Aids. Also, at the end of each chapter students can test their understanding by answering the questions.

Besides the supplemental information on aquatic species, the appendix contains helpful tables with information for converting units of measure, and for making contact with the aquaculture industry and agencies affecting aquaculture. Also, the appendix lists the web addresses (URLs) for agencies and other Internet sites.

SUPPLEMENTAL TEACHING

Besides the textbook other supporting material for teaching aquaculture includes the Instructor's Guide, a Lab Manual (correlated to the textbook chapters) Lab Manual Instructor's Guide, Lab Manual CD-ROM, and Class Master.

The Instructor's Guide provides an Overview and Summary of each chapter; Chapter Objectives; Suggested Lesson Developments; and Study/Review Answer Key. In addition, it includes three appendices: Use of the Internet; Supply Companies; and Suggested Resource Books.

The Lab Manual consists of 20 laboratory exercises written to enhance learning of the material in the textbook *Aquaculture Science, Third Edition.*

The Lab Manual Instructor's Guide contains a description of the Purpose of the lab, discussion of the Preparation needed to perform the lab; a Table Value and Notes portion describing potential outcomes for the labs, and finally an Answers to Analysis Questions section.

The Lab Manual CD-ROM to Accompany Aquaculture Science contains pdf's of both the Lab Manual and Lab Manual Instructor's Guide. This product was designed so that instructor's can pick and choose what labs to incorporate into their class. It also allows the teacher the flexibility to print up as many copies of the labs as they need during the lifetime of their edition.

The ClassMaster is a new resource for the Aquaculture Science text. Contained on the ClassMaster is the Instructor's Guide in pdf form; over 500 instructor slides in PowerPoint; a 600 question ExamView testbank; student worksheets to accompany each chapter; an Image Library of all the photos contained in the text and a correlation guide to Delmar's Introduction to Agriscience DVD Series.

All are available from Delmar Cengage Learning.

ACKNOWLEDGMENTS

THROUGH THREE EDITIONS THIS BOOK would still be a dream or idea without the help and support of Marilyn, wife to the author, mother, and now grandmother. As a true friend and partner she critiques ideas, types parts of the manuscript, writes questions and answers, organizes artwork, takes photographs, and checks format. She is a full partner in all aspects of the author's life.

Appreciation also goes to the author's immediate and extended family who are understanding and realize that "we (the parents) aren't much fun during a book revision!"

Unless otherwise noted, the photographs in the book were taken by the author or by Marilyn Parker.

The author appreciates the support, help, and encouragement of Ben Penner, Chris Gifford, and the rest of the Delmar team.

Delmar and the author also wish to express their thanks to the content reviewers. Their input and expertise added greatly to this new edition.

Joe Rasberry
Florala High School
Florala, AL

Hans Toft
Cape May County Technical School
Cape May, NJ

Chris Clemons
Highland High School
Highland, IL

Alan Godbey
Mercer High School
Harrodsburg, KY

Bryan Duncan
Coeur d'Alene High School
Coeur d'Alene, ID

Nathan Papendorf
Westby Area High School
Westby, WI

Darren Farmer
Richmond High School
Richmond, MD

ABOUT THE AUTHOR

R.O. (RICK) PARKER GREW UP on an irrigated farm in southern Idaho. His love of agriculture guided his education. Starting at Brigham Young University, he received his bachelor's degree and then moved to Ames, Iowa, where he finished a Ph.D. in animal physiology at Iowa State University. After completing his Ph.D., he and his wife, Marilyn, and their children moved to Edmonton, Alberta, Canada, where he completed a post-doctorate at the University of Alberta. The next move was to Laramie, Wyoming, where he was a research and teaching associate at the University of Wyoming.

After a stint as a co-author with M.E. Ensminger, he served as a division director and instructor at the College of Southern Idaho (CSI) in Twin Falls for 19 years. As director he worked with faculty in agriculture, information technology, drafting, marketing and management, and electronics. Dr. Parker also taught computer classes, biology, and agriculture classes at CSI. As an educator his motto is: "I hear; I forget; I see; I remember; I do; I understand."

Dr. Parker is the editor for the peer-reviewed NACTA Journal, a journal dedicated to the scholarship of teaching and learning (www.nactateachers.org). Additionally, he serves as the director for AgrowKnowledge (www.agrowknow.org), the National Resource Center for Agriscience and Technology Education, a project funded by the National Science Foundation (DUE #0802510) and located at Kirkwood Community College in Cedar Rapids, IA.

Dr. Parker is also the author of numerous online lessons, booklets, and these other Delmar/Cengage texts: *Introduction to Plant Science, Fundamentals of Plant Science* (with Marihelen Glass), *Plant & Soil Science: Fundamentals and Applications, Introduction to Food Science* and *Equine Science* (3rd edition).

He and Marilyn, his wife of 41 years, live in southern Idaho on his great-grandfather's small farm of 20 acres. They are the parents of eight children and grandparents to 22.

Aquaculture is arguably the only way to satisfy an increasing demand for fish and seafood products. Fortunately, as in the past, aquaculture can maintain its status as the fastest growing agricultural industry in the United States by continuing to successfully meet challenges

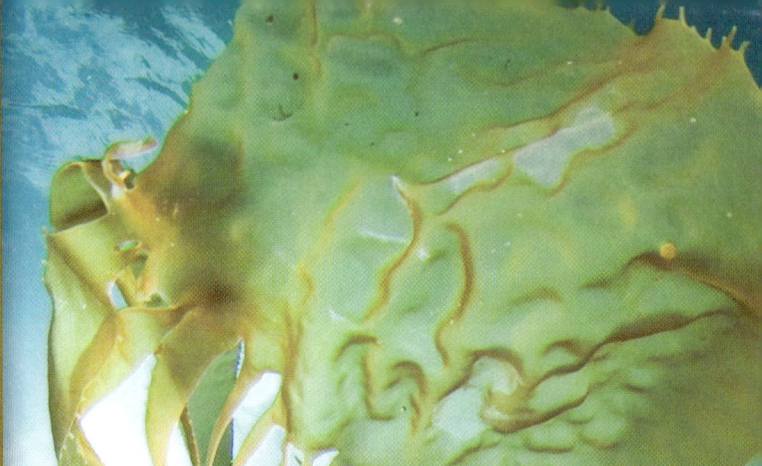

CHAPTER 1

Aquaculture Basics and History

OBJECTIVES

After completing this chapter, the student should be able to:

➤ Explain the development of aquaculture as a part of agriculture

➤ Name three civilizations that practiced aquaculture more than 200 years ago

➤ Define aquaculture

➤ Compare traditional farming to aquaculture

➤ Discuss why aquaculture evolved from fishing practices

➤ Discuss how the catfish industry developed and why Mississippi leads in catfish production

➤ Explain why Idaho leads in trout production

➤ List five main activities that are a part of aquaculture but often become a separate industry

➤ Discuss how aquaculture is expanding and what the future holds for aquaculture

➤ Identify significant scientific events or people contributing to the development of aquaculture

➤ Explain the National Sea Grant Program and its role in scientific research

➤ Discuss the role of science and technology in the development of aquaculture

➤ Indicate the role of scientific research in the future of aquaculture

Understanding of this chapter will be enhanced if the following terms are known. Many are defined in the text, and others are defined in the glossary.

KEY TERMS

Agriculture	Husbandry
Aquaculture	Incubate
Aquifer	Larvae
Brackish water	Mariculture
Broodstock	Monoculture
Coldwater	Polyculture
Culture	Processing
Eggs	Salinity
Fingerlings	Seed
Freshwater	Self-feeders
Grow-out	Spawn
Harvesting	Warmwater
Hatchery	

INTRODUCTION AND DEFINITIONS

From prehistoric times to the present, two primary needs of humans persist—food and shelter. (See Figure 1-1.) Through time, however, the means of obtaining food and shelter change. As societies moved from hunting and gathering to the culturing of plants and animals for food, their shelter needs changed from temporary to permanent. Also, as societies learned to **culture** plants and animals for food, they generated food surpluses that allowed society members to pursue other priorities and stimulated the need for preserving and marketing surpluses.

Agriculture is the art, science, and business—the culture—of producing every kind of plant and animal useful to humans. Agriculture is the oldest and most important of all industries. It continues to evolve in conjunction with the evolving knowledge and needs of civilization.

Typically, agriculture evolves through four stages—
1. A hunting-gathering activity
2. An object of husbandry
3. A craft
4. A science and business

Agriculture includes not only the cultivation of the land but also dairy production, beef production, sheep production, swine production, and all other farming activities, including aquaculture. Examples of aquaculture include catfish farming, crawfish farming, trout farming, salmon ranching, and oyster culture.

Aquaculture is a relatively new word used to describe the art, science, and business of producing aquatic plants and animals useful to humans. Aquaculture is a type of agriculture. Fundamentally, aquaculture means farming in water instead of on land. Often, agriculture and aquaculture include all of the activities involved in producing plants and animals, the supplies and services needed, the **processing** and marketing, and other steps that deliver products to the consumer in the desired form.

The Food and Agriculture Organization (FAO) of the United Nations defines aquaculture as "the farming of aquatic organisms, including fish, mollusks, crustaceans, and aquatic plants. Farming implies some form of intervention in the rearing process to enhance production, such as regular stocking, feeding, and protection from predators. Farming also implies individual

FIGURE 1-1 Shelter and food have always been primary human needs. (Theodor de Bry engraving after a John White watercolor. Indians (Native Americans) fishing with weir and spears in a dugout canoe. The drawing was made somewhere in the region of the colony of Virginia, which included the modern state of North Carolina. published 1590 from 1585 drawing.)

or corporate ownership of the stock being cultivated." Aquaculture and farming have some similarities and some differences. Table 1-1 compares traditional farming to aquaculture.

Aquaculture occurs in these general environments—
- Warmwater aquaculture
- Coldwater aquaculture
- Mariculture or marine culture (saltwater).

Warmwater aquaculture is the commercial raising of stock that thrives in warm, often turbid (cloudy or opaque), **freshwater** with temperatures exceeding 70°F. Examples of warmwater species include catfish, crawfish (crayfish), baitfish, and many sport fish. **Coldwater** aquaculture involves the commercial production of stock that thrives in cool, clear freshwater with temperatures between 50° and 65°F. Trout and salmon are examples of coldwater aquaculture. Warmwater and coldwater are also generally

TABLE 1-1 COMPARISON OF TRADITIONAL FARMING TO AQUACULTURE	
Farming	**Aquaculture**
Occurs on land	Occurs in water
Limited by water supply	Limited by oxygen dissolved in water
Many plant and animal crops	Many plant and animal crops
Domesticated plants and animals	Wild and/or domesticated plants and animals

considered freshwater—no **salinity**. Shrimp, oysters, and seaweed cultures are examples of **mariculture** (marine culture) where the crop thrives in saltwater of various temperatures. The salinity of saltwater ranges from 30 to 35 parts per thousand (ppt) and the salinity of **brackish water** is 1 to 10 ppt. (See Figure 1-2.)

Aquaculture, like agriculture, involves controlled culture and an individual or individuals who own the crop. Fisheries differ from aquaculture but are involved in aquaculture. Fisheries involve hunting and general public access to the crop—fish—being hunted. Aquaculture enhances fisheries by providing fish to restock streams, lakes, and oceans. This makes sportfishing more enjoyable and stable and it helps ensure the economic success of commercial fisheries. (See Figure 1-3.)

Historical events that made aquaculture a viable, growing, and profitable enterprise are not always easy to identify. Aquaculture probably evolved through a combination of human observation and serendipity in several areas of the world at different times. Perhaps aquaculture developed from fishing practices that involved trapping fish and holding them for freshness, which led to trapping, holding, and feeding to maintain a food supply for a longer time. Once people saw that fish could be fed and held, they refined techniques to ensure a more constant supply of fish. Possibly, cage culture developed when fishers realized that their surplus catch could be held in baskets in the water. Pond culture likely developed when some fishers observed fish trapped in pools of water formed by a flood. Some aquaculture likely developed in conjunction with farming and irrigation, since irrigation provided structures and a source of water.

In the United States, this relatively new business continues to grow in production and value (Figure 1-4). Catfish production dominates the U.S. aquaculture output, accounting for about half the total production. As the demand for aquaculture products increases and technology is developed for different species, aquaculture will grow worldwide.

FIGURE 1-2 Three-fourths of the Earth's surface is covered with salt water. Surf fishing the outer banks of North Carolina.

FIGURE 1-3 Family fishing —father, daughter, and son enjoy an afternoon of fishing.

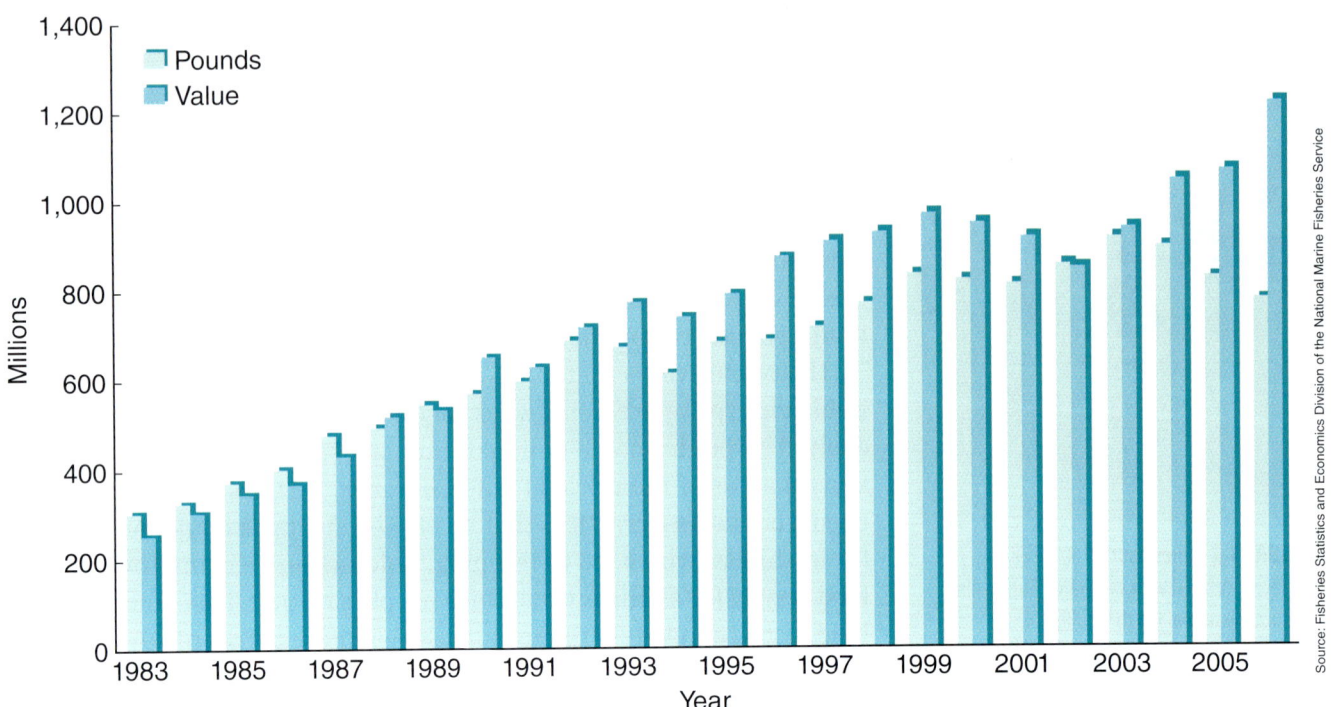

FIGURE 1-4 US Aquaculture increased rapidly in the decade from 1983 to 1993.

HISTORICAL PATTERNS AND PRACTICES

Aquaculture seems like a fairly new agricultural endeavor. But actually, many ancient civilizations developed some form of aquaculture.

Chinese Aquaculture

Aquaculture in China began around 3500 B.C. with the culture of the common carp. These carp were grown in ponds on silkworm farms. The silkworm pupae and feces provided supplemental food for the fish. Carp are hardy and easy to raise in freshwater ponds, and, because fish were an important part of life in ancient China, their culture developed very early. In Chinese, the word for fish means "surplus." Indeed, fish were equated with a bountiful harvest.

In 475 B.C., Fan-Li, a politician and administrator, wrote the oldest document on fish culture. Fan-Li was renowned for his self-taught expertise in carp culture. His document described methods for pond construction, **broodstock** selection, stocking, and managing ponds.

Emperor Li of the Tang Dynasty (A.D. 618–906) banned the culture of common carp because the word for carp was "Li." Emperors were considered sacred. Apparently, anything that shared the emperor's name was sacred and should not be eaten. This ban led the Chinese to develop

TABLE 1-2	POLYCULTURE WITH FOUR SPECIES OF CARP	
Species	**Location**	**Feed**
Grass carp	Topwater	Large vegetation near shore
Bighead carp	Midwater	Minute animals known as zooplankton
Silver carp	Midwater	Minute plants or algae known as phytoplankton
Mud carp	Bottom	Wide variety of plants and animals

polyculture—growing more than one species in the same water. Realizing that water is a three-dimensional habitat, all of the productive portions of a pond will not be used by just one species. Different species occupy different locations in the pond and feed on different food. Using polyculture, the Chinese cultured four species of carp, as Table 1-2 indicates.

China is generally considered to be the cradle of aquaculture. Chinese aquaculture evolved from providing food for the elite to supplying a food staple for the common people. (See Figure 1-5.)

In the Zhujiang Delta of South China, a dike-pond system of agriculture still exists after more than 500 years. Mulberry, sugarcane, fruit, forage crops, vegetables, silkworm breeding, and pig rearing integrate with fish rearing. Crops and crop residues are fed directly to bighead carp, grass carp, silver carp, mud carp, common carp, black carp, bream, and tilapia.

FIGURE 1-5 China is home to over a billion people, and aquaculture is still important in helping feed them all.

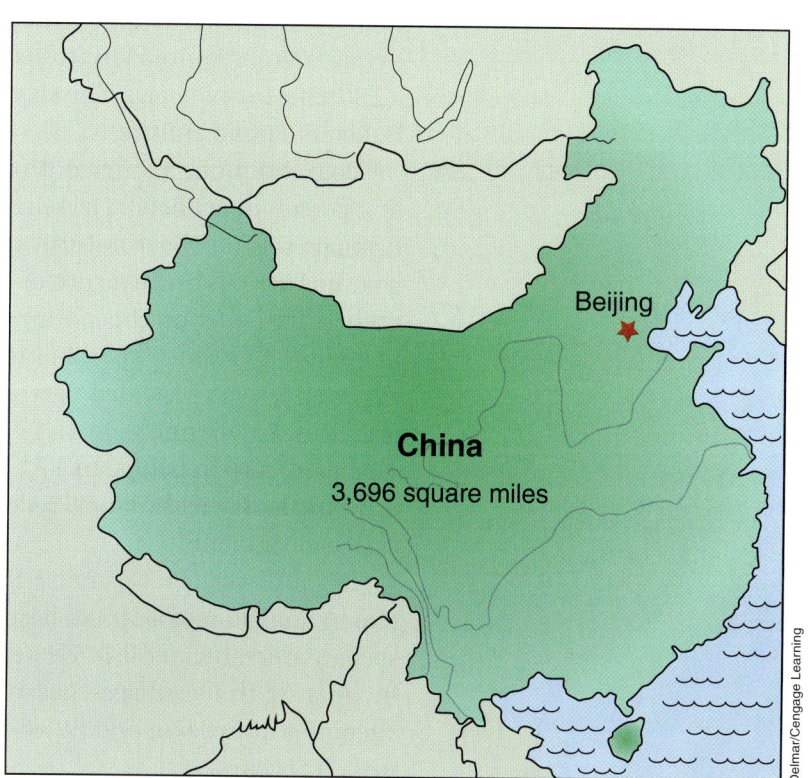

Common carp raised in the United States since the 1870s never gained a broad acceptance as a food fish. Since 1963, three species of Chinese carp—the grass carp, silver carp, and bighead carp—have been introduced into the United States for testing as farm fish.

Egyptian Aquaculture

For the ancient Egyptians, aquaculture seems to have evolved in tandem with the development of irrigation systems. Aquaculture in Egypt focused on tilapia, and developments seem consistent with those of carp in China. No written documents from early Egyptian aquaculture exist, but drawings in tombs, dated about 2000 B.C., show tilapia.

Roman Aquaculture

During the Roman Empire, fish were kept in ponds called "stews" next to the manors of the wealthy. Roman aquaculture focused on mullet and trout. Pliny the Elder recorded that saltwater and freshwater fish culture was practiced in Rome in the first century B.C.

During the Middle Ages, stew ponds became important for both monks and lay people, providing a source of fresh fish.

English and European Aquaculture

In central Europe, the history of pond-fish culture began at the close of the eleventh century and the beginning of the twelfth. Pond management in Bohemia, a part of the Czech Republic, peaked in the fourteenth century. Bohemia had about 185,000 acres of ponds for carp. From spawning to marketing required four to six years. Each acre was stocked with about 120 two-year-old carp. The sixteenth century became the golden era of Bohemian pond culture.

Dom Pinchon, a fourteenth-century French monk, possibly was the first person to artificially fertilize trout **eggs**. At the very least, he was the first person to collect natural **spawn** and **incubate** them in a hatching box. In 1600, John Taverner of England presented the first known comprehensive paper on the management of carp, bream, trench, and perch in ponds ("Certaine Experiments Concerning Fish and Fruite"). From his experiences, he provided very accurate details that mesh with today's practices. Furthermore, in 1613, Gervais Markham described in detail the raising of carp in ponds. In 1713, Sir Roger North presented a treatise on fish **husbandry** techniques, reflecting the ideas of Taverner but without acknowledgment.

In southwestern Germany, Stephan Jacobi published a series of articles in English and German describing his results in propagating several species of freshwater fish. He perfected the technique of spawning trout, incubating the fertilized eggs, and raising the fish. His methods did not become widespread, and little was written about fish husbandry for the next 100 years.

For a couple of reasons, many consider France as the birthplace of modern aquaculture. Two commercial fisherman, Joseph Remy and Antoine Gehin, became concerned by the decline of trout in French streams. Using their observations of trout, natural habits, and with the help of two well-known scientists, M. Miline Edwards and M. Coste, the first fish **hatchery** was established in 1852 in Huningue. With Coste as the director, the hatchery became well known and supplied trout eggs for most of central Europe.

During the 1800s, one more contribution added to the body of knowledge of aquaculture. In 1856, V. P. Vrasski developed the dry, or Russian, method of fertilizing trout eggs. Unfortunately, this method was not published until 1871, the same year that G. C. Atkins perfected the American method of dry fertilization.

Native Americans and Aquaculture

In the United States, almost every young student hears the story of how Indians instructed the pilgrims to include a fish with each corn seed planted, in order to improve the harvest. Native Americans knew a great deal about aquaculture, but much of that knowledge was not recorded. Today, we must surmise from things they left behind.

Hawaii

Hawaiian society centered around the ocean, agriculture, and aquaculture. By A.D. 400, an organized system of aquaculture existed in Hawaii. Extensive pond systems were developed, and the chiefs controlled aquaculture by leasing tracts of land to governors, who ensured that the ponds produced fish and were maintained.

Four types of agriculture/aquaculture existed in Hawaii:
1. Freshwater fish ponds fed by canals from streams
2. Taro ponds that irrigated agricultural plots
3. Brackish water fish ponds located near the shoreline
4. Seawall fish ponds along the shoreline, walled off from the ocean by human-made walls

Hawaii's integrated aquaculture existed until 1778, when Europeans arrived, disrupting the ancient religion of the people and removing the chiefs from ruling the ponds. A few of the old ponds are still used, and Hawaii maintains a prominent role in modern aquaculture.

America

In southern California, near the Salton Sea, the Cahuilla people built fish ponds. The Maya in Mesoamerica developed irrigation systems. Some evidence suggests that the Maya trapped or cultured fish in ponds or canals around 500 to 800 B.C.

U.S. Aquaculture Development

Theodatus Garlick collaborated with H. A. Ackley in working with brook trout. Their work inspired pioneers like S. H. Ainsworth, T. Norris, Seth Green, and Livingston Stone. Supposedly, Seth Green established the first public hatchery at Mumford, New York, in 1864. Early emphasis in Europe and the United States was on restocking depleted streams and lakes, not on the culture of food crops.

Soon, several New England states established fish and game commissions. As a result, private facilities increased. The combined interest of the state and private agencies lead to the formation of the American Fish Cultural Society in 1870. In 1885, this organization changed its name to the American Fisheries Society. The formation of this society is credited with providing the push to establish the U.S. Commission of Fish and Fisheries, which eventually became the U.S. Fish and Wildlife Service.

Whereas aquaculture developed in several areas in the United States, two species dominate U.S. aquaculture development—catfish and trout. In turn, two states dominate this area of development—Mississippi and Idaho.

Catfish and Mississippi

Commercial warmwater fish farming began in the late 1920s and early 1930s, initiated by a few individuals who raised minnows to supply the growing demand for baitfish for sportfishing. Shortly after World War II, the demand for minnows increased as the result of the boom in farm pond and reservoir construction and the many water conservation projects inspired by the dust bowl years of the 1930s. By the early 1950s, the number of producers increased enormously, and farmers also began to raise food fish such as buffaloes, bass, and crappies. Many of these early attempts at fish husbandry failed because the operators were not experienced in fish culture, ponds were not properly constructed, and low-value species were being raised.

From 1955 to 1959, the U.S. Fish and Wildlife Service, with funds from the Saltonstall-Kennedy Act for Commercial Fisheries, sponsored research on channel catfish at the University of Oklahoma. The purpose of the research was to develop better production methods in national fish hatcheries and to develop a basis for commercial fish farming. Other agencies and universities also became interested in channel catfish as a commercial and sport species. During the next few years, the service established three warmwater fish cultural research facilities: the Southeastern Fish Cultural Laboratory, Marian, Alabama, in 1959; the Fish Farming Experimental Station, Stuttgart, Arkansas, in 1960; and the Fish Farming Development Station, Rohwer, Arkansas, in 1963. These stations began research mainly with buffaloes, catfishes, and baitfish, but other species were added later.

Land grant universities, in cooperation with the U.S. Department of Agriculture, made substantial contributions to warmwater aquaculture research through Agricultural Experiment Stations in Alabama, Arkansas, California, Florida, Georgia, Hawaii, Louisiana, Mississippi, Puerto Rico, South Carolina, Tennessee, Texas, and the Virgin Islands.

Farmers received the technology through concentrated extension education efforts involving the Fish Farming Experimental Station, state cooperative extension services, and the Extension Service of the U.S. Department of Agriculture (http://www.usda.gov). Various research laboratories, national fish hatcheries, and fish cultural development centers of the U.S. Fish and Wildlife Services throughout the country also provided technical assistance to fish farmers, as have the agricultural experiment stations, Sea Grant Programs (http://www.seagrant.noaa.gov/), the U.S. Soil Conservation Service, universities, the Tennessee Valley Authority, the National Marine Fisheries Service, state departments of conservation, and various private foundations. (Refer to the Appendix for more detail on the agencies and services that supported and continue to support the development of aquaculture.)

The National Sea Grant Program, established in 1966, provides grants to U.S. universities that are designated sea grant colleges. The program, administered by the National Oceanic and Atmospheric Administration (NOAA) in the federal Department of Commerce, encourages those schools to provide education, research, and advisory programs in such areas as ocean engineering, aquaculture, pollution studies, environmental studies, seafood processing, coastal management, and mineral resources. In its concern for the marine environment, the program parallels the one that established land grant colleges to develop the agricultural environment.

Warmwater aquaculture blossomed during the 1960s. The channel catfish industry was originally limited to south-central Arkansas, but now it is centered in the delta region of northwestern Mississippi. Mississippi accounts for about 55 percent of U.S. catfish sales. (See Figure 1-6.)

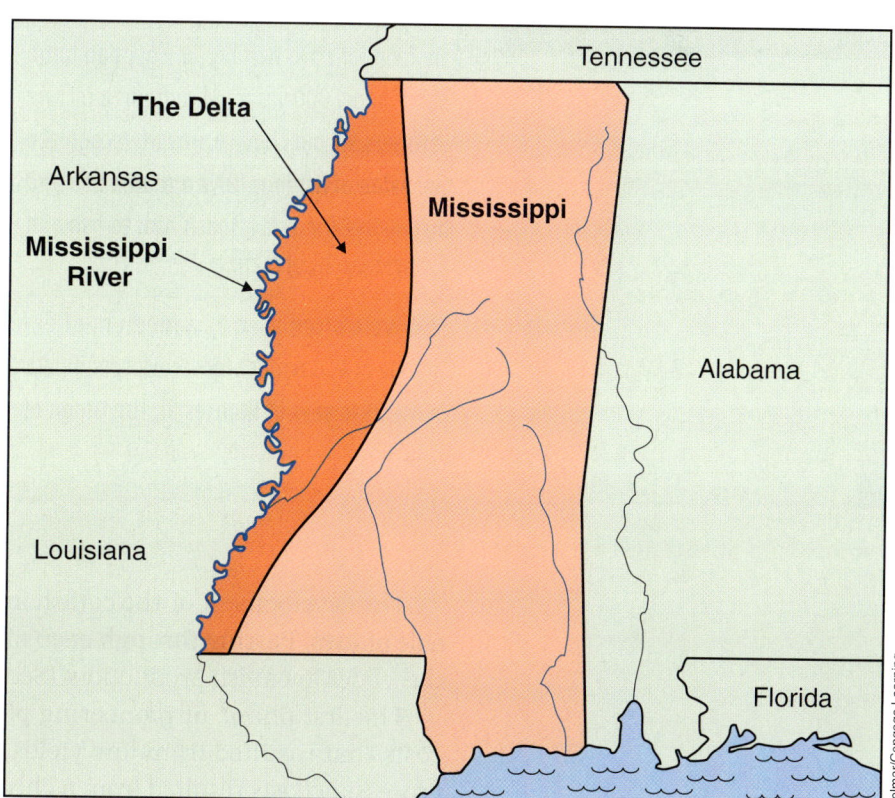

FIGURE 1-6 Most of the catfish production in Mississippi is located on the Delta.

A BRIEF HISTORY OF FISHING

Fishing is one of the oldest and most important activities of humankind. Ancient remains of spears, hooks, and fishnets have been found in ruins of the Stone Age. The people of early civilizations drew pictures of nets and fishing lines in their art. Through the ages, people wrote about fishing, used fish in exchange for services, and even learned to fish farm.

Early hooks were made from the upper bills of eagles, as well as from bones, shells, horns, and thorn plants. Spears were tipped with the same materials or sometimes with flints. Lines and nets were made from leaves, plant stalks, and cocoon silk. Ancient fishing nets were rough in design and material, but they were amazingly like some of the nets currently in use.

Many examples of fishing remain in art or are mentioned in writing. An Egyptian tomb more than 4,000 years old contains a picture of fishers. An old Chinese proverb recognized the value of fishing: "Give a man a fish and he will live for a day; teach him to fish and he will have food for life."

Fish were often used as a medium of exchange or as payment for services rendered. In the 29th year of Ramses III, the Union of Grave Diggers in Egypt filed a petition with the royal authorities for higher wages. As part of their wages, these workers received large amounts of fish four times each month. The petition requested a pay increase, pointing out that the petitioners came to the authorities without clothes and ointments—and even without fish.

The herring industry grew up around the Baltic Sea in the twelfth century and was controlled by the Hanseatic League, a group of German cities whose merchants traded all over northern Europe in fish, timber, cloth, salt, and many other goods.

A fourteenth-century discovery by a Dutchman named Beukelszoon helped the Dutch fishing industry. Buekelszoon pickled herring in brine instead of preserving them in dry salt.

In the fifteenth century, the herring mysteriously disappeared from the Baltic Sea. Fishers had to seek their herring in the North Sea and the Atlantic Ocean. The Dutch took over the herring fisheries and led commercial fishing of all kinds until the end of the seventeenth century.

Commercial fishing on the North American continent started more than 300 years ago

The development of the catfish industry moved through three identifiable phases. Passing through each of these phases, Mississippi emerged as the leader in catfish production. (See Figure 1-7, page 14)

The first phase, or pioneering phase, saw relatively high production costs that resulted from low yields and inefficiency. Additionally, high processing costs resulted from a chronic underuse of processing capacity.

with the arrival of the first colonists. So many fish were close to shore that the colonists did not need to build large sailing vessels as the Europeans did. Instead, the colonists followed the Indians' example and fished from small boats. Some fish were caught in traps and weirs (brush fences) set in the mouths of rivers and harbors. Shore fishers used nets or, when the tide had gone out, searched the rocks and sand for shellfish.

As colonization progressed, fishers began sailing farther out to sea to find enough fish for a good catch. They sailed for months, working the fishing banks off Canada and northeastern United States. Many early houses along the coast in colonial America featured a walkway around the roof so that the family could watch for returning ships. Because of the many hazards of sailing the sea, many ships did not return, and this walkway became known as a "widow's walk."

As ships grew larger and fishing methods were developed and refined, the success of fishing voyages and the types of fish and seafood increased. Like other commercial operations, the fishing industry became mechanized. With new technology, ships sailed to new fishing grounds by sailing farther from port and returning safely, loaded with fish. During the years between 1900 and the late 1960s, the world fish catch increased 27-fold.

For quite a while, a country's fishing rights have been a source of concern and agitation. As early as 1377, records indicate lawsuits against fishers who used a large net called the "wondrychoun" that fishers dragged through the water. The net caught little fish as well as big, and some people were afraid that soon there would be no fish left.

In the 1860s, individuals and groups acknowledged that fishery resources were finite, and that they must be managed through international agreements. In 1902, the International Council for the Exploration of the Sea (ICES) was formed by the major European fishing countries. In the mid-1960s, other nations joined, including the United States. The formation of ICES led to several conventions for the regulation of fisheries by the mesh size of nets and by quota, in order to obtain the highest yields consistent with the maintenance of fish stocks.

Farm-raised catfish faced severe competition from channel catfish caught in rivers by commercial fishers and from imported fish. The markets were fragmented. Low product acceptance outside the principal market areas of the Deep South and the lack of an effective marketing strategy combined to limit expansion.

FIGURE 1-7 Roadside sign along highway in Mississippi proclaiming Humphreys County as the catfish capital of the world.

During the second phase, from 1971 to 1976, production improved and unit costs declined. Average annual yields increased from 1,500 to 2,000 lbs per acre to 3,000 to 4,000 or more lbs per acre. Processing, typically limited to the fall, became less seasonal. Unprofitable and marginal producers quit the business when feed costs rose as a result of a scarcity of fish meal. Competition from river fish and imports continued, but supplies of these fish stabilized as the total demand for catfish rose. Marketing strategy improved, and Mississippi emerged as the clear leader in channel catfish production, processing, and related activities.

In the third phase, from 1977 to 1982, productivity continued improving, acreage increased, and production costs declined. The processed fish market became the major sales outlet. These developments were coupled with a more sophisticated marketing approach that led to single companies being involved in culture, processing, and marketing. This vertical integration started because processors needed to handle a nearly constant volume of fish throughout the year. A fall production peak is a built-in feature of catfish farming because most of the fish stocked as **fingerlings** in spring reach harvest size in fall.

The 1973–1975 shakeout period provides a lesson for all agribusinesses. Unfortunately, many farmers constructed ponds and started producing fish without considering two critical factors—management expertise and identifiable, dependable markets. Even when catfish farmers in areas like Georgia and South Carolina produced fish, they often had no ready markets. Local oversupply was especially critical when high feed prices reduced profit margins. In Mississippi, processing technology grew with the industry to provide a market for the crop. The catfish industry continues to thrive (Figure 1-8).

In 2008, the catfish industry faced major obstacles as producers dealt with sky-high feed prices, declining acreage, and fierce competition from imported fish. This led to a reduction in the number of producers and in

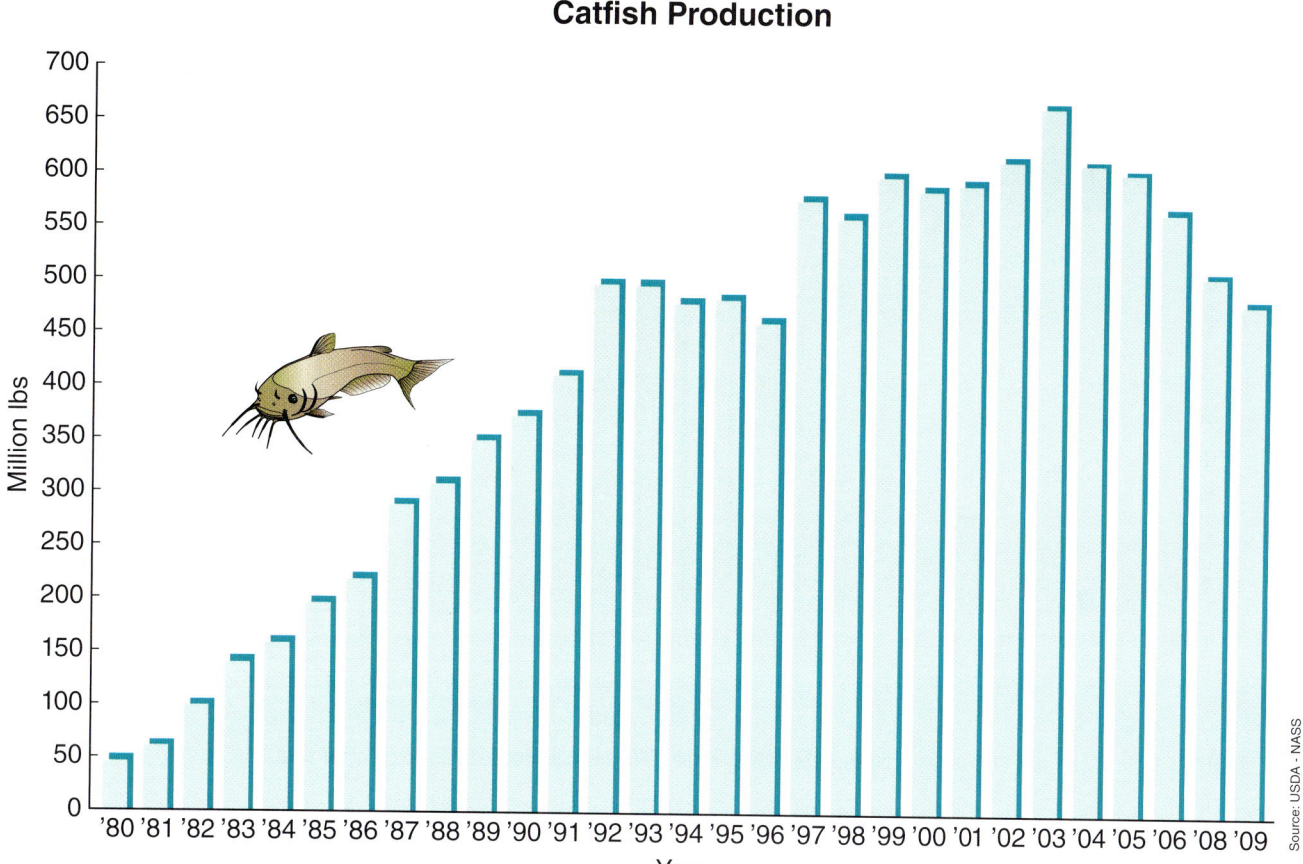

FIGURE 1-8 Catfish production soared from 1980 into 1990s.

the amount of acres of catfish ponds. For example, in 2009, Mississippi had just 70,000 acres of catfish ponds. This was down from a high of 113,000 acres in 2001. Feed prices that only a few years ago were about $240 a ton are now $330 a ton. Also, increasing imports of catfish and whitefish such as tilapia are putting pressure on catfish sales. Growers in countries like China have low-cost labor, favorable currency rates, and support from their governments.

Trout and Idaho

Rainbow trout were introduced into commercial fish farming in the early 1900s. Beginning in 1906 and continuing to 1947, the state of Idaho built 14 hatcheries located mainly in the southern part of the state. These hatcheries produced mostly rainbow trout to maintain productive fishing in rivers, lakes, and reservoirs. From the early 1920s until the end of World War II (1945), private trout-hatchery development proceeded slowly because of the easy availability of sport-caught fish and low demand.

The first commercial trout farm began operation in 1909 at Devil's Corral Spring, near Shoshone Falls in the Snake River Canyon. By 1914, Warren Meader started brood-stock production. By 1940, he was supplying up to 60 million eggs to public and private hatcheries around the country. Another early innovator was Jack Tingey, the former

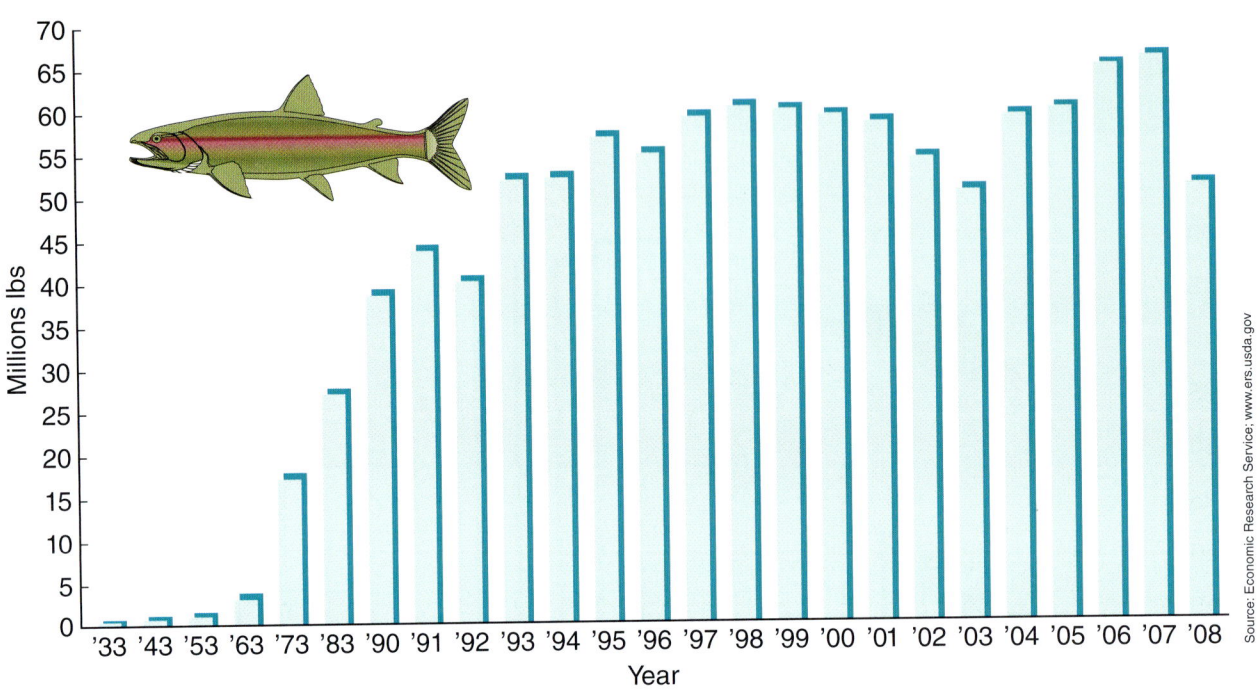

FIGURE 1-9 Trout production began slowly but increased rapidly in the 1960s. Idaho produces 75-80 percent of the trout.

commissioner of the Utah Fish and Game agency. In 1928, Tingey and his wife, Selma, started the first commercial hatchery near Buhl, Idaho. In the late 1940s, the trout industry began to grow. From the 1960s through the 1980s, trout production rapidly increased. (See Figure 1-9.) Most of the expansion occurred in the processed fish segment of the industry.

Today, the industry in Idaho is dominated by the world's largest trout production facility—Clear Springs Food Company in Buhl, Idaho. Idaho's trout industry is complete with feed mills and processing plants and a training program at the College of Southern Idaho in Twin Falls. Recent surveys indicate that Idaho trout producers—only about 30—produce about 75 percent of the commercial trout in the United States, a farm value of about $37 million.

Clear Springs Foods is the world's largest producer of aquacultured rainbow trout in the world. Producing in excess of 20 million pounds annually in its U.S. facilities, Clear Springs is a vertically integrated company with farms, feed manufacturing, cut-trout processing plant, specialty products processing plant and fish health and product-research facilities. In keeping with the trend of sustainability and reduction of carbon footprints, Clear Springs added a by-product production facility. This operation allows the use of 100 percent of the trout viscera, fish scraps, and farm mortality, which is turned into Clear Organic™ liquid fish fertilizer, sold to organic farmers.

Although trout is cultured in 45 of the 48 contiguous states, Idaho leads in trout production. Idaho leads because of an abundant supply of water at an ideal year-round temperature and because of entrepreneurs who use knowledge from various agencies and from their own observations.

History credits the Idaho trout industry with the development of many technological advances. Dry diets developed in the 1950s allowed the industry to expand. In 1956, the Snake River Trout Company started the first processing plant, allowing the opportunity for product diversity and distribution. By 1970, selective breeding for off-season spawn provided a year-round supply of eggs.

The idea of **self-feeders** (demand feeders) developed at the College of Southern Idaho's fish technology training facility. This idea was modified into many types of self-feeders and spread throughout the industry.

Water in Idaho comes from the Southern Idaho **Aquifer**. (See Figure 1-10.) The water in this aquifer enters the vast and extremely porous lava plain in southern Idaho. Eventually, water emerges from the aquifer at the ideal temperature of 55° to 58°F.

Some of the water in Idaho comes from warm and hot water wells. This water is being used to produce catfish, tilapia, and prawns on a limited basis. Some individuals are even trying to raise alligators.

Oysters, crawfish, clams, and shrimp are also important to the history of American aquaculture.

Oysters

About A.D. 43, Roman settlers in England harvested oysters along the seacoasts. In the winter, they packed the oysters in cloth bags and sent them to Rome. Eventually, to satisfy their taste for the delicacy, ancient Romans learned to farm oysters in the water off the Italian coast.

FIGURE 1-10 Water coming out of the Southern Idaho Aquifer into the Snake River Canyon near Buhl, Idaho.

When Europeans first came to North America, they found Indian tribes along the coast who depended upon oysters as part of their diet. Large piles of oyster shells existed around Indian settlements. The new Americans developed a taste for oysters and harvested the natural supplies. In 1894, the harvest of Chesapeake Bay oysters peaked at 15 million bushels and then began to decline. Soon, Americans developed culture methods to supplement the natural supply. Oyster culture in the United States is more than a century old. Worldwide, South Korea, Japan, the United States, and France lead in the culture of oysters.

Crawfish (Crayfish)

Culture of crawfish developed as a simulation of the creatures' natural life cycle in ponds. Now, some crawfish culture is tied to agricultural practices such as rice fields in the south.

Clams

Shortages and increasing prices for clams are creating more interest in aquaculture, either as an investment venture or to replace over-harvested stocks in public areas. Methods of spawning and growing hard clam **larvae** were described as early as 1927 and patented in 1929. Interest in culturing clams remained low until the early 1950s. The first commercial aquaculture operation, including a hatchery, began in 1957 near the town of Atlantic, Virginia. A short time later, another project was started in Sayville, New York, by Joe Glancey, who later patented a new method of growing clams.

During this period, a number of other companies formed, tried various methods of growing clams, and generally failed. The major problems involved in raising clams from **seed** (developmental form suitable for transplant) to market size are the ability to culture large enough numbers and to culture them at a reasonable cost. A few companies managed to survive as producers of clam seed for experimental planting and for replenishment programs carried out by various state agencies. By 1970, new technology and new materials contributed to the development of several promising methods for aquaculture of hard clams.

Shrimp

Shrimp are widely cultured in Asia, where, historically, the culture occurred almost by accident in brackish water ponds. Culturing shrimp alone, or **monoculture**, is a fairly recent occurrence in the United States and even Asia.

AQUACULTURE ACTIVITIES

When any industry such as aquaculture develops, the functions or activities performed to produce the product become identified in groups. Often, these become separate industries. In aquaculture, five main activities are performed:
1. hatchery
2. grow-out

3. harvesting
4. marketing
5. processing

Hatcheries produce the seed or young fish used to stock growing facilities. Seed are obtained by capturing wild seed or raising from broodstock—adults kept for reproduction.

Grow-out facilities produce crops (fish) from the seed. Like any agriculture venture, these can be intensive or extensive production systems. Intensive systems involve a very dense population of fish in relatively small spaces and require careful management. Extensive systems involve lower populations and less stringent management. Grow-out facilities may be land-based, such as ponds, tanks, and runways. Or, they may be water based, such as pens, cages, and ranching.

Harvesting involves the gathering or capturing of fish for marketing and processing. Aquaculture harvesting is typically topping (partial) or total harvesting.

Marketing connects producers with consumers. (See Figure 1-11.) The purpose of marketing is to provide a consumer with desired products and to provide the producer with a price to cover production and make a profit. Fish are the major aqua crop in the United States. Five markets, depending on the reason for production, are associated with fish:

1. food for human consumption
2. bait for sportfishing
3. pets or ornamentals for home or office aquaria
4. sport fish for release into lakes and steams
5. fish for feed ingredients

Processing changes the form of the product into something more desirable to consumers. Processing occurs in three forms: minimal, medium, and value-added.

FIGURE 1-11 Marketing is an important activity of aquaculture.

All the activities and functions of aquaculture are covered in more detail in other chapters.

THE FUTURE OF AQUACULTURE

The National Aquaculture Act of 1980 established aquaculture as a national priority. It consolidated federal support for aquaculture and the development of national planning for policy and cooperation by federal and state governments, universities, and industry. The purpose of the act was to support aquaculture as an industry that makes major contributions to the nation. The National Aquaculture Development Plan in 1996 (http://aquanic.org/publicat/govagen/usda/dnadp.htm) continues to emphasize aquaculture in the United States. Additionally, the U.S. Joint Subcommittee on Aquaculture (JSA) promotes a strategic plan that will take aquaculture into the future (http://aquanic.org/jsa/Strategicplan.htm).

In the United States, most traditional fisheries are being harvested at or near maximum sustainable yields. About half of the fishery products consumed in this country are imported to meet the high demand. Thriving aquaculture industries improve the balance of trade, increase the stability of seafood industries and markets, and provide more jobs for U.S. workers.

At present, the FAO estimates that world aquacultural production represents about 50 percent of the world aquatic food production by fisheries. For food fish, 50 percent of the total world supply is derived from aquaculture. Worldwide, aquacultural production in the 43 countries that have such industries produces more than 110 million tons of fish and fish products. Exclusive of the aquaculture of sport, bait, ornamental organisms, and pearls, this production includes more than 152 species, including finfish, species of shrimp and prawns, crawfish, diverse marine plants, oysters, clams, and other mollusks.

According to the FAO, world aquatic plant production by aquaculture was 15.1 million tons in 2006. The culture of aquatic plants has increased consistently, with an average annual growth rate of 8 percent since 1970. In 2006, aquaculture contributed 93 percent of the world's total supply of aquatic plants. Some 72 percent originated in China, at 10.9 million tons. Virtually all of the remaining production also stemmed from Asia. Japan is the second-most important aquatic plant producing country in terms of value, owing to its high-priced Nori production. Japanese kelp *(Laminaria japonica)* showed the highest production, followed by Wakame *(Undaria pinnatifida)* and Nori *(Porphya tenera)*.

Major species cultured in the United States are shown in Figure 1-12. Several technological breakthroughs have increased the potential of aquaculture in the United States:

- ▶ Development of net/pen culture and ocean ranching in the Pacific Northwest
- ▶ Establishment of abalone culture in California
- ▶ Introduction of Malaysian prawn culture to Hawaii and South Carolina

FIGURE 1-12 With thousands of species to culture, catfish make up the majority of finfish and shellfish cultured in the United States.

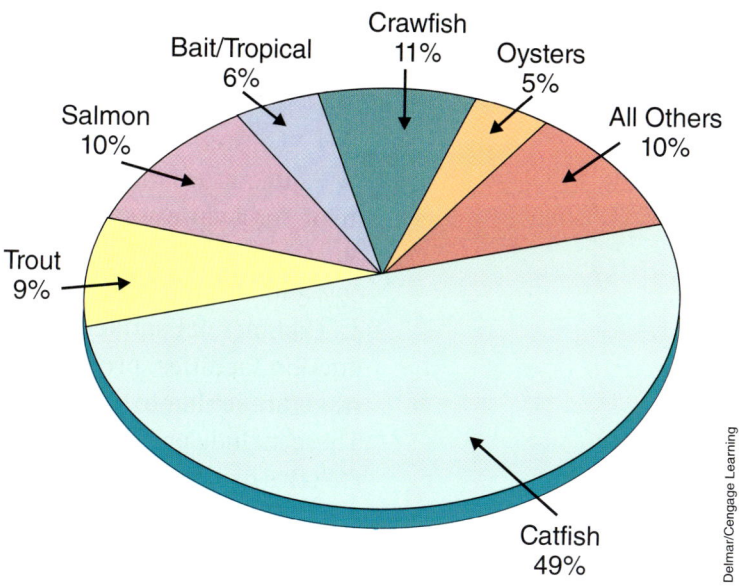

- Improvement of raft culture of blue mussels and oysters in New England
- Development of oyster hatcheries in the Pacific Northwest and the Atlantic States
- Establishment of marine shrimp farms in Central America by U.S. firms

The original Sea Grant plan included development of these aquaculture of species:

- Baitfish
- Channel catfish
- Crawfish (Crayfish)
- Rainbow trout
- Penaeid shrimp
- Prawns
- Salmon (net/pen rearing and ocean ranching)
- Yellow perch
- Oyster (hatchery/nursery production)
- Mussels
- Abalone
- Striped bass
- Scallops
- Clams
- Eels
- Bait leech
- Channel bass
- Scallops
- Red drum
- Sturgeon

- Southern flounder
- Speckled trout
- Red snapper
- Pompano
- Milkfish
- Lobster

Aquaculture throughout the world exists at different levels of development, for a variety of reasons. Levels of development include commercial aquaculture, infant industries, pilot scale or partially developed technology, and major lack of technology.

Commercial aquaculture represents enterprises with established production facilities, profitable markets, and continuity of sales. Research needs are similar to those that support established agricultural enterprises. These include product improvement, increased production efficiency, and effective marketing.

Infant industries may require research on several aspects of production, marketing, and creation of an acceptable institutional framework.

Pilot scale includes promising organisms for which proof of concept is established and basic breakthroughs in production technology have been achieved. Pilot scale aquaculture requires refinements to solve scale-up problems and ensure reasonable prospects for making money.

Major lack of technology represents those species of high market potential for which many major problems (such as reproduction, larval survival, domestication, strain selection, nutrition, and production systems) must still be solved.

Aquaculture is now considered a significant part of U.S. agricultural food production. Several factors suggest that the role of aquaculture will continue to grow: increased demand, new marketing and processing, and the culture of new species. Continual research on the problems facing aquaculture will ensure its future.

Demand

Aquaculture is the only known mode for increasing domestic fish production. The world's capture fisheries—wild-caught fish—are harvested at close to the maximum sustainable level. The demand in the United States for fish increases. A more health-conscious public consumes more fish each year. Recent marketing breakthroughs in several national fast food and restaurant businesses have extended sales of the southern tradition, catfish, into nontraditional regions.

Every 1-lb increase in per capita consumption requires 700 million more lbs of fresh fish. Over the past 40 years, per capita consumption of fish increased from 11 lbs to about 16 lbs. Some experts predict the per capita consumption of fish and shellfish could reach 25 lbs by the year 2025. Even if the per capita consumption remained constant, the U.S. population continues to increase. Fish raised to replenish dwindling wild stock also increases the demand for fish.

Marketing and Processing

The success of aquaculture depends on how the product meets the demands of the market—different products for different markets. Food service, retailers, and food processors market fish. The trend is toward more value added, fresh-refrigerated products, including bone fillets, seasoned and marinated products, smoked products, and vacuum-packed prepared fresh products that are ready to bake or broil.

Techniques and Technology

New techniques and technology continue to improve the profitability of aquaculture. Feeding represents 40 to 50 percent of the costs associated with aquaculture production. New feeding techniques and technology will improve feed conversion and use. Biotechnology, genetic engineering, genetics, and selective breeding will increase aquaculture production. New rearing methods such as cage culture and closed systems will open the door for more people to try aquaculture.

New Species

Scientists recognize about 21,000 kinds of fish, but only a few of these are widely used as food. In the United States, aquaculture is dominated by catfish production. This will continue, but the culture, technology, and marketing for many other species are being developed. Some of these species include carp, tilapia, hybrid striped bass, alligators, buffalofish, red drum and shrimp, prawns, and some aquatic plants. Chapter 2 discusses the potential species for aquaculture.

Research and Problems

Although aquaculture is generally successful, it is still several decades behind traditional livestock husbandry in research and development. Virtually every aspect of aquaculture can still be improved. Hundreds of thousands of acres of land are still available for expansion of fish farming. The water supply, if properly used, is adequate. The cooperative efforts of federal and state governments, private agencies, universities, and industry will be necessary to overcome the barriers that prevent the development of that acreage.

Research needs identified by members of the aquaculture industry touch every aspect of aquaculture. (See Figure 1-13.) General topics needing research include:

▶ Life history and biology
▶ Genetics and reproduction
▶ Nutrition and diet
▶ Environmental requirements
▶ Effluent (waste) control and water availability
▶ Control of diseases and parasites
▶ Predation and competition

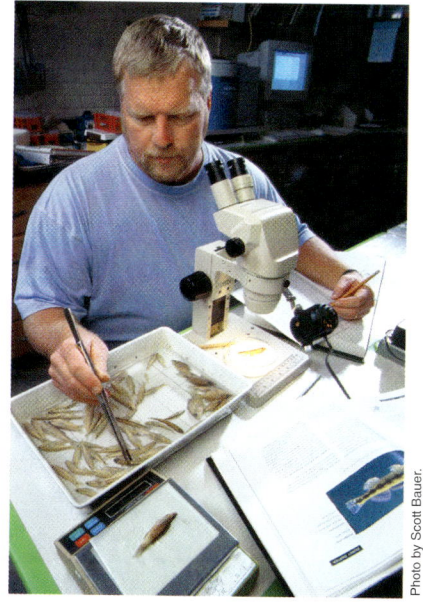

FIGURE 1-13 Biological technician Terry Welch identifies fish and measures their lengths and weights to evaluate changes in fish communities resulting from new watershed management practices.

- Harvesting, processing, and distribution
- Transportation
- Introduction of nonnative species
- Drug and chemical registration
- Production of rare seed stocks
- Sustainable fisheries and aquaculture
- Safe handling of fish and seafood
- Combating of aquatic nuisance species
- Educating the public
- Discovering new drugs

Environmental issues will continue to receive attention for all aspects of agriculture. This will present some problems—challenges—to the aquaculture industry. For aquaculture, these environmental issues include waste feed and excretory products, reduced water resources, endangered species, multiple uses of water, and water pollution from other sources.

SUMMARY

A thriving and developing aquaculture industry is important for several reasons. Aquaculture supplies a quality, healthy food source to a growing human population and does so through the efficient use of resources. Aquaculture creates jobs and stimulates economic activity. It provides valuable nonfood items such as eel skins, alligator hides, and by-products from the processing of finfish and shellfish. The feed demand of aquaculture increases the demand for other agricultural products such as corn, soybeans, wheat, oats, and barley. Finally, aquaculture contributes to recreation by providing fish to stock lakes, streams, and ponds for sportfishing and fee-fishing.

For U.S. aquaculture producers, future markets will grow, but producers will be faced with increased environmental regulations, the need for new and better technology through research, and competition from foreign producers as aquaculture expands worldwide.

STUDY/REVIEW

Success in any career requires knowledge. Test your knowledge of this chapter by answering these questions or solving these problems.

True or False

1. Aquaculture is a form of agriculture.
2. Modern Americans were the first to practice aquaculture.
3. Trout were one of the first fish involved in polyculture.
4. Aquaculture helps fisheries by providing fish to restock streams, rivers, oceans, and lakes.
5. Aquaculture is a minor part of U.S. food production.
6. Aquaculture developed rapidly in the United States from the 1960s to the 1980s.

Short Answer

1. Scientific _____ of the problems in aquaculture helps the industry grow.
2. _____ connects the aquaculture producer with consumers.
3. The state of _____ leads in trout production; the state of _____ leads in catfish production.
4. _____ and _____ are examples of coldwater fish.
5. _____ in water limits aquaculture.
6. List four stages of evolution for all agricultural activities.
7. What are the significant aquatic species cultured in the United States?
8. List five activities that are a part of aquaculture and often become separate industries.

Essay

1. Define aquaculture.
2. Explain how aquaculture may help maintain a traditional fishing industry and sport fishing.
3. Compare farming the water to farming the land.
4. Define freshwater, saltwater, warmwater, and coldwater.
5. Describe the importance of aquaculture to two ancient civilizations.
6. Why is the history of aquaculture in England and Europe important to U.S. aquaculture?
7. How did Mississippi emerge as the leader in catfish production and Idaho emerge as the leader in trout production?
8. Why is the National Sea Grant Program important to U.S. aquaculture?
9. List and describe five areas that will determine the future of aquaculture.

KNOWLEDGE APPLIED

1. Visit local grocery stores and survey the fish and seafood sold in each store. How much does a freshwater product typically cost? How much does a saltwater (marine) product typically cost? How are the products sold—fresh, frozen, canned? Are any of the products produced locally?

2. The history and development of aquaculture can be used to teach geography. Obtain a world map and identify the locations discussed in the chapter. The history of agriculture may be used as a springboard to other history lessons.

3. Visit a local aquaculture production site. Have questions ready to ask about the operation. For example:

    How did the production facility get started?

    What problems have they encountered with environmental concerns, diseases, feeding, processing, and marketing?

    How are prices for the product established?

    What type of training is required to be successful? What legal regulations are involved?

    If a local aquaculture production facility is not available, arrange a teleconference using a speakerphone with one in another part of the state or a different state.

4. Using the list of Regional Aquaculture Centers listed in the Appendix Table A-12, visit the Web sites and find information on one of the species listed in this chapter.

5. Although carp farming never really became important in the United States, it has become very important in China and parts of Europe. Develop a series of reports on the culture of carp in either China or Europe. Include in the reports the development of carp culture, **Polyculture** with other species, integration with other forms of agriculture, and any recipes for carp or ways of serving carp. Several books in the Learning/Teaching Aids section will be helpful.

LEARNING /TEACHING AIDS

Books

Bardach, J. E., Ryther, J. H., and McLarney, W. O. (1995). *Aquaculture: The farming and husbandry of freshwater and marine organisms.* New York: John Wiley & Sons.

Food and Agriculture Organization of the United Nations. (2004). *Global Aquaculture Outlook In the Next Decades: An Analysis of National Aquaculture Production Forecasts to 2030.* Rome.

Kirk, R. (2003). *A history of marine fish culture in Europe and North America.* Ames, IA: Iowa State Press.

Michaels, V. K. (1991). *Carp farming.* New York, NY: John Wiley & Sons.

Stickney, R. R. (2000). *Encyclopedia of Aquaculture.* Malden, MA: Wiley-Interscience.

Wittwer, S., Youtai, Y., Hans, S., and Lianzheng, W. (1987). *Feeding a billion: Frontiers of Chinese agriculture.* East Lansing, MI: Michigan State University Press.

Internet

Internet sites represent a vast resource of information. The URLs (uniform resource locator) for World Wide Web sites can change. Using a search engine such as Google, find more information by searching for these words or phrases: aquaculture, warmwater aquaculture, coldwater aquaculture, mariculture or marine culture, freshwater aquaculture, saltwater aquaculture, salinity, brackish, polyculture, fish hatchery, catfish, trout, monoculture, and wild-caught fish.

For some specific Internet sites, refer to Appendix Table A-14.

Aquaculture includes the art, science, and business of cultivating plants and animals in water (Figure 2-1). Aquaculture in the United States involves only a few successful species of fish and plants. This quantity could change as aquaculture evolves. Future aquaculturists need to have an awareness of potential plants and animals and a basic knowledge of aquatic plants and animals.

CHAPTER 2

Aquatic Plants and Animals

OBJECTIVES

After completing this chapter, the student should be able to:

- Name the major aquatic species in the United States
- Name five aquatic animals that hold potential for aquaculture in the United States
- Explain why aquatic crops may be more productive than terrestrial crops
- Briefly describe the general water and feeding characteristics of five aquatic animals
- List three aquatic plants that potentially could be cultured in the United States
- List three other uses for aquatic plants besides human food
- Give examples of aquatic animals and plants that could be used in polyculture
- Recognize the scientific names for some common aquatic species
- List and describe important biological characteristics in selecting a species for aquaculture
- Explain how aquatic species save energy when compared to terrestrial species
- List and describe the major characteristics of aquatic plants and animals
- Discuss the morphology, anatomy, and physiology of common aquatic animals

- Name and describe the nine body systems of aquatic animals
- Identify and describe the internal and external anatomy of a fish
- Identify and describe the basic structure and internal anatomy of crustaceans
- Identify and describe the basic structure and internal anatomy of an oyster or mussel
- Describe the basic morphology of aquatic plants

Understanding of this chapter will be enhanced if the following terms are known. Many are defined in the text, and others are defined in the glossary.

KEY TERMS

Adductor	Homocercal
Antennae	Incubation
Appendages	Inorganic
Asexually	Macrophytes
Assimilation	Mantle
Bivalve	Molting
Bloom	Omnivores
Calcareous	Phycocolloid
Carnivores	Phytoplankton
Chlorophyll	Polysaccharide
Decapods	Protandrous
Diffusion	Regeneration
Ectothermic	Rotifers
Fusiform	Semipermeable
Gametes	Siphon
Gastropods	Spores
Herbivores	Terrestrial
Hermaphroditic	Zooplankton
Heterocercal	Zygote

U.S. AQUATIC PLANT SPECIES

Aquatic plants are important components of aquaculture in other parts of the world, particularly in Asia. Europe and North America rank dead last in aquatic-plant production worldwide. Some aquaculture producers cultivate aquatic plants for (1) food, feed, and chemical products, (2) wastewater treatment, and (3) biomass production for conversion to energy. Table 2-1 lists aquatic plants and their potential uses.

Of the chemicals or products obtained from aquatic plants, a **phycocolloid** called carrageen is one of the most widespread. Carrageen is used in foods for gelling, thickening, and stabilizing. It is a **polysaccharide**.

Phytoplankton

A list of potential aquatic plants for culture should not overshadow the important role of **phytoplankton** in aquaculture. They are the primary producers, forming the first link in the aquatic food chain. Through photosynthesis, phytoplankton use sunlight to produce food energy and contribute oxygen to the water. Phytoplankton serves as a food source for **zooplankton** and for some fish and produces a **bloom** that helps shade out unwanted rooted aquatic plants. Pond fertilization encourages the production of phytoplankton.

U.S. AQUATIC ANIMAL SPECIES

Catfish and salmonids (trout and salmon) dominate U.S. aquaculture. Other species of freshwater finfish, marine finfish, mollusks, and crustaceans hold promise for the future of U.S. aquaculture. Tables 2-2 and 2-3 (page 31 through 33) categorize the important points of U.S. aquaculture species.

Ornamental Fish

Ornamental/hobby, tropical, and aquarium fish represent several families and over 100 species of small, colorful, and unique fish. These fish occur naturally in tropical, semitropical freshwater, saltwater, or brackish water. The major ornamental/hobby fish industry is located in central Florida, but hobby fish are raised in most of the other states. Water-temperature management is a prime concern because hobby fish are sensitive to cool temperatures. Culturalists specialize in the production of colorful varieties that are easy to propagate.

FIGURE 2-1 Aquaculture requires a good supply of clean water.

TABLE 2-1	AQUATIC PLANTS FOR AQUACULTURE			
Common Name	**Scientific Name**	**Water Type[1]**	**Uses**	**Notes**
Spirulina	*Spirulina* spp.	F	Food	Protein content of some species can be 70 percent; collected and dried into patties for human consumption in some Asian countries and Mexico; nutritious supplement; distinct taste.
Brown algae or Kelp	*Undaria pinnatifida* macro *Macrocystis pyrifera* *Macrocystis integrifolia*	S	Food Mulch Fertilizer Phycocolloids	Called "wakame" in Japan; dried, chopped, and used in salads; brownish color comes from xanthophyll; giant kelp may grow to 200
Green algae	*Monostroma macro* *Enteromorpha* *Chlorella* *Chlorophyceae* spp.	S,F	Food Mulch Fertilizer Biodiesel	Least cultured of three macroalgae; called "nori" in Japan; can occur as single cells or as colonies
Red algae or Laver	*Porphyra* spp. *Gelidium* spp. *Gracilaria* spp.	S,B	Food Feed Mulch Fertilizer Phycocolloids	Cultured in Japan as early as 1570; dried, high in protein; some harvested for livestock feed; United States leads in carrageen production—a phycocolloid.
Duckweed	*Lemna* spp. *Spirodela* spp. *Wolffia* spp. *Wolfiella* spp.	F	Feed Waste water treatment	Favored food of herbivorous fish and water fowl; harvested and used for livestock feed; one of least expensive to produce.

(Continued)

TABLE 2-1 AQUATIC PLANTS FOR AQUACULTURE (Continued)

Common Name	Scientific Name	Water Type[1]	Uses	Notes
Water spinach	*Ipomoea reptans*	F	Feed food	Commonly cultured in Thailand, Malaysia, and Singapore; often in polyculture; low protein and carbohydrate content.
Water hyacinth	*Eichhornia crassipes*	F	Waste water treatment Fuel source	Effectively removes waste from water and easy to harvest; possibly used for methane gas production.
Chinese waterchestnut	*Eleocharis dulcis*	F	Food	Small-scale production in United States compared to Asia; corm consumed; each corm produces about 20 lbs. of new corms in about 220 days; labor intensive; useful in polyculture.
Watercress	*Nasturtium officinale*	F	Food	Primarily freshwater aquatic plant produced in United States; requires abundant, continuous-flowing water; many people harvest wild crop.
Cattail	*Typha latifolia* *T. angustifolia*	F	Ornamental	Grown in aquatic gardens, used in dried flower arrangements; edible parts but not cultured for food.
Arrowhead	*Sagittaria* sp.	F	Ornamental	Grown in aquatic gardens; edible parts but not cultured for food.

[1] Freshwater (F), saltwater (S), brackish water (B).

TABLE 2-2 FINFISH FOR AQUACULTURE

Common Name	Scientific Name	Water Temp.[1]	Water Type[2]	Diet[3]	Notes
Atlantic Salmon	*Salmo salar*	C	A	C	Important as rod catalyst, sport fish, and commercial netting; fishing regulated by national, international, and local laws.
Bighead Carp	*Aristichthys nobilis*	W	F	C	Excellent food animal; suited for polyculture; acceptance increasing in United States.
Black Bullhead	*Ictalurus melas*	W	F	O	Susceptible to disease; tolerant of adverse water conditions; demand low.
Blue Catfish	*Ictalurus furcatus*	W	F	C	Some culture work; silvery white to light blue color.
Brook Trout	*Salvelinus fontinalis*	C	F	C	Used in hybrid crosses with Lake Trout—Splake.
Brown Trout	*Salmo trutta*	C	F (A)	C	Naturalized populations on every continent except Antarctica.
Buffalofish (Largemouth)	*Ictiobus cyprinellus*	W	F	C	Technology for spawning and rearing available; possible polyculture species.
Channel Catfish	*Ictalurus punctatus*	W	F	O	Principal farm-raised species in United States; oxygen depletion major problem.
Chinook Salmon (King)	*Oncorhynchus tshawytscha*	C	A (F)	C	Coastal species; researched and cultured in New Zealand; may live in fresh water.
Chum Salmon	*Oncorhynchus keta*	C	A	C	Most cold tolerant of Pacific salmon; widest distribution; hatchery techniques developed in Japan.

(Continued)

TABLE 2-2 FINFISH FOR AQUACULTURE (Continued)

Common Name	Scientific Name	Water Temp.[1]	Water Type[2]	Diet[3]	Notes
Coho Salmon	Oncorhynchus kisutch	C	A (F)	C	Grow rapidly second year when feeding on other fish; introduced into Great Lakes to feed on alewife, smelts, and sea lampreys.
Common Carp	Cyprinus carpio	W	F	O	Deep yellow body; member of minnow family.
Crappie	Pomoxis spp.	W	F	C	Member of sunfish family, centra-chidae; spawn readily.
Cutthroat Trout	Salmo clarki	C	F	C	Possible to propagate artificially; hybrid potential.
Fathead Minnow	Pimephales promelas	W	F	O	Baitfish; short-lived; seldom reach 3 in. or 3 years.
Flathead Catfish	Pylodictis olivaris	W	F	C	Predator species; not economical to raise on large scale.
Golden Shiner	Notemigonus crysoleucas	W	F	C	
Goldfish	Carassius auratus	W	F	H	Baitfish; very hardy; used as feeder fish or forage fish.
Grass Carp	Ctenopharyngodon idella	W	F	H	Slim carp; feeds on aquatic plants but accepts pelleted feed when cultured; cultured in Asia.
Lake Trout	Salvelinus namaychus	C	F	C	Used in hybrid crosses with Brook Trout—Splake.
Largemouth Bass	Micropterus salmoides	W-C	F	C	Large bass eat small ones; spawn in gravel nest; jaw extends beyond eye.
Milkfish	Chanos chanos	W	S-B	H	Very disease resistant; popular in tropical Pacific; will not spawn in captivity.
Mullet, Striped	Mugil cephalus	W-C	F-B-S	H	Commonly cultured; tropical and semitropical; possible polyculture.
Muskellunge	Esox masquinongy	C	F	C	Some cannibalism; prefer temperatures warmer than trout but cooler than catfish.
Northern Pike	Esox lucius	C	F	C	Wild stock usually captured for egg-taking; requires forage fish.
Pink Salmon	Oncorhynchus gorbuscha	C	A	C	Attempts to extend range not very successful; ranched in Alaska.
Pompano	Trachinotus carolinus	W	S	C	Naturally not very abundant; commercial production expensive.
Rainbow Trout	Oncorhynchus mykiss	C	F	C	Tolerant to relatively high water temperatures and low oxygen levels; fast growth.
Red Drum	Sciaenops ocellata	W	S-B	O	Popular in Cajun-style restaurants; popular sport fish; some successful culture.
Smallmouth Bass	Micropterus dolomieui	W	F	C	Special equipment and techniques to collect fry.
Sockeye Salmon	Oncorhynchus nerka	C	A (F)	C	Landlocked form called kohanec; crustaceans diet; pigments flesh red.

(Continued)

TABLE 2-2 FINFISH FOR AQUACULTURE *(Continued)*

Common Name	Scientific Name	Water Temp.[1]	Water Type[2]	Diet[3]	Notes
Steelhead	*Oncorhynchus mykiss*	C	A	C	Anadromous form of Rainbow Trout.
Striped Bass, Hybrid	*Morone saxatilis* x *Morone chrysops*	W	F	C	Cross of female striped bass and male white bass; approved for aquaculture late 1970s.
Sturgeon	*Acinpenseridae* spp.	C	F	O	Cultured to increase numbers; some culture for roe.
Sunfish (Green, Bluegill, Redear)	*Lepomis* spp.	W	F	C	Spawn readily; hybridize easily; female drab.
Tilapia	*Tilapia* spp.	W	F	H	Controlling reproduction is a major problem to culture; feed on algae, detritus, and waste feed.
Walleye	*Stizostedion vitreum vitreum*	W-C	F	C	Wild stock captured for egg-taking; requires long, slender forage fish.
White Catfish	*Ictalurus catus*	W	F	C	Determined inferior to channel catfish for aquaculture; hardy; stocked for fee-fishing ponds.
White Sucker	*Catostomus commersoni*	C	F	C	Forage fish; adapt to formulated feed as a supplemental diet.
Yellow Perch	*Perca flavescens*	C	F	C	Famous in the Midwest; cultured in United States.

[1] Warmwater (W) temperature exceeds 70°F or coldwater (C) temperature of between 50° and 65°F
[2] Freshwater (F), saltwater (S), brackish water (B), or anadromous (A).
[3] Herbivorous (H), carnivorous (C), or omnivorous (O).

Some common ornamental/tropical fish include sailfin mollies, guppies, clown barbs, black tetras, angelfish, and blue gouramies. Ornamental fish are covered in more detail in Chapter 6.

Bullfrogs

In the United States, most bullfrogs (*Rana catesbeiana*) for consumption come from the wild. Demand for food frogs and live frogs for biological research is greater than the supply, and the availability is seasonal. All of this makes the possibility of commercial production appealing. Unfortunately, frog culture is very complex because of the complicated life cycle and demanding feeding habits of the bullfrog. The Japanese and Taiwanese practice open pond culture of bullfrogs from eggs to adults.

Alligators

Alligators (*Alligator mississippienis*) are large aquatic reptiles valued for their meat and hides. They were once abundant in the lower South before overhunting and habitat destruction reduced their numbers. Extensive conservation efforts have restored alligators where the habitat has permitted and have led to the development of alligator-culture techniques. Presently, alligators are commercially cultured in Texas, Georgia, South Carolina,

TABLE 2-3   MOLLUSKS AND CRUSTACEANS FOR AQUACULTURE

Common Name	Scientific Name	Water Type[1]	Notes
Abalone, red	*Haliotis rugescens*	S	The only gastropod (snail) of significance cultured in United States; largest hatchery in California; prolific spawners.
Clams Hard clam Soft clam	*Mercenaria mercenaris* (hard clam)	S	More culture of hard clam; not widely cultured around the world; United States has most advanced culture; two to seven years to market size depending on location.
Crabs Blue crab	*Callinectes Spidus*	S	Primarily a fisheries product; aquaculture techniques produce soft-shelled crabs.
Crawfish (crayfish)	*Procambarus clarkii* *P. blandingi acutus*	F	About 300 species in United States; harvested from wild and cultured; found on every continent except Africa and Antarctica; six to fourteen months to reach market size.
Mussels	*Mytilus edulis*	S	New to U.S. culture; easy to raise; grow faster than other shellfish.
Lobster	*Homarus americanus*	S	Farming from egg to market size not profitable; minimum of five years to reach market size.
Prawns (Malaysian prawn)	*Macrobrachium rosenbergii*	F-B	High demand; started in Hawaii.
Oysters (American oyster)	*Crassostrea virginica*	S	Culture over 100 years old in United States; larvae swim free, then attach to something for rest of life.
Shrimp	*Penaeus* spp.	S	Widely cultured in Asia but new to United States; great demand for shrimp.

[1]Freshwater (F), saltwater (S), brackish water (B), or anadromous (A).

Louisiana, and Florida. Some producers in Idaho with access to warmwater wells are also considering raising alligators. The demand for alligator meat and hide keeps prices high and production profitable. Strict regulations govern intra- and interstate commerce in alligators and alligator products.

Eels

Eels are considered a gourmet food in Japan, Taiwan, and most European countries. The commercial production of food-sized eels for export has captured the interest of some U.S. aquaculturists. The life cycle of eels is complicated. They spawn at sea, and seed stock must be captured from the wild when the elvers—small eels—migrate upstream from the sea. Captured eels are raised in ponds, and they need to be trained to eat artificial feed. Eel culture is risky business without a stable supply of elvers, and there are few markets.

Zooplankton

Discussing the potential of aquatic animals tends to overshadow the minute animals important to aquaculture. Zooplankton, primarily copepods (very small crustaceans) and **rotifers** serve as vital food sources for all fish fry, and they feed on the phytoplankton. They are primary consumers in the food chain, as Figure 2-2 indicates.

FIGURE 2-2 An example of the food chain. Zooplankton feed on the phytoplankton. Crawfish eat the zooplankton, and a bass eats the crawfish.

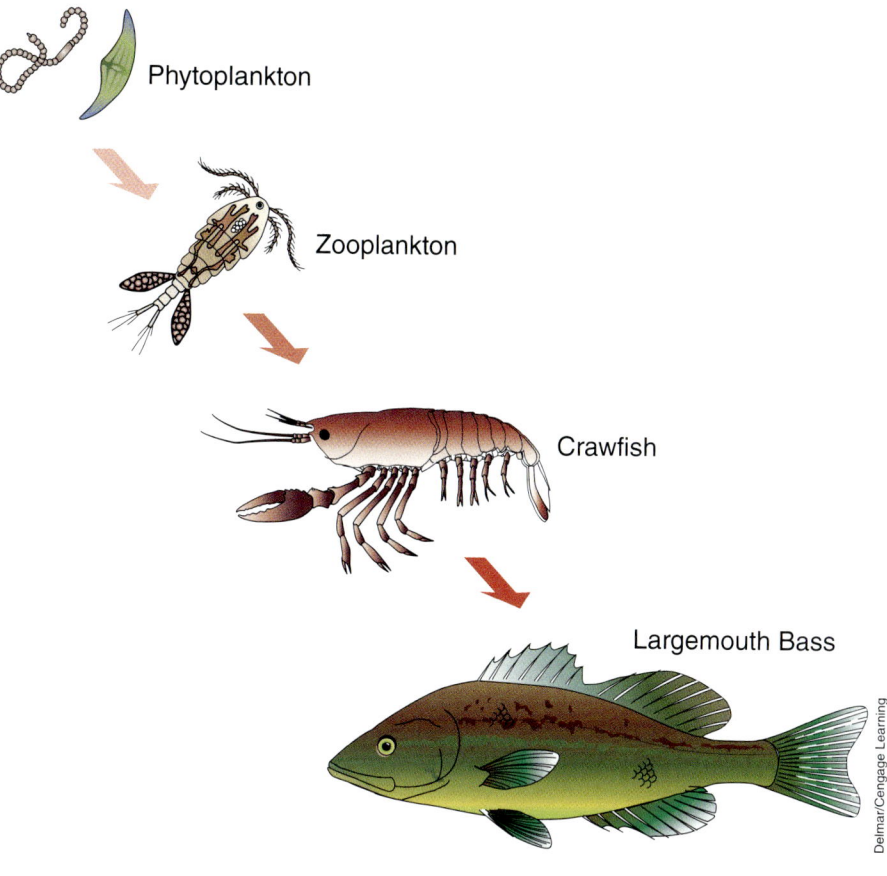

COMMON CHARACTERISTICS OF AQUATIC SPECIES

Aquatic plants and animals hold a greater productive potential than terrestrial plants and animals. Reasons for this include:
- Body temperature about same as environment (**ectothermic**)
- Body density similar to habitat
- Reduced energy required for getting food
- Efficient feed conversion
- Rapid growth
- Live in multidimensional environment

Since the body temperature of aquatic animals is near that of their environment, energy normally required to regulate body temperatures can be directed toward growth. Because their body density is near that of their habitat, energy normally reserved for overcoming gravity can be used for growth.

In land animals, the search for food requires energy. In aquatic species, this energy expenditure can be minimized. For example, filter feeders such as clams filter surrounding water through their bodies in order to find and use particles of suspended food.

Compared to livestock such as beef cattle and hogs, some aquatic species efficiently convert feed to growth. For example, beef cattle and hogs require 4 to 8 lbs. of feed for 1 lb. of gain. Catfish and trout produce 1 lb. of gain from 1.5 to 2 lbs of feed. The less feed used, the more profit made.

Some aquatic organisms, known as "primary producers." can grow rapidly. Some species of algae and plankton represent the best examples of this, growing at a rate of almost 10 percent per day. Figures 2-3a–c (pages 36–38) show some examples of algae and plankton.

Different species inhabit various spaces and positions within the aquatic environment. This variety expands the aquaculture options available at any single site. Polyculture with different species of carp is a good example. Fish, crustaceans, and mollusks all occupy different spaces. Structures such as floating cages, pens attached to the bottom and extending above the surface, and strings on poles extending into the water create a dimensional variety (see Figure 2-4, page 38).

FIGURE 2-3a Types of algae.

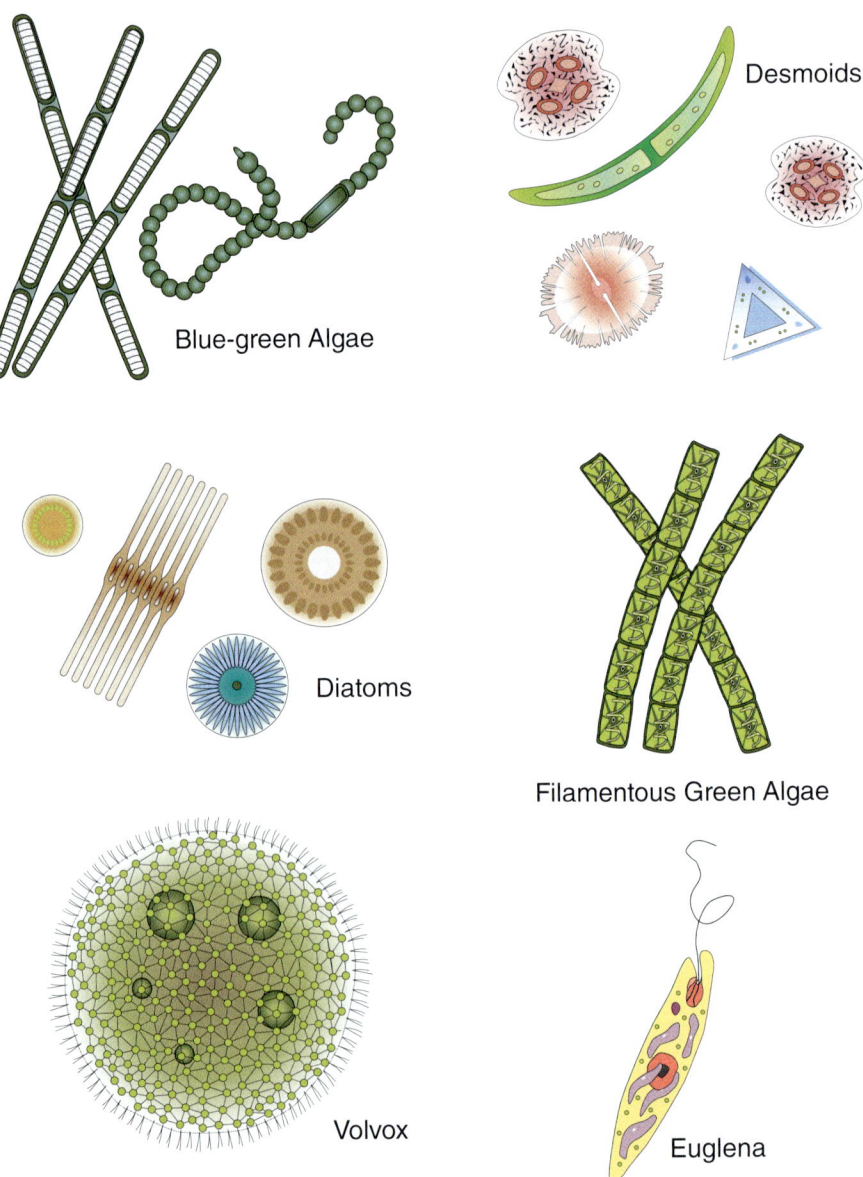

FIGURE 2-3b Pond plankton.

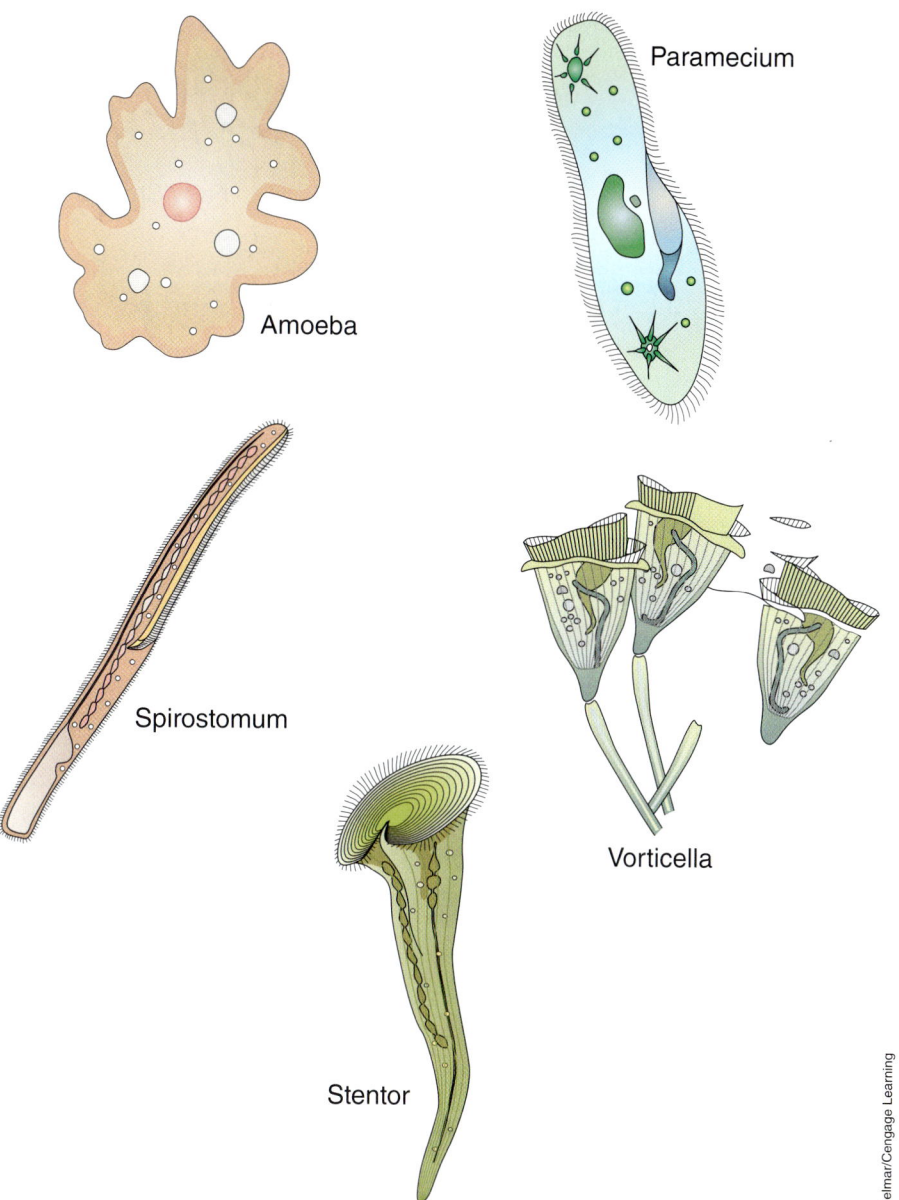

Choosing a species for aquaculture is similar to choosing any crop or livestock for culture. Successful culture means considering:
- Reproductive habits
- Egg and larvae requirements
- Nutritional needs and feeding habits
- Polyculture possibilities
- Adaptability to crowding
- Disease resistance
- Market demand

The ability to reproduce easily is a primary requirement. For successful culture, a stable supply of seed (young) must be available. Also, the reproductive processes of the species must be understood, and genetic selection and improvement must be possible.

FIGURE 2-3c Pond plankton metazoa.

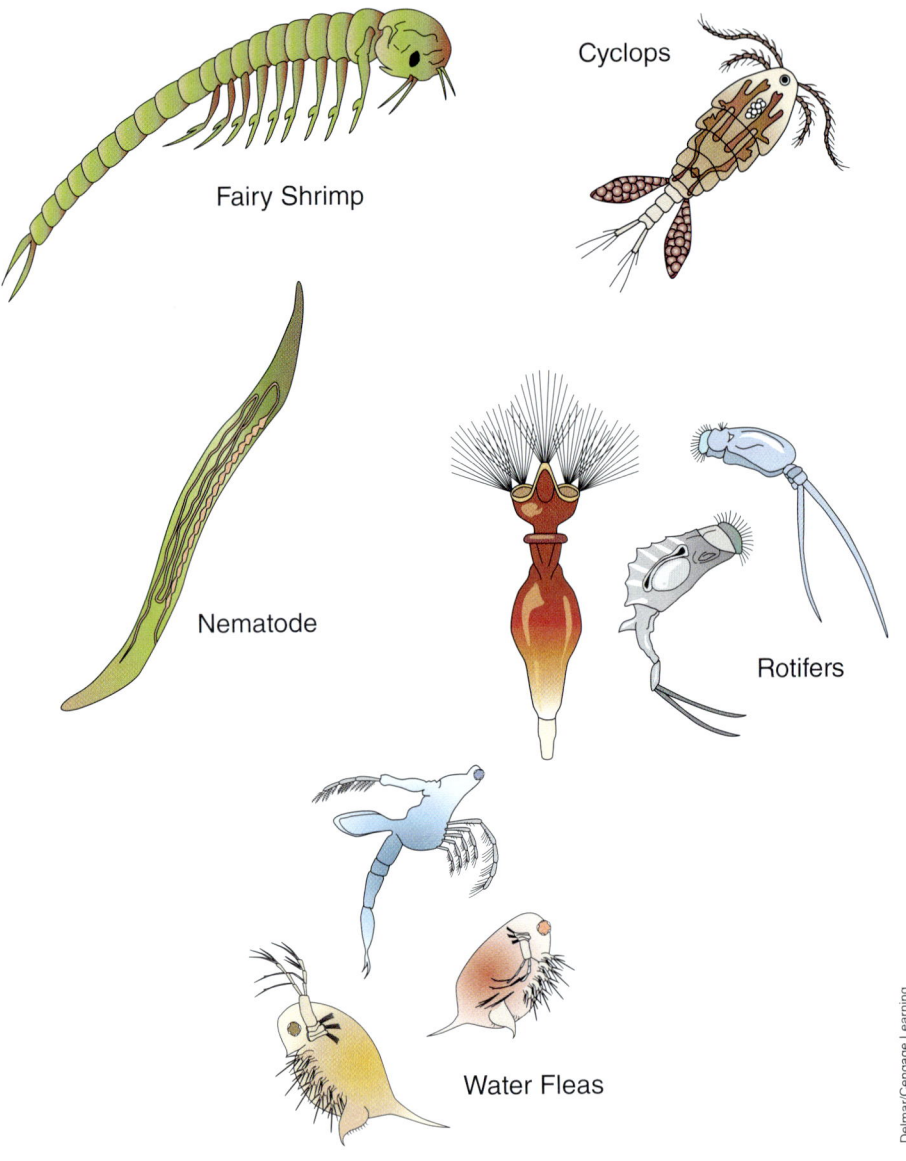

FIGURE 2-4 Structures such as cages create dimensional variety.

Reproduction should produce massive quantities and occur frequently. Eggs and larvae need to be large, hardy, and easy to culture. Table 2-4 (page 40) indicates the wide variation in finfish and their ability to produce eggs.

One female oyster may produce 500 million eggs per year. Among crustaceans, crawfish produce 100 to 500 eggs, and shrimp produce 500,000 to 1 million eggs.

After reproductive ability, the next factors to consider when selecting an aqua species to culture are nutritional needs and feeding habits. For feeding habits, a species can be selected low on the food chain or high on the food chain. An aquaculture species low on the food chain consumes low-cost vegetable matter or by indirectly consuming primary foods within the pond. Examples of species low on the food chain include carp, tilapia, and crawfish. Species high on the food chain include shrimp, trout, and bass. These species require a more expensive high-protein diet.

Aquatic animals, like terrestrial animals, require protein, carbohydrates, fat, vitamins, and minerals. Unlike terrestrial animals, however, some of aquatic animals' nutritional needs are met directly from the aquatic environment. Research on the optimal amounts and forms for each species continues. The more completely nutritional needs are understood, the more efficiently other aquatic animals can be produced. Complete information about nutritional needs is covered in Chapter 9, Fundamentals of Nutrition in Aquaculture.

Selecting polyculture as a criterion for determining which species to produce depends on the type of production system. In an intensive culture system, such as trout or catfish production, growth rate could be more of a concern than efficiency in the use of water space and nutrients. Polyculture increases the total aquatic production in a volume of water by using species that occupy different dimensions of the water and feed on different feedstuffs.

Aquaculture crowds species that are not used to crowding. Crowding increases the productivity of a space while increasing management for the space. Aquatic species selected for culture exhibit adaptability to withstand crowding.

Species vary widely in their ability to resist disease. Aquaculturalists select species for disease resistance based on the conditions at their production site.

Production of an aqua crop can be successful and efficient, but, without a market, production efforts are wasted. A market for a product consists of:

- Desire by consumers
- Price that consumers can afford
- Prepared, easy-to-use forms of the product
- Storage to reach consumer
- Desired flavor

Chapter 3, Marketing Aquaculture, contains more information about marketing.

TABLE 2-4 SPAWNING FREQUENCY AND EGG PRODUCTION IN VARIOUS FINFISH[1]		
Species	**Spawning Frequency**	**Eggs per Lb[2] of Fish**
Chinook salmon	Once per life span	350
Coho salmon	Once per life span	400
Sockeye salmon	Once per life span	500
Atlantic salmon	Annual-Biennial	800
Trout	Annual	1,000–1,200
Northern pike	Annual	9,100
Walleye	Annual	25,000
Striped bass	Annual	100,000
Channel catfish	Annual	3,750
Largemouth bass	Annual	13,000
Smallmouth bass	Annual	8,000
Bluegill	Intermittent	50,000
Golden shiner	Intermittent	75,000
Goldfish	Intermittent	50,000
Common carp	Intermittent	60,000

[1]Source: Fish Hatchery Management.
[2]1 lb = 0.45 kg

STRUCTURE AND FUNCTIONS OF AQUATIC ANIMALS AND PLANTS

A study of aquaculture requires some information about the structure and form or morphology and anatomy of aquatic animals and plants, and the function of aquatic animals and plants or their physiology. The suitability of an organism for culture depends on its morphology, anatomy, and physiology.

Animal Surfaces

Any discussion of the structure and function of animals begins with an understanding of the concepts of dorsal, ventral, anterior, and posterior. Dorsal pertains to the upper surface of an animal. Ventral relates to the lower or abdominal surface. Anterior applies to the front or head of an animal. Posterior pertains to the tail or rear of an animal. These are easy to understand in many species, but in species like clams and oysters, these positions can be a little confusing. Figure 2-5 shows the dorsal, ventral, anterior, and posterior of fish, crawfish, and clams.

FIGURE 2-5 A perspective of the dorsal, ventral, anterior, and posterior of a fish, crawfish, and clam.

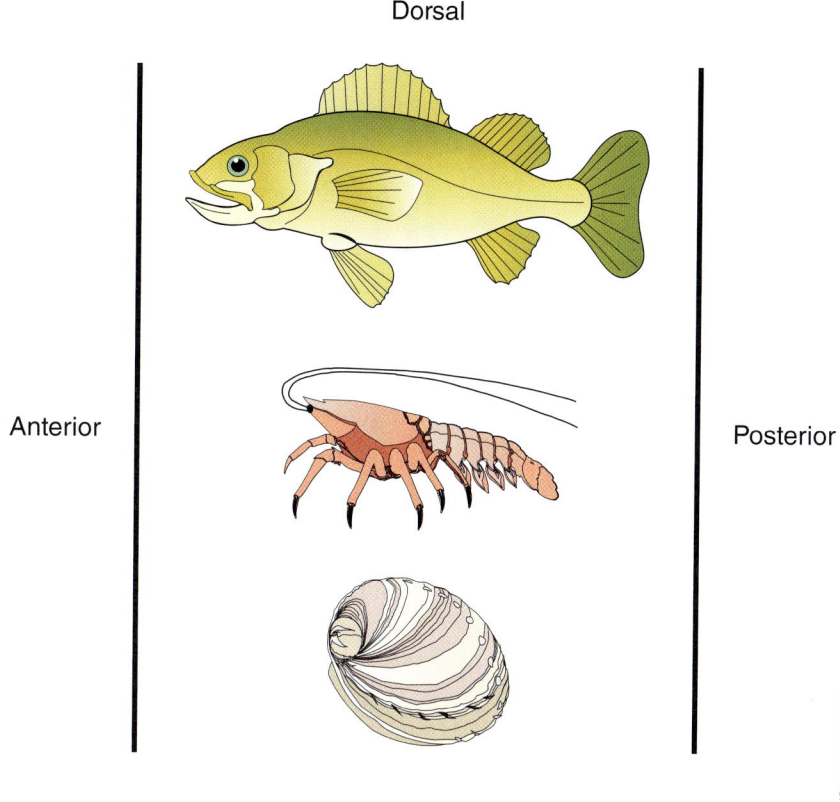

Morphology

Morphology, or the structure and form of fish, can affect feeding and the type of culture facility. For example, fish with small, upturned mouths generally are **herbivores** and/or surface feeders like tilapia. Fish with downturned mouths are generally bottom feeders like catfish.

Single-lobed or **homocercal** tail fins suggest that fish are slow swimmers and survive well in water that is free of much movement. Fish with forked or **heterocercal** tail fins are fast swimmers and prefer flowing water.

The body shapes of fish also suggest the type of culture facility. Fish like salmon, with long bodies that are tapered toward the ends (**fusiform**), are the best swimmers and need water space. Fish that are wide and flat or ventrally compressed tend to stay on the bottom and require lots of bottom space for growth. Laterally compressed fish are rounded and thin from side to side. These fish tend to hover in the water and are not particularly fast swimmers.

Physiology

Physiology is the function of the body systems of aquatic animals. These systems in aquatic species are adapted to the water environment. Nine body systems are found in animals, including aquatic animals. These systems are:

1. Skeletal
2. Muscular

3. Digestive
4. Excretory
5. Respiratory
6. Circulatory
7. Nervous
8. Sensory
9. Reproductive

Skeletal System

The skeletal system is a rigid framework that gives the body shape and protects its organs. This system is composed of bony or hard material and cartilage. Tissues and organs attach to the skeletal system. In aquatic animals, the skeleton can be internal or external. Fish possess an internal skeleton, or endoskeleton. Oysters, shrimp, and crawfish possess an external skeleton, or exoskeleton.

Muscular System

The muscular system provides movement internally and externally. Muscles vary in strength and function. Muscles contract and relax to cause movement. Organisms require movement for such functions as obtaining food and oxygen and eliminating wastes.

Digestive System

The digestive system converts feed into a form that can be used by the body for maintenance, growth, and reproduction. It consists of all the parts of an organism that are involved in taking food into the body and preparing it for **assimilation**, or incorporation into the body. In its simplest form, the digestive system is a tube extending from the mouth to the anus with associated organs. In most species, this includes the mouth, esophagus, stomach, intestines, anus, and other associated organs like the liver. Digestive systems vary according to whether the animals are herbivores eating only plants, **carnivores** eating only animals, or **omnivores** eating plants and animals.

Excretory System

Life processes produce waste products. The excretory system eliminates wastes from the body. Typically, it consists of the kidneys, urinary ducts, urinary bladder, and urinary opening. Kidneys filter wastes from the blood. The urinary bladder holds these wastes until they are excreted through the urinary opening.

Respiratory System

The respiratory system takes in oxygen from the environment, delivers it to the tissues and cells of the body, and picks up carbon dioxide from the tissues and cells, delivering it to the environment. Gills are the respiratory organs of fish, shellfish, and crustaceans. Water that these creatures take in is forced over the gills, where oxygen is removed by **diffusion** into the blood.

Circulatory System

The circulatory system distributes blood throughout the body. Generally, this system consists of a heart, veins, and arteries. The pumping action of the heart causes blood to flow through the arteries to the gills, where it picks up oxygen and carries it to the rest of the body. Oxygen is necessary for all cells of the body. As the blood delivers oxygen to the body's cells, it picks up carbon dioxide, a waste product, which is then carried in the blood back through the veins to the heart and gills. The gills release the carbon dioxide to the environment and pick up more oxygen.

Nervous System

The nervous system supplies the body with information about its internal and external environment. This system conveys sensation impulses—electrical-chemical changes—between the brain or spinal cord and other parts of the body. The nervous system consists of the brain, spinal cord, many nerve fibers, and sensory receptors. It is a complex system. Sense organs or receptors receive stimuli and convey these by nerve fibers to the brain or spinal cord, where they are interpreted. The brain or spinal cord may send responses to the stimuli back through the nerve fibers.

Sensory System

The sensory system includes the five senses—sight, touch, taste, smell, and hearing. The sensory system relays information through the nervous system. Fish use eyes to find food and identify predators. Ear bones in the skull pick up water vibrations as sound. The sense of taste is important to the aquafarmer when selecting and preparing feed for fish. Some species have an enhanced sense of touch through organs like the barbels on catfish. Lateral lines in fish contain nerves that detect water vibrations and motion. This helps keep groups of fish together in schools.

Reproductive System

Sexual reproduction is the process of creating new organisms of the same species through the union of the male and female sex cells—sperm and eggs.

Males and females exist in most species. Testes in the males produce sperm. Ovaries in the females produce eggs or ova. Fertilization occurs when the sperm unites with the egg forming a **zygote**. After a period of **incubation**, the zygote develops into a new organism. An understanding of the reproductive process is important to the success of the culture of a species. Some aquatic species reproduce **asexually**.

Anatomy

An understanding of the anatomy, or the internal and external structure, of aquatic animals is essential for the successful aquaculturalist. Understanding external anatomy helps distinguish between the sexes and spot abnormalities caused by disease.

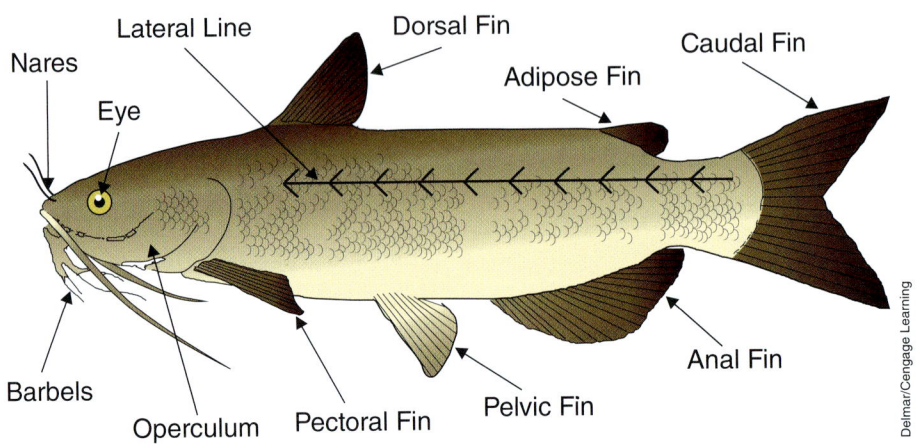

FIGURE 2-6 External anatomy of a finfish.

Anatomy of Finfish

Almost all fish used in aquaculture are considered bony fish with hard calcium-based endoskeletons. The skeleton gives the fish form and protects the internal organs, such as the digestive system, nervous system, and reproductive system. Figure 2-6 illustrates the external anatomy of a typical finfish.

Exterior coverings of fish vary. Bony plates or scales cover the skins of many fish, such as trout and carp. Scales grow as the fish grows. A few species, such as the catfish, have skin without scales.

Figure 2-7 shows the typical internal anatomy of a finfish and the location of major organs in the body system. Depending on the species, organs vary in size and shape.

The digestive systems of fish vary depending on the type of food eaten. Fish consuming algae and detritus have small stomachs and long intestines. Carnivorous fish possess large stomachs and short intestines.

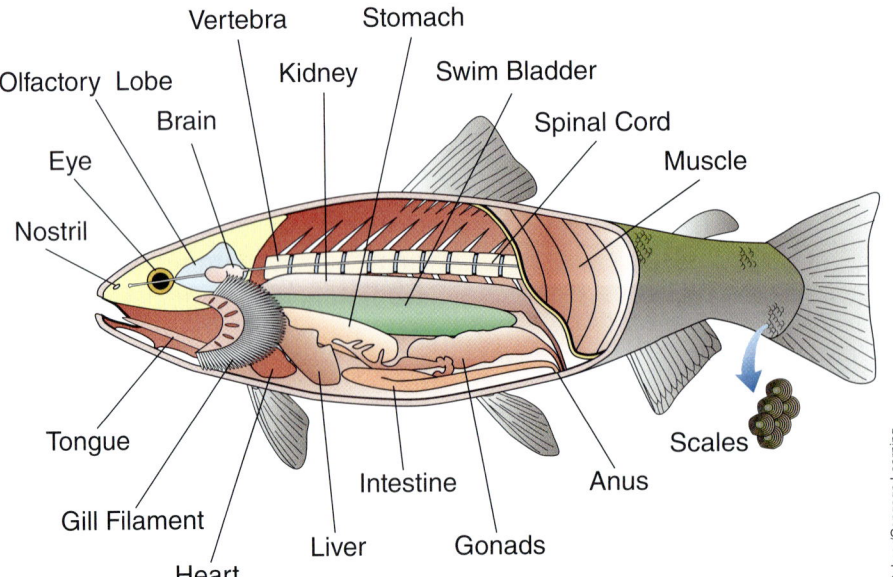

FIGURE 2-7 Basic internal anatomy of a finfish.

FIGURE 2-8 Gills are the lungs for fish and shellfish. Gills of fish are located on each side of the head (view A). They are covered by a protective movable flap of skin called the operculum. Four gills on both sides of the head each have a double row of slender gill filaments supported by a flexible white gill arch. Each side of the filament has many small cross plates called lamellae (view B). It is across the gill lamellae that the oxygen and carbon dioxide gases are exchanged. The lamellae have spaces through which blood rapidly percolates. Oxygen that is picked up at the gill lamellar surface is carried throughout the body in the blood. Waste carbon dioxide is also carried in the blood for release into the water at the lamellar surface.

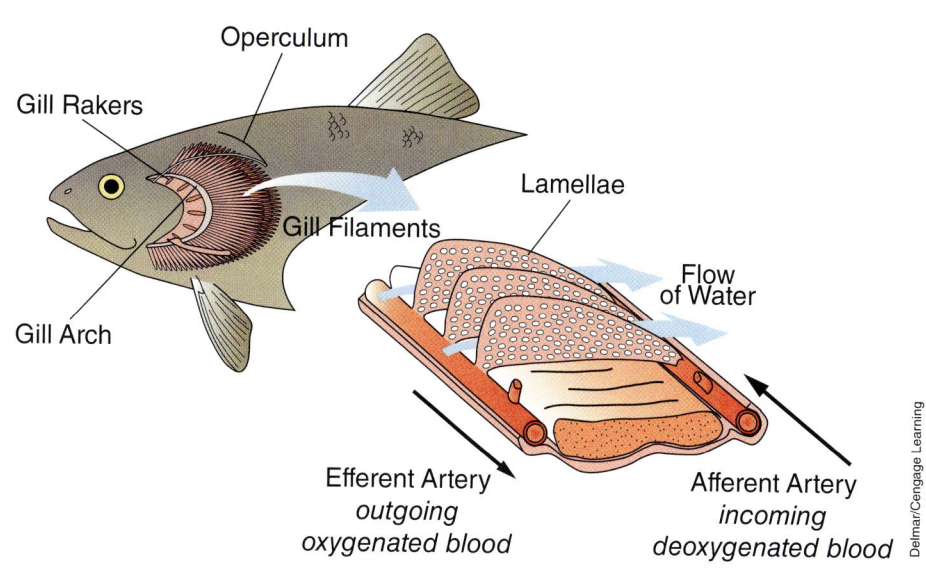

The nervous systems of fish are well-developed, with brains and spinal columns. The lateral line plays an important part as a sensory organ. It helps the fish maintain balance and position in the water.

Male and female fish are fairly easy to distinguish. The reproductive organs are located in the body cavity, but reproduction involves the fertilization of eggs laid in the water.

Gills remove oxygen from the water. Transfer of oxygen from the water to the bloodstream occurs by diffusion in the gill cells. Deoxygenated blood from the body that is pumped to the gills picks up oxygen and releases carbon dioxide. The higher concentration of oxygen in the water causes the oxygen to enter the blood, moving to an area of lower concentration (diffusion). Carbon dioxide in the blood is higher than in the surrounding water. Thus, it diffuses out of the blood through the gills. Cell membranes of gill cells are very thin and **semipermeable**, allowing gases to pass through. Figure 2-8 shows two views of fish gills.

Anatomy of Crustaceans

Crustaceans include shrimp, prawns, lobsters, crabs, and crawfish (crayfish). They all possess an exoskeleton made of chitinous material. Chitin is a polysaccharide—a carbohydrate—of a hexose (sugar) and also contains some tightly bound noncarbohydrate material, including proteins and **inorganic** salts. Most of the crustaceans considered for aquaculture are known as **decapods** (ten legs). The exoskeleton protects and supports the soft body, because all the muscles are attached to the inside of the exoskeleton. As a crustacean grows, the shell is cast off in a process called **molting**. When crustaceans molt, they are known as softshell animals. No more than a day is usually required to regrow the shell. During molting, crustaceans are subject to attack by other aquatic animals, including their own species.

FIGURE 2-9 External anatomy of a crawfish, dorsal view.

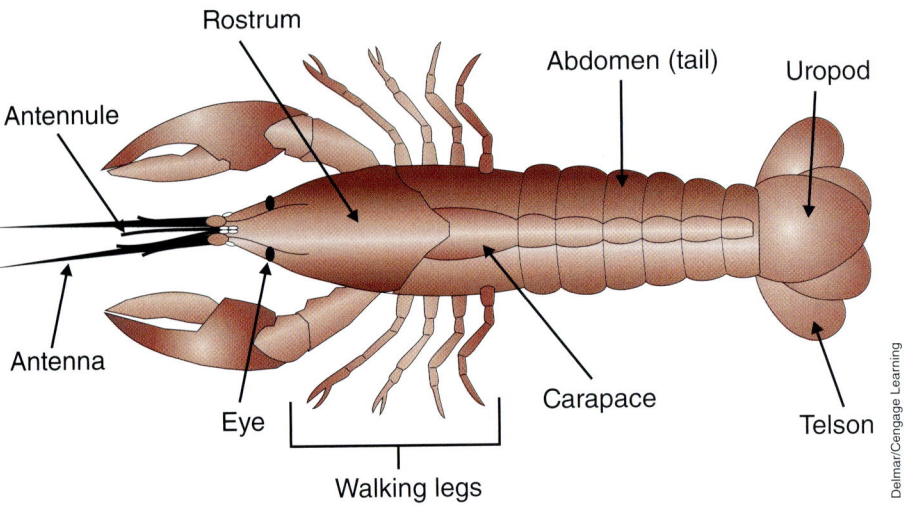

Figure 2-9 shows the external anatomy of the crawfish, a representative crustacean. The bodies of crustaceans are divided into three sections:

1. Head
2. Thorax (carapace)
3. Abdomen

Each segment has a pair of **appendages**. The head has two pairs of **antennae**. In crawfish, next to the antennae are the mandibles, or true jaws, and then two pairs of maxillae, or little jaws, that aid in chewing food. The jaws work from side to side, not up and down. The first appendages of the thorax are three pairs of maxillipeds or jaw feet. These hold food during chewing. Next come the large claws, for protection and food getting. The last four pairs of legs on the thorax consist of two pairs with tiny pincers at the tip and two more pairs with claws. The abdominal appendages of the crawfish are called "swimmerets" and are small on the first five segments. During reproduction, the female's eggs attach to her swimmerets. The sixth swimmeret develops into a flipper or uropod for locomotion.

Through a process known as **regeneration**, crustaceans regrow limbs that have broken off. This process is used to produce crab legs. One claw is removed, and the crab is returned to the water to grow another.

Figure 2-10 illustrates the internal anatomy of a crawfish. Crustaceans possess simple circulatory, nervous, and excretory systems. In the crawfish, the colorless blood is pumped by a very simple heart into several large arteries, which pour the blood over the major organs. Then the blood collects in spaces called sinuses and eventually returns to the heart. The nervous system of the crawfish consists of a brain and a ventral nerve cord. The acute senses of smell and touch are located in the antennae, maxillae, and maxillipeds. The compound eyes are on movable stalks, and sight is probably not keen. Hearing is poorly developed, but ear sacs located at the base of the antennules probably aid in balance.

Crustaceans use gills to breathe. In the crawfish, gills are located at the base of the legs or maxillipeds and protected by the thorax or carapace. Thus, the gills are exposed to water every time the legs or maxillipeds move.

FIGURE 2-10 Basic internal anatomy of a crawfish.

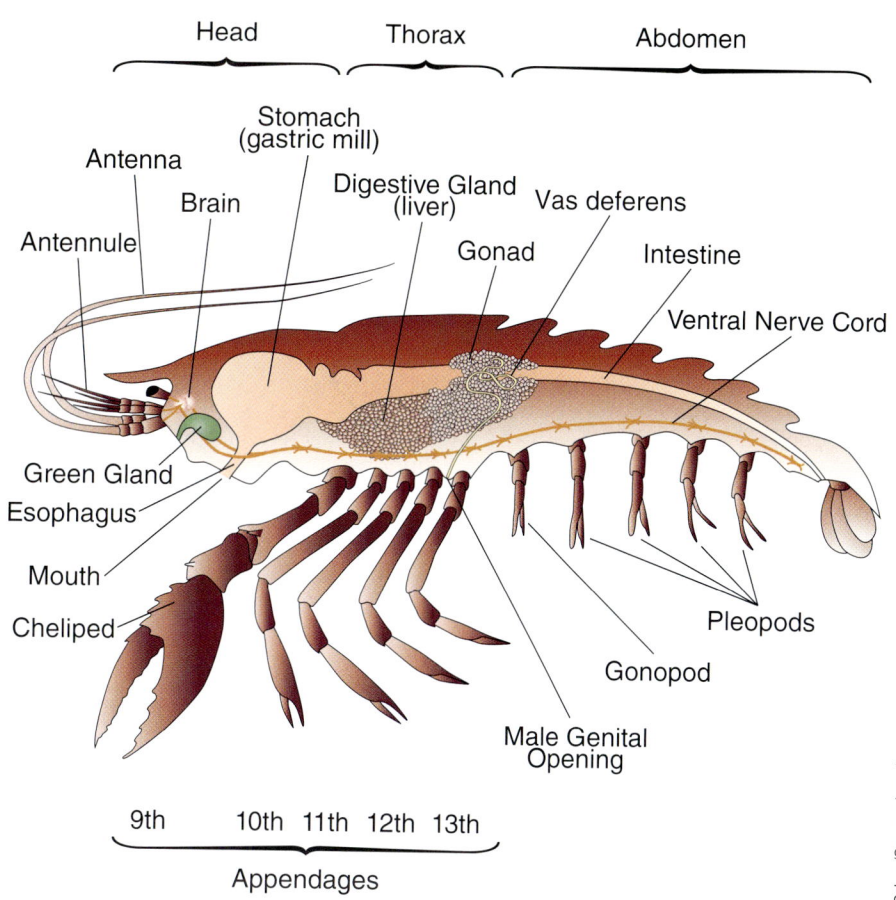

The crustacean life cycle and reproduction cycle is quite complex. Testes of males and ovaries of females are located inside the exoskeleton. A duct from the testes or ovaries leads to the outside for the release of sperm or eggs. Using one of the appendages, the male deposits sperm into a receptacle on the abdomen of the female.

Members of the pandalids group of shrimp all begin life as males. At about two years of age, they change to females.

Anatomy of Mollusks

In the United States, the most commonly cultured mollusk species include oysters and clams. These are **bivalve** mollusks. Two shells completely enclose the animal. These shells are made of a **calcareous** material that is very hard and resembles limestone. Anterior and posterior **adductor** muscles clench the shells together.

A muscular, hatchet-shaped foot can extend from between the shells. Clams use this for digging. The **mantle** lays over the internal organs and secretes the hard shell. The digestive system and nervous system are simple. A simple heart pumps the colorless blood to all parts of the body. Figure 2-11 shows the basic internal anatomy of a clam.

Gills not only serve as a respiratory system, but they filter material from water that is then either consumed or discharged. Small particles of matter stick to a thin mucous layer on the gills. The gill surfaces feature cilia, or small hairlike structures, which continually beat back and forth, carrying

FIGURE 2-11 Basic internal anatomy of a clam.

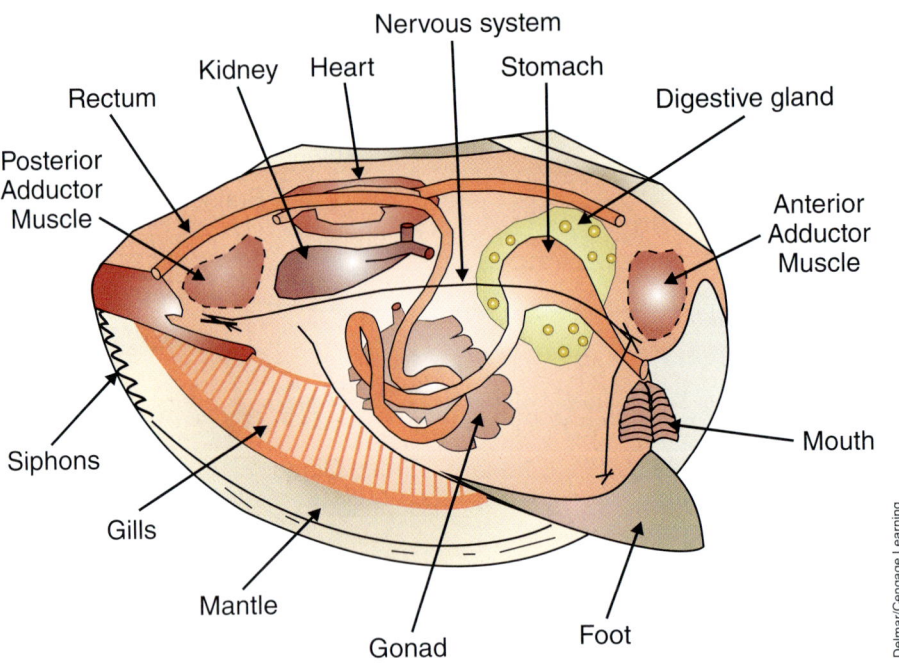

trapped material to the mouth. Water enters the mollusk through a **siphon** and passes over the gills. Next, water exits the mollusk through another siphon passing to the anus, where undigested matter is excreted.

Mollusks reproduce via egg and sperm. Typically, mollusks release their eggs into the water, which are then fertilized by waterborne sperm. Some bivalves are **protandrous**, meaning that they may change their sex one or more times during their lives. Some mollusks, such as scallops, are **hermaphroditic**, meaning individual organisms have gonads (testes and ovaries) for both sexes.

Univalves or **gastropods** include the snails, conches, and abalones. They have the same general anatomy as the bivalves, but only one shell.

Aquatic Plants

Aquatic plants share many characteristics of terrestrial (land) plants but are unique in other ways. Like land plants, aquatic plants make their own food through the process of photosynthesis. Photosynthesis requires light and **chlorophyll** to convert carbon dioxide and water to sugar (carbohydrates), oxygen, and water. Chlorophyll allows photosynthesis. Plants use the sun energy stored in the carbohydrates through respiration. The respiration process employs carbohydrates and oxygen to produce carbon dioxide, water, and energy. Plants take advantage of this energy for growth and reproduction. Figure 2-12 shows the complete chemical formulas for photosynthesis and respiration.

Microscopic algae such as diatoms, desmoids, blue-green algae, euglena, volvox, and filamentous green algae represent the smallest of the aquatic plants. Giant kelp reaching 200 ft. (61.0 m) or more in length represent the largest of the aquatic plants.

FIGURE 2-12 Chemical formulas for photosynthesis and respiration.

Photosynthesis	Respiration
$6CO_2 + 6H_2O + \text{Energy} \longrightarrow C_6H_{12}O_6 + 6O_2$ (carbon dioxide) (water) (light) (glucose) (oxygen)	$C_6H_{12}O_6 + 6O_2 \longrightarrow 6CO_2 + 6H_2O + \text{Energy}$ (sugar) (oxygen) (water) (carbon dioxide)
• constructive process • food accumulated • energy from sun stored in glucose • carbon dioxide taken in • oxygen given off • complex compounds formed • produces glucose • goes on only in light • only in presence of chlorophyll	• destructive process • food broken down (oxidized) • energy released • carbon dioxide given off • oxygen taken in • simple compounds formed • produces CO_2 and H_2O • goes on day and night • in all cells

Aquatic plants reproduce sexually—through the fusion of sex cells—or asexually. Many algae reproduce asexually by forming **spores**. Some algae produce spores that produce **gametes** or sex cells that fuse. Asexual spores, called "monospores," that are produced by young plants eventually become new plants. All forms of algae reproduce by the simple splitting of a cell into two. Some aquatic plants produce seeds—for example, watercress. Like potatoes, the fleshy corms of Chinese water chestnuts planted below the ground produce more corms. Finally, like terrestrial plants, some aquatic plants can be propagated by cuttings.

Because aquatic plants obtain most of their nutrients from the water, such plants are very useful for removing ammonia and nitrite wastes from water. This property makes aquatic plants an ideal component for polyculture because the wastes produced by the aquatic animals provide sufficient nutrients to the plants.

Like aquatic animals, aquatic plants differ in structure and occupy different positions in their environment. The two main groups of aquatic plants, categorized based on structure, are algae and **macrophytes**. Algae are primitive plants without true roots, stems, or leaves. Macrophytes are vascular plants with true roots, stems, and leaves. Planktonic algae occupy the space between the pond bottom and the surface. Filamentous algae form floating mats or hairlike strands attached to underwater objects, often called moss or pond scum. Macrophytic algae resemble true plants. They are large and are attached to the pond bottom.

Free-floating macrophytes are tiny green plants floating on the water's surface that resemble algae. However, they possess small leaves and roots that hang down into the water. Examples include duckweed or watermeal. Emergent macrophytes root in the pond bottom, but their leaves float or extend above the water surface. Examples include water lilies or lotus. Submergent macrophytes, such as pondweed or hornwort, are rooted in the bottom and grow completely underwater. Marginal macrophytes grow in very shallow water or wet soil on the edge of a pond or ditch. Examples include cattails and bulrushes. Figure 2-13 illustrates a common aquatic plant.

CARRAGEEN

Carrageen is a natural food gum obtained from red seaweed (algae). Old recipes indicate that freshly gathered seaweed was used to make meat gels, puddings, and broths. Crude carrageen was extracted from algae by cooking it with water or milk. The resulting solution was used to prepare gelled foods.

Carrageen is a giant molecule whose molecular weights range from 100,000 to 500,000. It is a polysaccharide made of chains of the sugar galactose.

Early Americans obtained the seaweed or algae called "Irish moss" from Ireland until about 1835. Then, Dr. J. F. C. Smith discovered that the seaweed could be obtained along the coast of Massachusetts.

A patent for the extraction of carrageen was granted to a Frenchman in 1871. Commercial production did not begin until 1937. Shortly after World War II, and due to the reduction of overseas trade, the industry became established in the United States.

Carrageen is used in batters, doughs, pastas, dairy products, fish, meat, poultry, fruit products, gelled desserts, relishes, salad dressings, and sauces. Special dietary products also use carrageen as a low-calorie emulsifier, thickener, and stabilizer.

FIGURE 2-13 A cattail, genus *Typha*, also called reed-mace, used to make mats, chair seats, and other items.

SUMMARY

Although other countries rely more heavily on aquaculture for food, aquaculture is gaining prominence in the United States. Catfish, trout, baitfish, and crawfish dominate U.S. aquaculture. Other aquatic species are being tried, and some appear to hold great promise for aquaculture after more research and development of the technology required for successful culture has been completed. Potential aquatic species for culture include a long list of finfish, crustaceans, mollusks (shellfish), and some aquatic plants.

For culture, aquatic species possess some advantages over land species. Selecting and successfully culturing an aquatic species requires an understanding of its biology, morphology, anatomy, and physiology. Successfully meeting the demands of a growing human population for food, feed, and energy means finding the best aquatic animals and plants for culture.

STUDY/REVIEW

Success in any career requires knowledge. Test your knowledge of this chapter by answering these questions or solving these problems.

True or False

1. North America ranks first in aquatic plant production worldwide.
2. Carrageen is an important chemical obtained from aquatic plants.
3. It is against the law to culture alligators commercially.
4. More feed is required to produce one pound of catfish than to produce one pound of beef.
5. Aquatic animals require large amounts of energy to regulate their body temperatures.
6. Primary producers grow rapidly and form the first link in the food chain.

Short Answer

1. _____ is a group of minute plants, and _____ is a group of minute animals important as food sources.
2. As a group, the colorful small fish used in aquariums are called _____ .
3. After spawning, small eels called _____ migrate upstream from the sea.
4. Successful culture of any aquatic animal requires a stable supply of _____ .
5. Name three things that make up a market.
6. Give the scientific and common names for the two freshwater finfish and one crustacean that currently dominate U.S. aquaculture.
7. Name two aquatic plants used for food, two used for waste-water treatment, two used for food feed, and two used for phycocolloid production.
8. List five salmonids.
9. List three catfish besides channel catfish.
10. List six saltwater or brackish water aquatic animals that could be—or are being—cultured.
11. Name two aquatic animals and two aquatic plants that could be used in polyculture.
12. With a pond of 200 female channel catfish weighing an average of 4 lbs each, how many total eggs could a producer expect when each female spawns?
13. Name the nine body systems found in aquatic animals.
14. List three different methods of reproduction in aquatic species.
15. For which aquatic species are molting and regeneration important processes?
16. Give two examples of algae and four examples of macrophytes.
17. Name two plants without true roots, stems, and leaves and two plants with true roots, stems, and leaves.

Essay

1. Explain the difference between phytoplankton and zooplankton.
2. Explain the differences between mollusks and crustaceans.
3. Why do eels, bullfrogs, and alligators hold potential as aquaculture species?
4. Using specific plants and animals, describe a polyculture system.
5. In what ways do aquatic animals save energy compared to terrestrial animals?
6. When selecting a species for possible aquaculture, identify six factors to consider.
7. What is the function of the skeleton? Name two types of skeletons.
8. Give an example of how morphology can affect the culture techniques an aquaculturist chooses.
9. Briefly describe how gills work.
10. Briefly define the following external anatomical features: adipose fin, lateral line, operculum, uropod, thorax, maxillipeds, and foot.
11. Compare the location of the gills in finfish, crustaceans, and mollusks.
12. Why is a basic understanding of internal and external anatomy important to the future aquaculturalist?

KNOWLEDGE APPLIED

1. Obtain some common pond water. Using a microscope and a picture guide such as *Pond Life: Revised and Updated* (A Golden Guide from St. Martin's Press) by George K. Reid, identify the aquatic life found in a pond. Collect the water in a clean, one-quart jar. To increase the number of plankton in the sample, pass a plankton net or nylon stocking through the pond and put the contents into the sample. Prepare microscope slides using a drop of pond water and a drop of methyl cellulose to slow the plankton movement. Add a cover slip and examine the sample at different levels of magnification.

2. As an alternative to using pictures to identify microorganisms in a pond water sample, use the dichotomous key in Appendix Table A-16 Place the water sample under a microscope as described in Question 1. Once the microorganism to be identified is in view, read the first two descriptions in the key. Decide which one better describes the microorganism. Next, go to the description number indicated at the right of the best description. Continue the process until the name of the microorganism occurs to the right of the description. Draw the microorganisms observed.

3. Choose an aquatic species from Table 2-2 or Table 2-3. Learn the common and scientific names. Develop a more complete profile of the species selected. Include such items as description, natural food, habitat, distribution, behavior, reproduction, larvae, adult size, edible qualities, culture possibilities, yield in culture, feeding, predators, diseases, harvest, and marketing. If a whole class did this for a number of species, the information could be entered into a computerized database so that all species could be compared on a variety of characteristics. For many species, more information can be found in Chapters 4, 7, and 8.

4. Investigate the external and internal anatomy of a finfish. Obtain some fresh, frozen, or preserved finfish. Identify all the external structures as well as the dorsal, ventral, anterior, and posterior areas. Lift the operculum and identify the gill arch, gill rakers, and gills. Dissect the fish and identify parts of digestive system,

reproductive system, and circulatory system. Open the skull and find the brain. Identify any sensory organs located internally or externally. Several books listed in the Learning/Teaching Aids can help you understand the anatomy of finfish.

5. Obtain fresh, frozen, or preserved crawfish. Identify the following external features: abdomen, carapace, head, swimmerets, maxilliped, and uropod. Dissect the crawfish and identify the internal structures, including the gonads, hind gut, gills, abdomen muscle, and any other parts of the body systems.

6. Obtain fresh, frozen, or preserved clams or oysters. Before opening the shells, identify the dorsal, ventral, anterior, and posterior surfaces. Pry open the shell and cut the adductor muscle. Identify the mantle, gonad, stomach, gills, mouth, food, intestine, kidney, heart, and siphons.

7. Conduct a survey of food labels. Find out how frequently and in what foods a phycocolloid, namely carrageen, is used.

8. Conduct a taste test of some aquatic plants used for food, like brown algae (wakame), water spinach, watercress, and Chinese water chestnut. Investigate how these plants are used in recipes by different cultures.

LEARNING /TEACHING AIDS

Books

Avault, J. W. (1996). *Fundamentals of aquaculture: A step-by-step guide to commercial aquaculture.* Baton Rouge, LA: AVA Publishing.

Chapman, V. J. (1980). *Seaweeds and their uses.* New York, NY: Springer Publishing.

Falkowski, P. G., and Raven, J. A. (2007). *Aquatic photosynthesis,* 2nd Ed. Malden, MA: Blackwell Science.

Fichter, G. S. (1988). *Underwater farming.* Sarasota, FL: Pineapple Press, Inc.

Pillay, T.V. R., and Kutty, M. N. (2005). *Aquaculture principles and practices.* Ames, IA: Blackwell Publishing Ltd.

Riemer, D. N. (1993). *Introduction of freshwater vegetation.* Malabar, FL: Krieger Publishing Co.

Stickney, R. R. (2005). *Aquaculture: An Introductory text.* Cambridge, MA: CABI Publishing.

Stickney, R. R. (2000). *Encyclopedia of Aquaculture.* Malden, MA: Wiley-Interscience.

Stickney, R. R. (1991). *Culture of salmonid fishes.* Boca Raton, FL: CRC Press.

Swann, L., and Swann, L. (1999). *Getting started in freshwater aquaculture:* West Lafayette, IN: Purdue University.

Usui, A. (1991). *Eel culture,* second edition. Boston, MA: Blackwell Science, Inc.

White, S. K. (1992). *A field guide to economically important seaweeds of northern New England.* University of Maine/University of New Hampshire Sea Grant Advisory Program.

Internet

Internet sites represent a vast resource of information. The URLs (uniform resource locators) for the World Wide Web sites can change. Using a search engine such as Google, find more information by searching for these words or phrases: anatomy of aquatic animals; physiology of aquatic animals; morphology of aquatic animals; aquatic plants; algae (specific names); phytoplankton; anatomy of mollusks; anatomy of crustaceans; ornamental fish; aquarium fish; tropical fish; bullfrogs; alligators; eels; zooplankton; pond plankton; animal surfaces (dorsal, ventral, anterior, posterior); monocercal or heterocercal tail fins; body systems of aquatic animals such as skeletal, muscular, digestive, excretory, respiration, circulatory, nervous, sensory, or reproductive; gills; photosynthesis of aquatic animals; respiration of aquatic plants; or macrophytes.

For some specific Internet sites, refer to Appendix Table A-14.

Marketing is the process of getting a product from producer to consumer. It is the final step in food production but should rate top priority in the mind of an aquaculturalist. Even though the aquaculturalist may possess the skills and resources to grow a crop, those efforts are in vain without a place to sell the product.

CHAPTER 3

Marketing Aquaculture

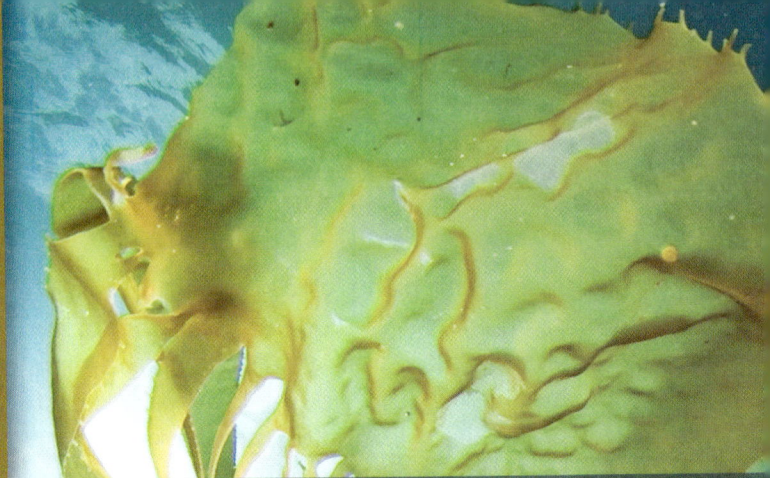

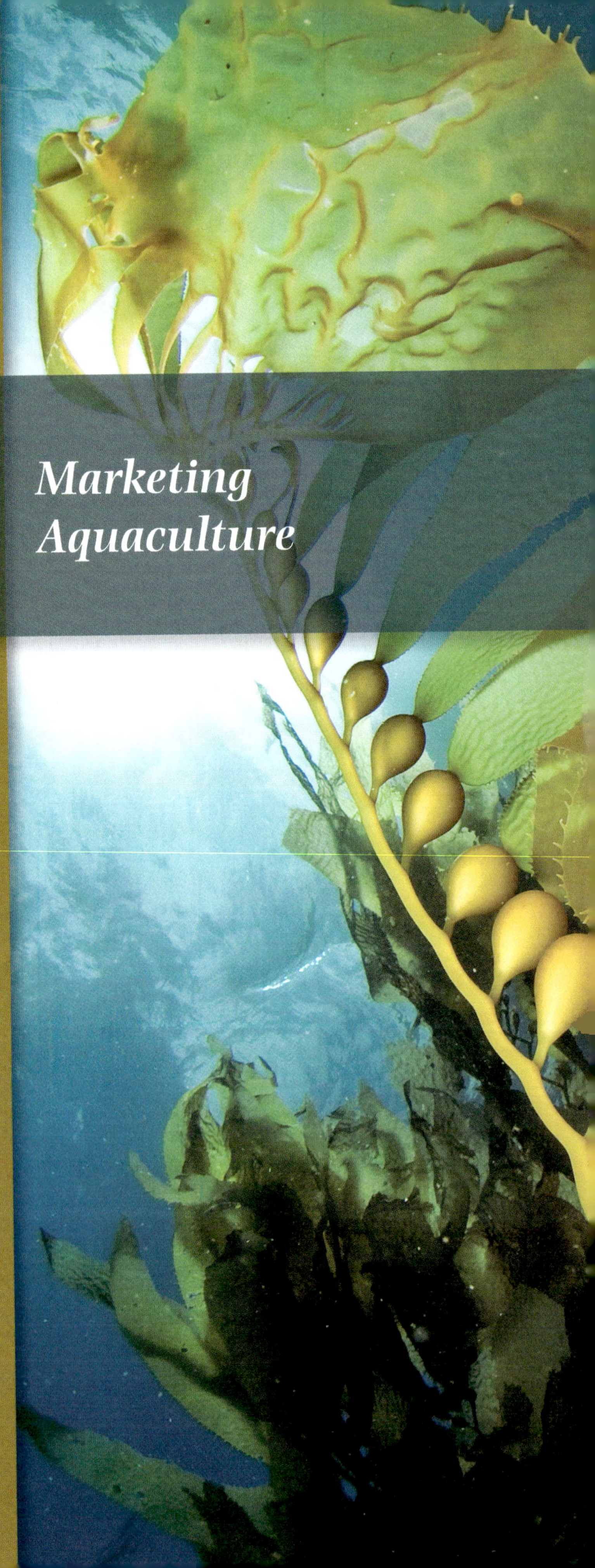

The first step in marketing is to understand the current production and consumption of a product. Next, producers need to understand marketing functions and strategies. Some of the specific details of marketing and processing for each species are covered in Chapters 4, 6, and 8.

OBJECTIVES

After completing this chapter, the student should be able to:

- Define marketing
- Describe the process of marketing aquaculture
- Explain the elements in developing a marketing strategy
- Explain the importance of developing a marketing plan
- Identify possible market outlets for aquaculture products
- Select an appropriate market
- Explain costs in marketing
- Describe the process of market promotion in aquaculture
- Identify terms related to marketing with their correct definitions
- Discuss quality control
- Describe some scientific skills required to maintain quality fish and fish products

- Recognize that development of a marketing plan and strategy requires research
- Describe processing
- Describe the grading process
- List factors to consider when exploring marketing alternatives
- Identify food fish processing cuts and forms with their correct descriptions

Understanding of this chapter will be enhanced if the following terms are known. Many are defined in the text, and others are defined in the glossary.

KEY TERMS

Assembling	Marketing
Branded	Off-flavor
Deheading	Offal
Demand	Processors
Distributors	Product pull
Enrobing	Product push
Eviscerator	Promotion
Fillets	Quality assurance
Grading	(or control)
HACCP	Shelf life
Inputs	Stunned
Inspections	Suppliers
Live-haulers	Value-added

INTERNATIONAL PRODUCTION

Fourteen countries or areas produce about 90 percent of all aquaculture products. As Figure 3-1 shows, the United States rates among the top ten. Aquaculture production in China, Japan, India, and Korea overshadows all other countries, as illustrated by Figure 3-2.

Table 3-1 (page 59) indicates the types of species traditionally cultured in these countries.

This chapter is about marketing aquaculture products produced in the United States in the U.S. market. Trout and catfish represent two success stories in marketing. Other species in the United States represent success stories for niche marketing.

CONSUMPTION

Marketing requires understanding the competition. Other meats compete with fish. Slowly and gradually, however, fish and seafood consumption in the United States has increased. Figure 3-3 (page 59) illustrates this increase in consumption.

Catfish, salmon, trout, and crawfish make up 90 percent of U.S. production. The other 10 percent includes striped bass, eel, alligators, and tilapia.

Although the per person consumption of fish and its increase look small compared to other meats, a small increase in per person consumption translates to big increases in production. For example, the per person consumption of catfish increased from 0.4 lb in 1985 to about 1.1 lb. in 2009. Assuming that the U.S. population is about 300 million, this translates to about 660 million lbs of live fish. Figure 3-4 (page 60) shows the historical trend of the per person consumption of fish only. This illustrates the gradual increase with consumption holding steady for the last few years.

MARKETING BASICS

Efficient aquaculture production matters little if the crop cannot be sold for a profit. Where and how an aquacrop will be sold should be the first concern of a producer. This means developing a marketing strategy or plan.

Marketing Plan

Depending on the operation, marketing plans can be long documents or informal plans of several pages. Developing a plan allows the producer to analyze

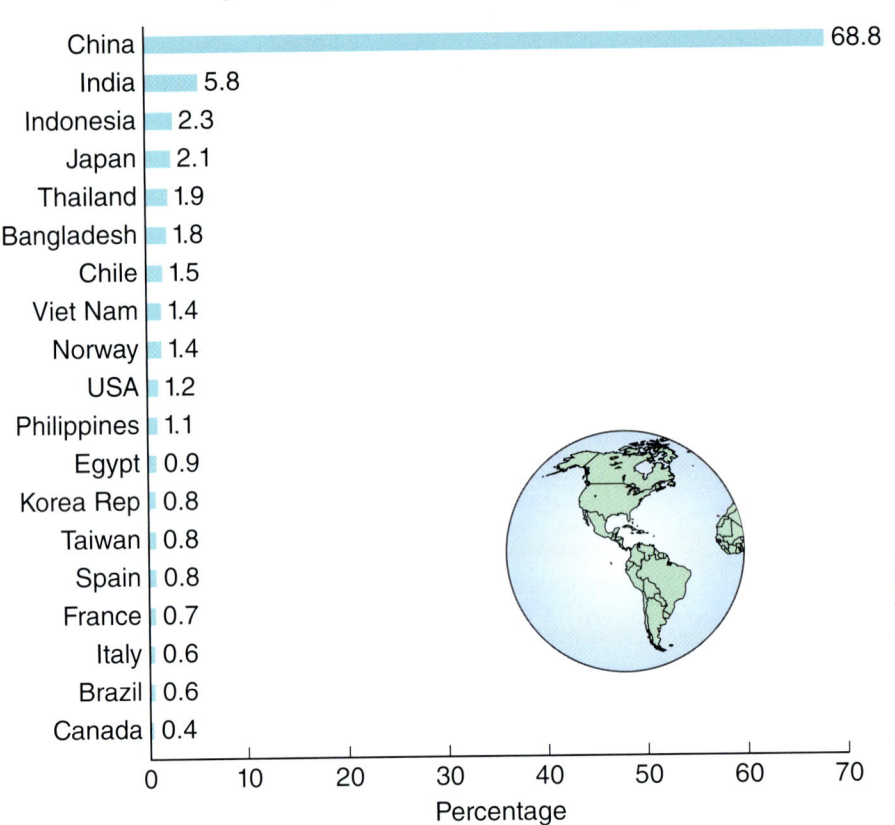

Figure 3-1 Top 14 aquaculture-producing countries based on the percentage of aquaculture's contribution to the national aquatic production.

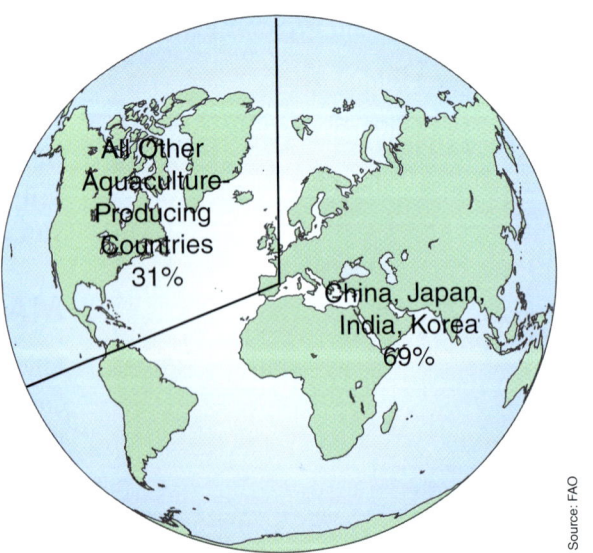

Figure 3-2 More than two-thirds of the world's aquaculture is produced in China.

TABLE 3-1 AQUACULTURE SPECIES CULTURED WORLDWIDE[1]

Country or Area	Species or Species Group Cultured
China	Carps, mollusks, shrimp
Japan	Amberjack, mollusks, algae
Taiwan PC	Eel, mollusks, shrimp
Philippines	Milkfish, shrimp, algae
USA	Catfish, pacific salmon, trout, mollusks
Russian Federation	Carps
Norway	Atlantic salmon, rainbow trout
Ecuador	Whiteleg shrimp
Indonesia	Carps, milkfish, shrimp, algae
Korea	Mollusks, algae
France	Rainbow trout, mollusks
Vietnam	Freshwater fish, crustaceans
India	Freshwater fish, crustaceans
Spain	Rainbow trout, mollusks
Thailand	Crustaceans, mollusks
Bangladesh	Freshwater fish
Italy	Rainbow trout, mollusks
Scotland	Atlantic salmon

[1]FAO, Rome, Italy.

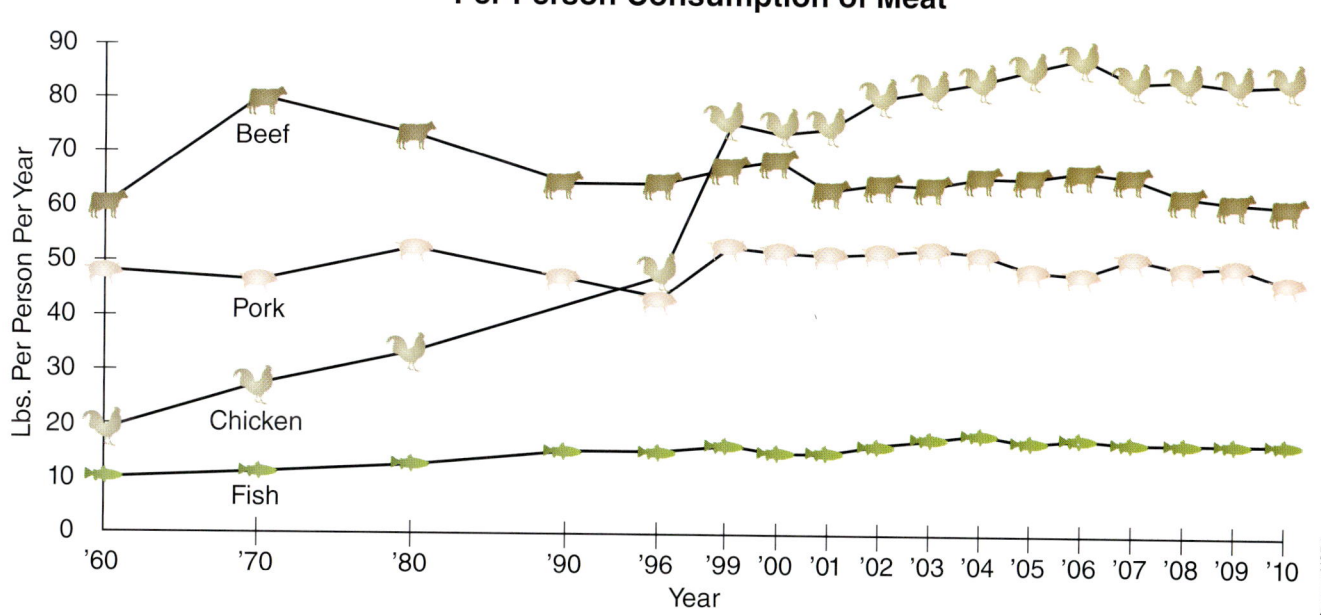

Figure 3-3 Trends in per person consumption of beef, poultry, pork, and fish.

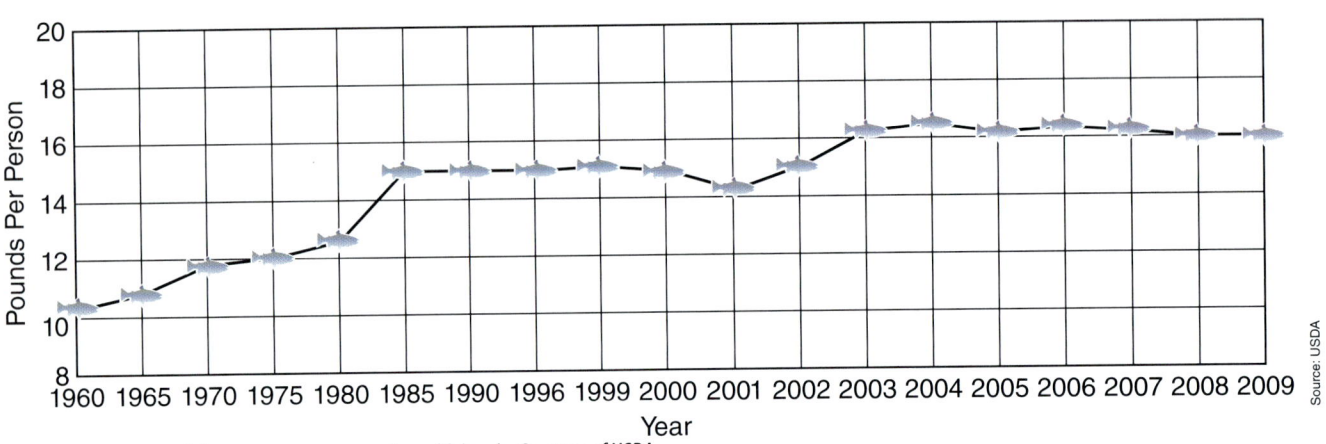

Figure 3-4 History of the per person consumption of fish only. Courtesy of USDA.

opportunities and needs. With this done, the producer can focus on production and make better decisions. Market plans or strategies contain three key elements:

1. Determination of the present situation
2. Determination of market goals
3. Developed plans to reach the goals

Often, separate marketing plans are developed for new products and for the continuing or annual plan.

Targeting the Buyer

New product introduction is a risky business. Successful new products can make a company and keep it competitive in its industry, while providing steady outlets for its input **suppliers**, such as farmers. Failure rates are high, so it is very important for suppliers, manufacturers, and **distributors** to understand the forces that affect new product success.

A new product's path from development to market acceptance depends on the type of buyer targeted. There are two basic buyer types: intermediate users, such as **processors** and manufacturers, and final consumers. Marketing channels and the prerequisites of success vary depending on which type of buyer is targeted.

Manufacturers—intermediate users—use new crops and new products from existing crops and foodstuffs as intermediate **inputs** in producing final goods. This market is made up of professional buyers who base purchasing decisions on strict price/quality specifications and who are highly knowledgeable about the availability of substitute inputs. Convincing them to buy a new product requires being responsive to their price, quality, quantity, and delivery needs.

Where the targeted buyer is the final consumer, the selling environment differs. Though consumers make the same type of price/quality comparisons as professional buyers do, they usually have less complete information, and factors such as brand name, advertising, packaging, coupons, convenience, and image play a bigger role. These elements make communication a crucial factor in successful new product introduction.

The consumer market also differs because often the producers are not in direct selling contact with the buyers, as they are in the intermediate goods market. The retail distribution chain links the two, so that the producer is faced with a double selling job—to convince the retailer to carry the product and the consumer to buy it.

In the two general areas—intermediate buyers or consumers—aquaculturalists sell or market their product to:
- Processing plants
- Live haulers
- Local stores and restaurants
- Backyard or pond bank sales
- Fee-fishing

A market should be selected based on the potential profits according to the scale of the operation. Each option should be carefully analyzed.

Intermediate Goods

Three key elements that largely shape the environment in which intermediate goods compete are:
1. The product
2. The buyer
3. The marketing system

Product characteristics

Intermediate agricultural goods are sold as inputs for further processing and distribution. Buyers need reliable information on the product's technical and functional characteristics. The producer must be able to demonstrate how the new product performs in its intended application. Buying decisions hinge on whether the new product contributes to the buyer's profit. Price is clearly an important consideration.

If the new product does not offer a price advantage relative to alternative inputs, then it must offer some performance edge in the manufacturing process.

Selling an intermediate good may also require the ability to customize the product to the buyer's specifications. Adapting to a particular buyer's needs may entail physical changes in the product, packaging changes, or changes in delivery methods.

Buyer characteristics

Intermediate buyers differ substantially from household buyers. Buyers of intermediate goods are well informed about prices and product characteristics. Consumers of final goods may be willing to buy a new product on impulse, but an intermediate buyer purchasing a vital input needs to know much more about a product before committing to a new supplier.

Compared to most household consumers, intermediate buyers face higher switching costs and risks in trying new products. Switching costs are one-time costs of changing to a new supplier. Such costs and risks for a consumer trying a new food or fiber product generally are small—an improperly cooked meal or an unenjoyable dining experience may be the only result of an unsuccessful experiment. For an intermediate buyer, switching costs and risks may be large.

In many cases, the seller of a new intermediate good needs to work with buyers to develop new product formulas. To provide assurance to buyers, the supplier may have to assume some of the financial risk that accompanies the switch to a new product.

Prices in producer goods markets adjust frequently to changing market conditions. Sellers of a new product must be prepared to negotiate prices with the buyer rather than simply offer a take-it-or-leave-it price.

Marketing system characteristics

Distribution networks play an important role in determining a product's success. For agricultural goods, shipping costs frequently are high relative to the product's value. Many agricultural products are bulky and costly to move long distances. Markets are likely to be local or regional in scope, presenting opportunities for entry by smaller scale operations—a niche market.

In addition to geographic market concentration, both buyer and seller concentration are important strategic considerations for the developer of a new product. Seller concentration helps determine the potential response by competitors to a new product's introduction. If concentration is high, with just a few large sellers, then existing sellers may compete vigorously to minimize sales lost to a new entrant. If the selling industry is competitively structured with many smaller producers, then a new product may be able to enter the market with little response from existing firms.

Product to the Consumer

Winning acceptance for new products is of key importance to food manufacturers because introducing such products is a centerpiece of their marketing strategies. The process has two stages: creating consumer **demand** (**product pull**) and encouraging distributors to give the product shelf space (**product push**).

Creating product pull

The ultimate success of a new product depends on generating strong consumer demand. Manufacturers of **branded** products seek to develop offerings with the price, quality, and convenience consumers will want, using advertising and coupons to make consumers aware of them. Advertising is a particularly important strategy for gaining new product acceptance because it plays a dual role. It builds consumer demand, and it signals to retailers and other manufacturers that the company is committed to spending the resources necessary to support the product in the early stages. Large advertising expenditures are routinely involved in introducing new branded products. This type of support builds the demand pull necessary to establish a new supermarket product.

Providing product push

Successful introductions in the retail channel also require the manufacturer to provide push for the new product. Push refers to incentives offered to wholesalers and retailers to carry the product. (See Figure 3-5.) Some are offered across the board, such as special introductory prices and free goods, whereas others are negotiated individually. Wholesalers and retailers frequently use buyers and buying committees to evaluate whether a new product is unique enough and has sufficient manufacturer support to merit shelf space. Recent research indicates that most products do not make it past this stage. This is not surprising, because about 90 percent of new products are extensions, for example, new flavors, of existing lines.

Push is necessary because retailers face restrictions in accommodating new products. Although average store size has increased, product numbers far outpace available shelf space, giving retailers a strategic advantage in choosing products to carry.

Additionally, new product introductions generate costs, such as establishing warehouse slots, resetting retail shelves, and changing store computer files. Product failures also generate costs. Given these strategic and cost factors, wholesalers and retailers have increasingly demanded more trade support (push) dollars for new products. One form this takes is charging slotting (and sometimes failure) fees to manufacturers.

Figure 3-6 summarizes how aquaculture products can be marketed.

Figure 3-5 Product push helps create space on the supermarket shelf for fish and seafood products.

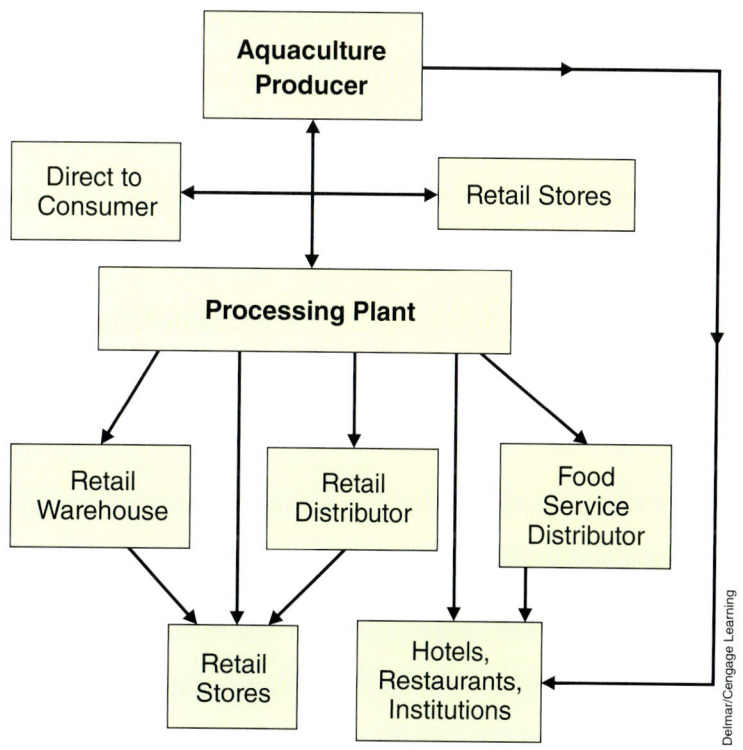

Figure 3-6 Marketing flowchart in aquaculture.

ADVERTISING

Advertising is any paid form of non-personal presentation of goods. It is a part of marketing and as old as recorded civilization. Ancient Greeks and Romans used advertising. The walls of ancient Pompeii and Herculaneum contain notices painted in black and red. A painted notice on a wall in Pompeii tells travelers about a tavern in another town. Another advertisement excavated in Rome offers property for rent. In the Middle Ages, merchants hired criers to walk the streets and cry the wares for their clients. Later, town criers became familiar figures on the streets of colonial America.

The first printed advertisement in English appeared in 1648. In 1704 colonial America, *The Boston News Letter* contained the first newspaper advertisement. By the middle of the twentieth century, advertising appeared in newspapers, magazines, direct mail, on billboards, cars, matchbook covers, radio, and television. Anywhere else an ad agency can think to place a piece of advertising, it will—T-shirts, public transportation, cigarette lighters, walls, and in movies. Advertising is big business.

United States companies spend billions of dollars each year on advertising. This money is spent mainly for advertisements in magazines, in newspapers, on television, on radio, and on the Internet. About $10 billion is spent advertising all types of food and beverages. Ad agencies know what they are doing. They employ a research staff, creative layout artists, copywriters, scriptwriters, graphic artists, and salespeople. Often they employ a recognizable face or voice to deliver their message. What they do works, and advertising is an important part of U.S. business.

Marketing Activities

Marketing of aquaculture products involves some characteristic activities. These include **assembling**, **grading**, transporting, changing ownership, processing, packaging, storing, wholesaling, retailing, and advertising. Not all marketing involves all of these activities.

Assembling

Assembling is collecting aquaculture crops from different production sites at a central location so that the volume to be processed will be large enough for efficient use of the processing facilities.

Grading

Grading is ensuring that the aquaculture crop batch is of uniform size and species. A grader may be used to screen out animals that are too large or too small. Some of this is done at harvest.

Transporting

Transporting means moving the aquaculture product to a location where it is to be processed. Most animals should be kept alive and in good condition until the time of processing. Specialized haul tanks with aerators and oxygen injection systems may be needed.

Changing Ownership

Most crops are sold several times between the farm and the consumer. Initially, the producer sells fish to the processor based on the weight of the fish at the time of delivery. The change of ownership involves the seller and buyer agreeing on the amount sold and on a price.

Processing

This involves a number of procedures to prepare fish for consumption. With fish, processing typically involves removing the skin and viscera, cutting into portions, pre-seasoning or cooking, and properly disposing of the waste products from processing.

Packaging

Consumers want to buy products that are packaged attractively and easy to use. (See Figure 3-7.) Packaging should also keep the food safe and wholesome. Package labels describe the product and how it is to be prepared.

Storing

Aquaculture products are stored several times between the farm and the consumer. Tanks are used at the processing plant to keep the fish alive until processing. Refrigeration and freezing are used with many fish and shellfish to preserve and store them. Canned products may be stored in large warehouses and at supermarkets.

Wholesaling

The processor sells the product to distributors (jobbers) or retail outlets. A price level is established so that the processor can make a profit.

Figure 3-7 Attractive packaging and brand names assure the consumer of quality.

Retailing

Selling to the consumer, restaurants, supermarkets, and fish markets may be involved. Attractive merchandising is needed.

Advertising

Consumers need to be aware of aquaculture products. Advertising develops awareness and encourages consumers to buy the product. Newspapers, radio and television, signs, and other means of advertising may be used. Grower associations, processors, and local stores may sponsor the advertisements. Advertising is also known as "product **promotion**." (See Figure 3-8.)

Selecting a Market

The producer must select a market. Markets vary according to the species produced, the location, and the amount returned to the producer. Some factors to consider when selecting a market for aquaculture products include profit, equipment, accessibility, species, quantity, size, and quality.

Profitability

Select the market that provides the greatest return on investment to the producer. The highest price per pound may not provide the largest profit if expenses are involved. Producers need to keep records that allow them to calculate the cost per pound to obtain or produce the product.

Figure 3-8 Depending on the location and the size of the market, any promotion is helpful.

Management is the secret ingredient to successful aquaculture. Management involves knowledge of the species being cultured—sources of the species, habitat, seed stock and breeding, accepted culture methods, stocking rates, feeding, diseases, processing, and marketing. With knowledge, the successful aquaculturalist uses good judgment.

CHAPTER 4

Management Practices for Finfish

No attempt is made in this chapter to cover all species that are, can be, or were cultured. Rather, this chapter covers a wide range of selected species. Management practices are similar for many species.

OBJECTIVES

After completing this chapter, the student should be able to:

- Describe the purpose and functions of a hatchery
- Describe the spawning facilities used in aquaculture
- Define harvesting
- Describe harvesting methods
- Arrange in order the phases of fingerling production
- Describe stocking rates for various stages of production and various species
- Describe trout culture
- Explain broodfish management
- Discuss egg management after fertilization
- Describe fry and fingerling management
- List general management guidelines for different species
- Describe the baitfish industry
- Explain the methods of pond preparation and fertilization for different species

- List control techniques for predators
- List guidelines for transporting fish to long-distance markets
- Describe the commercial culture of tilapia
- Describe different production systems used by various species
- Define terms related to harvesting and hauling
- Explain how sex is determined in fish
- Discuss methods of controlling reproduction in fish
- Describe the sexual reproduction processes of aquatic animals
- List salmonids that could be or are cultured
- Describe the commercial production of hybrid striped bass
- Identify popular baitfish species
- Demonstrate familiarity with scientific names of different aquatic animals
- Describe breeding systems and their purposes
- Understanding of this chapter will be enhanced if the following terms are known. Many are defined in the text, and others are defined in the glossary.

KEY TERMS

Anadromous	Microsporidean
Benthic	Milt
Catadromous	Mouthbrooders
Clarification	Nursery
Crossbreeding	Phenotypes
Density Index	Photoperiods
Detritus	Production ponds
Eyed stage	Ranching
Farming	Recruitment
Feeding chart	Rotational line crossing
Genes	Salmonids
Hormones	Sexing
Hybrid vigor	Spawning
Hybridization	Standing crop
Hybrids	Stocking rate
Inbreeding	Substrate
Inventory	Volumetric
Metabolites	

SPAWNING

Sexual reproduction involves egg production by the female ovaries and sperm production by the male testes. Reproduction is critical to successful aquaculture, as is an understanding of reproduction. **Spawning** is the act of obtaining eggs from the female and sperm or **milt** from the male.

In nature, most finfish are seasonal breeders. Reproductive cycles are controlled by **hormones** produced by endocrine glands. Figure 4-1 shows the approximate location of the endocrine glands in a fish. The production and release of the hormones is controlled by environmental stimuli—internal or external. Under natural conditions, climatic changes such as day length and temperature act as stimuli. Environmental stimuli interpreted by areas of the brain influence the release of hormones, as shown in Figure 4-2. Besides controlling the production of eggs and sperm cells, the reproductive hormones control secondary sexual characteristics, such as coloration and breeding behavior. Hormones and their actions are described in Table 4-1 (page 93).

Often, reproductive cycles are artificially controlled to ensure continuous seed production. Three approaches can be used to control reproduction—
- Genetic
- Environmental
- Hormonal

Genetic

Through genetic selection, early-maturing or late-maturing broodstocks can be developed. Such genetic management of the broodstock stretches the breeding season. This is a relatively difficult task.

Environmental

Controlled light periods—**photoperiods**—have been used with several species of fish to manipulate spawning time. Salmon exposed to shortened periods of light spawn appreciably earlier. Egg mortalities can be significantly higher. Light, not temperature, is apparently the prime factor in accelerating or retarding sexual maturation in this species. Artificial light has been used successfully to induce early spawning in brook, brown, and rainbow trout. Rearing facilities are enclosed and

92 AQUACULTURE SCIENCE

FIGURE 4-1 Location of endocrine glands in fish.

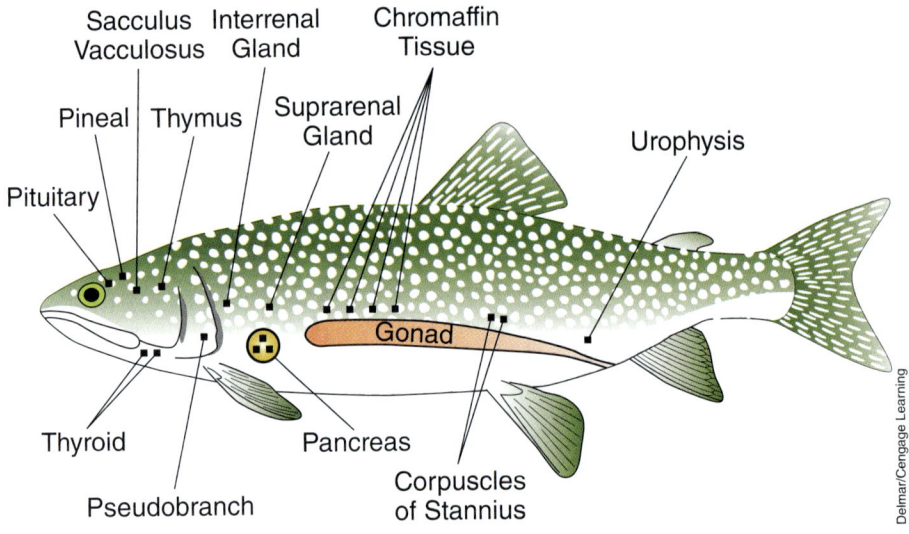

FIGURE 4-2 Environmental stimuli interpreted by areas of the brain influence the release of hormones.

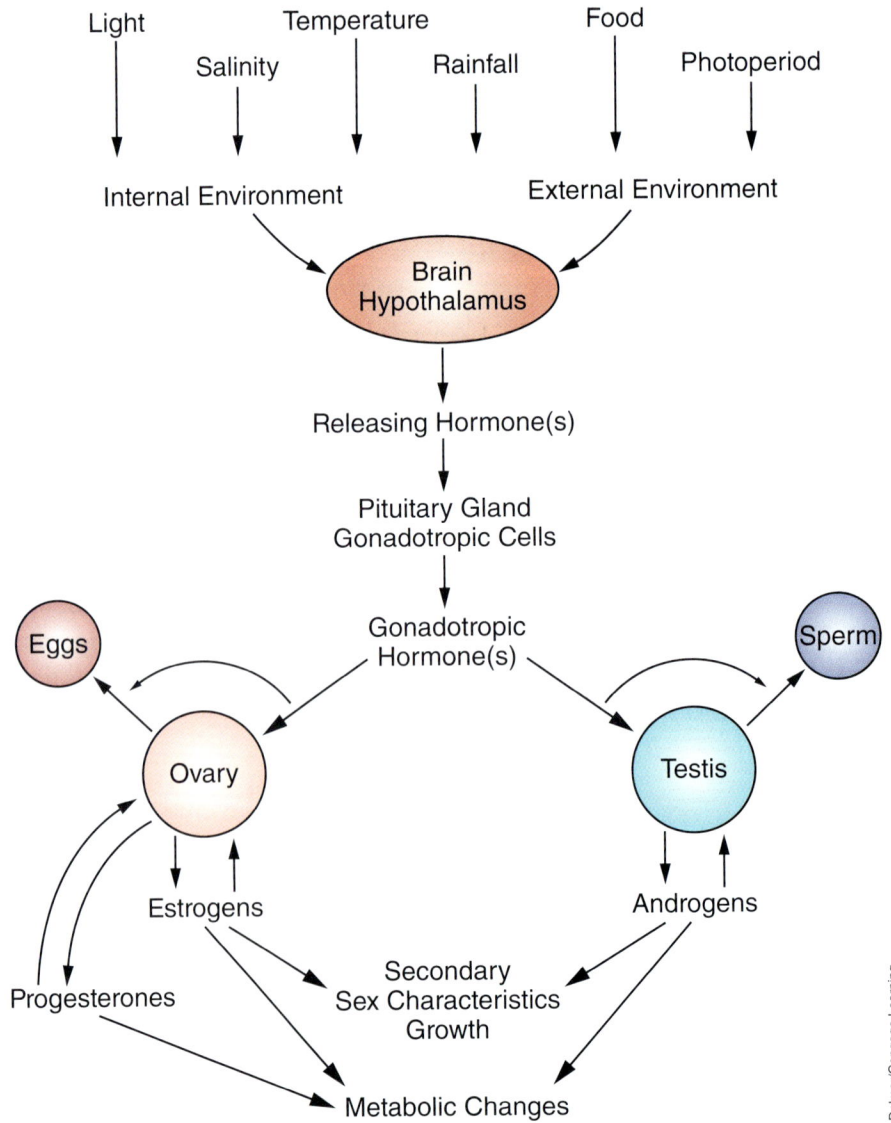

TABLE 4-1 MAJOR ENDOCRINE GLANDS AND HORMONES IN FISH

Gland	Hormone	Type	Function
Hypothalamus	Releasing hormones	Peptides	Controls the pituitary gland
Pituitary gland	Growth hormone (STH)	Protein	Controls growth
	Prolactin (LTH)	Protein	Controls ion balance
	Adrenocorticotropic hormone (ACTH)	Peptide	Controls inter-renal gland
	Melanocyte-stimulating hormone (MSH)	Peptide	Controls color change
	Thyroid-stimulating hormone (TSH)	Protein	Controls thyroid gland
	Gonadotropic hormone (GTH)	Protein	Controls reproduction and gonads
Thyroid gland	Thyroxine (T4) Triiodothyronine (T3)	Amino acids	Controls growth, reproduction, metabolism, and nutrient assimilation
Interrenal (adrenal gland)	Adrenaline	Amino acids	Counteracts stress
	Cortisol	Steroid	Controls ion balance
Testis	Testosterone	Steroids	Metabolic effects. Control secondary sex characteristics. Sperm production.
Ovary	Estrogen Progesterone	Steroids	Metabolic effects. Yolk and egg production.
Pancreas	Insulin Glucagon Somatostatin	Peptides	Protein metabolism and control of endocrine pancreas
Pineal gland	Melatonin	Peptide	Provides information about day/night and seasonal time
Ultimobranchial gland	Calcitonin	Peptide	Controls calcium levels
Stannius corpuscles	Hypocalcin	Protein	Controls calcium balance

lightproof, and all light is artificial. (See Figure 4-3.) Broodstock often have at least one previous spawning season before they are used in a light-controlled spawning program.

The light schedule used to induce early spawning in trout follows this scheme: An additional hour of light is provided each week until the fish are exposed to nine hours of artificial light in excess of the normal light period. The light is maintained at this schedule for a period of four weeks and then decreased one hour per week until the fish are receiving four hours less light than is normal for that period. By this schedule, the spawning period can be advanced several months.

FIGURE 4-3 Bins used to control artificial light, which controls spawning in trout.

Most attempts at modifying the spawning date of fish have been to accelerate rather than retard the maturation process. Artificial light periods that are longer than normal will delay the spawning activity of eastern brook trout and sockeye salmon. Temperature and light control are factors in manipulating the spawning time of channel catfish. Reducing the light cycle to eight hours per day and lowering the water temperature by 14°F (8°C) will delay spawning for approximately 60 to 150 days.

Hormonal

Fish spawning can be induced by hormone injection. Fish must be fairly close to spawning to have any effect, as the hormones generally bring about the early release of mature eggs and sperm rather than the promotion of their development. Both pituitary material extracted from fish, mammals, and human chorionic gonadotropin (HCG) have been used successfully. Recently, synthetic releasing hormone has also been used successfully on some species.

Use of hormones may produce disappointing results if broodfish are not of high quality. Under such conditions, a partial spawn, or no spawn at all, may result. Some strains of fish do not respond to hormone treatment in a predictable way, even when they are in good spawning condition.

Injection of salmon pituitary extract into adult salmon hastens the development of spawning coloration and other secondary sex characteristics, ripens males as early as three days after injection, and advances slightly the spawning period for females but may lower the fertility of the eggs.

Dried fish pituitaries from common carp, buffalo, flathead catfish, and channel catfish will all induce spawning when injected into channel catfish. The pituitary material is finely ground, suspended in clean water or saline solution, and injected intra-peritoneally. One treatment is given each day until the fish spawns. Generally, the treatment should be successful by the third or fourth day.

Channel catfish can also be successfully induced to spawn by intra-peritoneal injections of HCG. One injection of HCG normally is sufficient.

SEX DETERMINATION

Genes are the basic unit of inheritance. Genes are carried on the chromosomes in the gametes—the eggs or sperm. Genes contain the blueprint, or code, that determines how the animal will look and interact with its environment. The number of chromosomes varies from species to species but is consistent within a species.

Chromosomes also determine the sex of the fish. Although sex determination is not as well understood in fish as it is in mammals, the most common system of sex determination in commonly cultured fish is the XY system, like that of mammals. In this system, females carry the XX chromosomes and males carry the XY chromosomes. When females produce eggs, every egg will possess one X chromosome. When males produce sperm, half the sperm will carry the X chromosome and half will carry the Y chromosome. When the eggs and sperm unite, half the zygotes will be XX and the other half will be XY. On the average, in a normal population, half of the offspring are males and half are females. Figure 4-4 illustrates how sex is determined with the XY system.

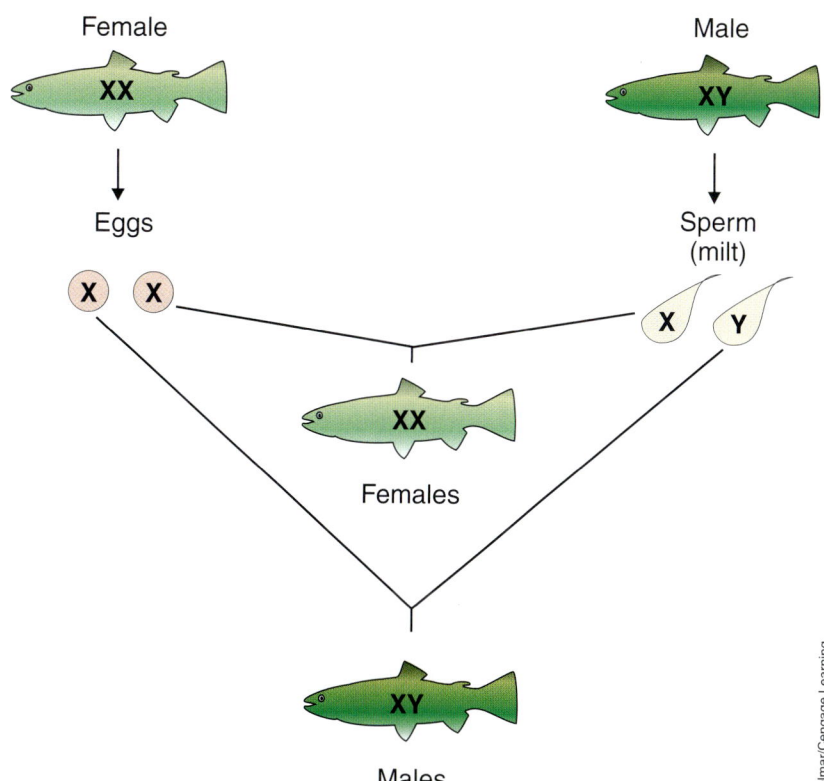

FIGURE 4-4 How sex is determined with the XY system.

The method for sex determination in fish is complex. At least eight chromosome systems may control the sex of different species. Also, while sex determination is primarily controlled genetically, environmental factors such as temperature, photoperiod, salinity, and crowding can help determine the sex of fish. Once sex determination is understood, this knowledge can be used to produce monosex cultures of fish.

FINFISH

Catfish and trout represent the major culture of finfish in the United States. Both are freshwater fish, but catfish is a warmwater species and trout a coldwater species. The culture of many other species of finfish is at various stages of development. Some of these are discussed in the sections that follow. For each species, the discussion includes sources of species, habitat, seed stock and breeding, culture method, stocking rate, feeding, diseases, harvesting and yields, processing, and marketing. Many of these topics are covered in more detail in other chapters, but not necessarily by individual species.

Channel Catfish

Channel catfish, Ictalurus punctalus, is the most important species of aquatic animal commercially cultured in the United States. It belongs to family Ictaluridae, order Siluriformes. Relatives of the catfish are found in fresh and saltwater worldwide.

In natural waters, channel catfish caught by fishers are usually less than 3 lbs., but the world record of 58 lbs. was caught in Santee Cooper Reservoir, South Carolina, in 1964. The size and age that channel catfish reach in natural waters depend on many factors. Age and growth studies suggest that, in many natural waters, channel catfish do not reach 1 lb in weight until they are 2 to 4 years old. The maximum age ever recorded for channel catfish is 40 years. Most commercially raised catfish are harvested before they are 2 years old.

Sources of Species

At least 39 species of catfish exist in North America, but only seven have been cultured or represent potential for commercial production. These include channel catfish, *Ictalurus punctalus*; the blue catfish, *Ictalurus furcatus*; the white catfish, *Ictalurus catus*; the black bullhead, *Ictalurus melas*; the brown bullhead, *Ictalurus nebulosus*; the yellow bullhead, *Ictalurus natalis*; and the flathead catfish, *Pylodictis olivaris*. (See Figure 4-5a and Figure 4-5b.)

Habitat

Channel catfish were originally found only in the Gulf states and the Mississippi Valley north to the prairie provinces of Canada, and to Mexico, but they were not found in the Atlantic coastal plain or west of the Rocky Mountains. Since then, channel catfish have been widely introduced throughout the United States and worldwide.

Figure 4-5a Channel catfish.

Figure 4-5b Other types of catfish.

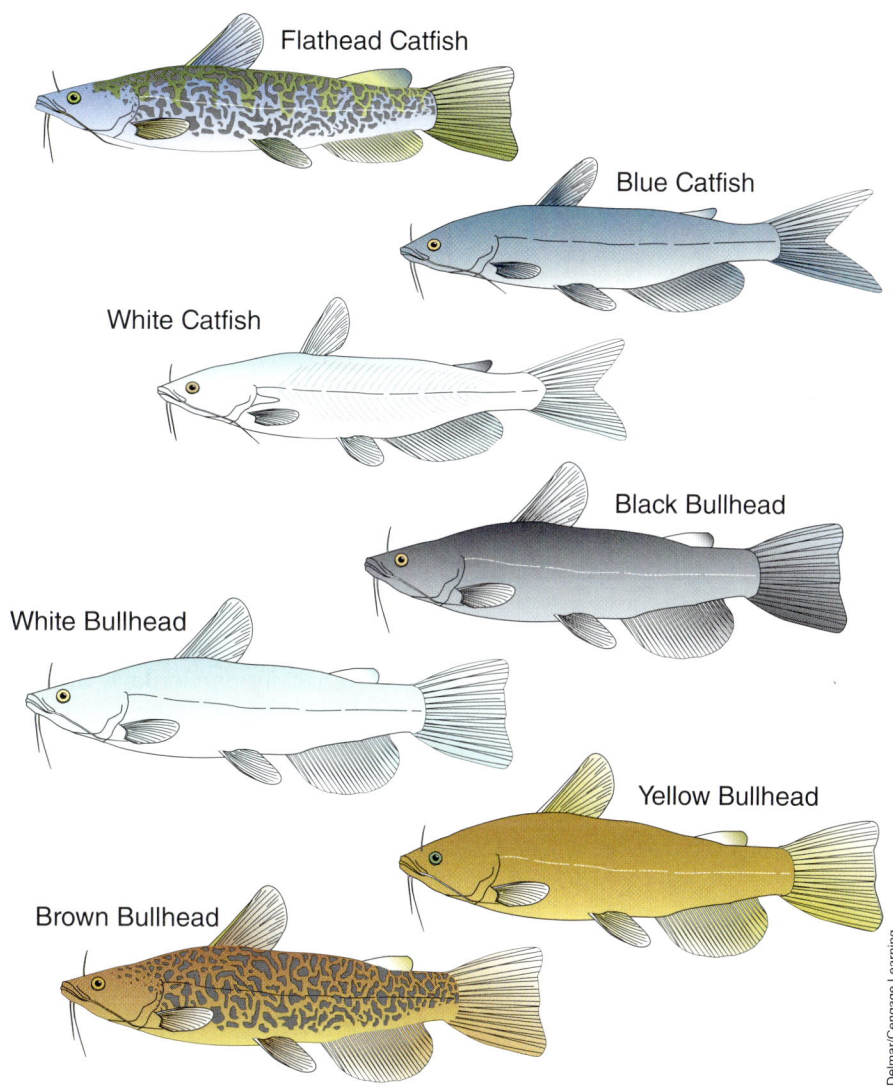

In natural waters, channel catfish live in moderate to swiftly flowing streams, but they are also abundant in large reservoirs, lakes, ponds, and some sluggish streams. They are usually found where bottoms are sand, gravel, or rubble, in preference to mud bottoms. They are seldom found in dense aquatic weeds. Channel catfish are freshwater fish, but they can thrive in brackish water. They can also be raised in raceways (Figure 4-6).

Channel catfish generally prefer clear-water streams, but are common and do well in muddy water. During the day they are usually found in deep holes wherever the protection of logs and rocks can be found. Most movement and feeding activity occur at night, just after sunset and just

Figure 4-6 Raising catfish in raceways.

before sunrise. Young channel catfish frequently feed in shallow river areas, whereas the adults seem to feed in deeper water immediately downstream from sandbars. Adults rarely move much from one area to another and are rather sedentary, whereas young fish tend to move about much more extensively, particularly at night when feeding.

Seed Stock and Breeding

Channel catfish spawn when the water temperature is between 75° and 85°F. About 80°F is optimal. Wild populations of catfish may spawn as early as late February or as late as August, depending on the location. The length and dates of the spawning season vary from year to year, depending on the weather and area, but peak spawning time in Mississippi usually occurs in May.

Channel catfish are cavity spawners and will spawn only in secluded, semi-dark areas. In natural waters, male catfish will build a nest in holes in the banks, undercut banks, hollow logs, log jams, or rocks. This behavior requires the use of spawning containers in order to successfully spawn channel catfish in commercial ponds.

The male selects and prepares the nest by fanning out as much mud and debris as possible. He will then defend this location against any intruder until spawning is completed and the fry leave the nest.

The female is attracted to the nest, and spawning occurs within the nest. Females lay eggs in a gelatinous mass on the nest bottom. After the female lays her eggs, the male takes over. He cares for the eggs by constantly fanning them with his fins to provide aeration and to remove waste products given off by the developing eggs.

Figure 4-7 Stages of the catfish life.

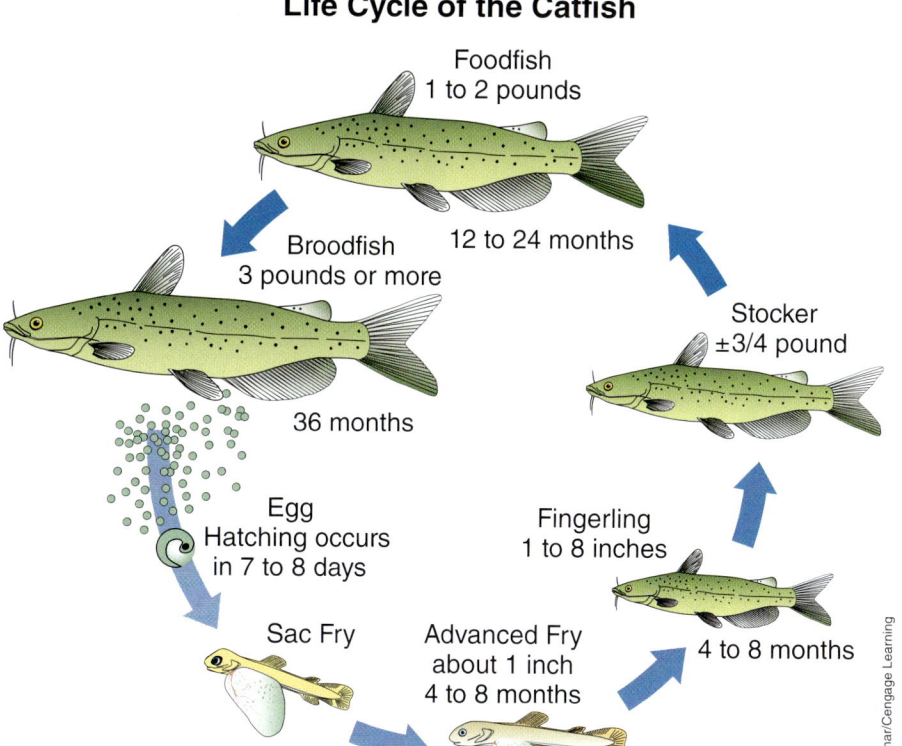

Females spawn only once a year, producing about 3,000 to 4,000 eggs per pound of body weight, whereas the males may spawn more than once. In wild populations, males seldom spawn more than once a year, but in hatcheries where the eggs are removed from the spawning container soon after being laid, males may spawn three or four times.

Channel catfish usually become sexually mature at three years, although some may spawn when two years old. (See Figure 4-7.) In wild populations they may not spawn until after age five. Channel catfish weighing as little as 0.75 lb may spawn if old enough. Farm-raised catfish usually weigh in excess of 2 lbs. when they spawn.

Eggs usually hatch in five to 10 days, depending on water temperature. At 78°F, eggs will hatch in about eight days. Each 2°F rise in temperature above 78°F requires one less day for hatching, and each 2°F fall in temperature below 78°F requires one more day. Water temperatures below 65°F and above 85°F will reduce hatching success. Newly hatched fry are born with a large yolk sac that contains the nourishment they need for the next two to five days until they are fully developed and ready to start feeding. After the yolk sac is absorbed, the fry take on their typical dark color and will begin to swim up, looking for food. At first swim up, fry will gulp air to fill their swim bladders, which helps them maintain and regulate their buoyancy.

Spawning containers used in commercial production can include milk cans, nail kegs, earthen crocks, ammunition cans, wooden boxes, and plastic buckets. The spawning container must be large enough to accommodate the brooding pair. The opening should be just large enough for them to enter. (See Figure 4-8.)

Figure 4-8 Old ammunition and milk cans used for catfish spawning.

The containers are placed in 1 to 2.5 ft of water, 1 to 10 yds apart, with the open end toward the pond center. Floats mark container locations. Enough containers are provided for 50 to 90 percent of the males.

Spawning activity sometimes diminishes for no apparent reason. Lowering the water level about a foot and rapidly refilling the pond may encourage additional spawning. Moving the spawning containers may also stimulate spawning.

Considering that not all females spawn and not all eggs, fry, and fingerlings survive, about 1,000 fingerlings will be produced per pound of healthy female brooder with the use of proper broodstock, hatchery, and rearing techniques.

Four methods are used in spawning channel catfish in ponds:

1. Spawning and rearing pond method. This approach requires the least skill, labor, and facilities. It is unreliable and not recommended for commercial operations. Spawning containers are placed in the pond, and the fish are allowed to spawn and hatch the eggs. The fry are left in the pond until ready for harvest.
2. Fry transfer method, open pond spawning. The fry transfer method is more productive than the spawning and rearing pond method but requires more skill and labor. The newly hatched fry are transferred from the spawning containers to previously prepared **nursery** ponds. Spawning containers are checked every three days.

   Males incubate the eggs. One day after the predicted hatching date, the fry are removed. The male catfish can bite hands and bare feet, so he should be chased from the spawning containers.

The fry are transferred to a bucket containing pond water by gently pouring them from the spawning container after counting. Next, fry are released into the nursery pond by slowly submerging the bucket, allowing them to escape into the pond near a shelter. If the water temperature or chemistry is not the same in both ponds, the fry must be slowly acclimated to the nursery pond water temperature before stocking. When temperature differences are more than 2 to 3 degrees, the water in the bucket is replaced slowly with nursery pond water until the temperature is equalized.

3. Egg transfer method, open pond spawning. Egg transfer is the most productive of the four methods, but it also requires the most skill, labor, and facilities. The fish are allowed to spawn in the containers as with the other methods, but the eggs are removed and incubated in a hatchery.

   Spawning containers are checked every two to four days. Late afternoon is the best time, because most spawning probably occurs at night or early morning. Checking at this time does not interrupt spawning activity and allows for timely removal of eggs. Eggs are removed immediately after finding them. Disturbed broodfish may sometimes eat eggs or dislodge them.

   The egg mass sticks to the container floor and must be gently scraped free. Egg masses are placed into a bucket and carried, immersed in water, to the hatchery. Eggs can be left in buckets in a shaded area for up to 15 minutes, but no longer, without aeration. Eggs must be shielded from sunlight. Egg masses near hatching must be taken to the hatchery immediately because they require more oxygen than young, or green, egg masses.

   Transporting egg masses in a cooler or other container can cause egg death due to suffocation. If a long time in transport is expected from the pond to the hatchery, aeration is required.

4. Pen spawning. (See Figure 4-9.) Each pen contains a pair of broodfish. The fish should be about equal in size. Daily checks ensure the welfare of the brooders and that the females are not being harassed or injured by the male fish. Females should be removed immediately after spawning to keep them from being injured or killed by the male. More than one female or male in the pen at a time can lead to fighting and injury to the females. Eggs can be left with the male, or they can be taken to the hatchery to incubate.

In maximum-production systems, eggs are transferred to a hatchery, incubated, and the fry started on food before they are moved into nursery ponds. The hatchery need not be elaborate. The critical ingredient is a water supply of the right quality and quantity.

Water temperature must be between 75°F and 82°F for proper hatching. Because eggs and fry have high oxygen requirements, oxygen levels should be maintained at a minimum of 6 parts per million (ppm). Water pH must be between 6.5 and 8.5 for best results. Risk of disease is less if

Figure 4-9 Pens for catfish spawning.

no fish are in the water supply. The best water is clean and free of organic matter such as algae and decaying leaves. A water flow of about 2 gal per minute is needed for a 100-gal hatching trough, or about one complete water change every 45 to 60 minutes.

Well water is probably best for the hatchery. It is usually clean and free of disease organisms. Well water is usually too cold for optimum hatching, but it can be warmed in a conventional water heater or stored and warmed in a small pond built specifically for this purpose. Some farmers keep two wells, one from a deep aquifer that contains warm water, and one from a shallow aquifer that contains cold water. A mixing valve is used to mix the two in the right proportions to provide uniform 80°F (26°C) water to the hatching troughs. The aeration tank should have a capacity of 25 percent of the hatchery's entire water volume. This will ensure at least a 15-minute retention time with a 60-minute exchange rate.

Total hardness and total alkalinity should exceed 20 ppm, and the pH should range between 6.5 and 8.5. Acidic or soft pond water can usually be corrected by adding agricultural limestone.

Eggs are commonly incubated in flat-bottomed, wooden, fiberglass, or metal troughs about 8 to 10 ft long, 18 to 24 in wide, and 10 to 12 in deep, holding about 100 gal. A series of paddles attached to a shaft are suspended in the trough. Paddles are spaced to allow wire-mesh baskets, holding the egg masses, to fit between them. The paddles should reach about halfway to the trough bottom and should extend below the basket bottoms. Baskets are made from ¼-in plastic-coated hardware cloth. An

Figure 4-10 Troughs in catfish hatchery. Paddles simulate fanning by male fish.

electric motor turns the paddles at 30 rpm. This motion gently rocks the egg masses and causes oxygen-rich water to flow through them. An 8-ft trough can hold six to eight egg baskets. A standpipe fitted into a drain at the other end controls water depth. A window screen over the standpipe prevents fry from escaping. (See Figure 4-10.)

Bacterial diseases and fungal infections are constant threats to eggs. The best disease control is prevention. A clean water supply of the proper temperature and frequent scrubbing and disinfection of troughs and equipment are essential. Debris and egg shells must be removed regularly with a siphon. Eggs are checked daily for bacterial egg rot or fungus. Bacterial egg rot appears as a milky-white dead patch, usually on the underside and in the center of the mass. Generally, bacterial egg rot occurs when water temperatures are higher than 82°F. The best preventative measure, besides maintaining good sanitation, is to keep water temperature at 78° to 80°F.

Fungus grows on infertile or dead eggs, usually when the pond water temperature is below 75°F or hatching water is below 78°F. It appears as a white or brown cotton-like growth made of many small filaments that can invade and kill healthy eggs. Formalin treatment controls fungus.

Bacterial egg rot or fungus seldom cause problems with good aeration and a proper water temperature, ranging between 78° and 80°F.

As the eggs hatch, sac fry emerge, swim through the screen baskets, and school together in a tight cluster on the trough bottom. For each pound of eggs, 10,000 to 11,000 eggs will be present. Approximately 95 percent of these will hatch.

Sac fry do not eat. They receive nourishment from the attached yolk sac. The yolk sac is gradually absorbed by the fry. After about three days, fry begin swimming up to the water surface searching for food. Their color at this time changes from pink to black. They are called "swim-up fry," and they begin feeding at this stage.

Sac fry can be left in the hatching trough for one to two days and then moved to rearing tanks or troughs. Many types of tanks can be used for holding fry, the most common in Mississippi being an 8 ft. × 2 ft. × 10 in. flat-bottomed trough, which will hold about 100,000 fry. If a large tank is used, a fry holding box is desirable. This is a 2 × 2 × 1-ft. wooden box made from boards or marine plywood and caulked with silicone. The bottom is made of 1/16 in. plastic window screen. One box can hold 20,000 to 30,000 fry, or the quantity obtained from a large egg mass. A tank that can hold 10 fry holding boxes should be supplied with 10 gal of water per minute.

Sac fry from the hatching trough can be siphoned into a bucket using a ½ in. hose and transferred to the rearing tanks. Oxygen in the rearing tank should remain about 5 ppm.

An estimation of fry number is crucial so that rearing ponds can be stocked correctly. A convenient time to do this is when fry are being transferred to the pond. Two acceptable methods are the **volumetric** and weight comparison methods.

Culture Method

Channel catfish grow best in warm water, with optimum growth occurring at temperatures of about 85°F. With each 18°F change in temperature, the metabolic rate doubles or halves. This means that, within limits, their appetites increase with increasing water temperatures or decrease with decreasing water temperatures.

Water quality preferences and limitations for wild channel catfish are not any different from those of farm-raised channel catfish. The lethal oxygen level for both wild and farm-raised catfish is about 1 ppm, and reduced growth occurs at oxygen concentrations of less than 4 ppm.

Stocking Rate

Initially, 4 to 6-in. fingerlings can be stocked at 3,000 to 4,000 per acre. New producers should not exceed a stocking rate of 3,000 to 3,500 catfish per acre for the first growing season. This allows new producers to gain experience in management procedures while reducing potential problems such as low oxygen (Figure 4-11). Exceeding this **stocking rate** increases the chance of substantial losses caused by water quality problems and diseases. In intensive pond culture systems, the stocking rate varies from 3,000 catfish per acre and upward. As the number per acre increases, management problems increase. In ponds with limited or no water available except run-off, stocking rates should not exceed 2,000 catfish per acre and a rate of 1,000 to 1,500 per acre would be better.

Figure 4-11 Aerators are frequently used in pond production of catfish.

Stockers 6 to 8 inches long are preferred when available because they will reach a size of 1.5 lbs. in about 210 feeding days when water temperatures are above 70°F. In order to help you determine the number and weight of catfish stocked, the average weights per 1,000 channel catfish and the number of catfish per pound for lengths from 1 to 10 in. are displayed in Table 4-2. These figures are averages, and they can vary a great deal, depending on the condition of the fish and when they were last fed.

Initial stocking is begun as soon as there is water in the pond and catfish of an acceptable size are available (see Figure 4-12, page 107). When a pond is clean cropped (all the fish are harvested at one time), restock the pond as soon as it is one-fourth to one-half full and stocker-sized catfish are available. When a pond is topped (multiple harvested), the pond is restocked as soon as possible after harvest with one 5 to 8-in. fingerling for each fish harvested.

Feeding

Feeding can occur day or night, and channel catfish will eat a wide variety of both plant and animal material. Channel catfish usually feed near the bottom in natural waters, however, they will take some food from the surface. Commercial channel catfish are fed complete diets of pellets sprayed over the pond surface. Chapters 9 and 10 provide more details on feeding catfish.

Diseases

Intensive catfish culture can set a producer up for problems with many diseases. The key to disease prevention is good management that does not stress fish or introduce disease-causing conditions. Chapter 11, Health of Aquatic Animals, discusses diseases affecting catfish.

TABLE 4-2 COMPOSITE LENGTH-WEIGHT CATFISH FINGERLING CHART

Length (inches)	Weight (lbs./1,000 fing.)	Length (inches)	Weight (lbs./1,000 fing.)	Length (inches)	Weight (lbs./1,000 fing.)	Length (inches)	Weight (lbs./1,000 fing.)
1.0	0.7	3.1	9.6	5.2	39.3	7.3	102.6
1.1	0.8	3.2	10.4	5.3	41.5	7.4	106.7
1.2	1.0	3.3	11.3	5.4	43.7	7.5	110.8
1.3	1.2	3.4	12.3	5.5	46.0	7.6	115.1
1.4	1.4	3.5	13.3	5.6	48.4	7.7	119.5
1.5	1.6	3.6	14.3	5.7	50.9	7.8	124.0
1.6	1.8	3.7	15.4	5.8	53.4	7.9	128.6
1.7	2.1	3.8	16.6	5.9	56.1	8.0	133.3
1.8	2.4	3.9	17.8	6.0	58.8	8.1	138.2
1.9	2.8	4.0	19.1	6.1	61.6	8.2	143.1
2.0	3.1	4.1	20.4	6.2	64.5	8.3	148.2
2.1	3.5	4.2	21.8	6.3	67.5	8.4	153.4
2.2	4.0	4.3	23.2	6.4	70.6	8.5	158.7
2.3	4.4	4.4	24.8	6.5	73.7	8.6	164.1
2.4	4.9	4.5	26.3	6.6	77.0	8.7	169.7
2.5	5.5	4.6	28.0	6.7	80.4	8.8	175.4
2.6	6.1	4.7	29.7	6.8	83.8	8.9	181.2
2.7	6.7	4.8	31.5	6.9	87.4	9.0	187.1
2.8	7.3	4.9	33.3	7.0	91.0		
2.9	8.1	5.0	35.3	7.1	94.8		
3.0	8.8	5.1	37.3	7.2	98.6		

Harvesting and Yields

In **production ponds** the growth rate of channel catfish is determined by water temperature, length of time held at different water temperatures, quantity and quality of food, palatability or taste of food, frequency of feeding, and water quality. Most farm-raised catfish are harvested at a weight of 1.25 lbs. at an age of about 18 months.

In a topping or multiple harvest production system, a pond is stocked initially and fed until about one-fourth to one-third of the fish are larger than 0.75 lb. Then the pond is seined with a seine having a mesh size of

Figure 4-12 Weight-length relationship in catfish.

1 3/8 to 1 5/8 in. The seine captures those fish that weigh 0.75 lb. or more and will allow smaller fish to escape. After partial harvesting, catfish fingerlings are restocked at a rate of one for each one harvested.

Processing and Marketing

Where and how the catfish will be sold should be the first concern of anyone thinking about raising catfish. Catfish farmers traditionally sell or market their catfish to:

- Processing plants
- Live haulers (See Figure 4-13)
- Local stores and restaurants
- Backyard or pond bank sales to local residents
- A fee-fishing operation

Obviously, some variations of these marketing schemes are used, but the following are the main outlets: In Mississippi, processing plants will not send a harvesting crew more than 50 miles from the plant, and they charge about three cents a pound for harvesting. In addition, plants charge from one to three cents per pound for transportation. The minimum load processing plants will take is 8,000 to 10,000 lbs. Arrangements for selling fish to a processing plant usually must be made between 7 and 60 days before harvest.

Like processing plants, most live haulers will not take less than 8,000 lbs per load. Also, they do not provide harvesting crews. This means that the farmers must harvest the fish. Live haulers want catfish only during a four- to five-month period, mid-April to mid-September. The farmer must set production and harvesting schedules to the live hauler's schedule.

Figure 4-13 Catfish being harvested from a pond in Mississippi and loaded into a live haul truck.

Local stores and restaurants usually want fish all year on a weekly basis. This means a farmer must be able to harvest fish weekly either by seining or trapping. One main problem is that many stores and restaurants will take only dressed fish, so the small catfish farmer must be willing to hand process fish.

Depending on location, area population, size of the catfish operation, the number and size of other catfish operations in the area, and other factors, the backyard sales method can be excellent or poor. Fish are available year-round and are sold live or dressed. Another method used is to harvest once a year and advertise by local radio and newspapers that fish will be available live at the pond bank on a certain date.

The fee-fishing method of marketing catfish allows the farmer to grow fish in one or more ponds and permits fishing in any or all the ponds for a fee, usually so much per day or fishing rod (to cover more than one fishing rod per person), and so much per pound. The pond may be open for fishing all year or just on certain days or weeks. In addition to the usual management problems, this system means that someone must be at the pond when it is open for fishing.

Trout

Salmonids include the members of the trout group and the salmon. Trout live in freshwater. Salmon hatch in freshwater, and then swim to saltwater, where they grow to maturity and return to freshwater to spawn. Salmon are **anadromous**.

Sources of Species

Table 4-3 lists cultured trout and chars. The most commonly cultured trout is the rainbow trout. (See Figure 4-14 and 4-15.)

TABLE 4-3 CURRENT STATUS OF CULTURED TROUTS AND CHARS OF THE GENUS SALMO AND SALVELINUS

Scientific Name	Common Name	Current culture status[1] Public	Private
Salmo aguabonita	Golden trout	LIM	LIM
Salmo clarki	Cutthroat trout	MC	LIM
Oncorhynchus mykiss	Rainbow trout	EC	EC
Salmo trutta	Brown trout	MC	LIM
Salvelinus aureolus	Sunapee trout	LIM	NA
Salvelinus fontinalis	Brook trout	MC	LIM
Salvelinus malma	Dolly Varden trout	LIM	NC
Salvelinus namaycush	Lake trout	MC	NC

[1]Status: EC—Extensively cultured; MC—Moderately cultured; LIM—Cultured on a limited basis; NC—Not cultured; NA—Data not available.

Figure 4-14 Rainbow trout.

Figure 4-15 Types of trout.

Habitat

Trout grow naturally in the streams and lakes of the northern half of the United States. They are coldwater fish, preferring water of 50° to 68°F. Growth slows below 50°F and above 68°F. Water temperatures above 75°F are lethal.

Seed Stock and Breeding

In commercial production of trout and other salmonids in the United States, eggs are typically produced on broodfish farms, which are separate from farms used for producing fish for food or for stocking. Producing good quality, disease-free eggs is a specialized activity requiring a high degree of skill and management.

Most eggs used in commercial trout production in the southeastern United States are produced in the Pacific Northwest region. Trout eggs are usually shipped when they reach the **eyed stage**, which is over halfway through the incubation period. (See Figure 4-16.)

Incubation time is temperature dependent. At 55°F, rainbow trout eggs will hatch approximately three weeks after fertilization, or within four to seven days after received as eyed eggs. Trout eggs arrive on ice. The first step is tempering, or gradually bringing them up to the hatchery incubation temperature. Any water loss in the eggs from shipping is replaced. Tempering of the eggs should be done in a clean bucket or other hatchery container by adding the eggs to water of identical temperature. Egg temperature can be increased to the hatchery water temperature over a 30 to 60 minute time period by adding small amounts of clean water. The eggs need to be gently stirred once or twice during the tempering process to ensure adequate water circulation to all eggs.

Figure 4.16 Trout eggs in the eyed stage.

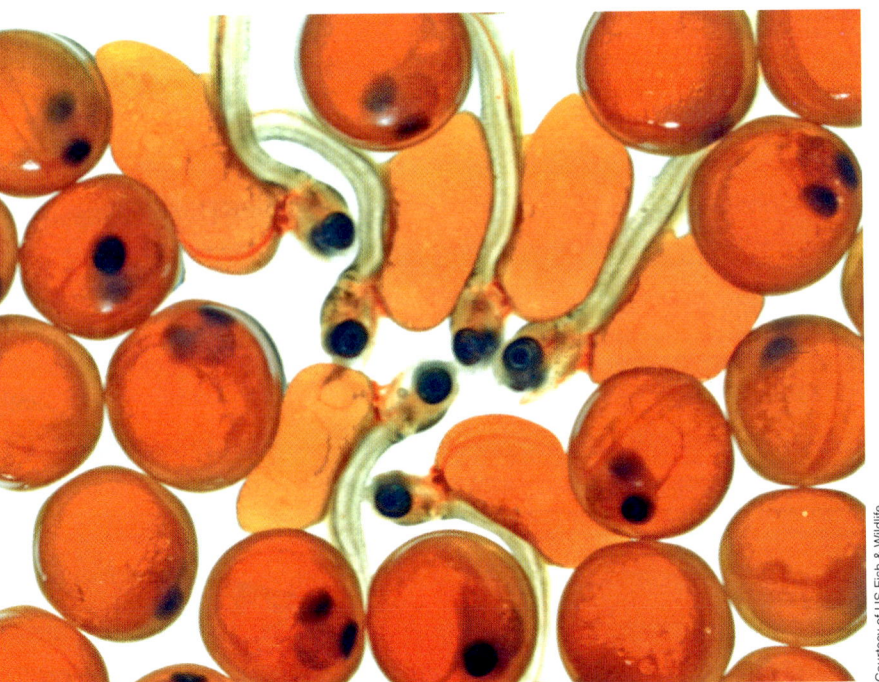

Figure 4-17 Types of incubator systems commonly used in trout hatcheries.

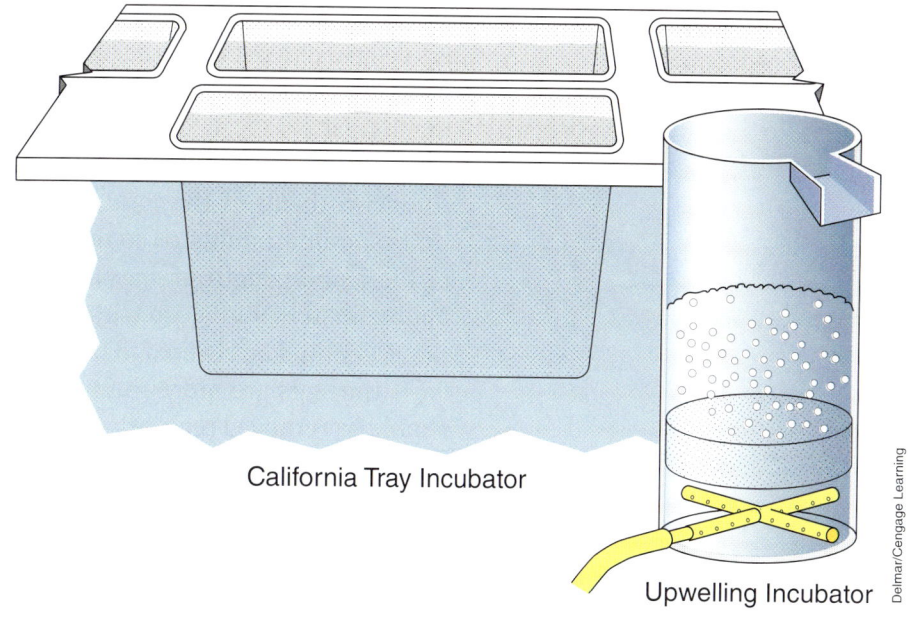

Figure 4-18 Typical small trout hatchery with hatching trays, fry troughs, and fingerling tanks.

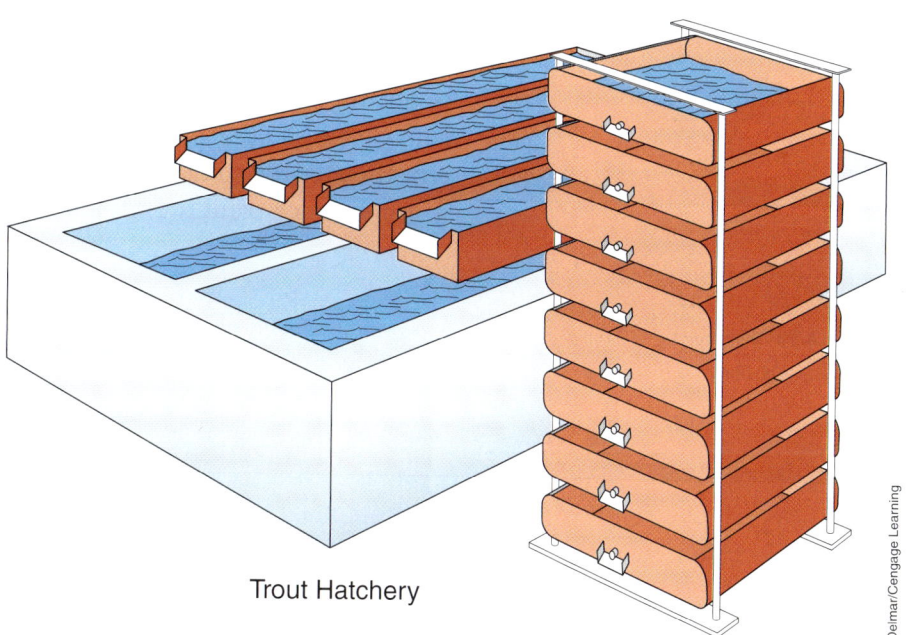

Three types of incubator systems are commonly used: California trays, vertical tray or Heath incubators, and upwelling incubators. (See Figure 4-17.) California trays are screened, flat-bottomed trays that fit inside rearing troughs, in series, horizontally. Between each tray, a partition extending to the trough bottom forces water through the eggs from below. Vertical tray incubators are essentially California trays arranged in stacks, having the advantage of requiring relatively little floor space to incubate large numbers of eggs (see Figure 4-18). Water is aerated as it flows down through the stack. Upwelling incubators are commercially available in several different models, or can be easily constructed from PVC or other materials.

To prevent smothering, trout eggs are placed no more than two layers deep in either California or vertical incubator trays. The recommended water flow in tray incubators is from 4 to 6 gal per minute (gpm). Upwelling incubators (Figure 4-19) maintain adequate circulation by using the water flow to partially suspend the eggs, but they should contain no more than two-thirds of the total volume in eggs. The flow rate in upwelling units should be adjusted so that eggs are lifted approximately 50 percent of their static depth. If eggs are 6 in deep with water off, they should be approximately 9 in. deep with water on. All types of egg-incubating containers should be covered to protect developing embryos from direct light.

If the eggs are more than three days from hatching, then dead eggs should be removed regularly to limit fungal infections. Siphoning off dead eggs is more effective than chemical treatment at controlling fungus but can be very time-consuming. Formalin added to the inflowing water controls fungus. Trout eggs should not be treated with formalin within 24 hours of hatching because the eggs will concentrate the chemical inside the shell and die. Once hatching, the start of the life cycle, (Figure 4-20) begins, the eggs and sac fry should not be treated with any chemicals.

Hatching rate depends on water temperature, but hatching will usually complete within two to four days. Empty shells should not be allowed to accumulate in the incubating units. If the eggs are incubated separately from the rearing troughs, the sac fry are transferred into troughs shortly after hatching is complete. Up to 30,000 fry can be stocked into a standard fry trough 10 ft long and 18 in wide. The water level in the

Figure 4-19 Trout eggs being incubated in upwelling incubators.

Figure 4-20 Life cycle of trout.

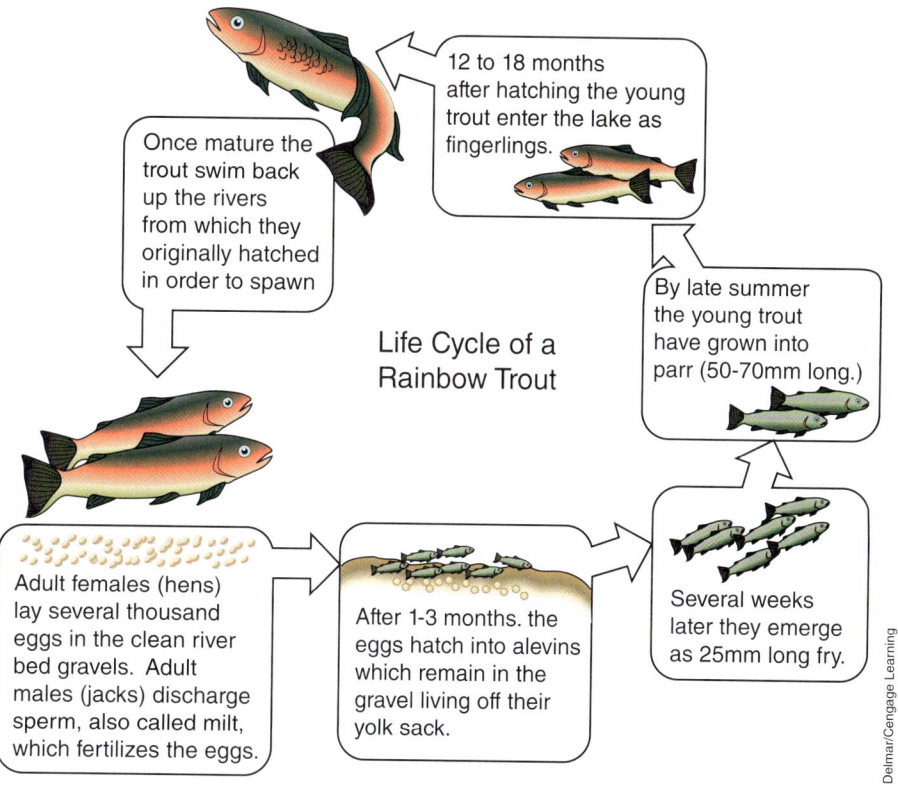

trough is kept fairly shallow (3 to 4 in.) until fry swim up, approximately two weeks after hatching at 55°F. Any mortalities or deformed fish are removed regularly.

When about 50 percent of fry swim up, feeding begins with small amounts of starter mash on the surface, three to four times daily, until most of the fish begin active feeding. Then, if possible, fry are fed every 15 minutes (but not less than every 60 minutes) at this stage. Automatic feeders usually are better and certainly are more convenient than feeding by hand.

After the fry have been actively feeding for two weeks, they are counted every week and the feeding rate and feed size are adjusted accordingly. Monitoring of dissolved oxygen levels is a good way to help determine when to reduce the density. Ideally, the oxygen level should not be allowed to be lower than 6 ppm. The fry will be ready to move into larger grow-out tanks when they grow to 200 to 250 per lb.

In areas where Yersinia Ruckeri, the causative agent of enteric redmouth disease (ERM) is present, the fish should be vaccinated seven to ten days before moving into a production facility.

Hatcheries avoid the possibility of disease being inadvertently introduced by restricting traffic and using a footbath.

All equipment used in the hatchery is reserved for hatchery use only. The hatchery and equipment are cleaned and disinfected regularly with a hypochlorite solution or an approved quaternary ammonium disinfectant. Troughs and floors are also disinfected between groups of fish. Additional ventilation prevents condensation from forming on walls or ceilings and spreading disease.

Culture Method

Raceways are generally constructed in a ratio of 5 to 1 (or greater) length to width, and with a depth of 3 to 5 ft. Water should flow evenly through the system to eliminate areas of poor water circulation where waste materials or sediment may accumulate. Raceways may be constructed above ground or in the ground from cement or fiberglass, and even wood has been used. Fish cultured in raceways require a large quantity of good-quality water, preferably supplied by gravity flow from artesian wells or higher elevations. If pumping is required, operating cost may be high, and risks may be increased due to possible failure of pumps or power supply.

On average, 1 to 3 gal per minute of flow should be available for each cubic foot of raceway volume at densities of 3 lbs. of fish per cubic foot. If supplemental aeration is used, the water requirement may be somewhat reduced. Water flow should be sufficient to keep solid waste material from accumulating in the raceway and to dilute liquid waste (primarily ammonia) excreted by fish.

To achieve good production and minimize problems of stress and disease, water quality should be sustained within desirable ranges at all times. Oxygen should be maintained above 60 percent of saturation. Ammonia levels should remain below 0.1 mg/l in the discharge. Water quality should be monitored frequently, especially oxygen and ammonia, to ensure that conditions remain suitable. This enables the producer to learn more about the production system and its operating characteristics.

Traditionally, raceways are considered to be single pass, flow-through systems. Some fish farmers have developed raceways that are joined with ponds and use the ponds to clean the water prior to reuse. If such a system is designed, the pond(s) should have a volume of at least seven times the total daily discharge volume of the raceway. This allows sufficient time for water quality improvement.

Recirculating systems are often proposed as a type of closed or semi-closed raceway. The water is reconditioned by **clarification**, biological filtration, and re-aeration so that most of the water is reused and only a fraction of the total daily flow is made up of new water. The productive capacity of this system depends on the filtration system's ability to remove wastes, as well as on the volume of replacement water used to improve water quality. Fish production in systems of this sort may reach levels similar to that achieved in raceways. Water quality should be monitored frequently in such a system because without high rates of water exchange, toxic **metabolites** may accumulate rapidly if the biological filtration system is not sufficient to handle the wastes. Chapter 15 describes recirculating systems. Trout may also be raised in earth ponds or cages.

Stocking Rate

The quantity of fish that can be grown intensively in a raceway depends more on water quantity and quality than on facility size. Small fish consume proportionally more oxygen per unit of body weight than larger fish and are normally stocked at lower densities. Densities of fish stocked in raceways may range from 1 to 10 lbs. per cubic foot of water, depending on a given system's capacity to support its population. In practice, stocking densities can be calculated based on expected harvest weight of fish to be produced, or based on the carrying capacity of the system. With the latter method, the number of fish is reduced as their sizes increase.

The carrying capacity of a trout-rearing unit is dependent upon fish size, as well as on several water-quality factors, principally oxygen content, temperature, water flow, and volume. Carrying capacity is usually expressed in terms of pounds of fish per cubic foot of rearing space or water flow. A number of different formulas have been devised to calculate carrying capacities, taking into account oxygen consumption, rate of increase in fish length, water volume and temperature, feeding rates, and other factors. As long as the appropriate limiting factors are monitored by the operator, the choice of a particular estimator is a matter of preference.

The easiest method for estimating maximum fish density for a rearing unit is to keep tank loadings within a level of 0.5 to 1 times the length of the fish (in inches) in lbs per cubic ft. ($ft.^3$), for example, 2 in. fish at 1 to 2 lbs. per ft^3, 4 in. fish at 2 to 4 lbs. per ft^3. The multiplying factor is referred to as a **density index**. Many trout farmers simply stock all sizes of fish at 4.5 lbs. per ft^3 as an upper limit for fish density. With proper management, the density can be much higher.

The density index estimates only the appropriate density of fish, without regard to system water flow. Water-flow rate will determine how quickly other water-quality characteristics become limiting in each unit. An estimate of the appropriate capacities of trout relative to water flow is to keep loadings within a range of 0.5 to 1 times the fish's length in pounds per gallon per minute (gpm) of water flow—for example, 2 in. fish at 1 to 2 lbs. per gpm, 4 in. fish at 2 to 4 lbs. per gpm. This factor is referred to as a "flow index," and it works on the assumption that inflowing water is at or near saturation of dissolved oxygen. In a properly designed facility, the estimate of carrying capacity obtained from the flow index and the density index will be nearly equal.

Grading

During the production cycle, fish should be graded periodically to maintain size uniformity. Trout usually are graded four times during the period from fingerling stocking, about 3 in, until they reach a marketable size of 12 to 16 in. Frequency of grading will vary according to individual circumstances but should routinely be done whenever loadings need to be decreased.

The simplest graders are wooden frames that measure as long as the tank is wide and slightly higher than the water is deep. Pieces of aluminum tubing, PVC pipe, or smooth wood are spaced at regular intervals across the frame to perform the grading (Figure 4-21). The grader is put in the top of the tank, and fish are crowded down toward the tail screen.

Figure 4-21 Bar grader for trout production.

Fish too large to pass through the bars remain at the tank bottom, where they can be moved to another tank containing fish of a similar size. The smaller fish swim through the bars and remain in the same tank, although 10 percent or more usually remain behind the grader. This method works best with fish larger than 2 to 3 in. long. Grading fish smaller than this is usually not necessary, and it will be stressful for the fish. Mechanized graders are available. These function by pumping fish onto a series of grading bars. These systems are very effective when properly sized for the fish to be graded, but they are difficult to justify economically for most smaller trout farms.

Inventory

Taking inventory is vital. Fish in each tank should be sample counted at least monthly to assure that they are growing as expected and to keep track of loading rates. Feeding according to a feeding-rate chart allows you to check daily ration amounts and adjust as necessary. When you are sample counting, the fish should be crowded starting from two-thirds of the way down the length of the raceway moving toward the head of the tank. The smallest, weakest fish, which will linger toward the tail of the tank, are not representative of the general fish population and will be left behind. With the fish loosely crowded at the head end of the tank, a sample of fish is netted into a bucket of water suspended from a spring-tension scale. The weight is recorded and the number of fish is determined as they are poured back into the tank. If fish are graded rather uniformly, three or four samples from different areas are sufficient. Fish size (expressed as number per pound) is calculated by dividing the number of fish in each sample by the total sample weight. The average for each tank is then used to estimate the weight of fish in the entire raceway.

Removing mortalities from each tank on a daily basis and recording their numbers is an important management detail. Dead fish left in tanks are a potential source of disease and indicate poor farm hygiene. Analyzing mortality rates in each tank may indicate developing fish health problems before they become severe. Also, mortalities, or morts, should be subtracted each month from the estimated population totals in order to maintain an accurate inventory.

Feeding

Research on trout nutrition has been conducted for more than 40 years. With the exception of the final sale price of the fish, the amount and suitability of feed used for trout farming will be the primary factor in determining production profitability. Digestive systems of trout and other salmonids are naturally equipped to process foods consisting primarily of protein (mostly from fish), and can obtain a limited amount of energy from fat and carbohydrates. Diets for fry and fingerling trout require a higher protein and energy content than diets for larger fish. Fry and fingerling feed should contain approximately 50 percent protein and 15 percent fat; feed for larger fish should contain about 40 percent protein and 10 percent fat. The switch to lower-protein formulations usually occurs at transition from a crumble feed to a pelleted ration, called a grow-out or production diet. Several brands of high-quality commercial trout diets are available. Although a farm could produce its own fish food, it is usually uneconomical to do so.

The primary goals in feeding trout are to grow the fish as quickly and efficiently as possible, maintaining uniformity of growth with the least degradation of water quality. The amount of feed trout require depends on water temperature and fish size. During normal production, trout should be fed seven days per week with a high-quality commercially prepared diet formulated for trout. Due to higher metabolic rates, smaller fish need more feed relative to their body weight than do larger fish, and fish in warmer water need more feed than fish in cooler water. Because fish are poikilothermic (cold-blooded), their body temperatures and metabolic rates vary with environmental temperatures.

The best way to determine the correct amount and sizes of feed needed for trout production is to use a published **feeding chart**, usually provided by the feed manufacturer. The chart should be used as a guide but may need adjustment to fit conditions on individual farms. Overfeeding will cause the fish to use the feed less efficiently and will not increase growth rates significantly. Chapters 9 and 10 provide more information on feeding limit.

Diseases

Trout are susceptible to a myriad of bacterial, viral, protozoan, metazoan, and mycotic pathogens, as well as to environmental alterations such as nitrogen supersaturation, free ammonia, low dissolved oxygen, and a host of environmental contaminants from industrial and agricultural sources. During the course of hatchery rearing, an estimated 50 percent of fish die. Losses are greatest in the yolk-sac and swim-up stages.

Most infectious and noninfectious diseases are easily diagnosed and their causes identified. Currently, oxytetracycline, sulfamerazine, salt, and formalin are approved for use with fish intended for human consumption.

Chapter 11, Health of Aquatic Animals, discusses diseases affecting trout.

Harvesting and Yields

Trout can be harvested by seining, trapping, netting, or draining the raceway or pond. Unless the processing plant is on-site, fish are transported in a live-haul truck.

Trout are harvested when they are 7 to 14 in long and weigh 0.5 to 1 lb. Depending on the culture conditions, food-size fish can be produced in 7 to 14 months.

Processing and Marketing

Trout are marketed at several stages in their production process. Culturalists specialize, depending on which of the following markets they intend to target:

- Broodfish marketed to hatcheries
- Eyed eggs
- Fingerlings
- Food fish for processing
- Fee for fishing or other recreational businesses
- Live haulers

Food fish are transported to a processor where they are killed, graded, dressed, boned, and packaged. Some are sold whole, and others are sold as fillets. Processed trout are sold fresh or frozen. Some trout are further processed into specialty products like smoked trout.

Salmon

The world supply of Pacific salmon (Figure 4-22), determined by commercial catches, declined from 16.9 billion lbs annually from 1935 to 1939 to 893 million lbs in the 1970s.

Source of Species

Table 4-4 briefly describes commercially important salmon species. (See also Figure 4-23.)

Figure 4-22 Salmon.

Figure 4-23 Types of salmon.

Habitat

Table 4-4 indicates the habitat for the various species of commercially important salmon.

Pacific salmon introduced into the Great Lakes now support a sport fishery. Hatchery-reared stocks must be used because natural reproduction contributes less than 10 percent of the **recruitment** to the fishery. Survival in the Great Lakes is substantially greater than in marine environments. Salmon in the Great Lakes form a self-perpetuating population in fresh water. Salmon help control the alewife fish (*Alosa pseudoharengus*).

Seed Stock and Breeding

Because salmon that are ready to spawn make their way back to their place of hatching, they are easy to catch. Once caught, they are moved to some hatcheries with tanks or some artificial spawning channels in natural streams or rivers.

The salmon and steelhead trout hatcheries of the Pacific states are among the largest and most technically sophisticated aquatic culture systems in the world. These hatcheries provide fish for a significant portion of the U.S. salmon fishery.

Culture Method

Pacific salmon are cultured by two methods—ranching and farming. California, Oregon, and Alaska allow private ocean salmon **ranching**. Companies rear salmon to migratory size in freshwater hatcheries, and then release them into a river or estuary. Fish swim to the ocean, where they graze on natural food. These fish are available to common-property fisheries in the ocean and during their return migration. After their return to the point of release, the ocean rancher processes

TABLE 4-4 CHARACTERISTICS OF MAJOR COMMERCIALLY IMPORTANT SALMON

Common Name	Scientific Name	Range	Time in Seawater	Mature Weight
Atlantic Salmon	*Salmo salar*	North Atlantic from New England-Ungava Bay on the west; Iceland, Greenland, from northern Portugal to the Kara Sea on the east	1 to 5 years	2 to 66 lbs
Sea trout, many local names for the "jack" form, which spends only a few months at sea (finnock, whitling, sewin)	*Salmo trutta*	Seagoing forms found in countries bordering the northeast Atlantic where sea temperatures are <70°F (<21°C) maximum	Months to 3 years, then 1 to 2 years between spawning	11 oz. to 22 lbs.
Steelhead	*Oncorhynchus mykiss*	Western North America from Mexico to the Bering Sea	<1 year to 4 years	7 oz. to 42 lbs.
Pink salmon, humpback, humphy, karafuto-maru, gorbuscha	*Oncorhynchus gorbuscha*	East and west Pacific	Always 2 years	2 to 11 lbs.
Chum salmon, dog, sake, keta	*Oncorhynchus keta*	East and west Pacific; widest distribution of all the Pacific salmon species	3 to 4 years, 5 for Yukon chum	8 to 9 lbs. 44 lbs. for Yukon chum
Sockeye salmon, red, blueback, beni-masu, nerka	*Oncorhynchus nerka*	East and west Pacific in rivers with lakes in the system; greatest numbers from Bristol Bay to the Columbia River on the east, Kamchatka on the west	3 years or more, males may mature as jacks	2 to 11 lbs.
Coho salmon, silver, gin-maru, kizhuch	*Oncorhynchus kisutch*	East and west Pacific from coastal California north to Norton Sound, Alaska, Hokkaido (rare) to the Anadyr River	2 years except for jacks	9 to 11 lbs.
Chinook salmon, king, spring, masunosuka, chavycha	*Oncorhynchus tshawytscha*	East and west Pacific, Ventura River California to Point Hope, Alaska, Hokkaido to the Anadyr River	1 to 5 years, 6 to 7 for Yukon females	Avg 22 lbs. max 121 lbs.
Masu, cherry, yamama, sima	*Oncorhynchus masou*	West Pacific, over the southern part of the range of the other Pacific salmons	1 to 2 years	Avg 9 lbs. jacks <1 year

them for spawning or marketing. Depending on the species, the return time can be two to five years. Figure 4-24 shows the life cycle of the salmon.

Salmon farmers use net pens or sublittoral enclosures. When the smolts, or fingerlings, are about 6 in. long, the farmer moves them from the hatchery to the enclosure or pen. Salmon **farming** provides a more reliable, year-round source of fish. Net-pen farming requires farmers to be concerned with the management of the dissolved oxygen, nutrition, wastes, temperature, and salinity.

Figure 4-24 Life cycle of salmon.

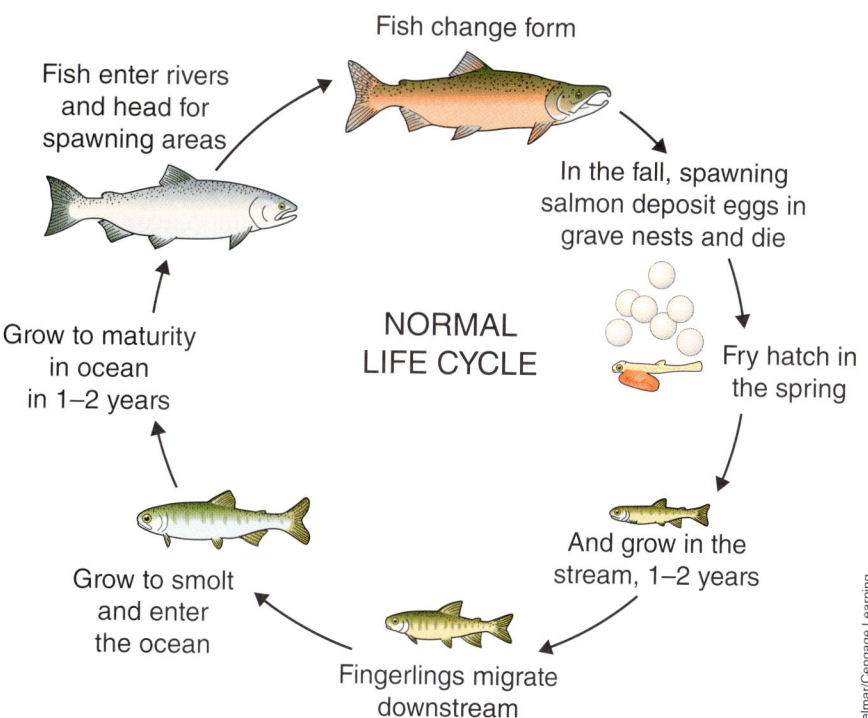

Stocking Rate

Large variations occur in numbers of returning adult salmon. Various climatic and oceanographic factors that control primary productivity most likely affect the size and abundance of returnees. A major problem involves gradual loss of wild populations as a result of heavy exploitation of hatchery stocks. Heavy fishing of hatchery stock in the ocean results in heavy fishing of wild stocks where they occur together.

Feeding

Nutritional requirements for growth of young chinook salmon, and coho salmon in fresh water, are established. Salmon's food requirements are similar to those of trout. As a carnivorous fish, salmon require a high-protein diet. Growers often feed the same diets to different species, although differences exist in their growth rates and length of time spent in fresh water.

Hatchery rearing may increase requirements of young salmon for certain nutrients that become depleted under stress conditions. Commercial diets for salmon are available.

Diseases

Prevention remains the most effective means of control. Conditions under which diseases occur vary among hatcheries, species, and stocks of fish. With salmon, the difficulty is to define conditions that increase or decrease the occurrence of disease. Requirements for drug certification are expensive and difficult to meet, and the small potential market does not motivate the pharmaceutical industry to develop new drugs. Commercial vaccines exist for vibriosis and enteric redmouth disease, but the cost and time involved limits their use.

Harvesting and Yields

In hatcheries, salmon growth can be enhanced by such means as varying water temperature, better diets, accelerated feeding, and genetic selection. Size at release seems to be a determinant of survival.

Farmed salmon are harvested between 0.5 and 1.5 lbs. At this size, the salmon compete and compare with the trout market. To harvest farmed salmon, nets are opened and fish removed with dip nets, or the net pen may be emptied into a boat. Seines harvest salmon from streams or sublittoral enclosures.

Processing and Marketing

Salmon products are highly valued and are in heavy demand throughout the world. Well-established markets and distribution channels exist for such products. The United States consumes 28 percent of salmon products; Japan, 25 percent; Russian Federation, 17 percent; Canada, 7 percent; and others, primarily European countries, 23 percent.

Commercial products include fresh and frozen steaks and fillets, canned, salted, pickled, smoked, kippered, and specialty items, and salted roe (caviar) for human consumption, as well as meal, oil, animal feed, and bait for industrial uses. The United States probably is the largest consumer of canned, fresh, frozen, and smoked salmon. Japan is the largest producer and consumer of salted salmon.

Demand for sport fishing has increased steadily, but outlook for the future remains unknown. Demand is a function of income, leisure time, cost, and availability of fish. Supply, not demand, represents the major problem confronting salmon fisheries worldwide. Limited supply aggravates conflicts over allocation and increases the prices that are paid by U.S. consumers.

Tilapia

In the United States, cultured species of tilapia include: the **mouthbrooders**: *Tilapia nilotica, T. aurea, T. mossambica, T. hornorum*, and the **substrate** spawners: *T. rendalli* and *T zillii*. Various **hybrids** between mouthbrooding species may also be important. For example, most of the reddish-orange tilapias are hybrids.

Sources of Species

Tilapia (Figure 4-25) belong to the cichlid family. They are native to Africa and the Middle East, but they have been distributed widely in tropical, subtropical, and temperate areas. Tilapia were introduced in the United States in the early 1950s. Tilapia established wild populations in parts of Texas.

Figure 4-25 Tilapia.

Potential tilapia culturalists in the United States first determine which species, if any, can be legally cultured in their state. Assuming no restrictions, the selection of a species depends mostly on growth rate and cold tolerance. Rankings for growth rate in ponds are:
1. *T. nilotica* (Nile tilapia)
2. *T. aurea* (Blue tilapia)
3. *T. rendalli*
4. *T. mossambica* or *T. hornorum*

Most tilapia hybrids tested grow as fast as their parent species. Cold tolerance may become an increasingly important criterion for selecting a species in more northerly latitudes. Tilapia aurea is generally recognized as being the most cold tolerant.

Habitat

In general, tilapia are extremely hardy creatures and can tolerate relatively poor water-quality conditions. One of the major constraints to development of the industry in the United States is the inability of tilapia to tolerate low temperatures. Tilapia will not survive at temperatures below 50°F. Activity is usually reduced at temperatures below 68°F. Tilapia exposed to low sublethal 50° to 60°F temperatures may develop fungal infections. Ideal temperature for growth is between 79° and 90°F.

Tilapia can withstand high water temperatures up to 95°F. Extremely high mortality rates, however, can result from handling and transporting fish at temperatures above 86°F, especially smaller fish.

Tilapia probably evolved from marine ancestors, and most species can tolerate a wide range of salinities. Changes in salinity greater than 10 to 15 ppt should be gradual. Most of the important species grow well at salinities up to 30 ppt. Spawning may be inhibited at salinities above 20 ppt, but this may be an advantage in production of food fish in brackish water.

Tilapia tolerate extremely low levels of dissolved oxygen in ponds. Tilapia can use atmospheric oxygen when dissolved oxygen levels drop in ponds. Oxygen levels reach near 0 ppm in heavily manured tilapia ponds without mortality. Although mortality is low, growth rates are reduced during these periods of low oxygen. Tilapia must have access to the water surface. If ponds are covered with duck weed or if fish are crowded in tanks, tilapia cannot survive the low dissolved oxygen.

Ammonia can be toxic to tilapia. As with most fish, ammonia toxicity in tilapia is related to pH. The un-ionized form of ammonia prevalent at higher pH is the toxic form. Chapter 12, Water Requirements for Aquaculture, describes water quality. Tilapia can acclimate gradually to increasing ammonia levels and can tolerate higher levels than most fishes.

Seed Stock and Breeding

Tilapia are mouthbrooders. (See Figure 4-26) The male establishes a territory and builds a round nest in the pond bottom. The nest's diameter correlates to the male's size. The female enters the nest and lays her eggs.

Figure 4-26 Diagram of reproduction in tilapia (mouthbrooders).

1. Male and female preparing nest.

2. Female lays eggs in the nest.

3. Male fertilizes the eggs.

4. Eggs picked up in mouth by female.

5. Fertilized eggs held in mouth until they hatch and yolk sac is absorbed—4 to 5 days.

The eggs are then fertilized by the male. The female then collects and incubates the eggs, which are yellow-colored, in her mouth. Tilapia eggs hatch in about four to eight days. After hatching, the fry remain in the female's mouth for another three to five days. Fry begin to swim freely in schools, but may return to the mouth of the female when threatened.

Females may spawn every four to six weeks but may spawn sooner if eggs or fry are removed during mouth brooding. The number of eggs produced per spawn is related to the female's size. A female of 100 gm can produce approximately 100 to 150 eggs, whereas a female weighing 1 kg (2.2 lbs) can produce between 1,000 and 1,500 or more eggs per spawning. Males in general will mate with more than one female.

Tilapia can spawn year round if maintained at a temperature between 77° and 86°F. Spawning activity is usually inhibited at salinities above 15 to 20 ppt, except for *T. mossambica*, which will spawn in seawater.

Tilapia can reach sexual maturity between 50 and 100 gm. If placed in ponds, tilapia will readily spawn, and, within a short time period, a pond can become overloaded with fingerlings. At high fingerling densities, growth of food fish is inhibited. This overpopulation of fingerlings is a major constraint to pond production of larger fish.

Culture Method

Pond culture is the most popular method of growing tilapia. One advantage is that the fish can consume natural foods. Management of tilapia ponds ranges from extensive systems, using only organic or inorganic fertilizers, to intensive systems, using high-protein feed, aeration, and water exchange. The major drawback of pond culture is the high level of uncontrolled reproduction that may occur in grow-out ponds. Tilapia recruitment, or the production of fry and fingerlings, may be so great that offspring compete for food with the adults. The original stock becomes stunted, yielding only a small percentage of marketable fish weighing 1 lb (454 gm) or more. In mixed-sex populations, the weight of recruits may constitute up to 70 percent of the total harvest weight. Two major strategies for producing tilapia in ponds, mixed-sex culture and male monosex culture, revolve around controlling spawning and recruitment.

Ponds can be any size, but for ease of management and economical operation, shallow (3 to 6 ft.), small (1 to 10 acre) ponds with drains are recommended. Draining is necessary to harvest all of the fish. A harvesting sump is needed to concentrate the fish in the final stage of drainage. Drying the pond eradicates any fry or fingerlings that may interfere with the next production cycle. Geographic range for culturing tilapia in ponds depends upon temperature. The preferred temperature range for optimum tilapia growth is 82° to 86°F. Growth diminishes significantly at temperatures below 68°F, and death occurs below 50°F.

In temperate regions, tilapia must be overwintered in heated water. In the continental United States, the southernmost parts of Texas and Florida are the only areas where tilapia survive outdoors year-round, with the exception of geothermally heated waters, most notably in Idaho. In the southern region, tilapia can be held in ponds for 5 to 12 months a year depending on location.

Mixed-sex culture

Mixed-sex populations of fry are cultured together and harvested before or soon after they reach sexual maturity. This eliminates or minimizes recruitment and overcrowding. A restricted culture period limits the size of the fish that can be harvested.

In mixed-sex culture, tilapia are usually stocked at lower rates to reduce competition for food and to promote rapid growth. One-month-old, 1-gm fry are stocked at 2,000 to 6,000 per acre into grow-out ponds for a -four-to five-month culture period.

Newly hatched fry should be used because older, stunted fish, such as those held over winter, will reach sexual maturity at a smaller, unmarketable size. Supplemental feeds with 25 to 32 percent protein are generally used. At harvest, average weight is approximately 8 oz, and total production is near 1,400 lbs. per acre for a stocking rate of 4,000 per acre. Expected survival is roughly 70 percent.

Species such as *Tilapia zillii*, *T. hornorum*, or *T. mossambica* are not suitable for mixed-sex culture because they reproduce at an age of two to three months and at an unmarketable size of 30 gm or less. Tilapia suitable for mixed-sex culture are *T. aurea*, *T. nilotica*, and their hybrids, all of which reproduce at an age of five to six months.

Male fingerling rearing

With male monosex culture, fry are usually reared to fingerling size in a nursery phase, and then male fingerlings are separated from females for final grow out. All-male fingerlings can be obtained by three methods: **hybridization**, sex reversal, and manual **sexing**. None of these methods is consistently 100 percent effective. A combination of methods works better. Hybridization may be used to produce a high percentage of male fish. The hybrids may then be manually sexed or subjected to a sex-reversal treatment. All three methods are sometimes used.

Hybridization and sex reversal reduce the number of female fingerlings that must be discarded during manual sexing. This saves time, space, and feed. Problems still exist with hybridization and sex reversal. Producing sufficient numbers of hybrid fry may be challenging because of spawning incompatibilities between the parent species. Sex reversal is more technically complicated and requires obtaining recently hatched fry and rearing them in tanks with high-quality water.

Manual sexing is commonly used by producers. Manual sexing (hand sexing) is the process of separating males from females by visual inspection of the external urogenital pores, often with the aid of dye applied to the papillae. Secondary sex characteristics may also be used to help distinguish sex. Reliability of sexing depends on the skill of the workers, the species to be sorted, and the species size. Experienced workers can reliably sex 15 to 50 gm fingerlings.

In temperate regions, fingerlings are produced during summer and stored in overwintering facilities for the next growing season. If manual sexing is used, it is done prior to overwintering. The best fingerling size for overwintering depends on the number of fingerlings that will be needed and the available storage capacity. Fingerlings that weigh less than 20 gm should not be overwintered because their survival rate will be low.

Overwintering facilities consist of geothermal springs, greenhouses, and heated buildings. Fingerlings can be held in cages located in geothermal springs or in small ponds or tanks through which warm spring water is diverted. In greenhouses and heated buildings, recirculating systems are used to hold large quantities of fingerlings. Fingerlings can be overwintered in long, narrow ponds that are covered with clear plastic if the winter is mild.

Male monosex culture

Males are used for monosex culture because male tilapia grow faster than females. Females expend considerable energy in egg production and do not eat when they are incubating eggs. Male monosex culture permits the use of longer culture periods, higher stocking rates, and fingerlings of any age. High stocking densities reduce individual growth rates, but yields per unit area are greater. With a longer growing season, fish weighing 1 lb. (454 gm) or more can be produced. Expected survival for all-male culture is 90 percent or greater. A disadvantage of male monosex culture is that female fingerlings are discarded.

The percentage of females mistakenly included in a population of mostly male tilapia affects the maximum attainable size of the original stock in grow out. For example, manually sexed *T. nilotica* fingerlings (90 percent males) stocked at 3,848 per acre will cease growing after five months when they average about 0.8 lb. because of competition from recruits. If larger fish are desired, females should make up 4 percent or less of the original stock and predator fish should be included.

Polyculture

Tilapia are frequently cultured with other species to take advantage of many natural foods available in ponds and to produce a secondary crop or to control tilapia recruitment. Polyculture uses a combination of species with different feeding niches to increase overall production without a corresponding increase in the quantity of supplemental feed. Polyculture can improve water quality by creating a better balance among the microbial communities of the pond, resulting in enhanced production. The disadvantage of polyculture is the need for special equipment, such as sorting devices and conveyors. At harvest, extra labor is needed to sort the different species. The role of natural pond foods is less important in the intensive culture of all-male populations, and it may not justify the expense of sorting the various species at harvest.

Tilapia can be cultured along with channel catfish with only a minor reduction in catfish yields. Also, catfish production does not decline when cultured in combination with tilapia, silver carp, and grass carp. With no additional feed, total net production increases over that for catfish cultured alone. The incidence of off-flavor catfish may be less in catfish/tilapia polyculture than catfish monoculture. Another promising polyculture system consists of tilapia and prawns. In polyculture, survival and growth of tilapia and prawns are independent. Feed is given to meet the requirements of the fish. Prawns, which are unable to compete for the feed, use waste feed and natural foods that result from the breakdown of fish waste.

Another type of polyculture involves the use of a predatory fish such as largemouth bass to reduce tilapia recruitment. Stocking predators with mixed-sex tilapia populations controls recruitment and allows the original stock to reach a larger market size. Predators must be stocked at a small size to prevent them from eating the original stock. Predators may be stocked when tilapia begin breeding.

The number of predators required to control tilapia recruitment in culture ponds depends primarily on the maximum attainable size of the predator species, the ability of the predator to reproduce, and the number of mature female tilapia. In general, as predators grow, they eat larger-sized tilapia recruits. More predators are required to control recruitment when there are larger numbers of mature female tilapia.

Stocking Rate

The stocking rate for male monosex culture varies from 4,000 to 20,000 per acre or more. At proper feeding rates, densities around 4,000 per acre allow the fish to grow rapidly without the need for supplemental aeration. About six months are required to produce 18-oz fish (500 gm) from 1.8-oz (50 gm) fingerlings. Total production approaches 2.2 tons per acre. A stocking rate of 8,000 per acre is frequently used to achieve yields as high as 4.4 tons per acre. At this stocking rate the daily weight gain will range from 1.5 to 2.0 gm. Culture periods of 200 days or more are needed to produce large fish that weigh close to 18 oz (500 gm). To produce an 18-oz fish in temperate regions, overwintered fingerlings should weigh roughly 70 to 100 gm and should be started as early as possible in the growing season. A stocking rate of 8,000 per acre requires nighttime emergency aeration when the **standing crop** is high.

Stocking rates of 12,000 to 20,000 per acre have been used in 1.2 to 2.5 acre ponds, but this requires the continuous use of two to four, 1-hp paddlewheel aerators per pond. Yields for a single crop range from 6 to 10 tons per acre.

As of 2005, 156 food-fish farms in the United States cultured tilapia. While the largest number of tilapia farms were located in Hawaii (19 farms) and Florida (18 farms), California (15 farms) ranked first in sales. Idaho ranked second, with sales from seven farms. Of the U.S. tilapia farms, the largest number (128 farms) reared food-size tilapia. Other tilapia farms specialized in stockers, fingerlings and fry, and broodstock.

As the tilapia industry has grown, so has the number of product forms. Today, fresh or frozen fillets are available in different sizes and packages, as skin-on, skin-off, deep skinned, individually quick frozen, smoked, and sashimi grade, and they are treated by carbon monoxide or ozone dipped. Interesting by-products have emerged, such as leather goods for clothing and accessories, gelatin from skins for time-released medicines, and flower ornaments made from dried and colored fish scales.

Feeding

With optimal temperatures, feeding rates depend on fish size and density. Commercial catfish diets can be used, but producers usually take advantage of the tilapia's ability to use the natural productivity of the pond. If densities are high, suboptimal feeding rates may have to be used to maintain suitable water quality, increasing culture duration.

ANCIENT FISH TODAY

Sturgeon and paddlefish are ancient fish surviving today. The two have some similarities. They both belong to the same order but to different families. Sturgeon belong to the Acipenseridae family, and paddlefish belong to Polydontidae.

Sturgeon are teleost (bony fish) fish that evolved about 250 million years ago. They are chondrostean fish with mostly cartilaginous skeletons. They are known for their longevity and large size. Sturgeon have protrusible mouths that draw in benthic organisms for food that are sensed by barbels beneath the snout.

Sturgeon and paddlefish are native in the temperate waters of Europe, Asia, and North America. Some anadromous species live exclusively in a marine environment and return to freshwater only to spawn. Semi-anadromous species, for example, the white sturgeon, live primarily in estuaries of large rivers and often form landlocked stocks in lakes and reservoirs. Freshwater species include lake sturgeon and paddlefish.

Sturgeon are iteroparous, meaning that they spawn several times in a lifetime. Females spawn at intervals from two to eight years. They usually spawn in areas of swift current, and the eggs are large and adhesive.

White sturgeon (*Acipenser transmontanus*) habitat is the Pacific coast from the Aleutians to Ensenada, Mexico. Some fish are semi-anadromous, and some live as landlocked freshwater fish. They swim upstream in fall and winter to spawn and migrate downstream in spring and summer.

Atlantic sturgeon (*Acipenser oxyrhynchus*) are found along the east coast of North America and in the Gulf of Mexico and north coast of South America. They are anadromous. Atlantic sturgeon stay in freshwater until age four to six, then migrate to sea.

Lake sturgeon (*Acipenser fluvescens*) inhabit lakes and rivers of Canada and the United States, south to Alabama, and west to the Missouri. This freshwater species migrates in early spring to smaller streams or shallow lakes to spawn.

Shortnose sturgeon (*Acipenser brevirostrum*) are considered an endangered species. Their habitat is from Canada to eastern Florida. Some migrate to saltwater on a seasonal basis. Others are anadromous.

Paddlefish (*Polyodon spathula*) inhabit the Mississippi River Valley and Gulf slope drainage, and they were known in Lake Erie before 1903. Another species, *Psephurus gladius*, inhabits the Yangtze River in China. Paddlefish migrate up river to spawn in the spring and summer. In the United States, populations of sturgeon and paddlefish are declining and, in many locations, are protected by federal and state resource agencies.

The most appropriate mouth-brooding tilapia for culture can feed low on the food chain on a diet of plankton and **detritus**. If the natural productivity of a pond is increased through fertilization or manuring, significant production of tilapia can be obtained without supplemental feeds. Although yields are not as high as those obtained with feed and fertilizers, animal manures can be used to reduce the quantity and expense of supplemental feeds.

Inorganic fertilizers are used less often because of their expense, but a single large application of an inorganic fertilizer high in phosphorus is frequently made prior to stocking fish to create an algal bloom. Tilapia productivity is stimulated mainly by an increase in phosphorus and to a lesser extent by an increase in nitrogen. Phosphorus is effectively increased through the application of liquid polyphosphate (13-38-0) at a rate of 20 lbs. per acre, or 2.4 gal. per acre.

Manuring, which is widely used for food fish production overseas, has not been practiced in the United States because of public perception. Manuring may have application in the production of tilapia as a source of fish meal for animal feeds. Manure's quality as a fertilizer depends on several factors. Pig, chicken, and duck manures increase fish production more than cow and sheep manure. Animals fed high-quality feeds (grains) produce manure that is better as a fertilizer than those fed diets high in crude fiber. Fresh manure is better than dry. Finely divided manures provide more surface area for the growth of microorganisms and produce better results than large clumps of manure.

Integrated systems

Collection, transport, storage, and distribution of manure involve considerable expense and are major obstacles to manured systems. These problems can be overcome by locating the animal production unit adjacent to or over the fish pond so that fresh manure can easily be delivered to the pond on a continuous basis. Effective and safe manure loading rates are maintained by having the correct number of animals per unit of pond surface area. Animal production units located adjacent to or over fish ponds in some areas include chicken, pigs, and ducks.

1. Chicken/fish farming. Maximum tilapia yields are obtained from the manure output of 2,000 to 2,200 chickens per acre, which deliver 90 to 100 lbs. (dry weight) of manure per acre per day. Several crops of chickens can be produced during a fish production cycle.
2. Pig/fish farming. Approximately 24 to 28 pigs per acre are required to produce a suitable quantity of manure, 90 to 100 lbs. of dry matter per acre per day for tilapia production. The pigs are usually grown from 44 to 220 lbs. over a six-month period. (See Figure 4-27.)
3. Duck/fish farming. Ducks are grown on ponds at a density of 300 to 600 per acre. The ducks are generally raised in confinement, fed intensively, and allowed access to only a portion of the pond where they forage for natural foods and deposit their manure. Ducks that are raised on ponds remain healthier than land-raised ducks. Also, by raising ducks on ponds, feed wasted by the ducks is consumed directly by the fish. Because ducks reach marketable size in 10 to 11 weeks, staggered production cycles are needed to stabilize manure output.

Although the growing season is adequate to produce marketable size tilapia, economical overwintering systems are essential for broodfish and fingerlings. In order to make the most of the limited growing season, it is essential to have fingerlings stocked early in the season. To do this, fingerlings must be produced in controlled systems during the winter or must be

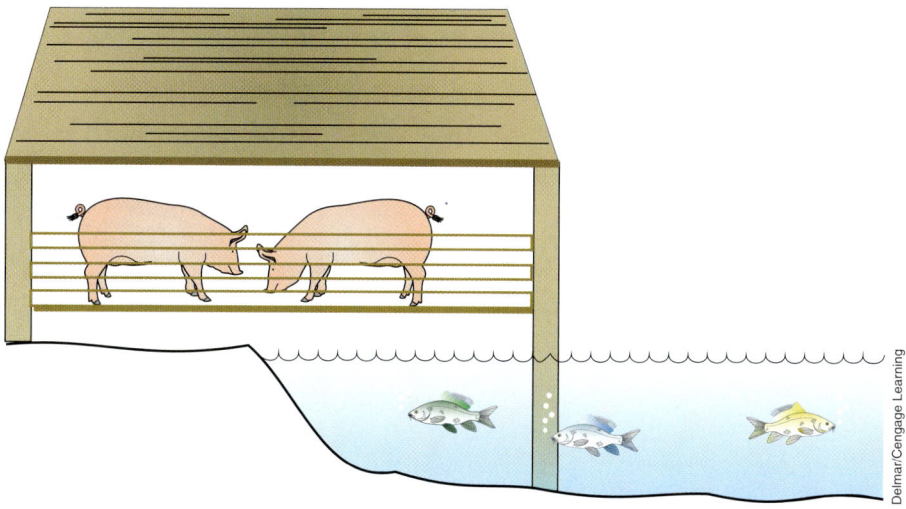

Figure 4-27 In other countries, fish and other animals are sometimes managed together. Stables and/or pens are built over fish ponds. Pig manure falls directly into the pond, providing food and fertilizer.

overwintered. This can be done using flow-through systems with warmer deep well water or in closed recirculating systems. Research and commercial applications use both systems.

Diseases

As a group, tilapia are very hardy and thrive under conditions that kill other fish. They have few diseases and parasites. At temperatures below 54°F, tilapia lose their resistance to disease and are subject to infections by bacteria, fungi, and parasites.

Harvesting and Yields

Tilapia are best harvested by seining and draining the pond. A complete harvest is not possible by seining alone. Tilapia are adept at escaping a seine by jumping over or burrowing under it. Only 25 to 40 percent of a *T. nilotica* population can be captured by seine haul in small ponds. Other tilapia species, such as *T. aurea*, are even more difficult to capture. A 1-in (2.54-cm) mesh seine of proper length and width is suitable for harvest.

Yields of male monosex populations in manured ponds have been modest, but production costs are very low if the manure is free. For example, all-male hybrids weighing 29 gm (*T. nilotica* x *T. hornorum*) stocked at 4,000 per acre will produce a net yield of 1,470 lbs. per acre of 200-gm fish in 103 days when given fresh cattle manure at an average rate (dry weight) of 46 lbs. per acre per day. In comparison, fish receiving a commercial high-protein feed will give a net yield of 2,370 lbs. per acre. Feeding costs per pound of production are 2 to 20 times higher for fish fed the commercial diet compared to fish receiving manure.

Processing and Marketing

Marketing constraints are always a stumbling block when a new product is introduced. Fish retailers indicate a demand for tilapia, but seasonal production is a problem. Ethnic markets hold a lot of potential. Introduction of red varieties definitely increased marketability in parts of the United States.

Figure 4-28 Striped bass.

Hybrid Striped Bass

The hybrid striped bass (Figure 4-28) has become a highly desirable substitute for the declining striped bass seafood industry. As a foodfish, the hybrid exhibits a mild taste and firm texture. Aquaculturalists find these hybrids well-suited to pond culture, and current research is helping to improve culture techniques. Hybrid fry are raised in rearing ponds until they grow to become 35- to 45-day-old fingerlings. Young fingerlings, when harvested, are graded by size and trained to feed on pelleted feed. Producing market-size fish requires 15 to 18 months.

Sources of Species

Hybrid striped bass generally refers to a cross between striped bass *(Morone saxatilis)* and white bass *(M. chrysops)*. This cross, sometimes called the "original cross," was first produced in South Carolina in the mid-1960s using eggs from striped bass and sperm from white bass. The accepted common name of this cross is the Palmetto Bass. The opposite cross, using white bass females and striped bass males, produces the Sunshine Bass.

Habitat

Hybrid striped bass are stocked into a variety of water types for recreational purposes. They do well in slow-moving streams, large reservoirs, lakes, and ponds. They are seldom found in extremely shallow areas or areas that contain dense aquatic-weed growth. Because they are pelagic in nature, they are generally found in open water areas. They are generally most active during periods of low light such as dawn and dusk. Beginning in late winter, they tend to concentrate in deep areas near inflowing streams and in the spring may undergo spawning migrations into upstream areas. Hybrids are fertile, and there are reports of successful reproduction in a few reservoirs.

Seed Stock and Breeding

Broodstock to produce fingerling striped bass and hybrid striped bass are collected from the wild during their spring spawning run. Spawning runs for striped bass species occur from late March to late May, depending on location. Spawning grounds for striped bass are usually found near deep, swift, and turbulent sections of a river, well upstream from lakes, reservoirs, and sounds. Males begin their spawning run one to three weeks before the females, when water temperature is less than 59°F. Female striped bass begin their spawning migration when water temperature is around 59°F. For any given population of striped bass, several periods of spawning may occur during a four to five week period when water temperatures are 61° to 68°F. Periods of spawning activity during this time frequently follow sudden temperature increases of 2° to 4°F.

White bass, although restricted to freshwater, also make spawning migrations from lakes and reservoirs to inflowing streams. They generally spawn in rocky areas where water flow is turbulent. Their peak spawning season usually occurs from late March to late May depending on location. As with striped bass, male white bass usually arrive at the spawning grounds before the females. Peak spawning activity occurs when water temperatures are 64° to 66°F. Usually more than one period of activity occurs for a specific population.

The production of hybrid striped bass must be accomplished by manually stripping the eggs and sperm from the ripe fish into a container. Sperm from two or more white bass or striped bass males is used to ensure fertilization of the eggs. Fertilization of striped bass eggs is accomplished by using either a wet or a dry method. Little difference exists in the percent of fertilization between the two methods. The most common method of incubating striped bass and white bass eggs is using a modified McDonald hatching jar. (See Figure 4-29.)

The incubation period varies inversely with water temperature. At 61° to 64°F, the incubation period is between 40 and 48 hours. Two hours after fertilization, percent fertilization should be determined by counting the number of eggs with dividing cells. At four hours, an estimate of total number of eggs should be determined volumetrically by letting the eggs settle to the bottom of the jar. The number of eggs per milliliter may be determined by counting the number of eggs in a known volume.

A hatch rate of 50 percent is acceptable and 60 to 80 percent is considered good. The fry are held in 30- to 75-gal. containers before pond stocking. Water exchange in these containers should be continuous. Newly hatched hybrids have no mouth opening, an enlarged yolk sac, and a large oil globule projecting beyond the head. At four to eight days post hatch, the yolk sac and oil globule are assimilated, the mouthparts develop, and fry begin to feed. Fry are stocked into fertilized ponds at two to ten days post-hatch. Fry held more than five days must be provided with live food, such as brine shrimp nauplii or wild-caught copopod nauplii and cladocerans. Fry should be fed frequently (at least every three hours) during the early rearing period.

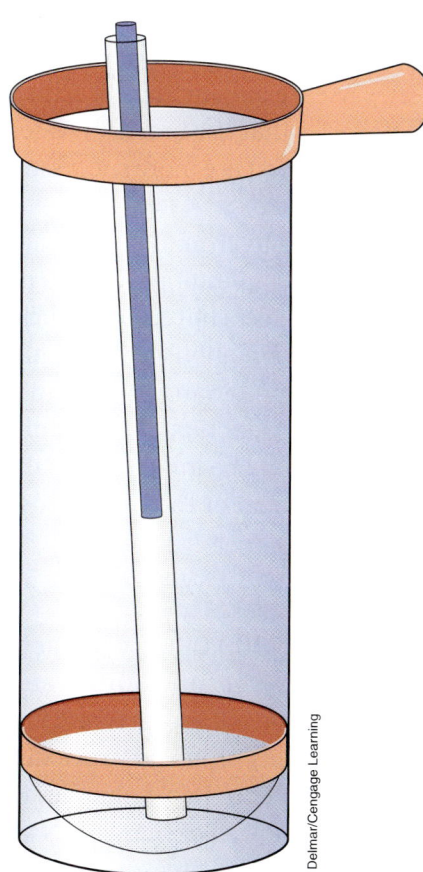

Figure 4-29 A modified McDonald.

Culture Method

Hybrid striped bass survive and grow well in a wide range of water-quality variables. Maintaining good water quality is a major part of all phases of production. Temperature and dissolved oxygen levels should be monitored daily—morning and evening—and aerators should be used to keep dissolved oxygen levels above 4 mg per l. Maximum growth occurs around 77° to 81°F. Hybrids can survive a temperature range of 39° to 90°F in culture systems. Below 59°F, feed consumption is reduced and growth slowed.

Dissolved oxygen is important in any culture operation, especially so for hybrid striped bass. Hybrids may survive dissolved oxygen levels as low as 1 mg per l for a short time, but these levels are very stressful. Dissolved oxygen levels below 4 mg per l reduce feed consumption and growth, while increasing amount of energy needed for respiration and increasing mortality.

Alkalinity, hardness, and pH levels are usually related, and hybrid striped bass grow well over a wide range of values. Alkalinity of 100 mg per l or above is desirable in culture situations. Mortality can be significant during transfer from water of high alkalinity per hardness to water with low alkalinity per hardness. Hybrids survive in a pH range of 6.0 to 10.0, although 7.0 to 8.5 is optimum for growth. Ammonia, the principal excretory product of fish, should also be monitored regularly in ponds. Concentrations should not exceed 1 mg per l.

Nursery ponds should be filled approximately two weeks prior to stocking fry. Ponds filled too early will develop large populations of predaceous insects that eat hybrid fry. Most hatcheries use freshwater, but brackish water is used in some areas. Generally, hatcheries that use brackish water or hard freshwater (more than 100 ppm Ca hardness) are more successful than those that rely on soft freshwater.

Ponds should be dried and disked prior to filling to promote the breakdown of nutrients in the pond bottom. Agricultural limestone may also be applied to the bottom at this time if necessary. Success in rearing hybrid striped bass depends on the presence of adequate populations of zooplankton. Nursery ponds are usually fertilized with a combination of organic and inorganic fertilizers to enhance the natural production of zooplankters. New ponds or ponds filled with well water may be inoculated with phytoplankton and zooplankton to foster development of the desired zooplankton populations.

Stocking Rate

Fry are generally stocked at a rate of 100,000 to 200,000 fry per acre at two to ten days of age. Food supply, dissolved oxygen, and other water quality factors are especially important to fish survival. Aeration and circulation of pond water help moderate water quality shifts, improve dissolved oxygen levels, and increase plankton production. Rapid changes in temperature, pH or hardness, and insufficient dissolved oxygen levels all affect survival of larval fish. Constant monitoring of water quality and food supply and remedying problems quickly improve fish survival.

Fingerlings are generally available from producers in the southeastern United States from May to July depending on location. They are stocked at a rate of 8,000 to 12,000 fish per acre to complete their first year of growth. Ponds of 2 to 4 acres are recommended for commercial production. Large ponds are more difficult to manage whereas small ponds are expensive to build.

Fingerlings (110 to 225 gm) are stocked into grow-out ponds at a rate of 3,000 to 4,000 fish per acre depending on the experience of the culturalist. With proper management these fish will reach marketable size by October or November. Survival rates for the second growing season are generally 90 percent or better.

Feeding

Hybrid striped bass is predaceous throughout its life. Survival and production of fingerlings depend upon the culturalist's ability to supply the young fish with live food of good quality and quantity. Striped bass female fry crossed with white bass male fry prefer large crustacean zooplankton as their first food. White bass female fry crossed with striped bass male fry must have an adequate supply of small zooplankters, such as rotifers, because the fry are smaller than fry hatched from striped bass eggs.

Pond management techniques are used to increase zooplankton populations in the nursery ponds. Pond fertilization requires culturalists to work out the techniques that work best for their situations.

At a size of 1 in. (25 mm), fish are introduced to prepared food. The transition to pelleted feed is begun at around 14 to 21 days old when fish are presented prepared food. Particle size of prepared food is critical to successful transition. By 28 days old, fish should be sustained on prepared feed and fed increasing amounts according to growth. Food particle size is increased as fish grow. Food should be offered daily with frequency depending on the amount of natural pond zooplankton.

Advanced fingerlings should be graded before they are stocked for grow-out to reduce the size variation in each pond. Feeding problems will be reduced, and all the fish in one pond will reach market size at about the same time.

Fingerlings are fed commercial feed at a rate of 1 to 3 percent of body weight per day. While temperatures are low and dissolved oxygen levels are high, fish can be fed at a rate of 3 percent of body weight per day. As temperature and biomass increase, dissolved oxygen levels become more difficult to manage. The feeding rate should then be around 1 percent of body weight per day. Food conversion ratios of 2:1 or less are expected.

Diseases

Infectious diseases pose few problems in the culture of striped bass even though many disease-causing agents are known. The few disease problems that do occur vary greatly from one cultural facility to another. Most of the variation in disease among hatcheries is accounted for by differences in intensity of culture, phase of culture, water quality, and management.

Figure 4-30 Seining a pond to harvest fish.

Harvesting and Yields

At the end of the 30- to 45-day nursery period, fingerlings are harvested by seining (Figure 4-30) and draining the ponds. Survival rates are extremely variable, and 0 to 80 percent of the larvae stocked may be harvested as fingerlings. Fish that are to be sold for aquaculture are held in tanks or raceways to be graded and trained to feed on pelleted food.

Grading, or sorting by size, is very important to prevent cannibalism (Figure 4-31). Losses of 50 percent or more can occur in one to two weeks if fingerlings are not graded every day or two. Cannibalism is prevalent because hybrid fingerlings would normally be switching to a fish diet when they are 2 to 3 in. long. Fast-growing fish should be graded out before they learn to cannibalize. Once trained to take pelleted food, fish are ready to be stocked into ponds.

Mortalities decrease when fish are transported in slightly saline water. Before transferring fish from one system to another, they should be gradually acclimated to the temperature and hardness of the new system.

By the end of the first growing season, fish may weigh an average of 0.5 lb.. Any fish from 3.8 oz. should reach a marketable size of 1.25 lbs. in the second year. Survival rates of 85 percent are common at the end of the first growing season. Fish are harvested after the growing season ends, usually beginning in December, when pond temperatures drop below 52°F and continuing through March. Handling fish at 52°F or above increases the likelihood of fungus and disease problems. The pond is seined, and the fish are herded through an opening in the seine into a holding net (live-car). The number and weight of fish are estimated by weighing several samples of a known number of fish and taking a total weight of fish. The fish should be weighed in water to reduce stress.

Figure 4-31 An auger-type grading device called a pescalator used successfully in pond culture of hybrid striped bass.

Processing and Marketing

Hybrid striped bass are processed or marketed at 1 to 3 lbs. They are sold whole in the round, dressed with the head on, and filleted with the skin on or off. Another marketing strategy involves the producer shipping the whole fish—no processing—on ice to distributors in large cities. After harvesting fish, they are chilled on ice and shipped in 50 to 25 lb. containers.

Hybrid striped bass production is rapidly expanding in the United States. Annual production has increased from about 400,000 pounds in 1987 to about 11 million pounds in 2005. According to the latest Census of Aquaculture (2006), 67 of the 87 farms raising hybrid striped bass in 2005 emphasized the production of foodsize striped bass. A much smaller number of farms (17), raised stockers. Production growth is a response to some reduction in natural fishery stocks of striped bass, generally increased urban-based market demand for seafood and the development of improved culture techniques for this species. Other countries, including Taiwan, Israel and Italy, have extensive, expanding production systems.

Carp

The Chinese cultured grass carp, silver carp, and bighead carp for food for several thousand years. In more recent history, other Asian and some European civilizations cultured carp (Figure 4-32).

Considerable information exists for the production and management of these fish for food throughout the world, and interest in production for food has increased in the United States. Interest in the grass carp centers on its use as a vegetation-control fish. The silver carp has been investigated principally as a nutrient removal species in sewage and animal-waste lagoons. Controversy surrounded grass, silver, and bighead

Figure 4-32 Common carp.

Figure 4-33 Types of carp.

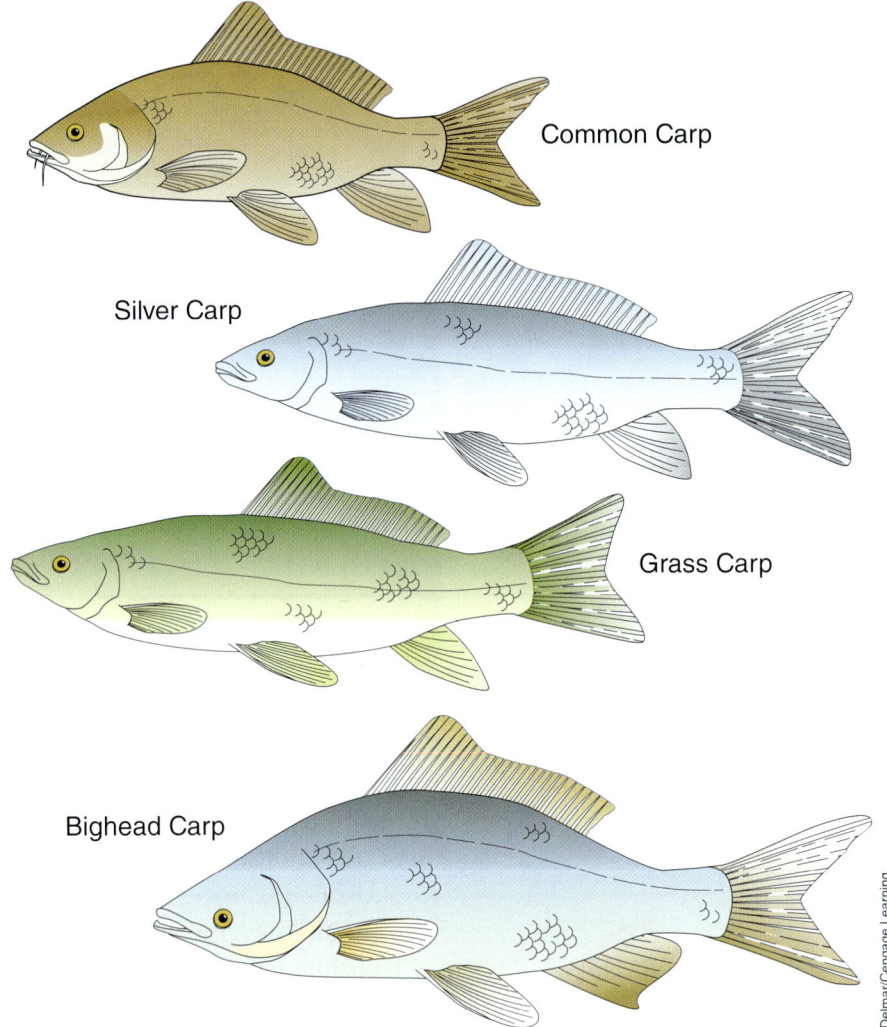

carp importation because people fear that they will become established in natural waters, displacing native species and possibly disrupting natural ecosystems, as the common carp is believed to have done. Most states outlawed these exotics. Anyone contemplating their culture must first become acquainted with local restrictions.

Sources of Species

The grass carp was introduced into the United States in 1963, and the silver carp and the bighead carp somewhat later. (See Figure 4-33.) Carp are native to Asia; they were introduced into Europe during the Middle Ages. They were introduced into the United States after the Civil War.

Habitat

Successful carp culture depends on the hardiness of the species, at all stages of life. Carp thrive in pond culture, and they adapt to acid or alkaline waters. They tolerate varying salinities, a wide temperature range, and turbid waters.

Seed Stock and Breeding

Grass carp, silver carp, and bighead carp for potential broodstock should be selected at one to two years of age and stocked in broodfish ponds. Ponds suitable for channel catfish broodfish are adequate. Generally, these ponds are from 1 to 10 acres in area and have adequate water supplies. Female grass carp and bighead carp mature at about four years, and the silver carp (both sexes), male grass carp, and male bighead carp at about three years. The different species of broodfish can be stocked alone, or some species can be stocked together. Males and females are generally stocked at ratios of 1:1 or 3:2.

In mid-April to mid-May in the southern United States, the broodfish should be seined and inspected. Because common carp spawn at about the same time as Chinese carp, they can be observed as readiness indicator. Female fish should have a distended, flaccid abdomen as compared with that of the males. A pinkish area around the egg vent indicates that a fish is ready to ovulate. The male grass carp should have protuberances on the head and operculum when ripe. Milt usually can be easily expressed from the vent of all mature male carp during the breeding season.

When the seined broodfish appear ready to spawn, fish that will be spawned immediately should be taken from the net and each put into a muslin or porous cloth bag, placed in water in a transport unit, and carried to the spawning shed. Water in the transport unit must be aerated, and the fish must be protected from rapid changes in water temperature. Anesthetizing the fish during transport prevents them from injuring themselves.

Broodfish should be stocked in separate tanks in water with sufficient dissolved oxygen and a water temperature near 77°F. The tanks must be covered to keep the fish from jumping out. Good quality water is essential for maintaining broodfish, hatching eggs, and culturing larvae and fry. Water obtained from wells is often preferred because it is usually free of pesticides and disease organisms. Like many fish, hormones can be used to artificially spawn carp. A number of hormone injection schedules are satisfactory.

About 10 hours after the 24-hour hormone injection, the females should be netted and inspected. When slight pressure is applied to the abdomen of a fish that is ready to ovulate, the eggs flow rapidly. If few or no eggs are obtained, the fish should be returned to the vat and examined again after another hour. When the eggs flow rapidly, the carp is ready to spawn. The following steps are typical for hand-spawning carp and some other fish.

1. Wrap the fish in a towel or place its head in a sack.
2. Dry the fish around the vent.
3. Holding the tail slightly lower than the head, allow the eggs to extrude into a wash pan.
4. Apply additional pressure to eliminate the remaining eggs.
5. Keep all water out of the egg-collecting pan.

6. Remove two males and press the abdomen to ensure that milt is flowing.
7. Dry the fish and press the abdomen to obtain enough milt to cover the eggs.
8. Stir the mixture of eggs and milt with a feather, paint brush, or dry finger.
9. Slowly add water and mix further.
10. Repeat until the water volume is twice the egg volume.
11. Continue stirring for about three minutes.
12. Decant and replace the water.
13. Repeat this process three or four times.
14. After about 10 minutes, the eggs are water-hardened.

At this stage a 1-qt. volume contains 225,000 to 250,000 eggs. Eggs of Chinese carp can be hatched in a variety of devices. For small quantities of eggs, the glass or plastic McDonald hatching jars used by the trout industry are adequate. Larger quantities can be hatched in fiberglass containers. Essentially, all that is necessary is to provide sufficient water exchange to maintain adequate dissolved oxygen and flush out waste metabolites, as well as sufficient agitation to move the eggs. Movement ensures no dead-water areas in the hatching container where the eggs might suffocate. The outlet must be screened to prevent loss of eggs and fry through the hatching-tank overflow. Fry can be held in the hatching tanks or removed to other tanks and kept for about five days before they are stocked.

Culture Method

Ponds suitable for the culture of Chinese carp resemble those used to culture channel catfish fry, bait minnows, and striped bass. Drain and dry the pond, and then disinfect any wet spots and puddles with hydrated or burnt lime. From one to two weeks before the fry are stocked, the pond should be filled with well water or filtered surface water. The surface water must be free of pesticides, and filtered to remove all fish and predaceous organisms. Any mixture of organic materials and chemical fertilizers that enhances the growth of zooplankton can be applied at filling time.

The fry can be poured or drained from the hatching device into tubs and immediately transported to the ponds. Because fry are delicate, netting them should be avoided, if possible. When transporting fry, you may use oxygenated plastic bags.

At first, the fish will feed primarily on zooplankton and other animal organisms. Supplemental feed can be offered. The growth of the fish must be observed. When growth rates diminish, either more supplemental feed must be offered, or density must be reduced by transferring some of the fish to other ponds. Water quality must be observed when the fish are fed.

Stocking Rate

Chinese carp are stocked initially in monoculture. The fry can be stocked at the rate of 100,000 to 500,000 per acre. Stocking them in the early morning reduces temperature shock. Vat water temperature can be gradually raised, or it can be lowered to that of the pond.

Fingerling grass carp, bighead carp, and silver carp can be produced in monoculture. Grass carp and bighead carp, or grass carp and silver carp, can be produced in polyculture. Date of sale and desired size at sale dictate the specifics of the production program. For example, relatively large numbers of fish can be stocked when only small fish are needed late in the growing season, but fewer fish must be stocked when the demand is for large fish early in the production period.

Feeding

Feeding and management techniques used depend on whether the fish are being produced alone or in polyculture. Polyculture species could also include some native U.S. fish such as catfish or black bass. Grass carp fingerlings and food-sized fish feed exclusively on aquatic plants in nature.

The fish in managed systems feed on floating pelleted fish feeds and terrestrial grasses such as millet and Bermuda grass. These plants can be fed green (freshly cut) at rates up to 200 lbs. per acre daily. Some producers feed smaller amounts of green forage to supplement a diet of pelleted feed. Most grass carp producers feed pelleted feed. When small fish first begin feeding, floating meals formulated for minnows may be necessary. The fish appear to feed best near sundown. During winter, the fish should be offered feed only on the warmer days.

Bighead carp feed on detritus and zooplankton in the natural environment. In managed environments, they feed on baitfish feeds and sinking channel catfish feeds. Fertilizers, plant materials, and animal wastes that result in the collection of pond bottom detritus can be used to indirectly support the growth of bighead carp. Silver carp feed on plankton in the natural and managed environment.

Chemical fertilizers and organic and animal wastes that stimulate plankton production probably provide the best conditions for growth. Plankton production varies with nutrient source and natural pond fertility. Production systems should be managed for maximum plankton production but without significant reduction in water quality. In polyculture systems consisting of grass carp stocked in combination with silver carp or bighead carp, the production system should be managed for grass carp. The uneaten feed, and other organic materials, provide detritus for bighead carp or nutrients for the phytoplankton that is eaten by silver carp.

Diseases

Carp are susceptible to the usual array of diseases and parasites, and they respond to the usual treatments and prevention.

Harvesting and Yields

Seining, handling, and transporting are more difficult for the Chinese carps—especially the grass carp—than for some other warmwater fish. Because grass carp are prone to jump over seines, a seine must be held up manually, or some other method must be devised to keep the fish inside the net. Some workers use floats to hold the cork line 1 to 2 ft above the water. Others use a floating seine inside the main seine. The simplest method is for workers to physically hold the net above the water.

Handling can cause self-induced injuries, some of which are fatal. The fish can be placed individually in porous bags (muslin or netting) to restrain them, and they can be anesthetized to reduce their activity in the transport tubs and tanks.

All species benefit from the addition of chemical and organic fertilizers that enhance the growth of food organisms. Annual yields of 2,000 to 5,000 lbs. per acre are common.

Processing and Marketing

People in many countries rely on carp as a food source. So far, the idea of carp as a food source has not been widely accepted in the United States. Some ethnic markets exist. An additional U.S. market is the sale of grass carp (white amur) to control vegetation.

Baitfish

Commercial development of baitfish began as early as 1915, but little progress was made until after World War II. During the 1950s and 1960s, rapid expansion occurred. Arkansas traditionally leads as the baitfish producer.

Sources of Species

Of about 100 species of fish used as bait in the United States, only 4 are raised in significant quantities. They include, in order of importance, the golden shiner (*Notemigonus crysoleucas*), the fathead minnow (Pimephales promelas), the white sucker (Catostomus commersoni), and the goldfish (*Carassius auratus*). (See Figure 4-34.) Many farms raise goldfish both for aquarium bait and the trade. Often, the most colorful go to aquarium dealers, and the others serve as baitfish. Some also serve as feeder fish for aquarium predators.

Other species that may prove profitable to baitfish farmers include killifish (*Fundulus* spp.), chub sucker (*Erimyzon* spp.), stoneroller (*Campostoma* spp.), tilapia (*Tilapia* spp.), Hawaiian topminnow (*Poecilia* spp.), and various other shiners (*Notropis* spp.). Hawaiian topminnows possess many behavioral and biological characteristics necessary as bait for the skipjack tuna (*Euthynnus pelamis*).

Habitat

The golden shiner, produced principally on fish farms in the Midsouth, reaches a size of 10 in. when mature. Most are marketed in smaller sizes. The fathead minnow is primarily reared in shallow lakes in Minnesota and the Dakotas, but it is also increasing in Arkansas. Fatheads also are produced on Midsouth and Midwest fish farms that use intensive culture systems. White suckers are produced principally in the upper Midwest as bait for bass, crappie, muskellunge, yellow perch, and northern pike.

Figure 4-34 Some common types of baitfish.

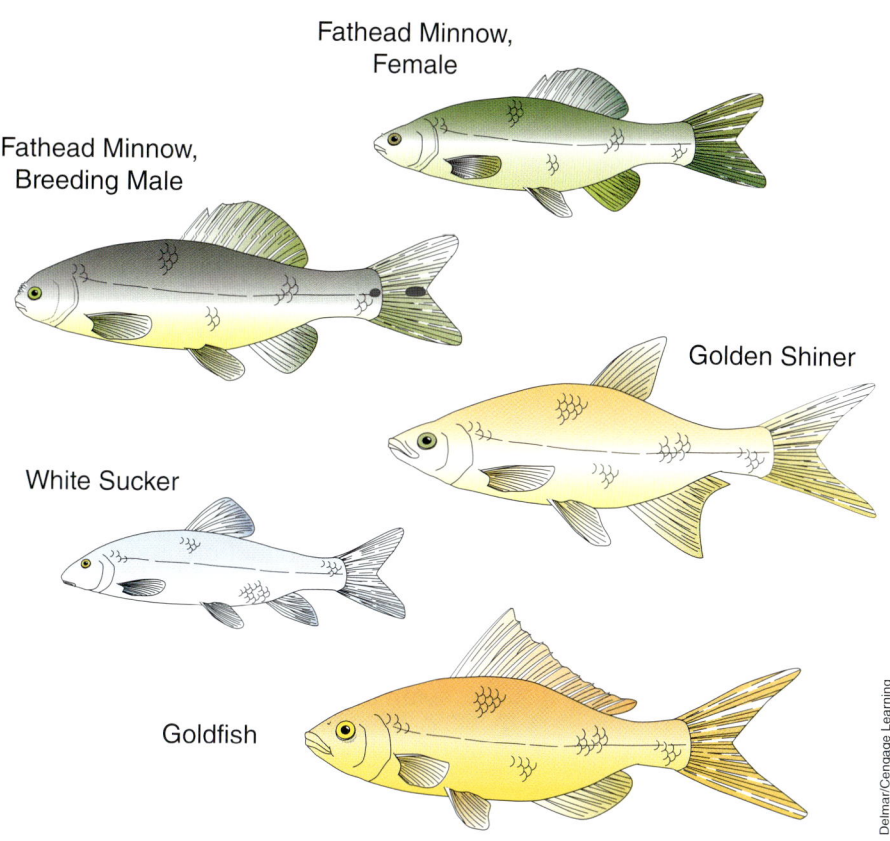

Seed Stock and Breeding

Almost no research has been done in this field. Selection of broodstock represents a major problem because of the large numbers of fish involved. Baitfish farmers often select broodstock from yearling baitfish. Baitfish begin spawning in the spring. The free spawning, egg transfer, or fry transfer methods are used to produce baitfish.

Culture Method

Most baitfish are cultured in enclosed ponds supplied with water from underground sources. Some small farms use water from nearby creeks. A combination of land, water, and climate favors concentration of the baitfish industry in the Midsouth. Close proximity to markets is also a factor.

A successful baitfish enterprise requires relatively flat land with good water retention. Rocky, gravelly, or sandy soils and rolling or steep terrain usually are undesirable. Adequate quantities of water that is neither too acid nor too alkaline must be available. Surface waters often contain excessive amounts of silt and pesticides, or undesirable fish species with the usual problems of disease and parasites. Absence of adequate underground water generally restricts development of a baitfish farm.

A host of animals prey on baitfish, including otters, diving ducks, egrets, herons, mergansers, bullfrogs, alligators, snakes, snapping turtles, back swimmers, dragonfly nymphs, and other aquatic insects. Baitfish farmers wage a constant battle against these predators.

Stocking Rate

Because baitfish are small and their size varies, so does the stocking rate. Even at market size of, often, no more than 2 in, a 1-acre pond contains about 286,000 fish. The number of fish per pound ranges from 250 to 300. If the young are to be left in the pond to grow, flathead minnows can be stocked at a rate of 2,000 brood females to 400 males per acre. If the fry are removed from the pond, 25,000 brood females and 5,000 males may be stocked per acre.

Feeding

Under certain water quality and algae conditions, golden shiners produce only 300 lbs. per acre per year. Farmers need to know what kinds of feed to use before, during, and immediately after spawning. Much of the knowledge about nutrition and diet comes from trout research. Artificial feeding increases production, and commercial diets are available.

Diseases

The nematode Capillaria can become a major problem now that more water is being reused. Also, the anchor parasite presents another problem. A **microsporidean**, Pleistophora ovariae, can progressively damage the ovaries of female golden shiners. One-year-old fish lack significant parasites, and farmers can use them as spawners because golden shiners reach sexual maturity within the same year as their birth. Blooms of large plankton, usually crustaceans or rotifers, compete for food and increase fry mortality in some ponds during the first month.

Harvesting and Yields

Baitfish harvesting techniques have remained constant over the years. Harvesting and grading fish requires considerable hand labor. Baitfish are seined using nets with 3/16 in. mesh. Next, they are dipped from the pond. Tanks or vats hold baitfish until they are sold (Figure 4-35).

Goldfish farmers produce yields as high as 4,000 lbs. per acre, and golden shiner farmers produce yields as high as 1,400 lbs. per acre. The average production industry-wide varies.

Processing and Marketing

When fishermen buy baitfish, they expect healthy, hardy animals. To produce fish that meet these requirements requires quality control throughout the growing and harvesting process. Marketing requires healthy fish graded into sizes suitable for different kinds of sport fishing. Growers use tanks or vats to hold baitfish until they are sold.

Red Drum

Major conflicts have occurred over red drum (*Sciaenops ocellatus Linnaeus*) (Figure 4-36) for 100 years. Recreational fishermen blamed inadequate catches on commercial overfishing, and commercial fishers accused sport fishers of having inadequate skills. Management of the conflict usually

Figure 4-35 Covered concrete holding vats for baitfish.

Figure 4-36 Red drum.

attempted to reduce the conflict by addressing recreational concerns without appreciably affecting commercial harvests. Size limits enacted in the 1920s to protect juvenile and adult fish, and seasonal and area net-closures were expanded to protect spawning adults.

Sources of Species

Broodstock are captured from natural habitats. Captured red drum can be spawned by controlling the day length and water temperature. Hormone injections can also be used to spawn red drum.

Habitat

The red drum is a quasi-**catadromous** fish that ranges from Tuxpan, Mexico, on the Gulf of Mexico to Massachusetts on the Atlantic Ocean. Fisheries for red drum have existed since the 1700s. In Texas, red drum are harvested primarily in estuaries where they once comprised as much as 35 percent of the commercial fish landings.

Juvenile red drum occur in all gulf and Atlantic estuaries from Chesapeake Bay to the Laguna Madre of Mexico. Their numbers vary, depending mainly on surface area. Salinity may also affect distribution, probably by affecting survival in early life stages.

Adult red drum permanently emigrate from estuaries to the Gulf when sexually mature. Sexual maturity is probably reached in three and one-half years. Adults rarely occur in the estuaries, but commonly occur in the Gulf of Mexico. Adult red drum in the Gulf are distributed to at least 70 miles from shore. Most occur within 10 miles off the shore of Texas.

Seed Stock and Breeding

Red drum males court females before spawning. Drumming and nudging intensifies prior to spawning and may be a major stimulus for spawning. Spawning occurs from August through January and peaks in September or October. Only simulated fall daylight and temperatures induced spawning in the laboratory. Females probably produce over 500,000 eggs annually. Hormone injections provide seeds at a time out of synchronization with natural spawning and can be conducted in small, inexpensive tanks.

Culture Method

Environmental requirements may change over the life span of the red drum. Red drum begin life in seawater, then passively drift or actively migrate into the brackish waters of bays and estuaries where they pass their larval and juvenile stages. As the fish grow toward adulthood, they migrate seaward and, in the process, encounter progressively increased salinity. In long-lived species such as red drum, the general seaward migration may involve a series of annual migrations, with fish moving into deeper, more saline water in winter and perhaps also during the warmest weather of summer and returning to fresher parts of the estuary during spring and fall.

The more serious obstacle to culture of red drum is their relative intolerance to cold, which limits outdoor overwintering in most of Texas and the other Gulf states. Experiments with juvenile (1 to 4 cm standard length) red drum suggest that cold tolerance is greatest in hard water with a salinity of 5 to 10 ppt.

Stocking Rate

As fry, red drum can be stocked in ponds at 296,000 to 492,000 per acre. After 30 to 40 days, the fish are harvested as fingerlings and show a survival rate of about 30 to 50 percent. Fish weighing about 0.5 lb have been stocked at 200 per acre.

Feeding

Red drum feed throughout the water column, but mainly at the bottom. Prey are sucked up from the bottom by a rapid expansion of the branchial region or captured by biting the bottom.

Red drum feeding in shallow water often exhibit tailing behavior. In this activity, their caudal and dorsal fins are out of the water as they feed on the bottom. Schools of tailing fish can be easily sighted, which increases their vulnerability to bay fishers.

Occasionally, red drum feed at the surface. Schools have been sighted close to gulf beaches feeding on other fish at the surface. Fish demonstrating this behavior are also easily sighted, which probably increases their susceptibility to fishers.

The most common components of the red drum's diet are crustaceans and fishes. Red drum are omnivorous feeders. Larvae begin feeding within four days of spawning and eat mainly zooplankton. The yolk sac provides nourishment for the first four days post-spawn. Juveniles eat mainly small bottom invertebrates and young fish. Fish of 3 to 4 in. rely heavily on amphipods for food. Shrimp, crabs, and fish (menhaden, gobies, mullet, killifish, and eels) predominate the diet in fish larger than 4 in. Adults eat mainly fish.

Diseases

Numerous parasitic organisms have been found on or in red drum, but copepods and cestodes are most frequent.

Harvesting and Yields

Like other fish in ponds or raceways, seines are used to harvest. The yield depends on survival, feeding, growth rate, and feed conversion. Feed conversion should be around 2:1.

Processing and Marketing

Estimated market for red drum is about 14 million lbs.

Because the Gulf is closed to commercial capture of red drum, aquaculture production of red drum should have a market. Processing plants on the Gulf processed red drum and provided a market for farm-raised fish. Most of these processors sell to wholesalers, who then transport the fresh fish directly to retailers. Inland production of red drum will need to find a different market.

Other possible outlets for red drum include fee-fishing operations.

Crappie

In some areas, the favorite sport fish of anglers includes crappie (*Promoxis* spp.). Anglers consider crappies to be excellent eating and excellent game fish.. Because of these qualities, many pond and lake owners request crappie fingerlings for stocking.

Sources of Species

The common names "black crappie" and "white crappie" (Figure 4-37 a and b), are not always a good key to distinguish between crappie species. The dorsal spines provide positive identification. Black crappie have seven or eight, whereas white crappie have five or six.

Figure 4-37a, b Some common crappies.

Habitat

Black crappie occur in clearer, slightly acidic waters. The white crappie is native to the more turbid waters of the South and West. Widespread stocking programs caused considerable overlapping of the species range.

Seed Stock and Breeding

Despite spawning only once annually, crappie are among the most prolific species of the sunfish family. White crappie are capable of producing 1,000 to 200,000 eggs per female as compared to 20,000 to 140,000 eggs per female for black crappie. The number of eggs produced depends on the size, condition, and age of females.

Most culturalists prefer broodfish ranging in size from 1 to 1.25 lbs.. Males become darker than females before spawning season, making them easily distinguishable. Females appear swollen with eggs. Broodfish are stocked at the rate of three to eight pairs per surface acre.

Black crappie begin spawning activity at 60° to 64°F. White crappie spawns at slightly higher temperatures (65° to 70°F). This difference probably reduces the incidence of natural hybridization. Males guard the nests and fry hatch in about five days. Once the yolk sac is absorbed, a ready supply of zooplankton is necessary to maintain growth.

Culture Method

A variety of culture techniques can be used to produce crappie. The extensive method (spawn and rear in the same pond) and intensive method (fry transfer) can be used. Crappie culture techniques are very similar to those used for largemouth bass and bluegill. Culture ponds need a bottom drainpipe, preferably, a concrete drain basin, and a reliable source of high-quality water.

If fingerlings are to be reared using the extensive method, the pond can be dried and planted with a cover crop such as rye grass (10 to 15 lbs. per surface acre) overwinter. The pond is filled two weeks before the onset of spawning. If a water source containing fish is used to fill the pond, water should be carefully filtered. Some ponds may require predacious insect control. To establish a bloom, many culturalists use both inorganic and organic fertilizers (cow manure or cotton seed meal, for example). Small but frequent applications of organic fertilizer maintain the bloom throughout the summer months. Use of inorganic fertilizers throughout the hot months often causes filamentous algae growth, which hampers harvest efforts.

When using the fry-transfer method, intensive culture, the spawning pond is not planted with a cover crop or fertilized. Success depends on water clarity in the spawning pond for fry transfer. One to two weeks prior to transferring fry, the grow-out pond is filled, treated, and fertilized as described for extensive production techniques.

Because of their high reproductive potential and forage requirements, crappie often stunt and become undesirable in many ponds. Crappie should not be stocked in ponds that are (1) less than 5 surface acres in size, (2) turbid (visibility less than 89 in throughout most of the year), and/or (3) forage deficient.

Stocking Rate

If the extensive method is utilized, broodfish should be removed from the pond once spawning is completed. A 1 to 2 in. mesh seine works well for broodfish removal. Broodfish are stocked at the rate of three to eight pairs per surface acre for extensive or intensive techniques.

If the intensive method is used, fry averaging 0.75 to 1 in. should be transferred to grow-out ponds. Fry can be concentrated by lowering the pond, then running fresh water to the drain basin or one end of the pond. Once the fry are in the drain basin, culturalists use a fine-mesh soft dip net for collection.

Avoid crowding fry. A tub containing pond water should be used to transfer the fry. Stocking rates vary between 25,000 to 50,000 per surface acre. Determination of fry numbers is normally based on volume displacement, weight of a known count, or estimation.

Feeding

A frequently fertilized pond and even a cover crop provide feed for crappies.

Diseases

The fingerlings are extremely delicate and must be handled carefully to avoid stress and bacterial disease.

Harvesting and Yields

Crappie fingerlings should produce 2- to 3 in. fingerlings by the start of cool weather. The water level in the grow-out pond is lowered. Next, fingerlings are dipped from the drain basin or seined with a soft, small-mesh seine. Fingerlings should be loaded into hauling boxes from tubs.

Many culturalists reduce mortality by (1) harvesting and hauling in cool weather only, (2) using additives in transport water, and (3) hauling with oxygen instead of agitators.

Processing and Marketing

Crappie are used to stock farm ponds and for some sport fishing. The person who catches the fish gets to process it—clean it.

Other Commercial Finfish

A number of marine finfish, including black drum, dolphinfish, flounder, groupers, rabbitfish, red snapper, sablefish, sea bream, sea trout, threadfish, tunas, and yellowtail, suggest important possibilities for aquaculture.

In countries other than the United States, many marine fish are reared in cages, pens, ponds, or other enclosures. These include sea bream and yellowtail in Japan; groupers, milkfish, and snappers in Southeast Asia; sea bream in Israel; and mullet in Asian and Mediterranean nations. Cage culture of high-priced fish such as groupers, sea bass, sea bream, and snappers appears to be a technique applicable to U.S. legal, social, and economic conditions. Other fish considered for aquaculture in the United States include the mullet, pike, muskellunge, pompany, sturgeon, sunfishes, walleye, white fish, and yellow perch.

Figure 4-38 Striped mullet.

Mullet

The striped mullet (*Mugil cephalus*) (Figure 4-38) inhabits coastal waters and estuaries throughout the tropics and subtropics. The commercial fishery plays an important role in the rural subsistence economics of many developing countries in Asia and the Pacific Basin. In the United States, mullet harvests constitute one of the largest fisheries in Florida.

Commercial mullet culture, as practiced today, is a low-intensity operation dependent on unpredictable natural supplies of fry. Desirable characteristics such as high-quality flesh, extreme tolerance to salinity and temperature, and low position on the food chain give mullet some aquaculture potential in the United States. For this to become a reality in most parts of the country, market development would have to accompany research on food technology, hatchery technology, and food-system dynamics.

Northern Pike and Muskellunge

The most commonly reared pike, the northern pike (*Esox lucius*), may weigh more than 44 lbs. (See Figure 4-39.) It ranges as far south as Iran in the Eastern Hemisphere and as far south as Missouri in the Western Hemisphere. North Americans cultivate it as a game fish, and Europeans stock it in carp ponds to control excess reproduction.

Figure 4-39 Just caught northern pike.

Figure 4-40 Pompano.

The muskellunge (*E. masquinongy*), a larger member of the pike family, also is cultured as a sport fish in the United States. Raising pike species for stocking programs will probably continue in the United States. Research possibilities include development of artificial feeds and studies on behavior and genetics.

Pompano

The Atlantic pompano (*T. carolinus*) (Figure 4-40), found in Atlantic coastal waters from Massachusetts to Brazil, is the only species cultured experimentally. Several commercial ventures were developed during the 1970s based upon the assumption that these fish could be reared profitably. All failed because of an inadequate technological base. Poorly understood factors, such as nutritional and environmental requirements, caused extensive mortalities. Pompano consume up to 20 percent of their body weight per day in feed-dry weight of feed to wet weight of fish, and fish may double in weight every two weeks. Because of the metabolic rate and the high rate of oxygen consumption, rapid rates of water exchange in ponds or cages are essential. At temperatures above 86°F, it becomes virtually impossible to supply enough oxygen to high concentrations of fish.

Sturgeon

The U.S. sturgeon fishery began with production of U.S. caviar in 1855. Smoked and fresh meat, oil, and isinglass also became important products, and a substantial fishery developed on both coasts by 1880. By 1915, overfishing reduced harvests to 10 percent of those in the 1880s. Attempts to replenish stocks through culture failed for both lake sturgeon (*Acipenser fulvescens*) (Figure 4-41) and Atlantic sturgeon (*A. oxyrhynchus*).

Soviet scientists, stimulated by the importance of sturgeon to that nation's economy, developed hormonally induced spawning techniques. Both Japan and Russia have established aquaculture centers for sturgeon, and the basic hatchery methods are considered routine. Extremely rapid growth rates and hardiness make the fish well suited for culture. In Japan, growth to a size of 7 to 9 lbs. has been recorded in 15 months. United

Figure 4-41 Shortnosed sturgeon.

Figure 4-42 Young Snake River sturgeon at the College of Southern Idaho hatchery.

States researchers have induced spawning of the sturgeon and reared wild juveniles in captivity on a variety of feeds. (See Figure 4-42.) Sturgeon do not spawn until five to six years of age.

Sturgeon products—caviar and meat—are of high quality and value and have well-established markets in the United States and worldwide. With over-fishing and habitat loss, wild sturgeon stocks are limited. Sturgeon aquaculture can help in restocking wild populations and in meeting consumer demand for products. Sturgeon have many positive culture attributes, including fast growth, good feed conversion, ability to accept crowding, and relatively good hardiness. Much of the necessary culture

Figure 4-43a Bluegill Sunfish. **Figure 4-43b** Redear Sunfish.

technology has been developed. Challenges in sturgeon culture are the long maturation period (5 to 10 years depending on species) and the limited availability of broodfish and seed. A consistent supply of juveniles needs to be developed in order to expand production. Prospective culturalists should do a thorough economic analysis of business potential and be prepared to make a significant capital investment.

Sunfish

Sunfish have little value in commercial fisheries, but they are popular sport fish. Some states forbid their sale as food. Natural reproduction maintains sport fisheries in most waters. Major U.S. programs of sunfish culture occur in conjunction with farm pond stocking activities. A U.S. farm pond program, which sought to conserve water and wildlife and to provide food and recreation for residents of rural areas, reached a peak in the 1950s. Popular species used in these pond stockings include bluegill (*Lepomis macrochirus*), and redear sunfish (*L. microlophus*). (See Figure 4-43 a and b.) No reliable data is available on the number and weight of bluegill reared, stocked, or harvested. According to the 2005 Census of Aquaculture, 217 farms in the United States raise sunfish, including fingerlings and fry (109 farms), stockers (63 farms) and food size (54 farms).

Experimental fish culturalists in this country have not neglected sunfish, but results are unspectacular. The principal obstacle to commercial culture involves their high rate of reproduction, which often leads to overcrowding and stunting. Sterile hybrids with exceptional growth potential can be produced, but intensive culture has not developed to any great extent. Hatchery propagation for stocking programs probably will continue in private and state hatcheries at the present rate. New aquaculture ventures probably will involve hybrids now cultured to a limited extent in some areas. Additional work is needed on genetics and hybridization.

Walleye

Walleye (*Stizostedion vitreum*) (Figure 4-44) have been artificially propagated since the late 1800s, but spawning and rearing remained inefficient until the 1960s. Cultured fingerlings are used to stock lakes and streams, and, in a few waters, to enhance commercial fisheries. State hatcheries in northern areas rear large numbers of walleye annually.

Figure 4-44 Walleye.

A significant culture problem involved maintaining dense populations of copepods and **benthic** organisms required by large fingerlings. Moreover, early fertilization methods did not work, and culturalists faced either a high percentage of cannibalism or premature stocking of small fingerlings, which are susceptible to predation. Intensive culture of walleye fingerlings is difficult as the percentage of fry that adapt to formulated feeds is low. Strong interest exists for development of intensive culture of fingerlings to market size. Strong demand, coupled with a drop in landings from Lake Erie, stimulated this interest.

Whitefish

Decreasing supplies from natural stocks caused by contamination of the Great Lakes created shortages of whitefish (Figure 4-45). Although little work has been done on cultivation of whitefish, an aquacultural potential exists, especially for pen culture in large lakes. Research is needed on nutrition, physiology, and culture systems engineering.

Yellow Perch

Traditionally, the Lake Erie fishery supplied 80 to 88 percent of the Midwestern yellow perch (*Perca flavescens*) (Figure 4-46) market. In 1969, this amounted to 33 million lbs annually. Threats to the fishery indicate that aquaculture may be needed to supply the market for yellow perch. According to the 2005 Census of Aquaculture (2006), the number of U.S. farms rearing yellow perch in 2005 was totaled 99, and their total sales that year reached $692 million. Wisconsin ranked first in number of farms, but not in sales. Ohio showed higher sales figures, but fewer farms.

Culture techniques for yellow perch are being developed. Two of the most important factors that currently constrain the expansion of the yellow perch aquaculture industry are that, compared to most aquacultured species, yellow perch are small and grow slowly. Research efforts aimed at improving the growth rate of yellow perch are high priority.

Figure 4-45 Whitefish.

Figure 4-46 Yellow perch.

Commercial culture of yellow perch depends on the economical production of fingerlings that accept an artificial diet. University researchers and private sector aquaculturists have developed a variety of methods for producing eggs, fry, and fingerlings. These range from the natural reproduction of adults with pond culture of fry and fingerlings to the induced reproduction by light period and hormone manipulation in tank systems. Additionally, nutritional requirements, disease control methods, and hatchery systems need improvement.

CONTROLS

Proper management of aquatic production requires control of water, disease, predators, and vegetation. Table 4-5 shows some the items controlled in a cycle of catfish production.

TABLE 4-5 DISEASE, PARASITE, AND WEED-CONTROL FREQUENCY OF OCCURRENCE; ACRES REQUIRING TREATMENT—DELTA OF MISSISSIPPI

Item	Expected Frequency of Occurrence	Acres Requiring Treatment (%)	Treatment
Parasite incidence	Once annually	26.6	8 ppm potassium permanganate per acre foot of water
Bacterial incidence	Four times annually	17.5	Maintain feeding schedule with medicated feed (Romet-30 or TM-50) for 5 or 10 days
Weed control	Once annually	75.0	1 ppm copper sulphate per acre foot of water
Fungus	Once annually	17.5	4 ppm potassium permanganate per acre foot of water
Nitrite	Once annually	100.0	30 ppm sodium chloride per acre foot of water
Ammonia	Once annually	100.0	Flush one-half of pond with fresh water and add 20 lbs of triple superphosphate per surface acre

Water Supply and Levels

Water quality and quantity are critical to a successful aquaculture venture. Requirements of many species are given in this chapter. Chapter 12, Water Requirements for Aquaculture, discusses water requirements for aquaculture in detail.

Disease

Disease represents a substantial source of lost income to aquaculturalists. Production costs are increased by fish disease outbreaks because of the investment lost in dead animals, cost of treatment, and decreased growth during convalescence. In nature, we are less aware of fish disease problems because sick animals are quickly removed from the population by predators. In addition, aquatic animals are much less crowded in natural systems than in captivity. Parasites and bacteria may be of minimal significance under natural conditions but can contribute to substantial problems when animals are crowded and stressed under culture conditions.

Disease is rarely a simple association between a pathogen and a host. Usually other circumstances must be present for active diseases to develop in a population. These circumstances are generally grouped under the umbrella term "stress." Management practices directed at limiting stress are likely to be most effective in preventing disease outbreaks. Complete details on diseases are contained in Chapter 11, Health of Aquatic Animals.

Bird Predation

Bird predation is a serious problem for producers. Most birds that cause problems are migratory, and these birds are protected under the federal Migratory Bird Treaty Act (MBTA). The birds' migratory nature complicates the problem, because predation varies greatly depending on migration patterns, time of the year, migratory concentrations, and the location of catfish ponds. Proximity to nesting or rookery sites can also compound the problem.

Besides eating fish, these birds can damage property. They are known to transmit fish diseases. Predatory birds consume the individual fish that are easiest to catch. Fish that are easily caught are often those that are diseased. So, the birds pick up diseases and transmit them to other ponds through their excrement and through simple body contact.

Catfish are preyed upon by cormorants, egrets, and herons throughout the year and by kingfishers and anhingas (water turkeys) during the warmer months. Ospreys and pelicans can sometimes cause problems, too. Frequent visits by flocks of anhingas, herons, egrets, pelicans, and cormorants can be devastating. The problem is generally most pronounced in fingerling ponds. The role protective role of the MBTA is often confused with the endangered species laws. Under the MBTA, migratory birds may not be killed or trapped without permits. But the species mentioned above can be harassed or frightened away from ponds, and habitat alteration and physical barriers are possible methods of control. Physical barriers can include hanging netting or wires over ponds and erecting fences around the edge of ponds. (See Figure 4-47.) These measures are expensive and may be in the way during harvest.

Figure 4-47 Wire netting used to keep birds out of trout raceways.

Figure 4-48 Propane gun to scare birds away from baitfish ponds.

The most common control measures are harassment techniques to frighten birds away from ponds. Birds can be frightened away by:

- Gunfire
- Fireworks
- Gas-powered noise cannons (Figure 4-48)
- Electronic noisemakers
- Flashing lights
- Reflecting material
- Repellents
- Bird distress calls
- Water fountains or cannons
- Scarecrows
- Electronic shocking devices

Each of these measures has had mixed success. Most methods appear to be effective at first, but they then become ineffective as the birds get used to them. The best approach is to use a combination of techniques and to frequently move the devices randomly around the ponds.

Producers may contact the U.S. Fish & Wildlife Service and the USDA's Animal and Plant Health Inspection Service (APHIS) for assistance. These agencies will recommend control measures. Permits can be issued to kill birds if producers keep good records of control measures and estimated losses of fish. These permits are only issued after other methods have proven ineffective as certified by the APHIS Animal Damage Control Office. More information about predators is given with the details of some of the species in this chapter.

Human Predators

Unfortunately, having a pond or raceway full of ready-to-harvest fish is a temptation for some individuals. Depending on the location, this can be a big problem. High numbers in ponds and raceways guarantee a potential thief a catch.

Control of Unwanted or Excess Vegetation

Some naturally occurring plants in ponds interfere with harvesting, shade out the more desirable ones, and cause water-quality problems when they decay. Measures must be taken to control them. Three types of control measures may be used to rid ponds of nuisance vegetation: mechanical, biological, and chemical.

In mechanical vegetation control, the farmer either physically removes the undesirable plants or alters the environment to create conditions discouraging the growth of nuisance plants. Hand removal of undesirable plants by pulling, raking, cutting, or digging may be accomplished in small ponds during the drying period. This may not be practical in large ponds. Some vegetation may also be mowed and disked under. Water drawdown is also an effective tool for controlling some aquatic weeds because it results in the drying out of underwater weeds and compaction of bottom mud. Prevention of the buildup of nutrients in the pond water by periodic exchange of the water from a nutrient-free source will reduce some aquatic plant growth. Rooted aquatic plants such as alligatorweed, smartweed, and primrose use nutrients in bottom sediments and will remain unaffected.

Biological vegetation control relies on the use of animals to consume the unwanted vegetation. Plant-eating fish such as grass carp and mouthbrooders (tilapia) are usually used to consume weeds and algae. Their introduction into ponds may cause some legal problems. The farmer should check state regulations.

Chemical control of nuisance weed and algae growths is generally an effective means, but it also involves certain risks. In many cases, an application of weed-killing chemicals kills a lot of plant material that can cause oxygen depletion. Some algaecides and herbicides used to control unwanted plants also kill desirable vegetation. Use is limited in crawfish ponds. If used, assistance from proper authorities must be sought, and care must be taken when handling and applying them by following the recommendations or directions on the manufacturers' labels. Whatever method is used to control unwanted or excess vegetation, weeds and algae are a continuing problem from year to year, and procedures must be taken regularly to control them.

GENETICS

Improving fish through genetic research is a relatively new activity. Much of the improvement in the last 40 years in all phases of agricultural production, both plant and animal, has resulted from genetic selection and hybridization. Faster growth, higher yield, better feed conversion, and increased resistance to disease can all be improved through genetic manipulation.

Several universities in the southeastern United States are involved in catfish genetic research. Scientists are doing work in selection, strain identification and evaluation, crossbreeding, hybridization, polyploidy, sex reversing, and gene splicing. The results of this research are encouraging, and research is expanding. Some of the more traditional genetic systems applied to aquaculture include:

- Selective breeding—the choosing of individuals of a single strain and species
- Hybridization—the crossing of different species
- Crossbreeding—the mating of unrelated strains of the same species to avoid inbreeding

Selective Breeding

Selective breeding is artificial selection, as opposed to natural selection. It involves selected mating of fish with a resulting reduction in genetic variability in the population. Criteria that often influence broodfish selection for selective breeding include:

- Size
- Color
- Shape
- Growth
- Feed conversion
- Time of spawning
- Age at maturity
- Reproductive capacity
- Past survival rates

These may vary with conditions at different hatcheries. No matter what type of selection program is chosen, aquaculturalists need an elaborate record keeping system in order to evaluate program progress.

Inbreeding occurs whenever mates selected from a population of hatchery broodfish are more closely related than they would be if they had been chosen at random from the population. The extent to which a particular fish has been inbred is determined by the proportion of genes that its parents had in common. Inbreeding leads to an increased incidence of **phenotypes**—visible characteristics—that are recessive and that seldom occur in wild stocks. An albino fish is an example of a fish with a recessive phenotype. Such fish typically are less fit to survive in nature. Animals with recessive phenotypes occur less frequently in populations where mating is random.

Problems after only one generation of brother-sister mating include reduced growth rate, lower survival, poor feed conversion, and increased numbers of deformed fry. Broodstock managers must be aware of the problems that can result from inbreeding and minimize potential breeding problems. To avoid inbreeding, managers should select their broodstocks from large, randomly mated populations. If inbreeding is avoided, selective breeding is an effective way to improve a strain of fish.

A system for maintaining trout broodstocks for long periods with lower levels of inbreeding requires the maintenance of three or more distinct breeding lines in a rotational line-crossing system. The lines can be formed by:

1. An existing broodstock arbitrarily subdivided into three groups
2. Eggs taken on three different spawning dates and the fry reared separately to adulthood
3. Three different strains or strain hybrids

Rotational line-crossing does nothing to reduce the level of inbreeding in the base broodstock but serves only to reduce the rate at which further inbreeding occurs. Consequently, a relatively high level of genetic diversity must be present in the starting broodstock. The use of three different strains or the subdivision of a first generation strain hybrid is the preferred method for line formation. Either of these methods tends to maximize the initial genetic diversity within the base population. After the three lines have been formed, the rotational line-crossing system can be implemented. At maturity, matings are made between lines. Figure 4-49 shows how a **rotational line-crossing** system works.

The rotational line-crossing system is flexible enough to fit into most broodstock operations. At least 300 fish—50 males and 50 females from each of the three lines—are needed for maintenance of the population, but this number could be set at any level necessary to meet the egg production needs of a particular hatchery operation. One potential problem with the system is the amount of separate holding facilities required for maintaining up to 15 groups if each line and year class are held separately. This problem sometimes can be overcome by using marks such as fin clips, brands, or tags to identify the three lines and then combining all

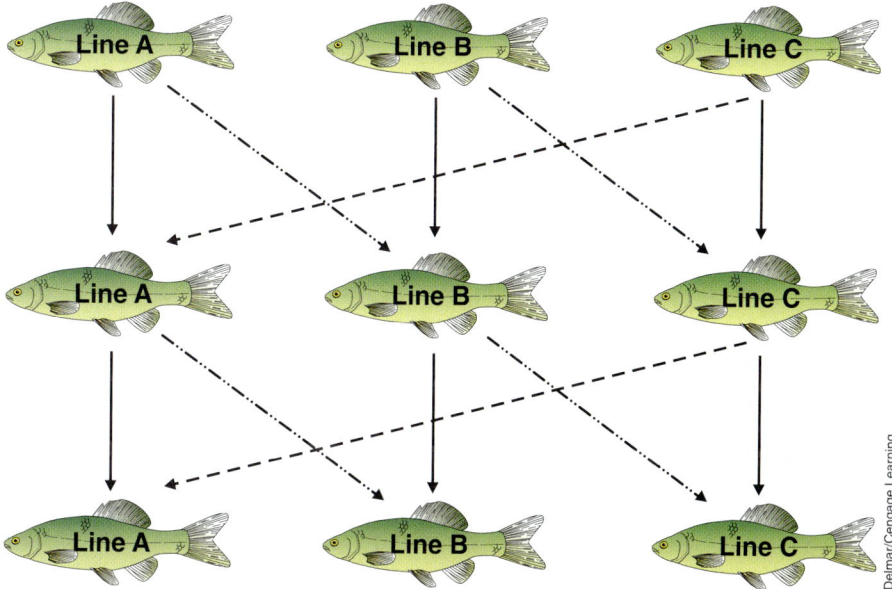

Figure 4.49 A line-crossing system. Rotational line-crossing based on three lines. Each fish represents a pool of fish belonging to a specific line. Each row of fish represents a generation of fish. The solid arrow lines show the females used to produce the next generation. The broken arrow lines shown represent the males used in the mating system.

broodfish of each year class in a single rearing unit. The total number of broodfish to be retained in each year class would be determined by the production goals, but equal numbers of fish should come from each line. This method will not only slow down inbreeding but will also make a selection program more effective.

Hybridization and Crossbreeding

Hybridization between species of fish and crossbreeding between strains of the same species have resulted in growth increases as great as 100 percent, improved feed conversions, increased disease resistance, and tolerance to environmental stresses. These improvements are the result of **hybrid vigor**, the ability of hybrids or strain crosses to exceed the parents in performance. Most interspecific hybrids are sterile. Those that are fertile often produce highly variable offspring and are not useful as broodstock themselves. Hybrids can be released from the hatchery if they cause no ecological problems in the wild. Several species of trout have been successfully crossed—for example, the splake, a cross between brook and lake trout. A cross between northern pike males and muskellunge females produced tiger muskie. The hybrid striped bass was developed by fertilizing striped bass eggs with sperm from white bass. The hybrids had faster growth and better survival than striped bass. The chief advantage of the reciprocal hybrid, from white bass eggs and striped bass sperm, is that female white bass are usually more available than striped bass females and are easier to spawn.

Both hybridization and crossbreeding of various species of catfish have been successfully accomplished at the Fish Farming Experimental Station, Stuttgart, Arkansas. Hybrid catfishes have been tested in the laboratory for improved growth rate and food conversion. Two hybrids, the white catfish crossed with channel catfish and the channel catfish crossed with blue catfish, demonstrated a greater growth rate than the parents.

Various hybrids of sunfish species also have been successful. The most commonly produced hybrid sunfish are crosses of male bluegill crossed with female green sunfish and male redear sunfish crossed with female green sunfish. They are popular for farm pond stocking because they do not reproduce as readily as the purebred parental stocks and grow much larger than their parents.

SUMMARY

Successful management requires the aquaculturalist to know about potential species that can be cultured. Managing the culture of a species involves understanding the nature of the species, where to obtain seed stock, the life cycle and biology, how to culture, stocking rates, feeding, potential diseases, harvesting, and processing.

Production systems and cultural methods may vary, depending on the aquatic animal and the stage of production. Many management concerns remain constant, such as water quality and prevention of stress and disease. Also, successful management demands a thorough understanding of the reproductive process, feeding behavior, and nutritional needs. Once managers understand the reproductive process, they try to enhance it. Once nutritional needs are identified, managers try to completely and economically meet these needs.

Some species may never be cultured from egg to adult without drastic changes in current cultural practices. Other species that are successfully cultured may be improved through genetic selection similar to the improvement of livestock and poultry.

STUDY/REVIEW

Success in any career requires knowledge. Test your knowledge of this chapter by answering these questions or solving these problems.

True or False

1. Sexual reproduction involves the production of eggs from the testes and sperm from the ovaries.
2. If the aquaculturalist has the appropriate permit, he or she can kill predatory birds around fish ponds.
3. Humans stealing fish from ponds can become a problem.
4. Management is the secret ingredient to successful aquaculture.
5. The color of the water controls spawning time in catfish.
6. Channel catfish spawn on spawning mats.
7. Oxygen levels of 5 ppm are lethal for fish.
8. Rainbow trout hatch in freshwater, swim to saltwater, mature, and return to freshwater.
9. Water temperatures above 75°F are lethal for trout.
10. Hatching rate depends on water temperature.

Short Answer

1. Give the scientific name for the majority of cultured catfish.
2. When do catfish spawn?
3. Besides fertilizing eggs, what role does the male catfish play?
4. What is the optimum temperature for growing catfish?
5. At what weight are catfish harvested?
6. Name three types of incubators used for trout.
7. In a trout hatchery, how many fry can be stocked in a trough 10 ft. long and 18 in. wide with 3 to 4 in. of water?
8. What two things should be considered when selecting a species of tilapia?
9. Tilapia are native to what parts of the world?
10. What is the major problem associated with the pond culture of tilapia?
11. At what size can tilapia reach sexual maturity?
12. Trout grow best in water temperatures of:
    a. 35° to 45°F
    b. 30° to 40°C
    c. 50° to 68°F
    d. 50° to 68°C
13. Name three types of incubator systems used in fish culture.
14. Trout fry are fed when 50 percent _____ _____ .

15. Most trout are commercially raised in _____, whereas most catfish are commercially raised in _____.
16. The process of adding oxygen to a pond or raceway is called _____.
17. Which of the following do not affect the carrying capacity of a trout production facility?
    a. Water quality
    b. Concrete thickness
    c. Temperature
    d. Water flow
18. How much time is required to raise market-size hybrid striped bass?
19. Grass carp, silver carp, and bighead carp are native to what continent?
20. Name three hormones involved in the natural spawning process.
21. Name two ways to increase spawning frequency.
22. What effect does temperature have on the development of fertilized eggs?
23. Name four baitfish.
24. What is the most common use for crappie?
25. Name four common ornamental fish.
26. Develop a list of fish that would be stressed if the water turned cold.
27. List three breeding techniques for improving any aquatic species.

Essay

1. Why would a producer consider raising tilapia in a male monosex culture?
2. Describe two polyculture systems using tilapia.
3. Describe three methods of fertilizing a tilapia pond for increased production.
4. How are hybrid striped bass produced?
5. Describe why an understanding of sex determination is important.
6. What does it mean to harvest fish by topping?
7. Give three reasons why culturing fish increases the concern for disease.
8. Define flow and density indexes.
9. Compare salmon ranching to salmon farming.
10. Give three reasons why tilapia are not widely available in the United States.
11. Explain why grading or sorting is necessary for hybrid striped bass culture.
12. Suggest two reasons why carp culture is dominant in countries other than the United States, especially in Asia.
13. Define a hybrid and give two examples.

KNOWLEDGE APPLIED

1. Using the information in this chapter and information from the Learning/Teaching Aids identified at the end of the chapter, construct a specie profile sheet for five finfish. The specie profile sheet should include the following information: common name, scientific name, brief description, habitat, distribution, reproduction, special features, problems, potential yield, feeds, and marketing potential. These Web sites might be useful: Agricultural Marketing Resource Center (http://www.agmrc.org/commodities__products/aquaculture/) and the FAO Fisheries and Aquaculture Database (http://www.fao.org/fishery/dias/en). Some ornamental fish are easily spawned in an aquarium. Obtain ornamental fish and research the requirements for spawning these fish. Compare the spawning of these fish with that of others described in this chapter.

2. Survey your community to determine whether any aquaculture products have either small niche markets or cultural markets.

3. Evaluate local water resources. Determine whether any of the water resources lend themselves to aquaculture. Determine which species could be raised in the water resources available. Perhaps the best way to do this is to invite a guest speaker who can discuss local water resources and possibly bring maps.

4. Catfish, trout, tilapia, and hybrid striped bass are most frequently cultured in the United States. Call local grocery stores and restaurants and ask if they sell trout, catfish, tilapia, or hybrid striped bass. If they do, find out how these fish are sold and where they are purchased.

LEARNING/TEACHING AIDS

Books

Beveridge, M. (2004). *Cage aquaculture.* Ames, IA: Blackwell Publishing, Professional.

Cabrita, E., V. Robles, P. Herraez. (2009). *Methods in reproductive aquaculture: Marine and freshwater species.* Sarasota, FL: CRC Press.

Conte, F., Doroshov, S., and Lutes, P. (1988). *Hatchery manual for the white sturgeon.* Oakland, CA: ANR Publications.

Halver, J. E., and Hardy, R. W., Eds. (2002). *Fish nutrition,* 3rd Ed. New York, NY: Academic Press.

Hepner, B. (2009). *Nutrition of pond fishes.* Cambridge, MA: Cambridge University Press.

Horvath, Laszlo, Tamas, G., and Seagrave, C. (2002). *Carp and pond fish culture,* 2nd Ed. Malden, MA: Fishing News Books.

Hutchinson, L. (2005). *Ecological Aquaculture: A sustainable solution.* Hampshire, England: Permanent Publications.

Lim, C. and Webster, C. D., Eds. (2006). *Tilapia: Biology, culture, and nutrition.* Binghamton, NY: The Haworth Press Inc.

Lucas, J. S. and Southgate, P. C. (2003). *Aquaculture: Farming aquatic animals and plants.* Ames, IA: Blackwell Publishing, Ltd.

Pillay, T.V. R., and Kutty, M. N. (2005). *Aquaculture principles and practices.* Ames, IA: Blackwell Publishing Ltd.

Stickney, R. R. (2005). *Aquaculture: An Introductory text.* Cambridge, MA: CABI Publishing.

Stickney, R. R. (2000). *Encyclopedia of Aquaculture.* Malden, MA: Wiley-Interscience.

Tucker, C. S., and Hargreaves, J. A., Eds. (2008). *Environmental best management practices for aquaculture.* New York, NY: Wiley-Blackwell.

Webster. C. D. (2006). *Tilapia: Biology, culture, and nutrition.* Boca Raton, FL: CRC Press.

Internet

Internet sites represent a vast resource of information. The URLs (uniform resource locator) for the World Wide Web sites can change. Using a search engine such as Google, find more information by searching for these words or phrases: finfish, freshwater fish, catfish, spawning, broodstock, seed stock, stocking rate, trout, salmonids, eyed egg stage, fish egg incubator, hatchery, grading of fish, harvesting fish, salmon ranching, salmon farming, tilapia, mouthbrooders, fingerlings, hybridization, polyculture, bass, carp, baitfish, red drum, crappie, ornamental fish, tropical fish, commercial finfish, mullet, northern pike, muskellunge, pompano, sturgeon, sunfishes, walleye, whitefish, yellow perch, or bird predation.

For some specific Internet sites, refer to Appendix Tables A-11, A-12, and A14

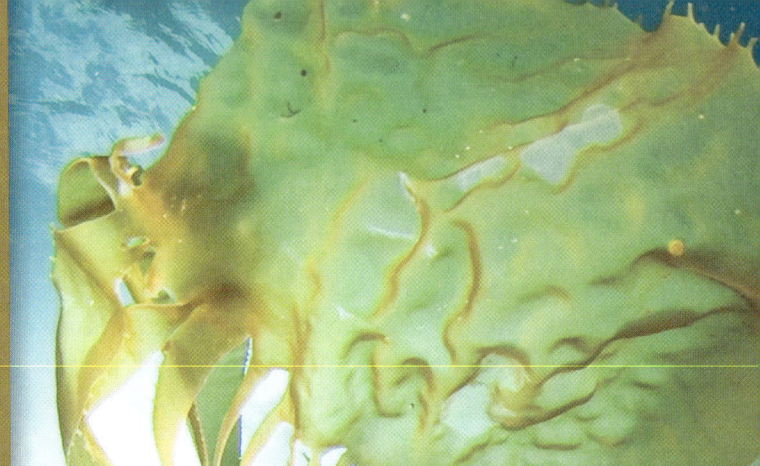

Fishing, as a method to provide food to sustain life, is probably as old as hunting. Sport or recreational fishing, also called "angling," maybe nearly as old. The sport of fishing has been practiced widely throughout the world for hundreds of years. Sport or recreational fishing is one of the largest and

CHAPTER 5

Recreational Fishing Industry

most profitable recreation businesses in the United States, as well as around the world. In the United States, about 36 million fishing licenses are issued each year—a number that does not include juvenile fishers or saltwater fishers, neither of whom require licenses. Freshwater recreational fishing depends on aquaculture to replenish the supply of fish.

OBJECTIVES

After completing this chapter, the student should be able to:

- Provide a brief history of recreational/sport fishing
- List five items that can be considered tackle
- Name three general method of fishing
- Identify eight freshwater fish caught for recreational purposes
- Discuss the economic impact of recreational fishing
- Describe recreational fishers based on the statistics
- Identify how recreational fishing supports conservation efforts
- Identify the four components of the National Fish Hatcheries System
- Explain the role of the 70 National Fish Hatcheries

- Name four states with National Fish Hatcheries
- Compare the role and location of Fish Technology Centers to Fish Health Centers
- Describe the importance of the Aquatic Animal Drug Approval Partnership to fisheries and recreational fishing
- Discuss the uniqueness of the life cycle of Pacific salmon to recreational fish and the work of the Fish & Wildlife Service

Understanding of this chapter will be enhanced if the following terms are known. Many are defined in the text, and others are defined in the glossary.

KEY TERMS

Bait

Cast

Line

Mitigation

Natal

Reel

Rod

Stocking

Tackle

Therapeutics

BRIEF HISTORY OF RECREATIONAL FISHING

Ancient Egyptian drawings show fishing scenes, and the sport is mentioned in writings or depicted in art of China, Greece, Rome, the Middle East, India, and Peru. Sport fishing that we recognize today began when people started writing about it.

The earliest English essay on recreational fishing (*Treatyse of Fysshynge wyth an Angle*) was published in 1496, shortly after the invention of the printing press. It was written by Dame Juliana Berners from the Benedictine Sopwell Nunnery. Many people read this work during the 16th century, and it was reprinted many times. The essay includes detailed information on fishing waters, the construction of rods and lines, and the use of natural baits and artificial flies.

Recreational fishing for sport or leisure became popular during the sixteenth and seventeenth centuries (Figure 5-1). The popularity corresponds with the publication of Izaak Walton's *The Compleat Angler, or Contemplative Man's Recreation* in 1653. This book is a classic work that supports those who love fishing for fishing's sake.

More than 300 editions of *The Compleat Angler* have been published. (The book is available today on Web sites like Project Gutenberg: http://www.gutenberg.org/etext/683.) The book contains writings on country fishing folklore, songs and poems, recipes and anecdotes, and meditations and quotes from classic literature. It is illustrated, including drawings of fish (Figures 5-2a and 5-2b). The central character, Piscator (meaning fisherman or angler), champions the art recreational fishing.

Equipment

As fishing equipment evolved the jointed rod, rods of different materials, the various types of reels fishing increased in popularity, and different styles of fishing developed.

The basic equipment used for fishing consists of a rod, a line, a **reel**, a hook, and a lure or **bait**, which can be live or artificial. The equipment, also called "**tackle**," can vary greatly with different types of fishing, and includes the additional pieces, such as weights or sinkers, floats or bobbers, and lures.

Figure 5-1 Drawing from *The Complete Angler* by Izaak Walton showing men heading out for some angling.

THE MEETING AT TOTTENHAM HIGH CROSS

From Hawkins's First Edition of Walton, 1760

Fishing Poles and Line

To begin with, fishing poles were somewhat short wooden rods. The development of the jointed **rod** enabled the use of longer fishing poles, and experimentation with different weight woods led to the split bamboo rod. Present-day fishing rods are made of graphite, fiberglass, or boron, although some expensive fly rods are still made of split bamboo. Along the length of the rod are a series of loops that act as a guide for the fishing line. The length and weight of the rod depends on the type of fishing; for example, fly fishing uses a long, light rod, while deep-sea fishing uses heavier rods and line.

Figure 5-2a Drawings from *The Complete Angler* by Izaak illustrating some of the fish.

The Trout

The Pike

The Carp

ILLUSTRATIONS FROM THE FIFTH EDITION, 1676

Early fishing **line** was made of natural fibers or animal products such as braided horsehair. Then fishing line developed through the centuries from gut and oiled silk, to modern products such as nylon, with specific weights and strengths for the type of fishing.

Reels

Fishing reels allow the fisher to **cast**, or throw out the line, and retrieve it untangled. A simple reel was in use by 1770. Development of reels through the years lead to the development of five general classifications: fly reel, bait-casting reel, spin-casting reel, opened-face reel, and larger reels for saltwater fish and powerful freshwater fish.

Bait

Fish are attracted to hooks by either natural or artificial baits. For freshwater fish, natural baits include worms, night crawlers, minnows, small baitfish (such as shiners and suckers), larvae, and crickets. Popular saltwater

Figure 5-2b Drawings from *The Complete Angler* by Izaak illustrating some of the fish.

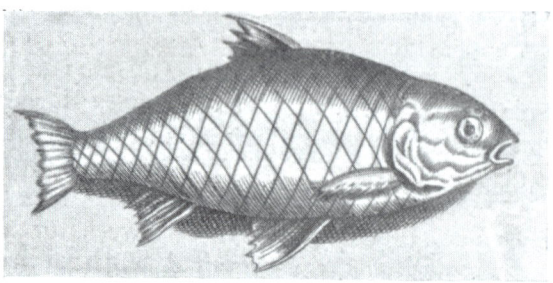

The Bream

The Tench

The Perch

ILLUSTRATIONS FROM THE FIFTH EDITION, 1676

baits are crawfish, clams, or small pieces of fish. Hundreds of artificial lures, shaped as small fish, worms, flies, or colorful objects, are also used. Both freshwater and saltwater bait have been in use for hundreds of years. The creation, or tying, of artificial flies for fly fishing has become an art. Interestingly, the *Treatyse of Fysshynge with an Angle* (1496), describes artificial flies that are still in use.

METHODS OF FISHING

Sport fishing consists of three general methods: trolling, casting, and still fishing. To troll, the angler lets out line attached to a baited hook or lure, which is pulled along as the boat moves through the water. To cast, the angler flips the baited hook toward an area where the fish are likely to lurk. For still fishing (Figure 5-3), the angler sits on shore or anchors the boat

Figure 5-3 Still fishing is popular with the very young.

and sinks a weighted line with bait and hook into the water. Sometime a bobber is attached to the line to indicate movement if the fish should nibble. From these methods, a variety of different styles of fishing have developed for freshwater ponds, lakes, and rivers, as well as saltwater bays and oceans.

HOW TO CAST, FROM *THE COMPLEAT ANGLER* BY IZAAK WALTON

zaak Walton's *The Compleat Angler, or Contemplative Man's Recreation*, first published 1653, is a classic. More than 300 editions of the book have been published. Here is an interesting excerpt from the book describing how to cast a fly and what do to if the wind is blowing. The wording is very different than our style of writing today:

> In casting your line, do it always before you, and so that your fly may first fall upon the water, and as little of your line with it as is possible: though if the wind be stiff, you will then, of necessity, be compelled to drown a good part of your line, to keep your fly in the water. And in casting your fly you must aim at the further, or nearer bank, as the wind serves your turn, which also will be with and against you, on the same side, several times in an hour, as the river winds in its course, and you will be forced to angle up and down by turns accordingly, but are to endeavour, as much as you can, to have the wind evermore on your back. And always be sure to stand as far off the bank as your length will give you leave when you throw to the contrary side: though when the wind will not permit you so to do, and that you are constrained to angle on the same side whereon you stand, you must then stand on the very brink of the river, and cast your fly at the utmost length of your rod and line, up or down the river, as the gale serves.

Figure 5-4 Ice fishing is popular with many recreational fishers.

Fly fishing, also called fly casting, is a popular method of fishing in mountain streams. Artificial lures designed to look like specific insects are attached to the end of a relatively heavy line, which is cast out by the angler. Fish such as bass, trout, or salmon are attracted to the flies, which land lightly on the water.

Bait casting, spinning, and spin casting are all similar in that the weight of the bait—either live or artificial—pulls out the light line. The bait may be weighted to sink into the water, or it may be made of lighter material to float on the surface. The equipment varies in weight, and different reels are used for different types of casting. Surf casting uses techniques similar to bait casting, but it is done at the ocean shorelines for larger game fish. The tackle and bait are heavier, to handle the rolling surf and the bigger fish.

Deep-sea fishing is for the larger ocean fish found in schools, such as marlin or tuna. Sport fishers go out on piloted boats and fish, either by trolling, drifting, or still fishing, depending on whether the location of a school of fish is known.

In wintertime, many anglers ice-fish by cutting holes in the ice; they sometimes fish from inside a small enclosure called a fish shanty (Figure 5-4).

TYPES OF FISH

Sport fish are raised in hatcheries to be released into lakes and streams to be caught by sport fishermen. These fish may also be sold to fee-fishing operations. Examples include largemouth and smallmouth bass, crappie, muskie, northern pike, salmon, perch, sunfish, trout, and walleye.

Among the most popular saltwater fish are flounder, bluefish, cod, striped bass, tuna, sea trout, porgies, red snapper, halibut, and haddock. The recreational freshwater fish species include: largemouth bass, rock bass, smallmouth bass, striped bass, whiterock bass, white bass, bluegills, catfish, carp, crappie, muskellunge, northern pike, perch, pickerel, pumpkinseed, Atlantic salmon, Pacific salmon, shad, smelt, suckers, brook trout, brown trout, cutthroat trout, lake trout, rainbow trout, walleye, and whitefish.

The most popular fish species among anglers who fish freshwater other than the Great Lakes is black bass. Following black bass in popularity are several species: panfish (edible game fish that has not outgrown the size of a frying pan), catfish, bullheads, trout, and crappie. On the Great Lakes, walleye, sauger (close relative the walleye), perch, salmon, lake trout, black bass, and steelhead are popular.

IMPACT OF RECREATIONAL FISHING

The economic impact of recreational fishing is huge. According to surveys of the U.S. Fish & Wildlife Service, about 30 million people in the United States enjoy a variety of variety of fishing opportunities. These anglers fish about 517 million days and take 403 million fishing trips per year. The anglers spend about $42.0 billion in fishing-related expenses during the year (Figure 5-5). Freshwater anglers represent about 25 million of these. They fish 433 million days per year and take 337 million trips to freshwater fishing locations. Freshwater anglers spend about $26 billion on freshwater fishing trips and equipment each year. Saltwater sport fishing attracts 7.7 million anglers who take 67 million trips on 86 million days per year. They spend about $9 billion per year on their saltwater trips and equipment.

Recreational fishers spend money on travel related costs (food, lodging, fuel, and airfare), specific fishing equipment such as rods, reels, tackle boxes, depth finders, and artificial, lures, and flies. They also spend money on auxiliary equipment like camping equipment, binoculars, and special fishing clothing and boats, vans, and cabins. Anglers

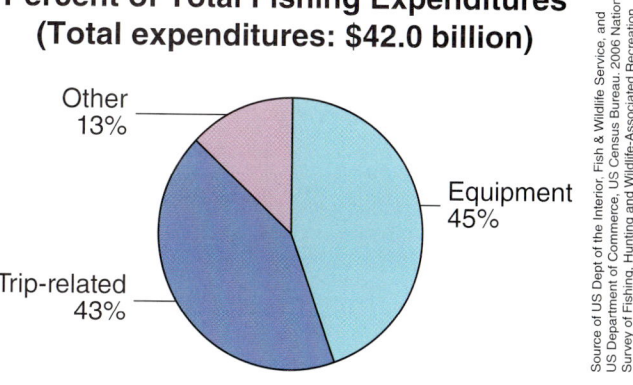

Figure 5-5 How sport fishers spend their money.

also spend a considerable amount on other fishing-related items, such as land leasing and ownership, membership dues, contributions, licenses, stamps, and permits.

Who Fishes?

The top 10 states in terms of money spent by anglers include:
- Florida
- Texas
- Minnesota
- California
- Michigan
- Pennsylvania
- Wisconsin
- South Carolina
- North Carolina
- Missouri

More than 90 percent of all sports fishers fish in the state of their residence, according to the American Sportfishing Association (www.asafishing.org).

Residents of metropolitan areas account for the majority of anglers (73 percent) but only 11 percent of metropolitan residents fish. Non-metropolitan anglers account for about 27 percent of those who fish. Most sport fishers are men, and most are older than 34 (See Figure 5-6).

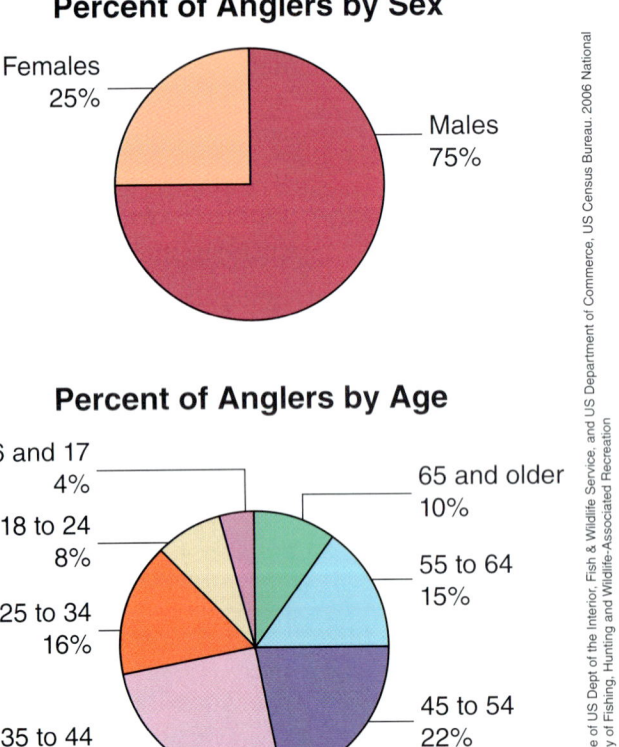

Figure 5-6 Percentages of sports fishers by gender and age.

Additional Benefits

Although many people recognize the recreational and economic benefits of fishing, its significant conservation benefits often go unnoticed. For each fishing-tackle purchase and each gallon of boating fuel consumed, a portion of the money is returned to state fish and wildlife agencies for conservation efforts. The success of the United States in restoring many species of fish and wildlife and protecting natural habitat can largely be credited to the billions of dollars generated by sport fishers for the Sport Fish Restoration and Boating Trust Fund (http://www.fws.gov/laws/lawsdigest/fasport.html).

Through the Federal Aid in Sport Fish Restoration Act, passed in 1950 at the request of the fishing industry, special excise taxes on fishing gear and boating fuel have contributed billions of dollars for fish and wildlife conservation. Added each year to this are nearly $650 million in annual fishing license sales, plus approximately $200 million in private donations from anglers for conservation efforts.

U.S. FISH & WILDLIFE SERVICE

The U.S. Fish & Wildlife Service works with other agencies and individuals to conserve, protect, and enhance fish, wildlife, and plants and their habitats for the continuing benefit of United States residents. The U.S. Fish & Wildlife Service is a bureau within the federal Department of the Interior.

The Service's National Fisheries Program has played a vital role in conserving United States fishery resources for more than 130 years, and today it is a key partner with states, tribes, federal agencies, other service programs, and private interests in a larger effort to conserve fish and other aquatic resources. Established in 1871 by Congress through the creation of a U.S. Commissioner of Fish and Fisheries, the National Fish Hatchery System's original purpose was to provide additional domestic food fish to replace declining native fish. Cultured fish were used to replace fish that were lost from natural (drought, flood, habitat destruction) or human (over-harvest, pollution, habitat loss due to development and dam construction) influences, to establish fish populations to meet specific management needs, and to provide for the creation of new and expanded recreational fisheries opportunities.

The National Fish Hatchery System (NFHS) has a unique responsibility in helping to recover species listed under the Endangered Species Act http://www.fws.gov/fisheries/nfhs/ restoring native aquatic populations, mitigating for fisheries lost as a result of federal water projects, and providing fish to benefit Tribes and National Wildlife Refuges (http://www.fws.gov/refuges/). The NFHS works closely with other service biologists and with states, tribes, and the private sector to complement habitat restoration and other resource management strategies for maintaining healthy ecosystems that support healthy fisheries.

The role of the NFHS has changed and diversified greatly over the past 30 years as increasing demands are placed upon aquatic systems. In recent years, the Service has integrated the work of fish hatcheries and fisheries management. This integrated effort has resulted in cohesive, more efficient national restoration programs, such as those for Great Lakes lake trout, Atlantic Coast striped bass, Atlantic salmon, and Pacific salmon.

The Fisheries Program consists of 70 National Fish Hatcheries (Table 5-1), seven Fish Technology Centers (Figure 5-7), nine Fish Health Centers and one Historic National Fish Hatchery (the D.C. Booth Historic National Hatchery), and the Aquatic Animal Drug Approval Partnership (http://www.fws.gov/fisheries/aadap/home.htm) program.

TABLE 5-1 NATIONAL FISH HATCHERIES[1]

Hatchery	State
Greers Ferry	AR
Mammoth Spring	AR
Norfork	AR
Alchesay-Williams Creek Complex	AZ
Willow Beach	AZ
Coleman	CA
Livingston Stone	CA
Hotchkiss	CO
Leadville	CO
Welaka	FL
Chattahoochee Forest	GA
Warm Springs	GA
Dworshak	ID
Hagerman	ID
Kooskia	ID
Wolf Creek	KY
Natchitoches	LA
Berkshire	MA
North Attleboro	MA
Richard Cronin	MA
Craig Brook	ME
Green Lake	ME
Jordan River	MI

(Continued)

TABLE 5-1	NATIONAL FISH HATCHERIES[1] *(Continued)*
Hatchery	**State**
Pendills Creek	MI
Sullivan	MI
Neosho	MO
Private John Allen	MS
Creston	MT
Ennis	MT
Edenton	NC
Garrison Dam	ND
Valley City	ND
Nashua	NH
Dexter	NM
Mora	NM
Lahontan	NV
Tishomingo	OK
Eagle Creek	OR
Warm Springs	OR
Allegheny	PA
Lamar	PA
Bears Bluff	SC
Orangeburg	SC
D.C. Booth (Historical)	SD
Gavins Point	SD
Dale Hollow	TN
Erwin	TN
Inks Dam	TX
San Marcos	TX
Uvalde	TX
Jones Hole	UT
Ouray	UT
Harrison Lake	VA
Pittsford	VT

(Continued)

TABLE 5-1 NATIONAL FISH HATCHERIES[1] *(Continued)*

Hatchery	State
White River	VT
Carson	WA
Entiat	WA
Leavenworth	WA
Little White Salmon - Willard	WA
Makah	WA
Quilcene	WA
Quinault	WA
Spring Creek	WA
Willard	WA
Winthrop	WA
Genoa	WI
Iron River	WI
White Sulphur Springs	WV
Jackson	WY
Saratoga	WY

[1]Complete contact information can be found on the Fish & Wildlife Service Web site: http://www.fws.gov/fisheries/nfhs/offices.html

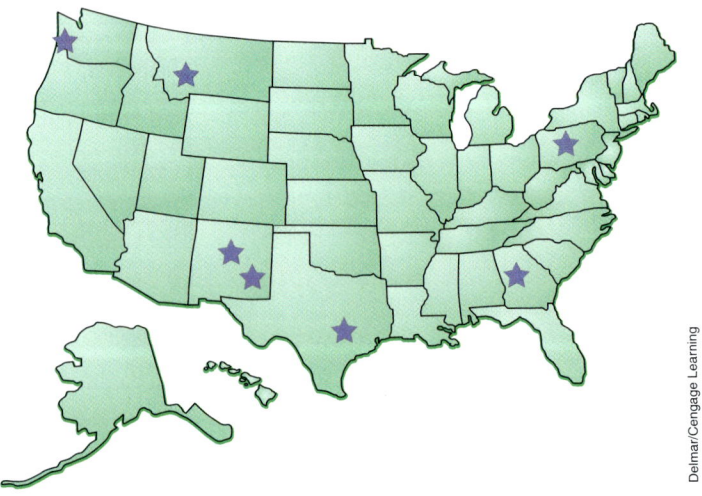

Figure 5-7 Locations of the seven Fish Technology Centers of the Fish & Wildlife Service.

The U.S. Fish & Wildlife Service Fisheries Program in the Mountain-Prairie Region helps conserve, protect, and enhance aquatic resources and provides economically valuable recreational fishing to anglers across the country. The program comprises 12 National Fish Hatcheries; a National

Fish Technology and National Fish Health Center; and eight Fish & Wildlife Management Assistance Offices serving Colorado, Kansas, Montana, Nebraska, North Dakota, South Dakota, Utah, and Wyoming.

The Region's National Fish Hatcheries produce millions of coldwater, coolwater, and warmwater game fish every year for **stocking** in public lakes, rivers, and streams; hatchery-raised fish meet legally mandated **mitigation** requirements, compensating for fish losses caused by federal water projects and associated dams. The hatcheries also raise native fish and other aquatic wildlife to help restore populations in the wild and to support recovery of threatened and endangered species. Two of the Mountain-Prairie hatcheries—one in Ennis, Montana and the other in Saratoga, Wyoming—are part of the National Broodstock Program, providing disease-free and genetically sound eggs to dozens of states, tribes, other hatcheries, and research facilities. These eggs support the production of millions of fish for recreational angling opportunities, species recovery and restoration, mitigation, tribal subsistence fishing, and other fisheries activities.

Fish Technology Centers (http://www.fws.gov/fisheries/nfhs/ftc/index.html) work nationwide to provide science and technology support and guidance to the National Fish Hatchery System and fish culture community. These centers continue to provide leadership in science and technology, especially for restoration and recovery of native species. With a focus on rapid turnaround and applied science, Fish Technology Centers assist fisheries field biologists with problem solving and new methods in areas such as genetics, nutrition, physiology, biostatistics, and fish-culture technology.

The U.S. Fish & Wildlife Service Fish Health Centers (Figure 5-8) (http://www.fws.gov/fisheries/nfhs/fhc/index.html) serve as resource centers that provide service, expertise and information supporting the service's mission to promote and protect aquatic animal health. Their work contributes to health, survival, restoration, and enhancement of fish and other aquatic species.

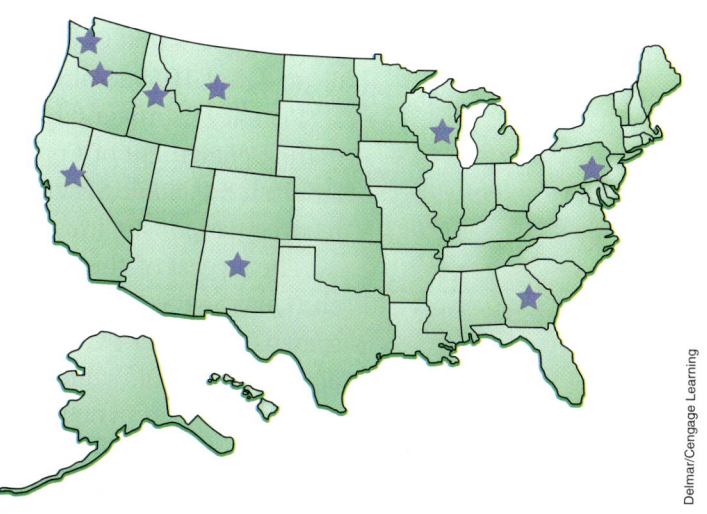

Figure 5-8 Locations of the nine Fish Health Centers of the Fish & Wildlife Service.

The Aquatic Animal Drug Approval Partnership (AADAP) program is located in Bozeman, Montana. The mission of the AADAP program is to work with partners to conserve, protect, and enhance the Nation's fishery resources by coordinating activities to obtain U.S. Food and Drug Administration (FDA) approval for drugs, chemicals, and therapeutics (methods of treatment) needed in aquaculture and fisheries management programs. This program is important to aquaculture because public and private aquaculture in the United States has struggled for many years due to a severe shortage of FDA approved drugs and therapeutics for use in aquatic species. This situation jeopardizes the health and fitness of aquatic species held in captivity, many of which are key to restoration, recovery, and management activities by the FWS and its many partners. New aquatic animal drug approvals benefit federal, state, tribal and private aquaculture programs alike throughout the United States.

PACIFIC SALMON

Pacific salmon encounter increasing human-caused hazards in their migrations to and from spawning grounds. Many salmon stocks are seriously threatened by habitat destruction, hydroelectric dams on migratory rivers, over-harvest of rare stocks, and competition with hatchery fish. Some stocks are so severely reduced that they have been listed as endangered or threatened species under the Endangered Species Act. Endangered means they are likely to become extinct. Threatened means they are likely to become endangered in the near future.

The winter-run chinook salmon, originating in California's Sacramento River, was listed as threatened in 1990, but was reclassified to endangered in 1994. In 1992, the Snake River stock of sockeye salmon was listed as endangered wherever found. The spring-summer and fall runs of chinook originating in Idaho's Snake River were listed as threatened in 1992. Others are being considered for listing, including the Columbia River (Washington) chinook and Oregon Coast coho salmon.

A 1991 report by the American Fisheries Society indicated that 214 of about 400 stocks of salmon, steelhead, and sea-run cutthroat trout in the Northwest and California are at risk of extinction. The report also indicated that 106 are already extinct.

Primary listing and recovery responsibilities for Pacific salmon belong to the Department of Commerce's National Marine Fisheries Service. The U.S. Fish & Wildlife Service and other federal and state agencies also have recovery responsibilities.

The largest of the Pacific salmon, chinook (Oncorhynchus tshawytscha) average about 24 pounds when they return to their **natal** river to spawn, most after two or three years at sea. The chinook is the least abundant of the Pacific salmon.

Coho salmon (Oncorhynchus kisutch), fourth in Pacific fishery abundance, is the number one sport fish. It spends only one winter at sea,

Figure 5-9 Male (top) and female (bottom) coho salmon, which become reddish as they mature.

Coho Salmon

returning the next fall to spawn. It averages about 10 pounds when full grown. (See Figure 5-9.)

Sockeye salmon (Oncorhynchus nerka) make up about 25 percent of the West Coast catch, and chum salmon (Oncorhynchus keta) make up about 13 percent. Both follow similar migration paths in the Pacific and reach a common weight of about 12 pounds before returning to their natal river to spawn.

Pink salmon (Oncorhynchus gorbuscha), the smallest of the Pacific salmon, average only about three to five pounds. However, they make up more than half the total West Coast commercial catch. Pink salmon seldom travel more than 150 miles from the mouth of their natal river.

SUMMARY

The sport of fishing has been practiced widely throughout the world for hundreds of years. Freshwater and saltwater sport or recreational fishing is one of the largest and most profitable recreation businesses in the United States. The direct economic impact of recreational fishing is over $42.0 billion in fishing-related expenses during the year. Freshwater recreational fishing depends on aquaculture to replenish the supplies of fish. The National Fish Hatchery System of the U.S. Fish & Wildlife Service consists of 70 National Fish Hatcheries, seven Fish Technology Centers, nine Fish Health Centers, and the Aquatic Animal Drug Approval Partnership, all of which support the recreational fishing industry.

STUDY/REVIEW

Success in any career requires knowledge. Test your knowledge of this chapter by answering these questions or solving these problems.

True or False
1. Discovery of ancient Chinese documents lead to an interest in recreational fishing during the sixteenth and seventeenth centuries.
2. The U.S. Fish & Wildlife Service is a part of the USDA.
3. Fishers create artificial flies and use them for catching fish.
4. Freshwater recreational fish often include flounder, bluefish, cod, red snapper, and halibut.
5. Most recreational fishers are men.

Short Answer
1. List five items that can be considered tackle.
2. Name three general methods of fishing.
3. Identify eight freshwater fish caught for recreational purposes.
4. Identify the four components of the National Fish Hatcheries System.
5. Name four states with National Fish Hatcheries.
6. Compare the role and location of fish technology centers to fish health centers.
7. Describe the importance of the Aquatic Animal Drug Approval Partnership to fisheries and recreational fishing.

ESSAY
1. Provide a brief history of recreational/sport fishing.
2. Discuss the economic impact of recreational fishing.
3. Identify how recreational fishing supports conservation efforts.
4. Explain the role of the 70 National Fish Hatcheries.
5. Discuss the uniqueness of the life cycle of Pacific salmon to recreational fish and the work of the Fish & Wildlife Service.

KNOWLEDGE APPLIED
1. Go online and read six different selections from *The Compleat Angler* (available on Web sites like Project Gutenberg: http://www.gutenberg.org/etext/683). Report on your readings.
2. Obtain rods, reels and line, learn to cast. As a team, demonstrate casting to the class. Or show or view two videos on YouTube that demonstrate casting.

3. Develop a display of lures and live bait used for fishing in your area.
4. Invite an avid fisher or a representative of the Fish & Game Service to visit the class.
5. Plan a presentation on the types of recreational fish common in your area. Include the common name, tackle used, and places found.
6. Search the World Wide Web and find out which National Fish Hatcheries are closest to your school. Visit the website for the Hatchery and report on the activities of the Hatchery.
7. Develop a report on the D. C. Booth Historical Hatchery (http://www.fws.gov/dcbooth/index.htm).

LEARNING /TEACHING AIDS

Books

Creative Publishing, Editors (2002). *The complete guide to freshwater fishing.* Minneapolis, MN: Quayside Publishing Group.

Garrison, R. (2003). *The everything fishing book: Grab your tackle box and get hooked on America's favorite outdoor sport.* Avon, MA: Adams Media.

Kaminsky, P. (1997). *Fishing for dummies,* Hoboken, NJ: Wiley.

Internet

American Sportfishing Association: http://www.asafishing.org/
Fish America Foundation: http://www.fishamerica.org/
Freshwater Fishing and Angling: http://fishresource.com/
National Fisheries Data Infrastructure: http://www.nbii.gov/far/nfdi/aboutnfdi.aspx
National Fish Habitat Action Plan: http://fishhabitat.org/
Take Me Fishing: http://www.takemefishing.org/
U.S. Fish & Wildlife Service: http://www.fws.gov/

Internet sites represent a vast resource of information. The URLs (uniform resource locator) for the World Wide Web sites can change. Using a search engine such as Google find more information by searching for these words or phrases: fishing, angling, sport fishing, recreational fishing, endangered fish, tackle, bait, lures, fishing flies, fly tying, and the common name of each recreational fish.

For some specific Internet sites, refer to Appendix Tables A-11, A-12 and A-14.

Most species of ornamental fish, also called "tropical fish" or "aquarium fish," occur naturally in tropical or semitropical fresh water, brackish water, or saltwater. Worldwide, the ornamental or hobby-fish industry is a multibillion dollar industry. The United States is the largest market,

CHAPTER 6

Raising Ornamental Fish

followed by Europe and Japan. United States' ornamental-fish imports come from Southeast Asia and Japan. Singapore, Thailand, the Philippines, Hong Kong, Indonesia, and South America (Colombia, Brazil, and Peru). Ornamental fish imported into the United States comprise about 1,500 species—about half are freshwater and about half are saltwater species. In total volume, however, freshwater species make up more than 90 percent.

No attempt is made in this chapter to cover all species that are, can be, or were cultured. Rather, this chapter covers a wide range of freshwater species. Management practices are similar for many species.

OBJECTIVES

After completing this chapter, the student should be able to:

- Identify six warmwater and three coldwater ornamental fish
- Describe the production of one warmwater and one coldwater ornamental fish
- Discuss ornamental fish production in the United States
- Compare spawning habits of ornamental fish
- Describe the feeding of fry and adults
- Explain the importance of water temperature to successful ornamental fish culture

Understanding of this chapter will be enhanced if the following terms are known. Many are defined in the text, and others are defined in the glossary.

KEY TERMS

Air stone
Broodfish
Carotenoids
Live bearers
Mouth-brooders
Nauplii

SOURCES OF SPECIES

Ornamental fish include representatives of several families and more than 100 species of small, colorful, and unique fishes. Table 6-1 describes some common ornamental fish. (See also Figure 6-1.)

Covering U.S. production, Table 6-2 summarizes the types of farms and their locations. Due to its mild to tropical climate, Florida is the leading state, but, depending on the species and resources, production of ornamental fish can be set up in any state because of its small scale.

Habitat

Habitats vary with the species selected. Table 6-1 provides some information.

Seed Stock and Breeding

Livebearers are fairly easy to propagate, but they hybridize so readily that great care must be taken to keep the individual species and strains separate. As a group, they are not as prolific as the egg layers. Livebearers can be propagated in earthen ponds. A producer may stock as many as 400 or as few as 25 broodfish in a pond. It is sometimes desirable to place all the broodfish in one pond, and net or trap the offspring and then remove them to other ponds. Specific hybrids of livebearers are usually bred in tanks or vats, and the fry are later transferred to earthen ponds for further growth.

Egg layers are generally difficult to propagate because their breeding requirements can be very exacting. Some species deposit adhesive eggs on vegetation, rocks, tiles, and other surfaces. Other species lay eggs in small nests that look like depressions in the pond or container bottom. Some species build floating bubble nests in which the embryos develop and hatch. A few popular species are **mouth-brooders** that gather fertilized eggs in their mouths and carry them until they hatch.

A few species of egg layers can be propagated in earthen ponds. Mats of Spanish moss may be used as a substrate for species that lay adhesive eggs on plants or rocks. The preferred method for propagating egg layers is inducing them to spawn in tanks or vats. Breeding pairs are stocked into 10- to 30-gal tanks. Some fish use flat surfaces, such as aquarium walls or vertical glass plates. Species that eat their eggs must be separated

TABLE 6-1 REPRESENTATIVE POPULAR ORNAMENTAL OR TROPICAL FISH[1]

Family	Scientific and Common Name	Characteristics
Poeciliidae (livebearers)	*Poecilia latipinna* sailfin molly *P. reticulata* guppy *Xiphophoroug helleri* swordtail *X. maculatus* platy	Often eat the living young they bear and need to be separated from them by using spawning cages or by providing cover for the young.
Cyprinidae (carps and minnows)	*Brachydanio rerio* zebra danio *Puntus everetti* clown barb *Rasbora* rasboras	Some of these egg layers scatter eggs in gravel; others scatter them among fine-leaved or broad-leaved aquatic plants, depending on species. Parents do not care for the young and should be removed after spawning has ended.
Cyprinodontidae (killifishes)	*Aphyosemion coeruleum* blue gularis	Eggs laid by these fishes adhere to plants or gravel. Parents do not care for young and should be removed from tank after spawning. The fish are very aggressive.
Characidae (dharacins)	*Carnegiella marthae* black-winged hatchetfish *Copeina gukttata* red-spotted copeina *Gymnocorymbus ternetzi* black tetra *Hemigrammus erythrozonus* glow-light tetra	Egg layers with varied breeding habits. Typically, adhesive eggs are scattered among thick plants. Parents eat the eggs if they are not removed after they are laid. Some species have very exacting water-quality requirements and will not spawn if the requirements are not met.
Cichlidae (cichlids)	*Symphysodon discus* discus fish *Cichlasoma meeki* firemouth cichlid *Pterophyllum scalare* angelfish	Includes the mouth-brooding tilapias. Some species build nests on the bottom or in crevices or natural cavities. Still others lay adhesive eggs that adhere to flat surfaces on rocks or vegetation. Parents guard eggs and young. The discus fish feeds newly hatched young with secretions in surface mucus.
Anabantidae (gouramies)	*Betta splendens* Siamese fighting fish *Trichogaster trichopterus* blue gourami	Labyrinth fishes with air chambers in heads. They require atmospheric oxygen and drown if denied access to the surface. Males of the more popular species build bubble nests at the surface for egg incubation and guard eggs and fry. Males of some species must be removed from the tanks because they eat the fry.
Callichthyidae (Corydoras catfishes)	*Corydoras aneneus* bronze catfish	Breeding pairs of species from this family join belly to belly, as eggs are extruded into a pocket created by the female's ventral fins, where they are fertilized. The female then deposits them on plants, and the process is repeated until all eggs are laid. Parents typically ignore offspring.

[1]Source: Third Report to the Fish Farmers.

from them almost immediately. Breeding traps or boxes with wire bottoms are commonly used to separate the eggs of free-spawning fish from the parents.

TABLE 6-2	ORNAMENTAL FISH PRODUCTION IN THE UNITED STATES[1]	
Type	**Number of Farms**	**Leading States**[2]
Goldfish	92	AR, OH
Koi	193	AR, CA, FL, OH, PA
Tropical fish (total)	158	FL, HI,
Tropical fish[3]	89	FL
Tropical fish[4]	129	FL, HI
Other ornaments	22	–
Total	358	AR, FL[5], HI, OH, PA

[1]USDA Census of Aquaculture (2005)
[2]Ten or more farms; many states have one, two or three
[3]Live bearers
[4]Includes egg layers and marine ornamentals
[5]Florida is by far the leading state, totaling 133

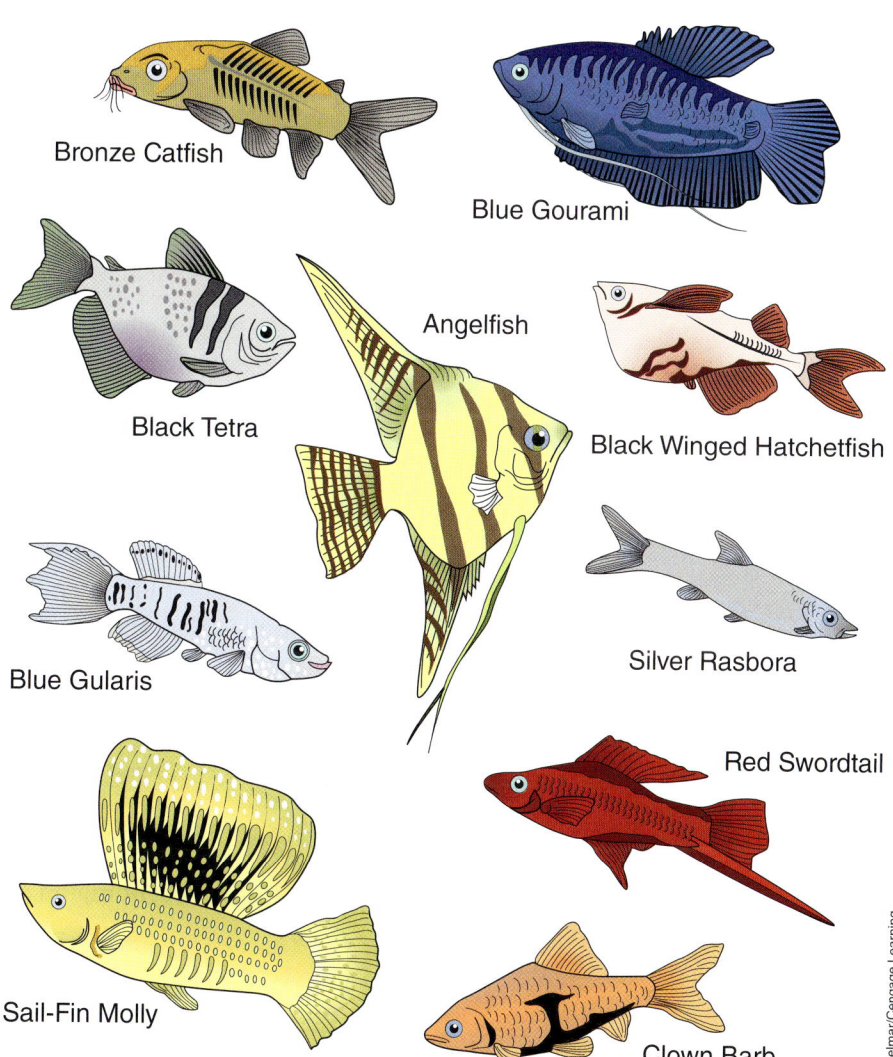

FIGURE 6-1 Some common ornamental fish or tropical fish.

ORNAMENTAL KOI

The word "koi" comes from Japanese, simply meaning "carp." Koi, or more specifically, "nishikigoi," meaning "brocaded carp," are ornamental, domesticated varieties of the common carp (*Cyprinus carpio*) that are kept for decorative purposes in outdoor koi ponds and water gardens. They are also called "Japanese carp." Koi are coldwater fish and thus benefit from being kept in the 59 F to 77 F temperature range. They do not thrive in prolonged cold-winter temperatures.

Koi were developed from common carp in ancient China during the Jin Dynasty and were later transferred to Japan in the 1820s in the town of Ojiya in the Niigata prefecture. By the 20th century, a number of color patterns had been established, notably the red-and-white Kohaku. The outside world was not aware of the development of color variations in koi until 1914, when the Niigata koi were exhibited in the annual exposition in Tokyo. At that point, interest in koi developed throughout Japan, eventually spreading worldwide. Koi are now commonly sold in most pet stores, with higher-quality fish available from specialist dealers. A variety of colors and color patterns have since been developed; common hues include white, black, red, yellow, blue, and cream.

In Japanese, "koi" is a homophone (a word that is pronounced in the same way as one or more other words but is different in meaning) for another Japanese word that means "affection" or "love." Accordingly, koi are symbols of love and friendship in Japan. An example of this is given in a short story by the Japanese author, Mukoda Kuniko, titled "Koi-san." Koi tattoos, incidentally, have become a popular trend in North America.

Since koi are domesticated common carp, culled for color, they are not a different species. If allowed to breed freely they will revert to the original carp coloration within a few generations.

Culture Method

Ornamental fish are either domestically produced or imported. Many of the domestically bred fish are cultured at the more than 100 farms in central Florida. Ornamental fish are also raised in most of the other states. Most culturalists specialize in the production of the varieties that are most colorful and easy to propagate. All operations, regardless of location, are limited by available space, cost of labor, and the cost of heating water during winter. Even in Florida's warm, semitropical climate, winter temperatures reduce production for a two- to four-month period in a warm-winter year, and virtually halt it in a cold-winter year.

Tanks, vats, and earthen ponds are used to culture ornamental fish. A typical farm may propagate 20 to 30 or more species or varieties, each of which requires separate accommodations. Ponds are usually 20 × 80 ft. or 80 × 100 ft.

Stocking Rate

Stocking rates differ depending on production scheme and species.

Feeding

Feeding of ornamental fish is a major consideration for the producer. Sometimes brine shrimp, or **nauplii** (free-swimming first stage of brine shrimp larvae), are fed to the fry of some species before they are stocked into ponds. Most of the varieties of fish must be fed in the ponds, and feeding is one of the major costs in production. Often, different formulations and feeding techniques are required for the various varieties of ornamental fish. **Carotenoids** (organic pigments) and other pigments are commonly included in the feed to promote bright coloration in the fish.

Diseases

Ornamental fish are susceptible to disease. Temperature and low dissolved oxygen levels are common stressors. For a complete discussion of diseases and fish health refer to Chapter 11 and water requirements are covered in Chapter 12.

Harvesting and Yields

Harvesting is a major consideration because fish that die in retail stores, or soon after being purchased by the aquarium owner, ultimately must be replaced by the producer. Livebearers can be caught with traps and seines. Because trapped fish suffer little trauma, they can be shipped soon after capture. Fish caught with seines must be held for several days before they can be shipped.

Processing and Marketing

Ornamental fish are normally shipped by air express to their market, like pet shops and aquaria. The fish are placed in plastic bags partly filled with water, and the rest of the volume of the bag is filled with pure oxygen. These bags are then placed in styrofoam or cardboard shipping containers. Consumers then buy the fish for homes or offices.

SPECIFIC ORNAMENTAL FISH

A brief discussion of the popular ornamental fish follows, providing more details on the commercial requirements for culture, feed, and spawning, and the production systems for these fish: angelfish, suckermouth, catfish, discus, goldfish, guppy, koi, molly, oscar, swordtail, and tetra (see Figure 6-1). Many of these fish are sold in pet stores, aquariums, and discount chain stores (see Figure 6-2).

Angelfish *(Pterophyllum scalare)*

Angelfish have a high market potential. They are 1/2 to 3 in. long. Water temperature requirement for growing is 75° to 84°F. For spawning, it is 82°F. A temperature of 60°F is lethal.

Feed requirements

Angelfish require a feed that is 40 percent crude protein and 10 to 15 percent fat, using a combination of flake, live, or frozen feeds.

FIGURE 6-2 A variety of ornamental fish and aquariums on display

Spawning requirements

Once broodfish start to exhibit courtship behavior, they are transferred to a 20 gal. spawning tank. Females spawn on a vertical substrate such as a slate tile. Eggs are adhesive and will hatch in two days at 82°F. Each female may lay up to 200 eggs every 7 to 14 days. First feeding using newly hatched brine shrimp can begin five days after hatching. After fertilization, the slate with attached eggs is placed in a 3 to 5 gal aquarium containing enough methylene blue, a dye, to give a dark-blue color. An **air stone** should be placed underneath the slate to provide circulation. After hatching one-half of the aquarium, water should be replaced each day so by the time the fry are free swimming, the water is only slightly blue.

Production systems

Most are produced in recirculating systems or ponds. When the fry are free swimming, they should be transferred to an aerated 15 gal aquarium at 300 fry per aquarium. The aquarium should have a water depth of approximately 4 in and be filtered with a sponge filter. The shallow water depth facilitates the feeding of the fry. When the fry are approximately 0.6 in in diameter, they should be transferred to a 30 to 55 gal aquarium with aeration and filtration. Fry should grow to a marketable size in 6 to 8 weeks.

Suckermouth Catfish *(Hypostomus plecostomus)*

The suckermouth catfish also has a high marketing potential. They are sold in many sizes, with ranges starting at 1 to 2 in. long. Suckermouth catfish require 71.6 to 82.3°F for growing and 75 to 79°F for spawning.

Feed requirements

Feed for suckermouth catfish should be 32 percent protein catfish feed, 1/8 inch in diameter.

Spawning requirements and production

Females are sexually mature after two years. Females burrow into the pond bank (cavity spawners) and lay around 250 eggs per spawn.

Production systems

Suckermouth catfish are commonly produced in ponds and tanks.

Discus (*Symphysodon discus* and *Symphysodon aequifasciatus*)

The discus have a high market potential. They range in size from 1/2 in. to 3 in. long and require 75 to 84°F for growing and 82°F for spawning. Temperatures 70°F or lower will set off disease outbreaks.

Feed requirements

Discus require a diet of 40 percent crude protein with a fat content of 10 to 15 percent, using a combination of flake, live, or frozen feeds.

Spawning requirements

Once **broodfish** start to exhibit courtship behavior, they are transferred to a 20 gal spawning tank. Females spawn on a vertical substrate such as a slate tile. Eggs are adhesive and will hatch in two days at 82°F. Each female may lay as many as 200 eggs per female every 7 to 14 days. For first feeding, newly hatched brine shrimp are used, and this can begin five days after hatching. After fertilization, the slate with attached eggs is placed in a 3 to 5 gal aquarium containing enough methylene blue to give a dark-blue color. An air stone should be placed underneath the slate to provide circulation. After hatching one-half of the aquarium, water should be replaced each day so by the time the fry are free swimming, the water is only slightly blue. Discus are extremely sensitive to poor water quality and require a near neutral pH and hardness levels less than 80 mg/l.

Production systems

Recirculating systems (tanks) and ponds are commonly used to produce discus. When the fry are free swimming, they should be transferred to an aerated 15 gal aquarium at 300 fry per aquarium. The aquarium should have a water depth of approximately 4 in. and be filtered with a sponge filter. The shallow water depth facilitates the feeding of the fry. When fry are approximately 0.6 inches in diameter, they should be transferred to a 30 to 55 gal. aquarium with aeration and filtration. Fry should grow to a marketable size in 6 to 8 weeks.

Goldfish (*Carassius auratus*)

The comet variety is the most common type of goldfish, but there have been many other varieties developed, such as black moors, calico, koi, and shubunkins (see Table 6-1). Goldfish are relatively easy to produce, and they have a moderate market potential as bait fish/feeder fish and ornamental fish. They range in size from 1 to 6 in long for ornamental fish and 1 to 2 in. for feeder fish. Goldfish require a growing temperature of 70°F. They spawn at a temperature above 60°F. Goldfish feed is 30 to 38 percent protein.

Feeding requirements

Spawning requirements

Goldfish spawn repeatedly from May to June, and the eggs hatch in two to eight days. They produce 50,000 eggs/lb body weight. The primary method used is the egg transfer method. In this method, the broodstock spawn on spawning mats placed in shallow water. When mats are covered with eggs, both mats and eggs are moved to rearing ponds.

Production systems

Small ponds, 0.25 to 1.0 acre, are used for spawning and larger ponds, 0.5 to 5 acres are used for rearing of fry.

Guppy (Lebistes reticulatus)

Guppy are easy to produce and they have a high market potential as ornamental fish. Guppies range in size from 0.25 to 1.5 in. long. The temperature for growing and spawning is 83°F.

Feeding requirements

Guppies feed on 40 percent crude protein flake food (8 percent fat), small zooplankton, or newly hatched brine shrimp.

Spawning requirements and production systems

Guppies are live bearers that can give birth to 200 young. Females become sexually mature in about three weeks.

Production systems

They are produced in recirculating systems (tanks) and ponds.

Koi (Cyprinus carpio)

Koi are relatively easy to produce and have a moderate market potential as an ornamental in the United States. They range in size from 3 to 12 in. long. The growing temperature needs to be 55° to 80°F.

Feed requirements

Feed should consist of a diet that is 31 to 38 percent protein and 3 to 8 percent fat.

Spawning requirements

Koi spawn at above 65°F. They spawn in the spring, and females produce 60,000 eggs/lb. body weight. The eggs hatch in two to seven days.

Production systems

Koi are commonly produced in ponds (see Figure 6-3).

FIGURE 6-3 A koi production facility in Niigata, Japan Prefecture.

Molly

The molly includes several species of live bearers in the family Poeciliidae, including *Platypoecilus mentalis*, the black molly, and *Poecilla velifera*, the sailfin molly. They are easy to produce and have a high market potential as ornamentals. Mollies range in length from 1 to 2 in. Mollies require a temperature of 77° to 86°F to grow and 80° to 84°F to spawn. A temperature below 60°F is lethal.

Feed requirements

Molly can be fed a diet of 40 percent protein flake food or 45 percent protein salmon starter.

Spawning requirements

Female mollies mature in 3 to 4 months and bear approximately 10 fry every two weeks.

Production systems

To prevent the adults from eating their offspring, producers provide cover for the fry. They are commonly produced in ponds and aquariums.

Oscar or Velvet Cichlid *(Astronotus ocellatus)*

Oscars are easy to produce and have a high market potential in the ornamental fish industry. They range in size from 2 in. and larger. Temperature requirement for growing is 79° to 86°F. They spawn when the temperature is 80° to 82°F, and 65°F is lethal.

Feed requirements

Oscars are fed a diet that is 32 to 38 percent protein as a pelleted fish food for adults. Broodfish and fry need supplements of live foods, such as brine shrimp.

Spawning requirements

Females will produce 1,000 to 2,000 eggs, which are laid onto a rock substrate.

Production systems

The most common production systems are ponds and recirculating systems.

Swordtail *(Xiphophorus hellerii)*

Swordtails are highly marketable as ornamentals and easy to produce. They range in size from 0.25 to 1.5 in. long. In order to grow, swordtails require a temperature of 83°F.

Feed requirements

A swordtail's diet should be 40 to 50 percent protein and 10 to 12 percent fat.

Spawning requirements

Swordtails are livebearers that can give birth to 200 young at one time. Females become sexually mature in about three weeks.

Production systems

Swordtails are commonly produced in ponds and recirculating systems.

Tetra

This group includes not only tetra but also other species in the family Characidae (Characins). *Paracheirondon innesi* is the scientific name for the common neon tetra, and Paracheirondon axelrodi is the scientific name for the cardinal tetra. Tetra are moderately difficult to produce, but they have a high market potential as an ornamental. Tetras range in size from 1 to 2 in. long. The temperature requirement for growing tetra is 77° to 82°F. Tetra spawn at 77° to 82°F. A temperature below 65°F is lethal.

Feed requirement

Tetra can be fed a flake fish food that is 40 percent protein.

Spawning requirements and production systems

Female tetra may lay up to 150 to 300 eggs per spawn onto a mat. Adults are removed. Fry hatch after 24 hours, and fry swim up after five days.

Production systems

Tetra are commonly produced in tanks and aquariums.

FLORIDA AND TROPICAL ORNAMENTAL FISH CULTURE

Ornamental fish comprise a varied list of species, each with their own peculiar requirements for commercial production, and markets for these fish are as varied as the fish themselves. Prior knowledge is recommended for anyone who wishes to enter the field. New producers may encounter difficulty in getting good information on how to produce tropical fish. Although many production techniques and management skills required for tropical fish are similar to those in a food-fish operation, the specific methods for producing a given species are usually held as closely guarded secrets by the people who are successfully producing tropical fish.

While not unique to Florida, warmwater ornamental fish production is concentrated there (see Table 6-2). This results primarily from the climate, but the historical strength of the industry is also a factor. Tropical-fish marketing depends on having a wide selection of fish to offer.

SUMMARY

Common ornamental fish include the angelfish, suckermouth, catfish, discus, goldfish, guppy, koi, molly, oscar, swordtail, and tetra. The United States is the largest market, followed by Europe and Japan. Ornamental fish imported into the United States comprise about 1,500 species—about half being freshwater and half being saltwater species. In terms of total volume, however, freshwater species make up over 90 percent. Florida is the leading production state, but, depending on the species, small-scale production of ornamental fish can be set up in any state. Culture methods vary with the species selected, spawning habits, temperature, and feeding requirements. Tanks, vats, and earthen ponds are used to culture ornamental fish. A typical farm may propagate 20 to 30 or more species or varieties. Ornamental fish are susceptible to disease. Temperature and low dissolved oxygen levels are common stressors that can lead to disease.

STUDY/REVIEW

Success in any career requires knowledge. Test your knowledge of this chapter by answering these questions or solving these problems.

True or False

1. All ornamental fish require warmwater temperature conditions.
2. Ornamental fish can be raised in ponds.
3. Brine shrimp nauplii are used to feed the fry of ornamental fish.
4. Ornamental fish are disease resistant.
5. Ohio ranks first in ornamental fish production.

Short Answer

1. Give the common names of six ornamental fish.
2. Identify a fish that often eats its young and so must be separated from its hatched young.
3. Why are carotenoids sometimes fed to ornamental fish?
4. What happens to a warmwater ornamental fish when the water temperature drops below 65°F?
5. The name "goldfish" includes what other common ornamental fish?

Essay

1. Discuss ornamental fish production in the United States.
2. Describe the production of one warmwater and one coldwater ornamental fish.
3. Compare spawning/breeding habits of ornamental fish.
4. If an individual planned to start an ornamental fish industry in Michigan, what resources would be needed?
5. If koi are carp, then why do koi have so many color variations?

KNOWLEDGE APPLIED

1. Visit or call a local pet shop or department store that sells ornamental or tropical fish. Make an inventory of the fish they sell and the price of each. Ask the manager who supplies the fish and how they are shipped. Compare your list with Table 6-1.
2. Some ornamental fish, such as guppies, are easily spawned in an aquarium. Obtain ornamental fish and research the requirements for spawning these fish. Compare their spawning with that of others described in this chapter.
3. Choose an ornamental fish discussed in this chapter that you would like to raise and market. Defend your choice. List the resources you will need. Describe how you will culture and market this fish. Report to the class and provide a color picture of the fish during your presentation.

4. Plan and set up an aquarium of 10 to 15 gal for ornamental fish. Describe your selection of fish, since some fish cannot be combined. Provide a checklist of the day-to-day maintenance of the aquarium. If possible, actually put the aquarium together as a class demonstration. (See Chapter 14, Aquariums.)

5. Develop data sheets on 10 ornamental fish. Include the following information on the data sheet: photo, common name, scientific name, water temperature requirements, feeding requirements of the adults, spawning/breeding behavior, type of production system, and any special needs or notes.

LEARNING/TEACHING AIDS

Books

Andrews, C., Exell A, and Carrington, N. (2003). *Manual of fish health: Everything you need to know about aquarium fish, their environment, and disease prevention.* Tonawanda, NY: Firefly Books, Ltd.

Cabrita, E., V. Robles, P. Herraez. (2009). *Methods in reproductive aquaculture: Marine and freshwater species.* Sarasota, FL: CRC Press.

Harper, D. (2006). *Aquarium fish.* New York, NY: HarperCollins Publishers.

Innes, W. T. (1994). *Exotic aquarium fish.* Surrey, England: T.F.H. Publications.

Ward, A. (2007). *Questions and answers on freshwater aquarium fishes.* Neptune City, NJ: TFH Publications, Inc.

Ward, A. (2008). *Questions and answers on saltwater aquarium fishes.* Neptune City, NJ: TFH Publications, Inc.

Wittenrich, M. L. (2007). *The complete illustrated breeders guide to marine aquarium fishes.* Neptune City, NJ: TFH Publications, Inc.

Internet

Internet sites represent a vast resource of information. The URLs (uniform resource locator) for World Wide Web sites can change. Using a search engine such as Google, find more information by searching for these words or phrases: aquarium, ornamental fish, tropical fish, feeder fish, aquarium fish, live-bearers, mouth-brooders, spawning mat, and aquarium feeds.

For some specific Internet sites refer to Appendix Tables A-11, A-12 and A-14.

Successfully managing the culture of crustaceans and mollusks requires a knowledge of the species being cultured—sources of the species, habitat, seed stock and breeding, accepted culture methods, stocking rates, feeding, diseases, processing, and marketing. Armed with this knowledge, the successful aquaculturalist is able to exercise good judgment.

CHAPTER 7

Management Practices for Crustaceans and Mollusks

OBJECTIVES

After completing this chapter, the student should be able to:

- Explain selection, management, and culture of broodstock
- Describe spawning methods and procedures
- Describe hatchery and grow-out operations and management
- Describe efficient harvesting procedures
- Describe management practices to encourage crawfish reproduction
- Describe the facilities needed to raise crawfish
- Explain design considerations with crawfish ponds
- Describe production systems used in crawfish farming
- Identify special considerations with soft-shell crawfish
- Describe how shrimp are cultured
- Discuss reproductive characteristics of shrimp
- Discuss how to manage a shrimp pond
- Describe how oysters and clams are cultured
- List commercially important species by common and scientific name
- Describe the life cycle of crawfish, shrimp, clams, and oysters
- Discuss environmental requirements for crawfish, shrimp, and mollusks

Understanding of this chapter will be enhanced if the following terms are known. Many are defined in the text and others are defined in the glossary.

KEY TERMS

Ablation	Metamorphosis
Byssus	Mysis
Chelator	Nauplius
Clutch	Selective breeding
Dredges	Spat
Hydrological	Zoeal

CULTURE STATUS

The culture of crustaceans and mollusks is not at the level of finfish. In the United States, most crustacean culturing is crawfish culturing. Shrimp and prawn culture is increasing in areas outside of the United States. Some potential exists to culture crabs and lobsters. Mollusk culture includes oysters, clams, mussels, snails, and abalone.

Crawfish (Crayfish)

The pond culture of crawfish is based primarily on the simulation of the natural **hydrological** (water) cycle to which the crawfish life cycle is adjusted in natural habitats. The crawfish farmer should establish, maintain, and manage a self-sustaining population of crawfish in the pond. This requires initial stocking, proper control of water and vegetation, and reasonable harvesting that will ensure adequate stock for the next season.

Sources of Species

Two species, the red swamp crawfish (*Procambarus* clarkii) (Figure 7-1) and the white river crawfish (Procambarus zonangulus) are cultured in the United States. The red swamp is native to the south-central United States.

Habitat

Crawfish are found in temperate freshwater throughout the world. Their natural habitats include swamps and marshes. Crawfish are low on the food chain, recycling decaying plant material.

Seed Stock and Breeding

Stocking of seed crawfish is necessary the first year, after which the population, if properly managed, should be self-sustaining. Ponds are usually stocked with adult crawfish from late April to mid-May. The ponds should be flooded for at least two weeks before stocking and have some vegetation or other form of cover for the crawfish. Freshly caught crawfish should be used for stocking purposes. Pond-spawned crawfish are preferable because they are fairly accustomed to pond systems and are less likely to migrate from the pond after stocking. The crawfish should be placed in the pond as soon as possible after being caught, or, alternatively, harvested and kept cool and damp until that time. To

FIGURE 7.1 Crawfish or crayfish.

reduce predation on newly stocked crawfish, the stock should be released in densely vegetated areas or in the deepest water far from the pond edge if little cover is present.

Culture Method

An important management procedure in crawfish culture is the manipulation of water level and quality. This involves the draining and flooding of the ponds at the right time to ensure reproduction by mature crawfish and production of young crawfish, respectively.

Every year, in late spring or early summer, ponds should be drained to simulate the summer drought of the hydrological cycle in the natural habitats of the crawfish, during which time the crawfish burrow and reproduce. Although crawfish may be able to reproduce even when there is water year round, draining accomplishes three functions. First, it forces all the crawfish to burrow close to the same time period, ensuring simultaneous reproduction and producing heavy recruitment of young crawfish during the flooding time. Second, draining and the subsequent drying out of the pond allow annual grasses and semiaquatic plants such as alligator weed, smartweed, and water primrose to grow and become established in the pond. This ensures enough vegetation for food and cover for the young crawfish after the pond is flooded. Finally, draining helps control unwanted vegetation and predators and allows work on pond or dike repair, if needed.

Water should be drained gradually in late June or early July. A quick method to determine when to start draining is to look for the burrows of early-burrowing females. If burrows can be seen, usually along the banks or under logs or heavy debris in the pond, draining begins. Slow draining allows young crawfish to seek hiding places for the summer and lets the adults have time to find suitable burrowing areas. Fast draining will strand some crawfish that are not ready to burrow and expose them to predators. Draining rate should proceed no faster than 2 to 3 in. per day.

After a period of drying, during which time the crawfish reproduce in the burrows, ponds are flooded. This ensures that the newly hatched crawfish will have ample water. Flooding softens the burrow plugs and allows the female and the young crawfish to escape from the burrow and start feeding and growing. Flooding should take place when the water is still warm enough to promote rapid growth and cool enough to hold more oxygen and to slow vegetative decay.

Generally, flooding takes place in late September and early October. Digging up a few burrows and observing the condition of the eggs or hatched young of the burrowed females help determine when to flood. If the females are in berry (carrying eggs) with dark eggs, flooding begins in two or three weeks. If the young crawfish have hatched out but still cling to the underside of the females, flooding begins within a week. If the young crawfish have left the females and are freely swimming in the burrow water, immediate flooding is needed. The burrows should be checked at least once a week, starting in early September.

Pond-water quality is a key factor to good crawfish production. Under certain conditions, oxygen depletion may occur. This usually happens during the warm fall and spring months when vegetation is decomposing rapidly. Good water circulation prevents this problem. The cheapest and easiest way to improve circulation is to exchange the pond water with good oxygenated water. This flushes out the deoxygenated water. Another way is to recirculate the water by pumping. In large ponds where water exchange may be a problem, mechanical aerators may be used.

Pond vegetation serves as food and cover for the crawfish as well as providing access to the water surface when dissolved oxygen levels are low. The crawfish will eat a variety of plants. The more tender plants are generally the most desirable, especially for young crawfish. Any animal matter that pond crawfish eat is the result of natural production and their own foraging. CThe crawfish farmers encourage the growth of suitable food plants and discourages undesirable vegetation. Plants used as crawfish food and cover should be capable of surviving and growing during the dry period and when the pond is re-flooded.

Plants generally considered desirable in a crawfish pond include—
1. Alligator weed (*Alternanthera* spp.)—An excellent food and cover, this plant can grow luxuriantly in many areas.
2. Water primrose (*Ludwigia* spp.)—Considered as good or even better than alligator weed, this plant does not grow as thick as alligator weed and is more tolerant to extremely cold weather.
3. Smartweed (*Polygonun* spp.)—A fair food and cover plant, it grows naturally and easily in most ponds and looks like water primrose.
4. Pondweed (*Potamogeton* spp.)—Another fair food and cover plant, it also grows naturally in ponds and is generally submerged.

A number of other plants occurring naturally in ponds, such as duckweed and Elodea, are also fair food and cover plants. Fertilization of the water is not necessary for crawfish culture, unless it is needed to get the food plants growing.

MOLTING

Growing crustaceans must shed their hard outer shell—the exoskeleton. This process is called "molting" or "ecdysis." The exoskeleton is made primarily of inorganic calcium carbonate (chalk) mixed with chitin and modified proteins. The crawfish provides a good example of the molting process.

The shell or exoskeleton prevents the crawfish from growing between molts. Crawfish grow in a series of abrupt steps, rather than in the uninterrupted progression of animals with skeletons. The animal eats for a period and builds tissues internally. Just before the crawfish molts, its shell steadily loses calcium carbonate. This is stored in the stomach in two small stones called "gastroliths." With the loss of calcium, the old shell weakens. It splits at a point between the back surfaces of the carapace and the abdomen. The crawfish crawls out, shedding all of its protective shell. The actual shedding lasts only a few seconds.

The body is soft at this time. It rapidly absorbs water and swells, often doubling in weight. Its new exoskeleton remains soft for about 12 hours. Next, it gradually hardens as it absorbs calcium carbonate from the supply stored in the gastroliths and from food and water. A crawfish may even eat its old shell for calcium and carbonate.

The molting processes are a complex physiological cycle, controlled chiefly by temperature, light, and hormones. The molt itself is hormonally controlled and occurs every 6 to 18 days depending upon the maturity of the crawfish and the water temperature. Great differences in growth rate are usually found among crawfish in a habitat or even in the same brood. These differences are probably due to varying activities and amounts of food consumed.

In the wild, a crawfish may grow about 0.25 in for each molt, which means a maximum size of 2.25 to 3 in. In well-managed culture ponds and wild areas where there is an abundance of food, crawfish often grow 0.5 in or more per molt and reach a maximum size of 3.5 to 5 in. Under conditions in the southern United States, fast growth occurs in spring. Crawfish require about 11 molts to reach full maturity.

Molting is used as a marketing tool for crustacean culturalists to produce soft-shell crawfish. Soft-shell blue crabs have been produced throughout the coastal areas of the south Atlantic and the Gulf of Mexico for decades. Currently, crawfish farmers in Louisiana produce soft-shell crawfish.

For a few hours after molting, crawfish shells are butter soft, and they are good to eat with very little preparation.

Immature, pond-raised crawfish are captured in ponds and transferred to indoor, recirculating aquaculture systems. Here the crawfish are fed and cared for in shallow troughs until they shed their shells. Then the live, soft crawfish are quickly frozen to stop the hardening process. Soft-shell crawfish may be fried or boiled using many different recipes to satisfy the consumer.

Too much vegetation may hamper harvesting and cause severe oxygen depletion when it decomposes. Too little vegetation may not provide enough food and cover. Generally, cover in the form of rooted, semiaquatic plants should cover at least 25 percent of the pond bottom during drawdown to provide protection for burrowing and young crawfish. At least a similar amount of vegetation, especially along the edges and shallower areas of the pond, is also needed during flooding to provide food and refuge for young crawfish.

Stocking Rate

Stocking rates vary with existing conditions of the pond. In ponds where crawfish are already present, 20 to 25 lbs per acre may be stocked. In a densely vegetated pond with no existing crawfish population, a stocking rate of 40 to 50 lbs per acre is recommended. Densely wooded ponds and open ponds with sparse vegetation should be stocked at 45 to 60 lbs per acre. Ponds with no or very little natural vegetative cover require a higher stocking rate because of higher predation losses. Up to 100 lbs per acre may be stocked in these ponds, depending on price and availability.

Stocking rates are based on medium to large crawfish. Sexually mature crawfish in a ratio of at least one female to one male should be used for stocking. A ratio of one male to three or four females will ensure heavier reproduction. In properly managed ponds, further stocking is generally not necessary. Enough adult crawfish present after harvest ensure adequate stock for the next season.

Feeding

A primary advantage of crawfish culture over traditional fish culture is that crawfish derive their nutrition from natural production of plants and organisms associated with decaying matter. If a good cover crop of desirable vegetation is present in the pond, no supplemental feeding is required. Studies indicate that the addition of pelleted and extruded commercial fish feeds in experimental ponds can produce significant increase in crawfish production. High cost and low feed conversions of the artificial feeds make them uneconomical as yet in commercial crawfish operations. Agricultural forages and by-products such as hay, sweet potato vines and trimmings, rice bran and stubble, and cottonseed cake, among others, serve as excellent sources of supplemental food and may be added to ponds. Experiments show that pelleted hay sold as rabbit or goat food can support molting and body maintenance requirements of crawfish. These products may be used as supplemental food in ponds with poor vegetation or as starter diets during the first few weeks of crawfish production in newly built ponds without properly established vegetation. Pelleted hay may also be used as maintenance ration for bait-sized crawfish until needed or sold.

Supplemental feed in crawfish ponds, particularly during warm periods, may cause water-quality problems. The farmer should watch for signs of low oxygen levels and be ready to circulate or exchange water and/or remove some of the feed.

The addition of lime in aquaculture ponds is practiced by many farmers to raise certain soil and water parameters to optimal levels. Liming increases the pH of bottom mud, increasing availability of nutrients, especially phosphorus, for use by plants. It raises the alkalinity of the water, increasing the availability of carbon dioxide needed by plants, as well as buffering against drastic daily changes in pH. Liming increases total hardness by the addition of minerals such as calcium required by crawfish for their molting and other metabolic processes. It also causes the precipitation of excessive organic matter in suspension in water.

Before adding lime, the farmer should have the soil and water in the pond analyzed. The exact amount of lime to be added depends on factors such as initial soil and water conditions and the chemical relationships among them.

Diseases

Few diseases cause problems in crawfish. Management of oxygen and temperature is critical for growing healthy crawfish. Cold water temperatures stress crawfish, predisposing them to disease. Currently, diseases and parasites are not a major concern to crawfish production. But as culture becomes more intensive, diseases and parasites could become a serious consideration, as water quality declines

When extreme environmental conditions are prolonged or intolerable, cultured crawfish can more easily become diseased. A bacterial infection known as shell disease could become a problem in intensive culture. Knowledge of crawfish disease lags behind that of finfish.

Crawfish grow well between 72° and 80°F, but they begin to die when temperatures exceed 91° to 95°F. Consumption of oxygen by crawfish increases with temperature. If low oxygen and high temperatures occur at the same time, this enhances the problems. The presence of some bacteria increase when temperatures exceed 83°F. The most reliable way to manage water is water circulation.

Nutritional diseases also occur when crawfish do not receive the proper diet or enough feed.

Harvesting and Yields

For maximum yield, a crawfish pond must be harvested intensively throughout the production season. Enough reproductively active crawfish are left to serve as broodstock for the next season. Harvesting ponds is quite tedious because it is generally done manually with the use of baited wire traps that should be run daily. (See Figure 7-2.)

Crawfish traps vary in size, shape, type, and construction, depending on the preference of the crawfish culturalist. The traps are cylinders of wire fencing or similar material that have funnel-shaped openings through which crawfish can enter. The mesh size depends on the size of crawfish to be caught. Traps made of 0.75 in. (19 mm) chicken wire or hardware cloth will trap only market-size—3 in. and larger—crawfish. Traps made of 0.5 in. mesh will retain bait-size—1.25 in. and larger—crawfish. Most crawfish farmers construct traps that are 24 to 36 in. long. Among the preferred types of traps (Figure 7-3) are:

1. Modified minnow traps with wider funnels and mesh sizes.
2. Stand-up traps with two or three bottom entrance funnels and open top; the traps are propped up on stakes to leave the open end above the water surface.
3. Pillow-shaped traps that are similar in conformation to the stand-up type, except that the non-funnel end is pinned shut. These traps may be set flat on the bottom in deeper water or propped up partly out of water.

FIGURE 7-2 Harvesting crawfish.

FIGURE 7-3 Crawfish are harvested with traps of various size, shape and type. Some are similar to this cylindrical trap.

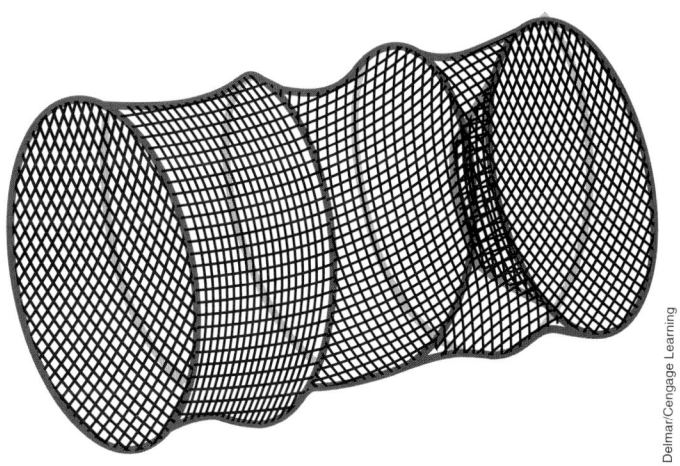

Daily trapping requires 15 to 25 traps per acre. For bait, cut fish and fish heads are usually used, with gizzard shad being the most preferred type. Other good baits are carp, alewife, and other oily-trash fish. Because crawfish prefer fresh animal matter, bait should be fresh or fresh frozen. When changing baits, the harvester should not throw the old bait or its remains into the pond. Added decayed matter will compound water-quality problems.

Crawfish may also be caught with minnow seines in ponds where absence of dense vegetation permits it. Seines are especially useful for collecting soft crawfish and those about to molt because these crawfish hardly move at all from their hiding places.

The crawfish crop may be trapped from as early as late November through approximately mid-June or just before water drawdown. The crawfish hatched in late September and early October could attain market size by mid-December, providing there is adequate food and still mild water temperature, above 50°F. While the young crawfish are growing, many adult crawfish, especially the spawned-out females, are still present. These holdover crawfish should be harvested in late November and early December to provide more space and food for the young of the year. Many of the adults will reach the end of their life span during winter and thus will be lost if not harvested during this period. This one-and-a-half to two months after re-flooding gives these adults in the burrows ample time to feed and fill out their tails. Harvesting one or two weeks after flooding will usually yield only hollow-tailed crawfish.

Harvesting may be spotty in January and February when water temperature is well below 50°F because crawfish are generally inactive during periods of low temperature. Intensive harvesting should commence when water temperature rises above 50°F, usually in mid-March or early April, and continue until just before draining. If, during the few weeks before drawdown, the average catch is less than 0.5 lb. per trap per day, it is better to stop trapping. This ensures that enough crawfish will be left as broodstock for the next season.

Grading and Marketing

Grading involves sorting crawfish for uniform size and quality. Damaged, dead, diseased, or off-color crawfish are removed. Crawfish are grouped by weight. Mechanical graders are sometimes used. Hand sorting is labor-intensive and requires a lot of time. Crawfish grades are uniformly established and are based on the number of whole crawfish required to weigh a pound. The common grades are as follows:

- Large or No. 1: 15 or fewer in a pound (weigh 1 oz or more each)
- Medium or No. 2: 16 to 25 in a pound (weigh 0.5 to 1 oz each)
- Small or No. 3: more than 25 in a pound (weigh less than .5 oz each)

The preferred size for crawfish is 1 to 1.5 oz., or the No. 1 grade.

Crawfish tend to be marketed by growers. Many crawfish producers use small, niche markets. Crawfish marketing tends to be seasonal. Crawfish growers may sell their catch on farm, to retail markets, to processors, to wholesalers, or to recreational stores. New product development with whole, cooked, and frozen crawfish and prepared frozen dishes has increased the distribution of crawfish sales.

Sometimes, incidental catches of small crawfish, less than 3 in. long, may also be sold as bait, which is another lucrative market. Bait-size crawfish are generally available in well-managed ponds from about six weeks after fall flooding into February and March. Because bait crawfish are usually in demand during spring and summer when most of the crawfish have grown to larger food-size ranges, or when ponds are dry, some modifications to the general production schedule for raising food crawfish have to be made.

FIGURE 7-4 Freshwater prawn.

Prawns

Worldwide, more than 300 species of prawns and shrimp exist. Only about 80 species are commercially important. Shrimp represent the most valuable U.S. fishery and the most widely cultured saltwater species. Culture of shrimp and prawns is attractive because of increasing demand. Shrimp culture in the United States is relatively new, whereas shrimp culture is typical to Asia.

Sources of Species

In the United States, commercial aquaculture of the freshwater prawn (*Macrobrachium rosenbergii*) (Figure 7-4) began in Hawaii and developed most rapidly there.

Habitat

Freshwater prawns, *Macrobrachium rosenbergii*, require warm temperatures of about 82°F and saline waters (10 to 20 ppt) for larval stages and fresh or slightly saline waters for grow out. Land and water resources suitable for pond-based aquaculture exist in Hawaii, Puerto Rico, the Southeast and Gulf states, California, and perhaps other areas of the United States, plus many foreign locations.

Shrimp live in the oceans from the Atlantic to the Pacific and the Gulf of Mexico.

Seed Stock and Breeding

Routine reproduction of *M. rosenbergii* in captivity poses no problem. Domestication is a selection process that favorably modifies economically important physiological, developmental, and behavioral traits. Although the freshwater prawn remains essentially a wild animal, it displays evidence of domestication. Because *M. rosenbergii* breeds readily in captivity, its domestication can be accelerated through **selective breeding**.

Shrimp are obtained by catching larval shrimp or catching gravid females and spawning them in hatcheries. Eyestalk **ablation**, removal of one of the female's eyes, is an effective method to induce spawners to early maturation. This method may help ensure a constant supply of larvae.

Culture Method

Facilities and equipment for shrimp and prawn aquaculture evolved mainly through trial and error, as well as through adoption of technologies from other types of operations. There is some need for improvement in mechanization and automation.

Prawn-culture technology includes three phases: hatchery, nursery, and grow out. Only the larval stage (hatchery phase) requires saltwater. Nursery production and grow out are effective in brackish water. Production facilities include ponds and tanks.

Shrimp aquaculture occurs in three phases: (1) maturation and reproduction for production of seed stock (larvae); (2) the hatchery for production of post-larvae; and (3) grow out to the adult stage in raceways or ponds. Seed stock produced in the maturation and reproduction phase supplies the hatchery phase, and the post-larvae produced in the hatchery supply the grow-out phase. Figure 7-5 shows the life cycle of a shrimp.

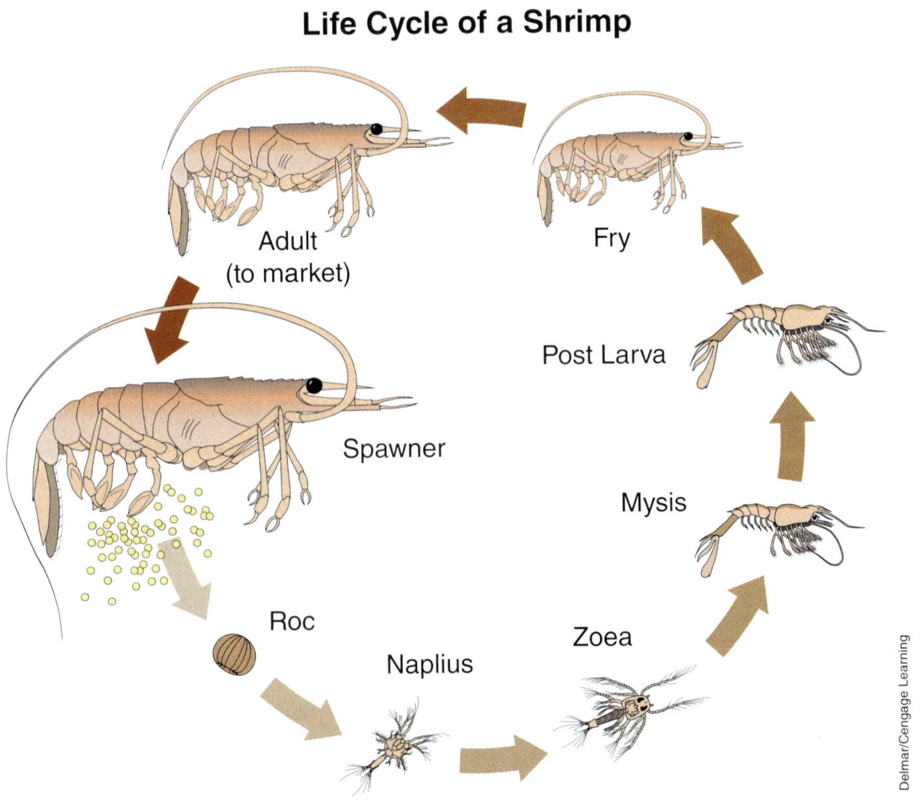

FIGURE 7-5 Life cycle of a shrimp.

Stocking Rate

Stocking rate varies with the species cultured, the method of culture, and the stage of production.

Feeding

Commercial feeds are available for the various stages of production. What and how much to feed depends on the culturalist's experience and continuous observation of the larvae, sediments, feces, and water conditions. Shrimp are fed on a variety of cultured plankton. Shrimp also eat small fish, other shrimp, soybean cake, peanut cake, and rice bran. Commercial, artificial, complete diets are the main food source for cultured shrimp.

A general lack of information exists on the nutritional requirements of *M. rosenbergii*.

A variety of commercial feeds like chicken-broiler starter and catfish and trout feed aids the start of experimental and commercial culture of prawns in ponds. Prawn obtain substantial nutrients from the natural life of the ponds. Because feed accounts for an estimated 20 to 40 percent of operating costs, it can affect the profitability of pond culture operations. The careful selection of an inexpensive, nutritious diet tends to improve profitability.

The principal expenditure in larval culture is for food, mainly brine shrimp (*Artemia nauplii*). Dependence on expensive and occasionally unobtainable cysts (eggs) of the brine shrimp troubles hatchery operators.

Diseases

Disease is an important factor in reducing shrimp numbers in natural populations. Natural mortality or death from old age is the potential fate of all shrimp, but the toll taken by predation, starvation, infestation, infection, and adverse environmental conditions is significant. A variety of viral, bacterial, fungal, and protozoan agents affect shrimp. Flukes, tapeworms, and nematodes also occur in shrimp populations.

Microsporideans parasitize crustaceans. In shrimp, microsporidean infections cause a condition known as "milk" or "cotton" shrimp. Microsporideans become abundant in the infected shrimp and cause the white appearance of tissues.

Cramped shrimp is a condition of shrimp kept in a variety of culture situations. The tail is drawn under the body and becomes rigid to the point that it cannot be straightened. The cause of cramping is unknown.

With the development of prawn culture, disease problems can be expected in the list of obstacles to successful production. High-density, confined rearing is unnatural and produces stress.

Harvesting and Yields

Shrimp and prawn are harvested by hand with seines or dip nets. In ponds, the water is drained and shrimp are caught at the drain or lower levels in the pond.

Shrimp and prawn growth depends on the stocking period, stocking density, feed, temperature, salinity, and pond management. Shrimp harvested at 1 to 2 oz. require 100 or more days of culturing from the postlarval period.

Perhaps the most prominent characteristic of a prawn population is its mixed growth rate. Highly variable growth affects the entire industry because it influences the production of large prawns, which command a higher price in the current market.

Processing and Marketing

As soon as the shrimp come out of water, they should be placed on ice. Shrimp are beheaded, graded, and frozen. Depending on the location and the market, some may be sold fresh as well as frozen.

As the volume of production increases, so does the need for processing control and improved quality. Currently, most prawns produced in Hawaii are sold fresh, on ice, or live. As this market fluctuates, producers lower prices or process their products to move them. On the mainland, nearly all shrimp are sold as frozen tails or in some processed form.

Shrimp

Shrimp exist naturally in brackish water in the tropical regions of the world. When they are cultured, it is most often through pond systems with extensive, semi-intensive, or intensive management techniques.

Worldwide, the most common way shrimp are cultured is in pond systems. After pond systems, tanks and raceways are used. Many producers and researchers are experimenting with indoor, intensive, closed recirculating systems.

Sources of Species

Penaeid species most widely cultured worldwide are *P. monodon, P. vannamei, P. chinensis, P. stylirostris, P. japonicus, P. penicillatus, P. merguiensis,* and *P. indicus.* Researchers and farmers also work with a wide range of other penaeid shrimp species.

In the Western Hemisphere, *P. subtilis, P. paulensis, P. setiferus, P. brasiliensis, P. duorarurn, P. occidentalis, P. schmitti,* and *P. californiensis* are cultured. In the Eastern *Hemisphere, P. semisulcatus, P. latisulcatus, P. kerathurus,* and others are cultured.

Tropical penaeid shrimp (Figure 7-6) look very much like local (endemic) species in the United States. Generally a biological key is required to separate the species and obtain a positive identification. Normal color patterns include white, brown, and blue.

Habitat

Tropical penaeid shrimp are naturally found in brackish water, but they can survive in freshwater. They do not grow well below 68°F and will experience stress below this temperature and will not eat. Shrimp die at 50°F and lower.

FIGURE 7-6 A young shrimp, *Penaeus vannamei,* feeds on a soy-based pellet.

Shrimp are omnivorous, but they are often described as primarily herbivorous. In the pond and in the wild, different species of shrimp occupy different niches. They feed or graze upon different organisms or plants (diatoms).

Seed Stock and Breeding

In most aquaculture enterprises, obtaining spawns is one of the most difficult tasks the producer faces. Shrimp will spawn more easily under the right conditions. Adhering to the right conditions separates the good hatcheries from the nonproductive ones.

Hatchery systems for producing saltwater shrimp can be grouped into two general categories:

1. Extensive. Animals are held in very low densities in tanks or ponds and allowed to spawn. Sometimes spawners are brought in from the wild or reared in ponds.
2. Intensive. Animals are held in higher densities for longer periods with high water exchanges, high-protein diets, and high water-quality standards.

Extensive hatchery systems are the least complex. They are also the least efficient, however, in terms of production for the volume of water used. Problems include low broodstock density, inefficient harvesting techniques, and possibly poor diet. Also, in temperate climates, tropical animals must be overwintered or brought inside during the winter.

Hatchery systems using tanks can be managed more intensively for higher production. Tanks in a controlled environment are the most productive and are more commonly used to propagate shrimp.

Intensive production is normally conducted with a 1:1 male-to-female ratio, or approximately 25 males and 25 females in a 12-ft. circular tank Environmental conditions, hormonal regulation, and nutrition are three major factors that affect shrimp maturation. Biological and environmental factors affect maturation results: Water temperature, photoperiod, light intensity, water quality, nutrition, age and size of broodstock, broodstock density, and sex ratio. Other management factors include the proper removal of the female and placement in spawning tank, return of the female to maturation tank after spawning, incubation of eggs, source of broodstock and stability of temperature, and salinity (28 to 36 ppt).

Hormonal regulation involves ablation or removal of one eye of female broodstock to induce maturation. Unlike fish, no known hormone treatments exist for shrimp used regularly in hatcheries.

Nutrition is the third major factor affecting spawning and maturation. Shrimp must be provided a high-protein diet, with high steroids and high unsaturated fatty acids (HUFA) for good reproduction and viable egg production. Additionally, a high source of carotenoid should be incorporated into the diet after females have been in intensive maturation systems for a period of time because they experience a carotenoid deficiency.

The number of eggs is positively correlated with size of female broodstock. The larger the shrimp, the more eggs are generally produced. Female shrimp can regenerate eggs and spawn again in approximately five days after a spawn, and males can regenerate the spermatophores in a similar period of time.

Fertilized eggs are allowed to remain in the spawning tanks until they hatch into the **nauplius** stage. They require aeration (high oxygen). Eggs normally sink, but they will remain in the water column as long as there is aeration. Over a three-month period, an average of 50 percent of eggs will typically hatch. Beginning hatches are generally in the 90-percent-hatch range and taper off over time.

Most drugs are not approved for use by the FDA in production systems. Researchers and growers are attempting to gain approval for some therapeutants for aquaculture. Check with your state aquaculture extension specialist for current recommendations. EDTA (ethylene diamine tetraacetic acid) is approved and is often used as a **chelator** (helps keep debris from adhering to eggs and larvae in the hatchery). A 2-ppm solution is used most frequently.

Culture Method

Shrimp larvae are staged and classified according to their stage of development. The larvae are then fed accordingly.

Microalgae are grown using standard algae culture procedures and fed to beginning stages of shrimp. Algae can be cultured in five-gallon plastic drinking bottles. Other larger containers may be used if greater quantities are required. Algae are fed to larval shrimp and maintained at certain critical levels (above 35,000 cells/ml and below 200,000 cells/ml; average 100,000 cells/ml).

At the late **zoeal** and early **mysis** stage, shrimp larvae are fed Artemia nauplii. Artemia cysts are placed in seawater (5 grams of cysts per liter of water). Aeration and a light source are provided and allowed to hatch (usually within 24 hours). Freshly hatched Artemia nauplii are then separated from empty shells and debris by removing the aeration and allowing settling to occur. Artemia cysts are usually disinfected with a dip into chlorine, or they may be decapsulated using chlorine.

Shrimp larvae may also be fed Artemia flakes, freeze-dried Artemia, or a dry microencapsulated diet. These diets are considered best used as supplemental diets only and should be given in combination with algae and live *Artemia nauplii*.

Larvae require approximately 18 days in the hatchery. This includes harvest, draindown time, tank dry-out time, and refill time.

From the standpoint of commercial production of shrimp ponds, a semi-intensive stocking density and management level are most common. The best post-larvae to stock in ponds, raceways, or tanks have proved to be the most active, with good color, observed with full guts, clean shells, and good muscle development. Pathogen-free or guaranteed healthy post-larvae should be obtained if possible.

Stocking Rate

Stocking rates for shrimp will vary according to the type of grow-out system in which they will be placed (Figure 7-7). In pond systems, stocking rates range from 2,300 to 220,000 shrimp per acre. Some producers stock nursery ponds at 1 million shrimp per acre, and after one month they transfer juveniles to grow-out ponds of considerably larger size. Most U.S. producers stock grow-out ponds at 150,000 to 200,000 shrimp per acre, and they manage the ponds using intensive management techniques (increased water exchange, aeration, etc.). In systems other than ponds, it is difficult to determine average or recommended stocking rates. Research continues on stocking rates for each system.

Some shrimp producers suggest that, in tanks, the stocking rate is about 100 pl/m3 up to 1,000 pl/m3. High stocking rates have been attempted, but the lower stocking densities have proven more successful. The problem is not really how many shrimp can be stocked in a given

FIGURE 7-7 An aerial view of a large shrimp-culturing facility in Malaysia.

system, but rather the issues of the most economical stocking density for the system and its level of management.

When a pond's stocking density is increased, the existing natural-food supplies are depleted faster and water quality is stressed. Shrimp do not react well to stress and often develop secondary infections as a result of stress. They do respond well to good management and good nutrition. The limiting factor of shrimp culture often is dissolved oxygen (DO), followed by ammonia toxicity.

Feeding

In a low stocking-density situation (300 lbs per acre) there often is enough natural food for shrimp to survive. They can eat diatoms, plankton, and/or detritus. On a dry-weight basis, natural food can contain about 55 percent protein. To encourage natural-food growth, culturists often fertilize ponds with organic and/or inorganic fertilizers.

As the stocking density is further increased, not enough natural food is available and supplementary feeding becomes necessary. Shrimp post-larvae can be weaned from Artemia to crumbled starter food (higher protein level than adult grow-out diet).

In other countries, producers have fed rice bran, broken rice, oil cakes, wheat flour, cornmeal, and a variety of plant refuse. In many countries of the world, excellent grow-out commercial feed formulas have been developed and proven effective, making feed the most expensive operating cost of a shrimp farm. New methods have been found to stabilize vitamin C in diets and improve shrimp growth. New binders and increased levels of wheat as a binder have all proven to be successful.

Some producers begin feeding post-larvae with a 40 to 50 percent protein feed at a rate as high as 18 to 20 percent of body weight per day. When post-larvae weigh 1 g (in about one month), they are fed 15 percent of body weight per day, gain 1 to 2 g per week. By the time they reach 18 to 20 g, they are consuming about 3 to 5 percent of body weight per day. Juveniles are fed commercial feeds (about 25 to 32 percent for *P. vannamei* and higher, 35 to 38 percent for P. monodon). Their feed conversion is about 2 pounds of feed per pound of gain. A typical feeding schedule based on percent of estimated body weight follows; semi-intensive systems manage feeding using trays and by feeding twice each day:

- 40 percent of daily ration in the morning
- 60 percent of the daily ration in the afternoon

At the end of the day, leftover feed is recorded and used to adjust the feeding level (see Table 7-1).

Shrimp do not tolerate low dissolved oxygen (DO) very well. Below 2.0 ppm DO begins to stress shrimp. A level of 0.1 to 1.5 ppm can be lethal to shrimp, depending upon species and other parameters such as salinity, pH, temperature, and the like. A chronic low DO level can cause shrimp to stop eating, can cause stress, and subsequently can cause the onset of secondary bacterial infections. Pond aeration and water-movement devices,

TABLE 7-1 FEEDING SHRIMP TO MAINTAIN WATER QUALITY[1]

Subjective Score[2] of Average Value on Feed Trays	Action
>2	Reduce previous day's ration by 30%
>1	Reduce previous day's ration by 20%
0.5 to 1	Feed same amount as the previous day
<0.5 for 3 days	Increase previous day's ration by 10%

[1]National Council for Agricultural Education, Aquaculture Curriculum Guide: Year Two. (1995).
[2]System for recording the presence of food left on tray: 0 = no feed remaining; 1 = small amount remaining, less than 12.5%; 2 = medium amount remaining, between 12.5 and 25%; and 3 = large amount remaining, more than 25%.

TABLE 7-2 GUIDE FOR ADJUSTING THE DAILY FEED RATION FOR SHRIMP[1]

Level of Dissolved Oxygen (DO)	Action
3.0	Reduce previous day's ration by 30%
2.5 and <3.0	Reduce calculated ration 50% and feed it all in the afternoon.
2.5 and 2.0	No feed that day.
<2.0	No feed that day and draw down level of water to 40 in. (90 cm) or lower and start a continual exchange until the morning DO level is above 3.

[1]National Council for Agricultural Education, Aquaculture Curriculum Guide: Year Two. (1995).

as well as pumping water are treatments for low DO. The level of DO can also be used to adjust the feed level (Table 7-2).

Culture systems should be designed and managed so that excretory products do not build up. In ponds, most excretory products will break down. In intensive systems, excretory products must be removed. Soluble metabolic by-products, such as ammonia, and by-products of organic materials breaking down to nitrites are a problem. Nitrites above 0.1 ppm may cause reproduction problems. Tolerance levels in grow-out are not well known, but much higher levels have been recorded (.75 to 2 ppm at 8.3 pH) without mortality. Some gill damage may occur when the level of un-ionized levels of ammonia go above .5 mg/l and when other stresses are present (low DO, handling, etc.). However, growth can be reduced at these higher levels.

Diseases

Diseases and parasites are somewhat less of a problem in shrimp grow-out than they are in the hatchery phase. Although there are few chemicals allowed for the treatment in the U.S. shrimp hatcheries, they are not as heavily restricted in other countries that culture shrimp. Antibiotics are allowed to be placed in shrimp grow-out feed in the United States, but this practice is heavily regulated. Most countries require that antibiotics be withdrawn from the shrimp feed 15 to 21 days before harvest if the shrimp are for human consumption.

Protozoans often attach themselves to shrimp. Some protozoans are Acimeta, Ephelota, Zoothamnium, Epistylis, and Lagenophrys. Bacterial diseases that affect shrimp include Vibro and filamentous bacteria.

The best treatment for diseases seems to be prevention. Water quality, temperature, low stress, and good nutrition are vitally important to preventing diseases. Purchasing post-larvae from a reputable source is also necessary. If possible, producers should obtain guaranteed healthy stock.

Viruses such as IHHN, Baculovirus, Parvovirus, and others have caused various negative effects—runting, size variation, and deformities—in penaeid shrimp. The development of guaranteed healthy shrimp helped in the control and hopefully the eventual eradication of these diseases.

Harvesting and Yields

Growth and yields of shrimp vary greatly with species, stocking density, and food supply. Other conditions, such as water quality and temperature, are also major factors. Under ideal conditions, P. vannamei can reach 20 g in 120 days, whereas *P. monodon* attains 35 g in the same period. The normal weight at which shrimp are harvested in the United States is about 16 to 18 g.

In pond culture, harvesting is usually accomplished by draining the pond. Sometimes cast nets and seines are used. Traps are used with P. japonicus. Usually they are caught in a net as the water passes out the sluice gate. In most places in the world where shrimp are cultured, they are processed (beheaded), frozen green headless, and then sold to the United States or Japan.

Processing and Marketing

The United States prefers frozen shrimp. Japan prefers fresh or secondly green headless (raw uncooked) frozen. A number of large food processors in the United States market frozen shrimp through grocery stores. Some shrimp farms market their own shrimp, but most are sold to a broker or processing house. The European market prefers a head-on product that is frozen mostly because fresh is not economical.

Crabs

While crab meat is high priced, ocean fisheries provide most of the crab (Figure 7-8) for the markets. Generally, crab are not considered a major aquaculture species. The blue crab (*Callinectes sapidus*) fills a special market niche, the demand for soft-shelled crabs or soft crabs, which sell for five to six times more than hard-shell crabs.

To produce soft-shelled crabs, blue crab that are just about to molt, are harvested. These crabs are placed in holding boxes, fenced pens, or indoor recirculating systems. Crabs are observed for signs of molting. After a split

FIGURE 7-8 Crab.

FIGURE 7-9 Lobster.

occurs along the back of the shell, molting takes place in a few hours. Within four or five hours after molting, crabs are removed and sold. If crabs are left in the water, their shells begin to harden.

Lobsters

Analysis of catch statistics shows that lobster fisheries are harvested at or near maximum sustainable yields. This is an especially severe problem for the American lobster (*Homarus americanus*) (see Figure 7-9), whose populations are on the verge of a major decline. Since 1955, annual U.S. landings have been fairly constant at approximately 30 million lbs. Fishing efforts quadrupled. Awareness of the limits of the fishery stimulated strong commercial interest in lobster aquaculture.

Hatcheries are used to supplement the number of lobsters available for commercial catch. Lobster farming from the egg to the market-sized adult is not profitable yet. One reason is the five to six years required to grow a lobster to market size.

Individualized containers will probably remain a requirement of lobster aquaculture and this is a major part of the large initial capital investment. Container size affects growth of the American lobster, but the reason is not entirely known.

Shell erosion, a chronic problem, becomes more critical as desirable broodstock is held for longer periods. Shell erosion also influences marketability and probably predisposes the animal to other infections.

The basic technology for lobster aquaculture is available, but a number of problems hinder rapid commercialization.

Oysters

The reasons for the decline in domestic production of oysters (Figure 7-10) includes overfishing, natural disasters, loss of habitat, diseases, economic factors, and pollution. Overfishing contributed to major declines after the turn of the century. Since the late 1940s, disease, pollution, and habitat losses have caused major declines.

Sources of Species

Four species of oysters are harvested commercially in the United States: the American oyster (*Crassostrea virginica*); the Pacific oyster (*C. gigas*); the European oyster (*Ostrea edulis*); and the Olympia oyster (*O. lurida*). The first two species make up the majority.

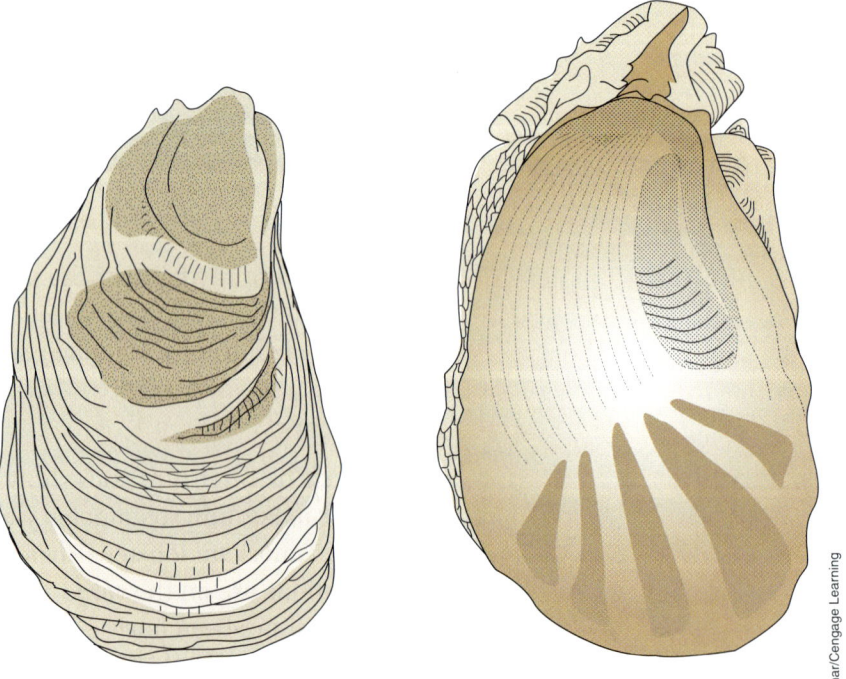

FIGURE 7-10 Oyster in shell.

Oyster

FIGURE 7-11 Production of oyster seed.

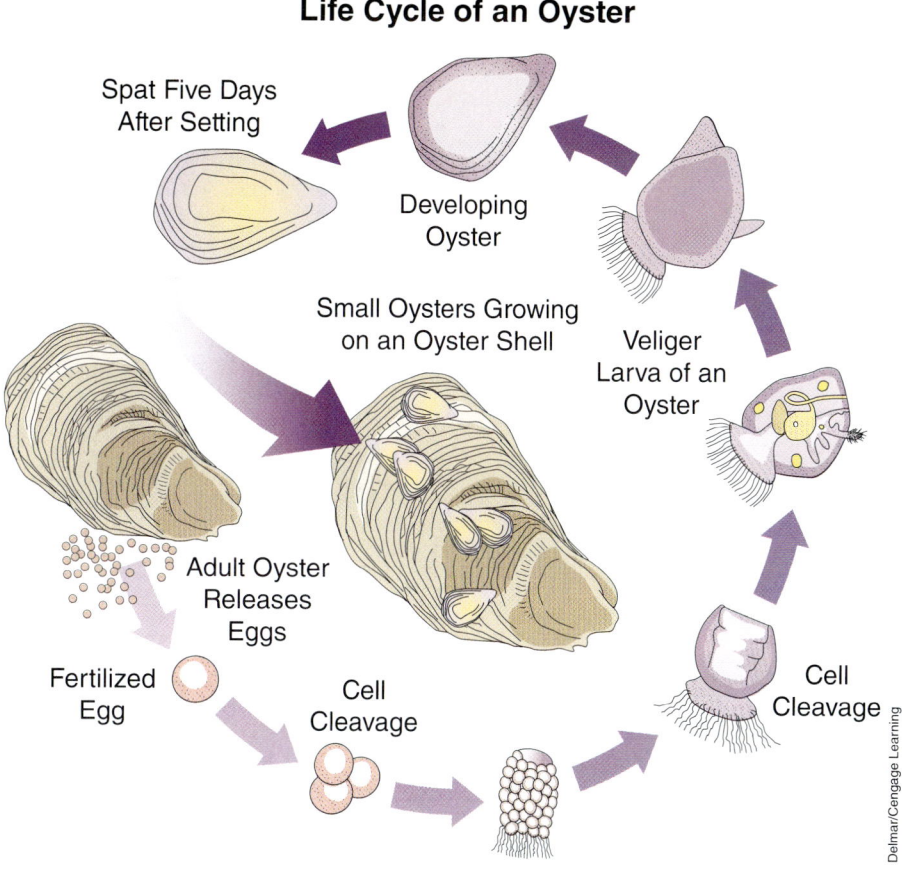

Habitat

Like other aquaculture species, oyster culture occurs in water free of toxic substances associated with industrial and domestic pollution. Bays and estuaries used for oyster culture require protection. Continuing research seeks to find ways to culture oysters in controlled environments. This should expand, not only to support controlled systems on a commercial scale but to help define water quality in the natural environment.

Seed Stock and Breeding

Figure 7-11 illustrates the production of oyster seed. In many areas of the United States, natural seed setting has become sporadic, and in some years it has been a total failure. The culturalist should take advantage of natural broodstock and setting areas in order to enhance recruitment. New or improved **spat** (young oysters) capture systems, including development of an economical artificial **clutch**, are helpful to culture. To increase natural sets, commercial hatcheries were constructed, especially on Long Island Sound and the West Coast.

On the West Coast, the availability of economically competitive seed is a major problem. Three sources of seed include Japan and commercial hatcheries and natural settings in Washington and Canada. No seed was available from Japan in 1978 and 1979 because of high local demand and poor setting.

FIGURE 7-12 Oyster culture.

Culture Method

The oyster industry is seriously constrained by pollution and continued loss of grow-out areas due to water-use conflicts. In many areas, it is almost impossible for a new grower to find high-quality grow-out areas. This difficulty may force the use of more efficient culture techniques, such as suspension and off-bottom relaying systems. Cost of these techniques may be higher because of materials and labor, and special permits must be obtained to place such systems in the water. Oyster culture requires space in bays and estuaries that often conflict with other uses.

Oyster culture requires water that is free from the toxicants that inhibit oyster growth and from the contaminants that reduce meat quality (Figure 7-12). Current culture techniques cause high mortalities. For example, the movement of seed from setting to grow-out areas can result in more than 50 percent mortality. On the West Coast, 75 percent of the oysters die during the first year. By increasing survival 25 percent, production would increase substantially and, in turn, lower culture costs.

Stocking Rate

Stocking rate depends on the culture system used.

Feeding

Oysters are filter feeders. A variety of algae species are needed to rear oyster larvae to **metamorphosis**. But to grow oysters from post-metamorphosis to seed size or to market size in intensive systems is too expensive because of the kinds and quantities of algae required. Growers need to know the nutritional requirements of oysters at every stage and how algae can be mass cultured to fulfill these requirements.

Diseases

A number of diseases and parasites plague oysters. The most serious ones that affect adult oysters are *Minchinia nelsoni* (MSX), *M. costalis* (SSO), and *Dermocystidium marinum* (Dermo). The first two organisms cause heavy mortalities in the mid-Atlantic area—Delaware, Chesapeake, and Chincoteague Bays. Dermo causes mortalities in the Gulf Coast states. Mortalities from Dermo also have been recorded as far north as the Chesapeake Bay. Stress conditions implicate viral and other microbial diseases in shellfish mortalities.

Major oysters predators include oyster drills, starfish, flatworms, crabs, and fish. Mortality varies with geographic area, salinity, and other factors.

Researchers and growers regard oyster drills as the most damaging predator. Successful research efforts to reduce or limit drill predation could save the industry millions of dollars.

Harvesting and Yields

With intensively managed bottom culture such as in Long Island Sound, it is possible to produce 4,450 lbs per acre per year. Public grounds, with little or no management, produce only 9 to 90 lbs per acre per year. Studies of off-bottom culture indicate that 0.25 acre covered by rack cultures could yield 2.9 tons of meats per year. Such yields depend on factors like productivity of the waters and total flow.

The size of an oyster culture operation varies from 2 to 3 acres in Maryland to 10 to 15 acres in Virginia to 200,000 acres in Long Island Sound. Many states limit the number of acres a person can lease.

Harvesting methods range from primitive tools, such as hand tongs, to modern hydraulic **dredges**. (See Figure 7-13.) In some cases, state laws dictate the type of gear that can be used. Also, the size of the operation limits the gear that can be used economically.

FIGURE 7-13 Oyster dredge for harvesting.

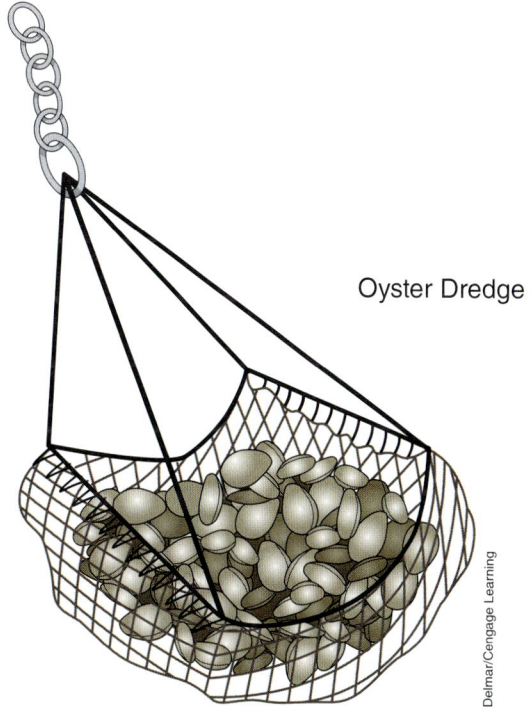

Oyster Dredge

Processing and Marketing

Oyster processing often begins with hand shucking. Oyster shucking machines that produce a fresh product need development. Oysters are sold whole, fresh, or canned.

Clams

Hard clam (Figure 7-14) harvesting traditionally provided supplementary employment for those in other seafood industries. For example, during closed seasons for oysters and crabs, oystermen and crabbers often go clamming. Because harvesting equipment is relatively simple and inexpensive, clamming also attracts many part-time and recreational clammers. Recently, clamming became more of a full-time occupation. Several trends encourage aquaculture of hard clams:
- A chronic shortage of smaller size clams—littlenecks and cherrystones—and a strong demand
- The increased price of small clams
- Declining harvests in New York, New Jersey, and Virginia—major producers in the past

Sources of Species

The hard clam (*Mercenaria mercenaria*) is harvested commercially in 16 states.

Seed Stock and Breeding

Gonadal development begins at 46° to 50°F 7. Spawning occurs between 72° and 82°F 7. Growth takes place in water between 46° and 82°F.

FIGURE 7-14 Clam in shell.

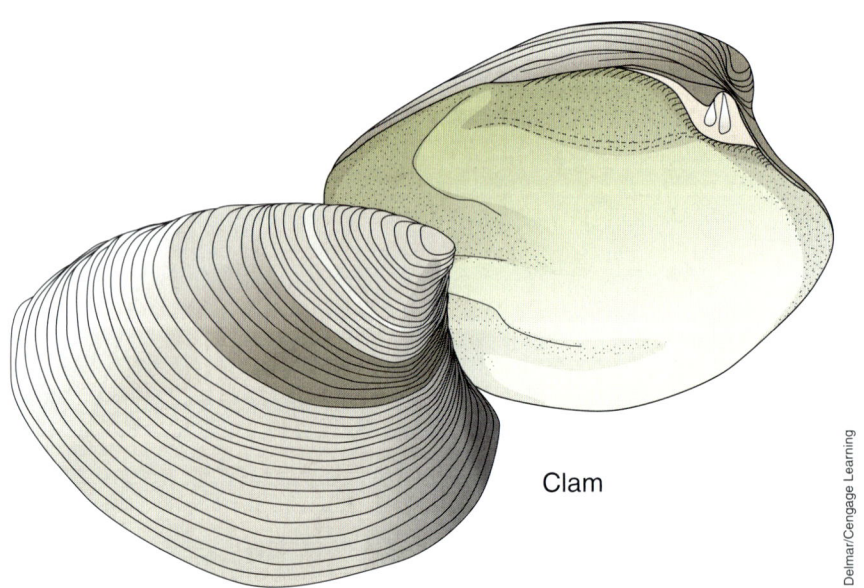

Clam

Culture Method

Clams require relatively clean seawater with a salinity range between 20 and 38 ppt. Salinities may go above or below that range for short periods without causing losses of larger seed or adult clams, but larvae and small clams are vulnerable beyond this range.

Clams grow intertidally and subtidally in virtually any bottom where they can burrow. They do well in soft mud, sand, shell, and rubble. Growth is maximum in coarse sand and inhibited in areas with high silt clay content. Shell or rubble offers some predator protection, and clams often are more abundant in these and other coarse substrates. Where currents run one knot or less, clams grow very well. Because they don't require a firm bottom, clams do not compete for oyster growing grounds.

Stocking Rate

They can be grown at relatively high densities of 1 to 2 million per acre. Clams reach littleneck size throughout most of their range.

Feeding

Clams are filter feeders, feeding on the phytoplankton in their environment.

Diseases

Bacteria are considered responsible for culture failures. The clam leech (*Malacobdella grossa*), other parasites, and commensals are found in clams.

Harvesting and Yields

Harvesting methods in use include rakes, hoes, patent tongs, modified oyster dredges, modified surf clam dredges—hydraulic dredges attached to cages—and hydraulic dredges attached to conveyors. Legal constraints usually dictate the type of harvesting dones in a given area. Although there is some justification for restricting methods of harvesting wild stocks, this is counterproductive when applied to aquaculture. No unreasonable restrictions should be placed on gathering cultured clams, provided the method used does not adversely affect the environment.

Processing and Marketing

Clams are usually sold in the shell, eliminating the need for packaging or processing. If the harvest increases because of aquaculture of larger numbers of individuals, some clams probably will be processed. Hard clams usually are sold in the shell, either by the bushel or individual count. Prices vary—smaller sizes or grades (littlenecks) have a higher value than larger clams. Most littleneck clams are consumed raw, steamed, or as specialty items, such as clam casino or deviled clams.

FIGURE 7-15 Mussels.

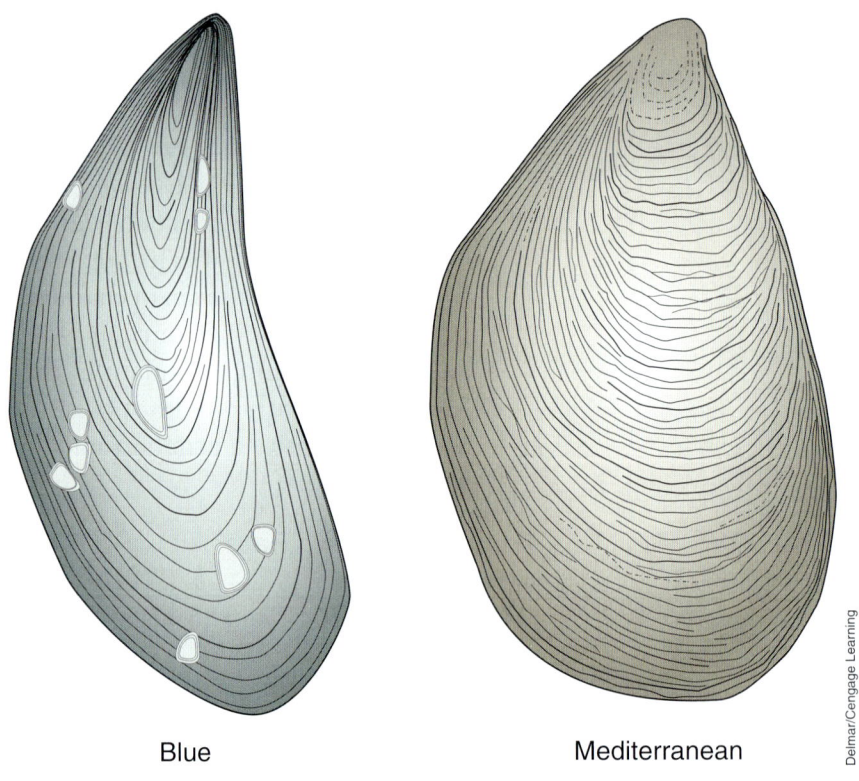

Blue   Mediterranean

Mussels

In the United States, the blue mussel (Mytilus edulis) (Figure 7-15) is the only indigenous mussel species cultured commercially. Mussels are commonly cultured in other parts of the world. Most of the mussels cultured in the United States are grown in suspended culture systems such as longlines or rafts. Bottom culture is also profitable.

Mussels are easy to raise, requiring little attention after being placed in the growing area. In the United States, the market for mussels is increasing. Most mussels are being cultured in the Northeast Atlantic states.

Seed stock for culture is obtained from wild stock. Mussels attach to almost any structure by means of a threadlike **byssus**, which they secrete. Unlike oysters, mussels can discard their byssus and move to a new location.

With good management and a good location, mussels will grow to market size of 2 to 3 in within one or two years, depending on the size of the seed stock and the climate.

Snails

Heliciculture is the process of farming or raising snails. Snail farming on a large-scale basis requires a considerable investment in time, equipment, and resources. Prospective snail farmers should carefully consider these factors, especially if their goal is to supply large quantities to commercial businesses. If you wish to raise snails, you should expect to experiment until you find what works best in your specific situation. Expect a few problems.

FIGURE 7-16 Snail.

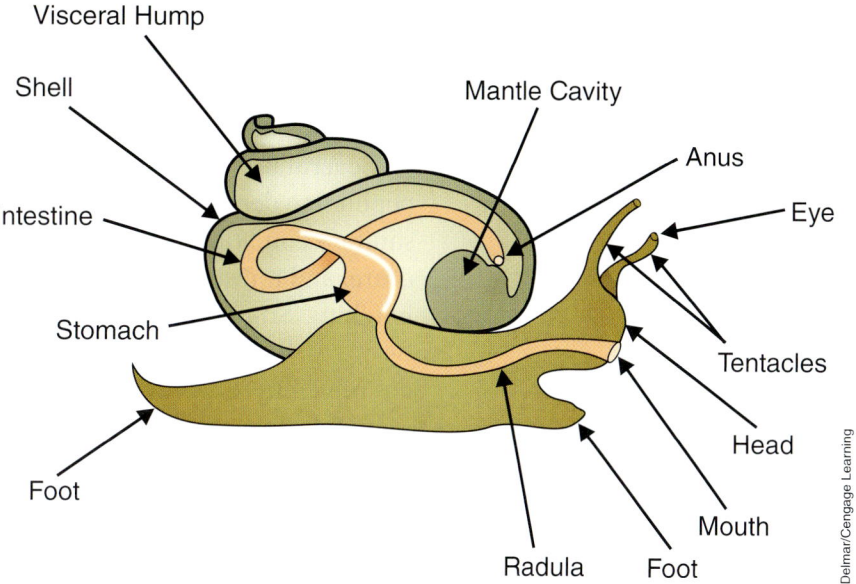

Whereas many different methods exist for rearing snails, reproduction and nursery take place indoors in climate-controlled areas, and growing takes place outdoors in pens.

Many species of snails can be farmed. The typical French escargot are *Helix aspersa aspersa*, the Petit-Gris, (about 10 g) or *Helix aspersa maxima*, the Gros-Gris, (about 20 g). Other species include: the apple snail (*Helix pomatia*), the vineyard or Spanish snail (*Otala lactea* or *Helix lactea*), the wood snail (*Cepaea nemoralis* or *Helix nemoralis*), the garden snail (*Cepaea hortensis* or *Helix hortensis*), and the banded snail (*Theba pisana*). Figure 7-16 shows the basic anatomy of a typical snail.

The production cycle in the Northern Hemisphere takes place from March to October in areas with a cold winter and a moderate, dry summer. In the Southern Hemisphere, the cycle is from September to May for countries with a wet and cool winter and a very hot and dry summer, for example, some Mediterranean regions.

Snails are fed a finely ground dry meal with a formula similar to the following:

- calcium carbonate 30 percent
- calcium phosphate 4 percent
- soybean meal 20 percent
- sunflower seeds 5 percent
- wheat flour 40 percent
- vitamin mix 1 percent

Breeding snails are put in cages at a density of 100 to 200 per yd.^{2} (or about 4 lbs. per square yd.), with food and water free choice, and some pots that are filled with a good soil—snails lay their eggs in soil. Ideally, these boxes are placed in a room that is kept at 68°F with a relative humidity of 95 percent, and a long photoperiod (16 hours light and 8 hours dark). Snails are hermaphrodites, but they have to mate before laying eggs.

When a snail lays eggs, the pot of soil is placed into an incubator at 68°F and covered. Snails hatch in about 3 weeks. Each breeder produces 70 to 100 young. These young are maintained in a nursery for 3 to 4 weeks before being placed in outdoor pens.

Several factors can greatly influence the growth of snails, including: population density, stress (snails are sensitive to noise, light, vibration, unsanitary conditions, irregular feedings, being touched, etc.), feed, temperature, and moisture. Snails should reach adulthood 10 to 15 weeks after hatching. Snails are mature when a lip forms at the opening of their shell. Before they mature, their shells are more easily broken, making them undesirable. After this time, snails are harvested, but the best specimens are saved for the next breeding season. Future breeders are placed in hibernation, in wooden boxes, well dried, and out of frost.

Parasites, nematodes, trematodes, fungi, and microarthropods may attack snails, and such problems can spread rapidly when snail populations are dense. The bacterium Pseudomonas aeruginosa causes intestinal infections that can spread rapidly in a crowded snail pen.

Predators include rats, mice, moles, skunks, weasels, birds, frogs and toads, lizards, walking insects (like some beetle and cricket varieties), some types of flies, centipedes, and even certain cannibalistic snail varieties such as *Strangesta capillacea*.

For market, snails are washed, steamed, shelled, then washed in a solution of vinegar (or lemon juice) and water before they are canned. To prepare live snails (escargot) for cooking, the membrane over the shell opening is removed. Snails are soaked in enough water to cover them. Salt and vinegar are added. Mucus turns the water white, so the water is changed several times during the three- to four-hour soaking. After rinsing several times, the snails are placed in cold water, boiled about eight minutes, and then drained and plunged into cold water.

A complete document on raising snails can be found on the National Agricultural Library Web site: http://www.nal.usda.gov/afsic/AFSIC_pubs/srb96-05.htm. The title of the document is simply "Raising Snails." It was developed for the Alternative Farming Systems Information Center.

Abalone

The largest and most commercially important species of abalone, the red abalone (*Haliotis rufescens*), occurs mostly in California waters. Fishermen harvest less than 1 million lbs annually from the wild, but demand and prices are high and are increasing.

Japanese scientists have developed hatchery-culture procedures for abalone, and Japan has large stocking programs. Wild and farmed abalone from Mexico, Australia, New Zealand, Taiwan, Japan, and China fulfill most of the demand for abalone throughout the world. In addition, a U.S. abalone aquaculture industry in central and northern California has been in production since the mid-1980s. Current production is minimal, with most of the product exported to Asia. Abalone farmers in California have faced significant regulations, disease and pest problems (sabellid worm and withering foot syndrome), and conflicts with interest groups such as kelp habitat conservationists and local fishermen.

Other problems associated with abalone culture include: slow growth rate, high post-larval mortality, design of tank-culture systems, protection from predation in open coastal waters, cost-effective feeds and feeding systems, and legal aspects of obtaining adequate space for production facilities. The time required to reach market size may be three or more years, depending upon species, water temperatures, and feeding rates. Research is needed on the physiology, nutrition, and genetics of abalone.

OTHER COMMERCIAL SPECIES

Depending on the market, location, and the entrepreneurship of an individual, many other species are or could be cultured. Some species require more research. Some species need markets developed for them. For some names of other species that hold potential, refer to Chapter 2, Aquatic Plants and Animals.

SUMMARY

Crawfish, freshwater prawns, shrimp, oysters, clams, snails and abalone are all actively cultured. Successful management of the culture of crustaceans and mollusks requires a thorough knowledge of the species being cultured. This includes where to obtain seed stock, the life cycle and biology, how to culture, stocking rates, feeding, potential causes of disease, harvesting, and processing. Culturing is a continual learning process. As observations are made, changes can be initiated and the culture method improved.

STUDY/REVIEW

Success in any career requires knowledge. Test your knowledge of this chapter by answering these questions or solving these problems.

True or False

1. The difference between soft-shelled and hard-shelled crawfish is management.
2. Shrimp eat other shrimp.
3. Crab meat is not high priced.
4. Snail reproduction takes place outdoors.
5. Abalone growth is very slow.
6. Crawfish have few disease problems.

Short Answer

1. Name the most commonly cultured crustacean in the United States.
2. What type of climate is necessary for a successful shrimp or prawn culture?
3. Name two markets for crawfish.
4. Shrimp can regenerate eggs and spawn again in about _____ days.
5. Oysters are _____ feeders.
6. Mussels attach to any structure by a threadlike _____ , which they secrete.
7. Another name for live snails is _____ .
8. What are the three phases of culture technology for prawns?
9. Spat is another term for a young _____ .

Essay

1. Explain why lobster culture from egg to adult does not occur.
2. Give some indication of the worldwide distribution of crawfish and shrimp.
3. What is molting?
4. Why is clamming a part-time employment?
5. Define heliciculture.
6. Describe the difference between extensive and intensive hatchery systems for saltwater shrimp.
7. What is the reason for the decline in domestic oyster production?

KNOWLEDGE APPLIED

1. Conduct a taste test of some different crustaceans and mollusks. Allow individuals to consume them cooked plain, without being a part of some recipe. Then find recipes for the preparation of these crustaceans and mollusks.

2. Using the information in this chapter and information collected from Internet searches, construct a specie profile sheet for crustaceans and mollusks. The specie profile sheet should include the following information: common name, scientific name, brief description, habitat, distribution, reproduction, special features, culture problems, potential yield, feeds, and marketing potential.

3. Conduct a survey and find out how many people actually eat crustaceans or mollusks, which ones, and how often. Compile this information and report on your findings.

4. Raise some snails in an aquarium and record your observations for a month and track the water quality during this time.

5. Search the Internet for successful changes being made in culture techniques for crustaceans or mollusks, or search for new species that are being cultured. Report on your findings.

LEARNING/TEACHING AIDS

Books

Dillon, R T, Jr. (2000). *The ecology of freshwater mollusks.* New York, NY: Cambridge University Press.

Dore, I., and Frimodt, C. (1997). *An illustrated guide to shrimp of the world.* Long Island City, NY: Osprey Publishing Inc.

Fallu, R. (1991). *Abalone farming.* Boston, MA: Blackwell Science, Inc.

Fast, A. W., and Lester, L. J. (1992). *Marine shrimp culture: Principles and practices.* Amsterdam and New York: Elsevier Publishing.

Gosling, Elizabeth, E. (2003). *Bivalve mollusks.* Malden, MA: Fishing News Books.

Groves, Roy E. (1991). *The crayfish: Its nature and nurture.* Boston, MA: Blackwell Science, Inc.

Leung, P and C R Engle. (2006). *Shrimp culture: Economics, market and trade.* Ames, IA: Blackwell Publishing.

Lee, E., and Wickens, J. (1992). *Crustacean farming.* New York: Halstead Press.

Manzi, J., and Castagna, M. (1989). *Clam mariculture in North America.* Amsterdam and New York: Elsevier Publishing.

Mattiessen, G. (2001). *Oyster culture.* Malden, MA: Fishing News Books.

Phillips, B. Eds. (2006). *Lobsters: Biology, management, aquaculture, and fisheries.* New York, NY: Wiley-Blackwell.

Shumway, S. E. (1991). *Scallops: Biology, ecology, and aquaculture.* Amsterdam and New York: Elsevier Publishing.

Stevens, M.M. (2003). *Cultured Abalone.* Seafood Watch, Seafood Report, Monterey Bay Aquarium (www.montereybayaquarium.org/.../MBA_SeafoodWatch_AbaloneFarmedReport.pdf)

Wichins, J. F., and Lee, D. O'C. (2002). *Crustacean farming: Ranching and culture, 2nd Ed.* Ames, IA: Iowa State University Press.

Internet

Internet sites represent a vast resource of information. The URLs (uniform resource locator) for the World Wide Web sites can change. Using a search engines such as Google, find more information by searching for these words or phrases: crustaceans, mollusks, crawfish (crayfish), shrimp, clams, oysters, molting, prawns, dredge, metamorphosis, crabs, lobsters, filter feeders, mussels, abalone, snails, or heliciculture.

For some specific Internet sites refer to Appendix Tables A-11, A-12 and A-14.

Depending on the market, location, and an individual's entrepreneurial spirit, many aquatic species can be cultured. An entrepreneur is always looking for new species and new culture methods. Many hold potential. The culture of alligators has removed them from endangered or threatened species status. Frog culture can be used to supply biological specimens or the delicacy of frog legs. Aquatic plants are more important in other parts of the world, but they could become important to the United States.

CHAPTER 8

Management Practices for Alligators, Frogs, and Plants

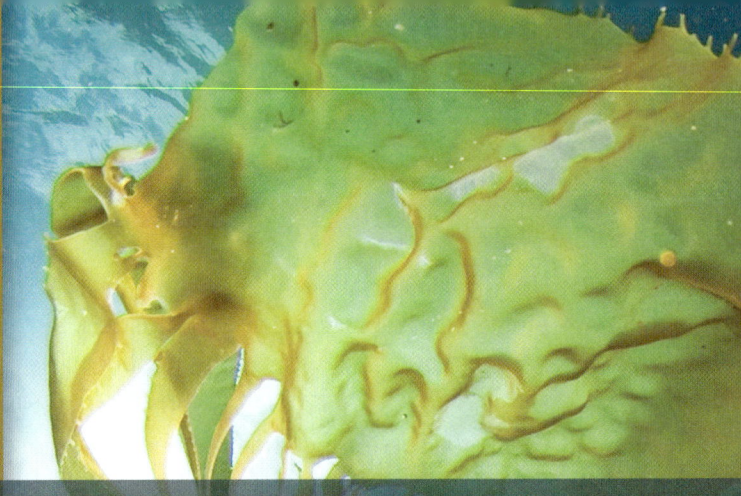

OBJECTIVES

After completing this chapter, the student will be able to:

- List problems of alligator culture
- Describe the uses of alligator products
- Identify the requirements for alligator culture
- Describe the dietary needs of alligators
- List problems of frog culture
- Describe the requirements for frog culture
- Compare the dietary needs of tadpoles to those of frogs
- Discuss the potential for aquatic plant culture
- Define plant aquaculture
- Describe economic and production considerations of plant aquaculture
- Explain the role of temperature and food supply in reptile and amphibian development

- Describe the amphibian life cycle
- Discuss the nesting behavior of alligators
- Identify disease-causing situations in reptile and amphibian culture
- Give the scientific names for alligators and bullfrogs
- Explain nutrient uptake and photosynthesis in aquatic plants
- List four reproductive methods used by aquatic plants

Understanding this chapter will be enhanced if the following terms are known. Many are defined in the text, and others are defined in the glossary.

KEY TERMS

Amphibian	Macrocystis
Bermed	Monospores
Carrion	Nutria
Clutch	Polliwogs
Fertility	Sporophytes
Hatchling	Tadpole
Hide	Territorial
Incubation	

ALLIGATORS

American alligators have been hunted for centuries in the southeastern region of North America, first by native peoples and subsequently by European settlers. Alligator harvesting has included a commercial component since the late 1800s. By the 1950s and early 1960s, alligators were widely harvested throughout the southeast, with no regulations to manage or protect their populations. Being rapidly overexploited led to serious reductions in breeding alligator populations, and most harvests were banned by the late 1960s. After protected by the 1973 Endangered Species Act, the alligator population recovered dramatically. As a result of this recovery, many states reopened limited alligator harvesting and have adopted conservative management of their alligator populations.

Production of alligators in a controlled indoor environment is relatively new. Advances in environmentally controlled production methods during the 1980s improved survival rates, allowed sex determination, and produced market-size alligators (4 ft) in less than two years. These advances, together with the high prices paid for alligator hides, led to rapid expansion of the industry in the 1980s. Production of farmed American alligators has expanded significantly over the past two decades. Currently, farms in Louisiana (14), Florida (11), Texas (3), and Georgia (2) sell alligator hides (Census of Aquaculture 2005). This number is down from in 1991. These trends illustrate the shifting of production and marketing to fewer, larger operations.

Sources of Species

The American alligator (*Alligator mississippiensis*) is a member of the order Crocodilia. This order includes alligators, crocodiles, and caimans (several species of reptiles related to alligators and found in Central and South America). The American alligator is native to the coastal plain and lowland river bottoms from North Carolina to Mexico. (See Figure 8-1). The American alligator can grow to 16 ft or more in length. The only other species of alligator is found in China.

Crocodile or caiman farming also is being attempted on about 600 farms in 47 countries around the globe. The total worldwide population of crocodilians on farms is more than one million.

FIGURE 8-1 Alligators are large reptiles native to North America from North Carolina to Mexico.

Habitat

Alligators inhabit fresh to slightly brackish aquatic habitats. These habitats include marshes, swamps, creeks, rivers, lakes, and ponds.

Seed Stock and Breeding

Alligator courtship and breeding correlate with air temperature and can occur between April and July depending on weather conditions. Courtship and breeding occur in open water at least 6 ft deep. Courtship behavior includes vigorous swimming and bellowing activity. Both males and females bellow, but the male bellow is much more bass and vocal than that of the female.

Adult alligators reared entirely in captivity/confinement behave very differently from wild stock. Farm-raised alligators accept confinement and crowding as adults much better than those captured from the wild. Also, adult alligators that have been raised together tend to develop a social structure, and they adapt more quickly and breed more consistently than animals lacking any established social structure.

In Louisiana, females in the wild usually reach sexual maturity at a length of 6.5 ft and at an age of 9 to 10 years. The alligator, like all reptiles, is a cold-blooded animal, and the difference in age to reach maturity is related to temperature-dependent growth. Optimum growth occurs at temperatures between 85° and 91°F. (See Figure 8-2.) No apparent growth occurs at temperatures below 70°F, and temperatures about 93°F can cause severe metabolic stress or death.

Captive alligators raised in temperature-controlled environments for 3 years reach sexual maturity at 5 to 6 years of age. Inbreeding can be eliminated by obtaining males and females from different clutches (eggs laid at one time).

FIGURE 8-2 Alligators require temperatures between 85° and 91°F in commercial grow-out facilities.

Courtship and breeding activity is extremely important. Courtship and breeding occur between April and July, depending on temperature/weather conditions. Courtship is a time of heightened activity by both sexes. This activity includes vigorous swimming and bellowing.

Nest building and egg laying occur at night. The nest is built from natural vegetation (e.g., broomsedge, bullwhip, cutgrass, wiregrass, and available annuals), hay (if provided), and soil. Nests are built into a round, mound-type structure. A female may start several nests before a single nest is successfully completed. Eggs are deposited at the top of the mound and then sink to the center, forming layers with vegetation. Finally, the eggs are covered with approximately one foot of vegetation. All nesting activity usually occurs within a two-week period.

Clutch size varies with age and condition of the female. Large and older females generally lay more eggs. Clutch size should average 35–40 eggs (range 2–58). Egg **fertility** can vary from 70 to 95 percent. Embryo survival can also vary from 70 to 95 percent. Hatching rate varies from 50 to 90 percent. Reproductive success of captive alligators has been the most difficult problem for producers to overcome.

Alligator embryos are very sensitive to handling (mechanical injury) from 7 to 28 days after they are laid. Many embryos will die if handled during this period. Current recommendations are either to collect eggs within the first week or wait until the fourth week of natural incubation.

Most alligator eggs cannot be turned or repositioned when taken from the nest (unlike bird eggs). Eggs should be marked with an "X" across the top before removing them from the nest, so that they can be maintained during transport and incubation in the same position as they were laid. Eggs that are laid upright in the nest (long axis perpendicular to the

THE TADPOLE: A CHANGELING

A frog seems nothing like a tadpole, just as a caterpillar seems nothing like a butterfly. The change from tadpole to frog is nothing short of a miracle. Tadpoles live in the water. They respire (breathe) by means of gills. They do not have arms or legs but they have dorsal or fin-like appendages and a tail with which they swim by lateral undulation, similar to most fishes. As a tadpole matures, it goes through metamorphosis by gradually growing limbs. Next, the tail is absorbed by apoptosis (a form of cell death necessary to make way for new cells). Lungs develop around the time of leg development, and tadpoles late in development will often be found near the surface of the water, where they breathe air. During the final stages of external metamorphosis, the tadpole's mouth changes from a small, enclosed mouth at the front of the head to a large mouth the same width as the head. The intestines shorten to make way for their new diet from herbivore to carnivore. The miracle is complete and the new animal carries on to start the cycle again.

ground) will die unless repositioned correctly (long axis parallel or lying on its side and not on end) before artificial incubation. This repositioning can only take place in the first few hours after nesting.

Artificial **incubation**, compared to wild nesting, improves hatching rates because of elimination of predation and weather-related mortality. The best hatching rates for eggs left in the wild is less than 70 percent. Hatching rates for eggs taken from the wild and incubated artificially average 90 percent or higher.

Hatchling alligators make peeping or chirping sounds after hatching. When 10–15 percent of the clutch has hatched, many producers carefully open unhatched eggs.

Culture Method

After 24 hours, the hatchlings are removed from the egg baskets, sorted into uniform-size groups, and moved into environmentally controlled grow-out facilities. Closely sizing the alligators is very important. Smaller, weaker individuals will not be able to compete with their larger siblings.

Hatchlings can be moved into small tanks, 2 ft by 2 ft or larger, that are heated to 86°–89°F. Maintaining hatchlings at 89°F for the first week helps increase yolk absorption. Usually the hatchlings will start to feed within three days.

Grow-out buildings are heavily insulated concrete block, wood, or metal buildings with heated foundations. The foundation is a concrete slab laced with hot-water piping or, less commonly, electric heating coils. A constant temperature is maintained in the building by pumping hot water through the pipes. The slab is poured over insulation board to reduce heat loss. Some grow-out houses are earth-**bermed** to reduce further heat loss. (See Figure 8-3.)

FIGURE 8-3 Alligator grow-out facility built into the ground to conserve heat.

Growth rates of young alligators can be as much as 3 in or greater per month when they are held at a constant temperature of 86°–89°F, fed a quality diet, and protected from stress. Many producers have been rearing alligators from hatchlings to 4 ft in 14 months.

Stocking Rate

Pens can be constructed in almost any size. In general, smaller pens are used for rearing small alligators, and, as the alligators grow, pens become progressively larger. Table 6-1 gives examples of pen size to alligator size and corresponding densities.

A commonly used stocking regime is:
1 sq ft per animal until 2 ft in length
3 sq ft per animal until 4 ft in length
6 sq ft per animal to 6 ft in length

TABLE 8-1 RECOMMENDED PEN SIZES FOR GROW-OUT OPERATIONS[1]				
Gator Length	Pen Size Sq. ft. (l × w)[2]	Gators/Pen	Sq. ft./Gator	Sq. ft. Needed 350 Gators
7–17"	9 (3 × 3)	20	0.45	158
15–30"	120 (10 × 20)	80	1.50	525
30"–4'	168 (12 × 14)	50	3.36	1,176
4–5'	192 (12 × 16)	50	3.84	1,344
5–6'	216 (12 × 18)	40	5.40	1,890

[1]Southern Regional Aquaculture Center (1993)
[2]Length times width

Feeding

Alligators eat almost anything. Research shows that young alligators primarily consume invertebrates like crayfish and insects. As they grow, fish are included in their diet. For adult alligators, mammals such as muskrats and **nutria** become more important. Larger alligators even consume birds and other reptiles (including smaller alligators). **Carrion** is consumed whenever available. In general, the alligator's natural diet is very high in protein and low in fat.

Most alligator farms used to be equipped with large walk-in freezers to store large quantities of meat. Meat sources used in the past included nutria, beef, horse, chicken, muskrat, fish, beaver, and deer. Today, however, artificial diets have been developed that provide adequate nutrition. These diets have eliminated much of the need to keep fresh-frozen meat products available.

Commercial feeds are approximately 45 percent crude protein and 8 percent fat. These feeds are a blend of fish meal, meat and bone meal, blood meal, and some vegetable protein. These feeds are also fortified with vitamins and minerals.

Diseases

Stress and/or poor water management may lead to brown spot disease. Although sores will heal, the spots are detectable and reduce the value of the skin.

Alligators are wild creatures that have been thrust into captivity. In the wild, alligators are relatively shy and reclusive creatures that do not normally congregate together except during the breeding season. Cultural conditions imposed upon them are unnatural and stressful.

Alligators crowded into pens appear to be very sensitive to light and sound. Many producers like to keep alligators in the dark, or with very reduced light conditions. They try to locate and insulate facilities to minimize external noise. Some producers, however, put lights on timers to simulate natural conditions and place radios in the grow-out houses, believing that the animals will grow accustomed to human voices and not be as stressed by daily feeding and cleaning routines.

Two antibiotics, oxytetracycline (OTC) and virginiamycin (VA) have been added to feed when bacterial problems occur. These antibiotics and any others can only be administered to alligators through a prescription from a veterinarian.

Harvesting and Marketing

Alligators are harvested when they are 4–6 ft long. All alligators must be tagged with tags from the state regulatory agency immediately after slaughter. Alligators can be skinned only at approved sites and by using specific skinning instructions issued by the state agency. Skinning, scraping, and curing must be done carefully to assure quality. Hides that are cut, scratched, or stretched, particularly the belly scales, have reduced value.

TABLE 8-2 PERCENT YIELD OF DEBONED ALLIGATOR MEAT ON A LIVE-WEIGHT BASIS[1]				
Tail	Leg	Torso	Ribs[2]	Jaw
16–17	4–5	6–12	7–10	1

[1]Southern Regional Aquaculture Center (1993)
[2]Ribs with bones

Producers processing alligator meat must comply with all sanitation requirements of federal, state, and local authorities. County or parish health departments can supply guidelines and assistance in complying with sanitation standards. Specific state laws regulate the size of meat cartons (for example, not larger than 5 lbs.), labeling of the cartons with the names of the seller and buyer, date of sale, and tag number that corresponds to the **hide**. Average deboned dress-out percentages of meat from alligators in the 4- to 6-ft range are given in Table 8-2. Prices for hides vary and are low, but the prices for meat remain relatively constant.

FROGS

Frogs are the best known of the **amphibian** family, and frog meat is one of North America's native aquatic gourmet foods. The most common frog in the United States is the leopard frog. The bullfrog (*Rana catesbeiana*) is so named because its sound resembles the distant bellowing of a bull. It is the most aquatic of all frogs (Figure 8-4).

The demand for frogs is created by their use as laboratory animals and their use as food. Also, supplies are reduced as prime frog habitats disappear.

FIGURE 8-4 Bullfrogs have been introduced into many areas of the United States.

Sources of Species

Because of its size and popularity as food, the bullfrog has been introduced outside its normal range. It is the most important of the frogs as a commercial food animal. Adult bullfrogs may measure 8 in long and weigh as much as 2 lbs.

Habitat

Bullfrogs are native to most of the United States and southern Canada east of the Rocky Mountains, and they have been introduced outside that range, especially in California. Bullfrogs live and forage mainly along the shorelines of ponds and hide in deep water when frightened.

Seed Stock and Breeding

Tadpole ponds can be stocked by introducing egg masses gathered from frog ponds. Frog ponds can be stocked with wild-caught or commercial frogs.

Bullfrogs lay their eggs in shallow water during March and June, depending on the latitude. Egg masses of bullfrogs are much larger than those of other frog species. Eggs hatch in four days to three weeks depending on the temperature (Figure 8-5). Water temperature is usually

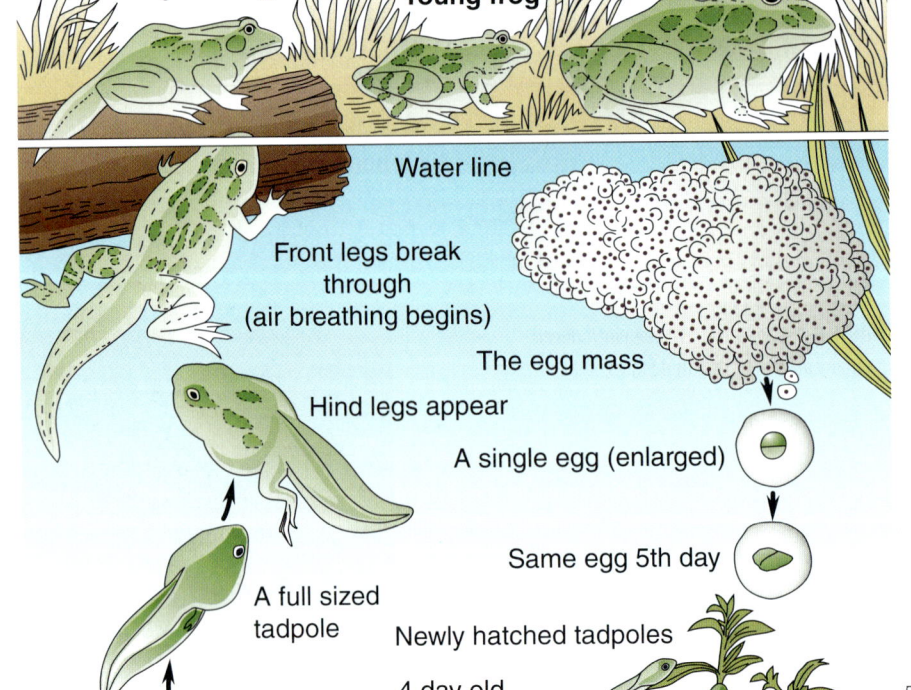

FIGURE 8-5 Life cycle of the frog.

not critical as long as it is reasonable (40°–80°F). Eggs in cold water take longer to hatch because the speed of the development of young frogs is directly related to the amount of heat they accumulate—sometimes called "accumulated temperature units."

Larval frogs or **tadpoles** (**polliwogs**) will eat almost any soft organic matter. Under culture conditions, tadpoles need constant aeration.

Culture Method

Metamorphosis from tadpole to frog (Figure 8-5) can take from five months to two years, depending on the temperature and food supply. The adult bullfrog is carnivorous and will accept only live, moving food. Frogs grow slowly, depending on the food supply and length of growing season. To reach a marketable weight of ¼ to ½ lb may require a year or more.

Some of the culture problems encountered by producers include:

- **territorial** nature of bullfrogs
- refusal to eat anything but live, moving food
- predators of frogs and tadpoles such as humans, fish, bird, cats, alligators, crawfish, and other frogs

Natural ponds or ponds constructed for fish culture, recreation, and irrigation will not support dense frog populations. For maximum production, special ponds should be constructed—one for tadpoles and one for frogs. Obviously, optimum culture conditions differ for tadpoles from that of frogs.

Frog ponds are shallow and do not need to be as deep as fish ponds, but they do need to provide food, shelter, and a place for hibernation (Figure 8-6). Tadpole ponds should have high algal production. Fencing the pond's sides and top will help eliminate predators such as birds and humans. Carnivorous fish should also be eliminated from ponds.

FIGURE 8-6 Underground frog pens supplied with geothermal water and used for culture of frogs.

Stocking Rate

Amphibians tend to release hormones that inhibit the growth of others in the tank. This results in pinhead-sized tadpoles sharing the same space with larger emergent frogs. To prevent this, producers should not overcrowd and should use the standard aquarium practice of one inch of fish per gallon of water.

Feeding

Tadpoles are fed indirectly by fertilizing the pond water to promote the growth of algae or fed directly by feeding any soft organic matter.

Adult frogs seem to insist on moving food. Some producers culture or mass harvest earthworms, flies, crickets, and other living frogs for food. Other producers create an environment around the pond that attracts living food sources.

Recently, some producers have been successful at getting frogs to accept an artificial diet. This could spark the development of more frog farming.

Diseases

Handling frogs during hot weather increases the chance of disease. Intensive frog culture will naturally create other disease problems not yet known. For now, the producer must rely on regular inspection, disinfecting of facilities, and culling of sick frogs.

Harvesting

Cultured frogs are harvested the same as those in the wild. Frogs can be caught by hand, with a net, a hook, or a spear. Night harvesting with a bright light can be effective.

Processing and Marketing

Whole live frogs are sold for laboratory research. Dressing the hind legs is done by simply snipping them off close to the body. Obtaining the other meat on the carcass is a labor-intensive process, so almost half the meat is wasted or becomes a by-product.

Live frogs are marketed to university research laboratories. Frog legs are marketed to restaurants.

AQUATIC PLANTS

Table 2-1 (page 30) provides a good overview of the aquatic plants. Three areas that are not discussed in this chapter are rice production, hydroponics, and aquaculture/irrigation systems. Rice is generally considered an agronomic crop that grows in water.

Hydroponics is generally considered a horticultural enterprise, with vegetables as the primary crop. Aquaculture/irrigation systems use nutrient-rich water from aquaculture systems to irrigate agronomic crops. This system gets two uses from the scarce water supplies.

Plant aquaculture is a very young enterprise in the United States. With the exception of some coastal states and Hawaii, little plant aquaculture can be found in this country. Plant aquaculture should increase as this potential is further developed.

Plant aquaculture is the culture of aquatic plants as food for humans or for other special purposes. Plant aquaculture is much more prevalent in other countries than in the United States.

Production may be in a monoculture or polyculture (with other plants or animals) system. Plant aquaculture is more prevalent in other countries for the following reasons:

- Other countries have been practicing aquaculture, in general, much longer
- Population density is much larger, with less area per capita for traditional agriculture
- A higher percentage of people live close to coastal areas
- People in these countries have established a taste for plant aquacrops over the years

Other types of production include phycocolloids, ornamental uses, animal feeds, mulches and fertilizers, and wastewater treatment.

Marine Plant Aquacrops

Commonly referred to as seaweeds, aquaculture contributes a significant percentage to the worldwide harvest of these attached algal forms. Seaweeds (Figure 8-7) have been harvested for food, fertilizer, and medicine for thousands of years. The Chinese used seaweed for medicinal purposes as early as 3000 B.C. One of the earliest records, the Chinese *Book of Poetry*, indicates that sea vegetables were considered a delicacy as far back as the time of Confucius. In Iceland, seaweed has been eaten for centuries; the oldest law book refers to the "rights and concessions involved before one might collect and/or eat fresh sol (*Palmaria palmata*) on a neighbor's land." Ancient Hawaiian nobility also kept limu (edible algae) gardens, where rare and choice varieties of seaweeds were cultivated to provide gourmet food for the royal family.

Other cultures have used seaweed for fertilizer and fodder. *Bellum Africanum*, written in 46 B.C., states, "The Greeks collected seaweed from the shore and having washed it in fresh water, gave it to their cattle." Throughout Europe and Great Britain, seaweed has been used for many years to replenish the soil and promote plant growth. In northern New England, rockweeds (*Ascophyllum* spp. and *Fucus* spp.) have been collected for use as fertilizer since colonial times. *Ascophyllum nodosum* is used as packing material for shipping live lobsters and clams, and for transporting sandworms in the marine bait industry. Dulse (*Palmariapalmata*) and Irish moss (*Chondrus crispus*) formed the nucleus of a cottage industry in which harvesters collect, dry, and sell the seaweeds for food and industrial use.

Seaweeds have many other important but low-volume uses. Because they concentrate trace elements, seaweeds historically have been a source of iodine, potash, and other minerals used in industry and medicine.

FIGURE 8-7 Seaweed.

A number have been used for drugs, including anticoagulants, antibiotics, anthelminthics (worms), antihypertensive (high blood pressure) agents, reducers of blood cholesterol, dilatory agents, and insecticides.

In the global market, seaweed is sold primarily for food. In the past decade, interest has increased in using seaweed as a health food. One seaweed used for food in the rapidly expanding Japanese cuisine market is nori or Porphyra tenera. China is the largest producer of nori, followed by Japan and the Republic of Korea.

Uses of Seaweeds

Japanese seaweed foods include kombu (Laminaria), aonori, nori (Porphyra), and wakame (Undaria). A Chinese seaweed food is called hai dai (Laminaria). Various Hawaiian seaweed foods, called "limu," are from different algal species (*Asparagopsis taxiformis, Codium, Grateloupia filicina, Ulva*). *Rhodymenia palmata* is used for a Scottish seaweed food called "dulse," or, in Irish, "dillisk," or "sol" in Iceland. Irish moss or carrageen (*Chondrus crispus*) is consumed in Europe.

Alginic acid is extracted from brown algae (Phaeophyta) and kelps (Macrocystis, Laminaria, Ascophyllum, Fucus, and Sargassum). Alginic acid (alginate) is a colloidal product used for thickening, suspending, stabilizing, emulsifying, gel-forming, or film-forming. About half of the alginate produced is used for making ice cream and other dairy products, the rest is used in other products, including shaving cream, rubber, or paint. In textiles, alginates are used to thicken fiber-reactive dye pastes, which facilitates sharpness in printed lines and conserves dyes. Dentists use alginates to make dental impressions of teeth.

Carrageen is extracted from red algaes (Rhodophyta) such as *Gigartina stellata, Chondrus crispus, and Eucheuma*. Carrageen (carrageenin, carragheen) is similar to agar, but requires higher concentrations to form gels. Carrageen is used for stabilizing in foods such as chocolate, milk, eggnog, ice cream, sherbets, instant puddings, frostings, and creamed soups.

Red algae (*Gelidium, Gracilaria, Pterocladia, Ahnfeltia*) are also a source of agar. Agar is another colloidal agent used for thickening, suspending, and stabilizing. It is best known for its unique ability to form thermally reversible gels at low temperatures. The most frequent use of agar is in food preparation and in the pharmaceutical industry. It is used as a laxative, or as an inert carrier for drug products where slow release of the drug is required. Agar is used in bacteriology and mycology (study of fungus) as a stiffening agent in growth media.

Agar is also used as a stabilizer for emulsions and as a constituent of cosmetic skin preparations, ointments, and lotions. It is used in photographic film, shoe polish, dental impression molds, shaving soaps, hand lotions, and in the tanning industry.

In food, agar is used as a substitute for gelatin, as an anti-drying agent in breads and pastry, and also for gelling and thickening purposes. It is used in the manufacture of processed cheese, mayonnaise, puddings, creams, jellies, and frozen dairy products.

Other major uses of seaweeds include as fertilizers or soil amendments and as filters in sewage treatment to remove inorganic nutrients and toxins.

Growing Seaweeds through Aquaculture

As the market for seaweed grows, many people are concerned about the effects of over-harvesting. Removing large quantities of seaweed from a rocky shore can upset the balance of plant and animal communities living there. It can also cause more rapid shore erosion.

Culturing economically important seaweeds has many advantages over harvesting wild stocks. When seaweed is cultured in suspended structures away from rocky substrata, an existing ecosystem is not substantially altered. In addition, harvesting cultured seaweed is many times simpler and more efficient. If seaweeds are cultured on nets or ropes strung horizontally on the surface of protected bays, harvesters do not have to wait for the proper tides and risk dangerous wave action to gather the crop. For example, commercial-scale culture in the United States involves seeding and transplanting the giant brown alga, **Macrocystis** (known as "kelp") along the California coast.

Given that the demand for marine algae is growing steadily, and with a finite natural supply, commercial culture will become a significant industry. Vast areas of the oceans may be devoted to algae farms. Problems of culture include predation and undesirable growth of competitive seaweeds with low commercial value.

Freshwater Plant Aquacrops

Freshwater plants, such as water chestnut and watercress, are grown widely for food. Water hyacinth, duckweed, and water spinach serve as feed and fuel. Use of freshwater plants such as water hyacinth in recycling wastes holds great potential, and investigations into this topic have just begun. Field testing of, and applied research on, plants in polyculture and waste recycling systems are needed. Cattail, arrowhead, and other deep-water plants are used as ornamental plants (Figure 8-8).

FIGURE 8-8 Deep water plants such as floating heart can be used as ornamentals.

Single-cell Algae

Culturists raise unicellular algae on both small and large scales as food for larval fish, oysters, and other invertebrates. Some species have value as human food. Spirulina is grown for this purpose in Mexico and Chad. Pigments, glycerol, and other hydrocarbons can be produced from unicellular algae, and pilot-scale production of several species is underway in several countries. Blue-green algae have nitrogen-fixing abilities similar to leguminous plants, and researchers are experimentally putting these algae on rice field soils.

Nutrient Uptake

Aquatic plants obtain most of their nutrients from the water in which they are grown. This factor makes them useful with polyculture with animal aquacrops because they use the various forms of nitrogen wastes (NH_3, NO_2, and NO_3) produced by animals. This ability also makes plants useful in wastewater treatment.

Like land plants, aquatic plants use photosynthesis and convert carbon dioxide (CO_2) and water (H_2O) into carbohydrates ($C_6H_{12}O_6$), which can then be used for energy for growth and reproduction. By-products of this process are water and oxygen (O_2).

Reproduction

Plants can reproduce sexually or asexually. The aquatic plants described in this chapter can be reproduced using either method. Sexual reproduction involves the fusion of gametes (sex cells) followed by meiosis (cell division). This fusion of gametes may result in a zygote (new plant) or a seed, which must germinate to produce a new plant.

Asexual reproduction includes any means other than sexual reproduction. This may include production of asexual spores, corms, shoots, or bulbs. It may also involve mechanical reproduction, such as cuttings.

Algae

Mature algae plants, called **sporophytes**, release spores that become microscopic plants. These plants then produce gametes, which fuse and produce new plants. Algae may also release asexual spores, called **monospores**, which then become new plants.

Watercress

In nature, watercress produces gametes that fuse to form a seed. It can also reproduce by sprouting shoots that result in new plants. In commercial production, however, mechanical reproduction is used. Terminal cuttings are used to start new plants.

Chinese Water Chestnuts

Chinese water chestnuts reproduce by developing corms, fleshy underground stems. Each corm can then be removed from the parent plant and planted to develop into a new plant.

One parent plant can produce hundreds of these new corms in a growing season. This is also the part of the plant that is eaten, so most corms are not used for reproduction.

Economic and Production Considerations

Before culturing aquatic plants, producers should consider the marketability of the crop. Marketability is the ability to sell a plant aquacrop. Several factors must be considered, including existing markets, market potential, market stability, and product development and delivery. Other factors to consider are site selection, seed stock availability, and labor.

Existing Markets

To consider the production of freshwater aquacrops and to develop a marketing and business plan, some questions need to be answered. How much of the aquacrop is currently being sold in the market area? Is the supply currently available to meet the demand? Can the product be grown at higher quality or at a lower price? Will the product be sold wholesale or retail?

Market Potential

Can advertising or promotional programs increase the existing market? How much promotion is needed? How much will it cost?

Market Stability

Does the market function year-round, or is it seasonal? Can you produce the amount needed when it is needed? Would market size increase if the product were consistently available? Is the existing market a niche market that may be taken over by larger suppliers if it develops into a larger market? How much storage is required, and for how long?

Product Development and Delivery

Does the product require further processing after harvest? How much will it cost to deliver the product to the market? Will the buyer pay a premium for extra service required?

Site Selection

Once a market has been identified, the aquafarmer must determine an appropriate site for production. Of course, if a polyculture is planned with one or more animal aquacrops, then the facility may already be in place.

When planning a new facility, several factors may need to be considered—for example, site, location, distance, structures, traffic, utilities, and access.

Seed Stock Availability

An aquafarmer must be able to locate seed stock for the initial start-up of a plant aquaculture operation. This supply should be close enough to offset shipping costs and loss of seed stock during shipping.

In cases where a source of seed stock is not readily available, the aquafarmer may start his or her operation on a small scale and keep the products for use as parent plants. Spores can be collected from algae, cuttings from watercress, and corms from Chinese water chestnuts for use during the next growing season. Almost all plant aquafarmers save seed stock each year for the next year's production.

Labor

Production of all plant aquacrops is particularly labor-intensive during the planting and harvesting stages. A source of seasonal labor is important in operations on a medium to large scale.

The broad potential of aquatic plants to produce protein, energy, chemicals, pigments, hydrocarbons, and other special products makes them deserving of further research. Areas of study should include basic physiology and reproduction, nutrition, environmental requirements, food chain dynamics, germ-plasm maintenance, and transfer.

SUMMARY

Successful aquaculture requires the identification of a market, proper location, an entrepreneurial spirit, and an understanding of the biology of the species to be cultured. Learning to culture alligators basically saved the species. Alligators are successfully cultured, and the methods for culturing them are continually being refined. Frogs represent a potential species for culture, but with many challenges. Aquatic plant culture also holds potential, but entrepreneurs need to identify plants that have marketability.

STUDY/REVIEW

Success in any career requires knowledge. Test your knowledge of this chapter by answering these questions or solving these problems.

True/False

1. Temperature determines when frog eggs will hatch.
2. Tadpoles and frogs have the same dietary needs.
3. Alligator eggs must be laid with their long axis parallel to the ground.
4. Aquatic plants reproduce by only asexual methods.
5. Phycocolloids and mulch are by-products of alligator culture.
6. Frogs are territorial.
7. Health and growth of alligators and frogs is independent of their environmental temperature.
8. Frogs have few predators.

Short Answer

1. Two main products of alligator culture are _____ and _____.
2. Alligators and crocodiles are both members of the order _____.
3. The average clutch size for alligators is 35–40 _____.
4. Name two sources of seed stock for alligators, frogs, or aquatic plants.
5. How are bullfrog eggs identified when comparing them to other species of frogs?
6. In general, an alligator diet is high in _____ and low in _____.
7. On an alligator farm, an alligator that is 2 ft long requires _____ sq. ft.
8. How are cultured frogs marketed?
9. Where do alligators lay their eggs, and when should a producer gather the eggs to put them in an incubator?
10. Alligator farmers expect _____ to _____ percent of alligator eggs to hatch.
11. Name two freshwater and two marine aquatic plants that could be commercially grown.
12. Give the scientific names for the alligator and for the bullfrog.

Essay

1. Discuss the conditions that create stress for alligators under culture (farming) conditions.
2. Before raising aquatic plants as a commercial venture, what items should be considered to ensure the success of the enterprise?
3. Discuss the conditions that create stress for frogs under culture (farming) conditions.
4. Compare the dietary needs of tadpoles to those of adult frogs.
5. Describe a grow-out unit for alligators.

KNOWLEDGE APPLIED

1. Collect tadpoles and raise them until they develop into frogs. Record observations of their daily progress and record the water temperature.
2. Find recipes that use alligator meat or recipes on how to cook frog legs.
3. Create a list of local items being sold that are made of leather from alligator hides. Report on the price of these items.
4. Purchase some seaweed, watercress, and Chinese water chestnuts. Conduct a taste test of these aquatic plants and determine how they are normally eaten.
5. Obtain catalogs from some biological supply companies or visit their Web sites on the Internet. Find out how frogs are sold through these companies and how much they cost.

LEARNING/TEACHING AIDS

Books

Chapman, V. J. (1980). *Seaweeds and their uses*. New York, NY: Springer Publishing.
Engle, C. R. and Quagrainie, K. (2006). *Aquaculture marketing handbook*. Ames, IA: Blackwell Publishing.
Falkowski, P. G., and Raven, J. A. (2007). *Aquatic photosynthesis, 2nd Ed*. Malden, MA: Blackwell Science.
Hutchinson, L. (2005). *Ecological Aquaculture: A sustainable solution*. Hampshire, England: Permanent Publications.
Pillay, T.V. R., and Kutty, M. N. (2005). *Aquaculture principles and practices*. Ames, IA: Blackwell Publishing Ltd.
Riemer, D. N. (1993). *Introduction of freshwater vegetation*. Malabar, FL: Krieger Publishing Co.

Associations/Organizations

American Alligator Farmers Association, 5145 Harvey Tew Road, Plant City, Florida 33565.
Florida Aquatic Plant Management Society: http://www.fapms.org/
Food and Agriculture Organization of the United Nations (FAO), Fisheries and Aquaculture Department: http://www.fao.org/fishery/dias/en
National Aquaculture Association: http://www.thenaa.net/
Southern Southeast Regional Aquaculture Association, Inc.: http://www.ssraa.org/
World Aquaculture Society: https://www.was.org/

Internet

Internet sites represent a vast resource of information. The URLs (uniform resource locator) for World Wide Web sites can change. Using a search engine such as Google, find more information by searching for these words or phrases: alligators, frogs, aquatic plants, tadpoles, plant aquaculture, amphibian, hatchling, nutria, caimans, crocodiles, wild stock, cold-blooded, phycocolloids, marine plants, seaweed, marine algae, kelp, wakame, nori, laver, aonori, macrocystis, water chestnuts, spirulina, polyculture, unicellular algae, or sporophytes.

For some specific Internet sites refer to Appendix Tables A-11, A-12 and A-14.

The primary purpose of aquaculture is the efficient conversion of feed into meat for consumption (Figure 9-1). Fish convert feed to food for human consumption very efficiently, especially compared to other meat-producing animals. An understanding of the digestive system and the nutrition of fish is essential for successful aquaculture.

CHAPTER 9

Fundamentals of Nutrition in Aquaculture

OBJECTIVES

After completing this chapter, the student should be able to:

- Find the protein, energy, vitamin, and mineral requirements for fish
- Understand the role of nonnutritive factors in feed
- Know what toxic substances to watch for in fish feed
- Identify the parts of the digestive system
- Explain the role of the digestive system in absorption
- Explain how anatomy and behavior affect feeding
- List factors that influence energy requirements
- List three sources of energy
- Identify factors that affect the digestibility of fat
- Explain the role of essential fatty acids and essential amino acids
- Name 10 essential amino acids
- Name two essential fatty acids
- List the fat-soluble and water-soluble vitamins
- Describe 10 effects of vitamin-deficient diets
- Name the macrominerals and the microminerals
- List 10 functions of minerals

Understanding of this chapter will be enhanced if the following terms are known. Many are defined in the text, and others are defined in the glossary.

KEY TERMS

Absorption	Intestine
Amino acid	Lipids
Anemia	Liver
Antinutrients	Macrominerals
Antioxidants	Metabolic rate
Binders	Microminerals
Breakdown	Nutrition
Carbohydrates	Omnivores
Carnivores	Osmoregulation
Carotenoids	Oxidation
Coenzymes	Parasites
Connective tissue	Peptide bond
Diet	Peroxide
Digestion	Pharynx
Dry feed	Phytin
Enzyme	Predators
Essential amino acids	Protein
Essential fatty acids	Pylorus
Fines	Saturated
Gall bladder	Sedentary
Gastrointestinal	Strainers
Grazers	Suckers
Heme	Synthesize
Herbivores	Vitamins

NUTRITION OF FISH

The science of **nutrition** draws heavily on findings of chemistry, biochemistry, physics, microbiology, physiology, medicine, genetics, mathematics, endocrinology, cellular biology, and animal behavior. To the individual involved in aquaculture, nutrition represents more than just feeding. Nutrition becomes the science of the interaction of a nutrient with some part of a living organism, including feed composition, ingestion, energy liberation, wastes elimination, and synthesis for maintenance, growth, and reproduction. Feeds and feedstuffs contain the energy and nutrients essential for the growth, reproduction, and health of aquatic animals. Deficiencies or excesses can reduce growth or lead to disease. Dietary requirements set the necessary levels for energy, **protein**, **amino acids**, **lipids** (fat), minerals, and **vitamins**.

The subcommittee on Fish Nutrition of the Committee on Animal Nutrition of the National Research Council (NRC) examines the literature and current practices in aquaculture. The NRC publishes recommendations on fish nutrition. The latest NRC publication on fish nutrition was issued late in 1993. (This book can be accessed online at The National Academies Press: http://www.nap.edu/catalog.php?record_id=2115) Many of the recommendations in this chapter are based on this NRC publication.

To make money in aquaculture, transforming feed to food must be done efficiently and economically. Fish can do this, as Table 9-1 shows, but the principles of nutrition must be applied. Successful nutritional practices also depend on breeding, health, and management.

Digestion and Absorption

Figure 9-2 represents the fish digestive system. The digestive or **gastrointestinal** tract is described as a continuous, hollow tube extending from the mouth to the anus with the body built around it.

The digestive system of fish includes the mouth, **pharynx**, esophagus, stomach, **pylorus**, **intestine**, **liver**, and **gall bladder**. It acts like an assembly line in reverse, taking the feedstuffs apart to their basic chemical components so that the fish can absorb them and rearrange them into its own characteristic body composition. Table 9-2 summarizes all structures of the digestive system in **digestion** and **absorption**.

FIGURE 9-1 A truck driver delivering catfish feed to farmers on the Mississippi Delta.

TABLE 9-1 FEED CONVERSION COMPARISONS

Species	Unit of Production	per Lb. of Product Average Lb. of Feed
Broiler	1 lb. chicken	2.4
Dairy cow	1 lb. milk	1.1
Turkey	1 lb. turkey	5.2
Layer	1 lb. eggs (8 eggs)	4.6
Rabbit	1 lb. fryer	3.0
Hog	1 lb. pork	4.9
Beef Steer	1 lb. beef	9.0
Lamb	1 lb. lamb	8.0
Fish	1 lb. fish	1.6

Other considerations of digestion and absorption in fish include the type of eaters, the anatomy of the mouth, and feeding behavior.

Feeding Type and Anatomy. Fish can be divided into three types of eaters:

- **Carnivores** consume primarily animal material. Foods consumed by this type of fish may be as small as a microscopic crustacean or insect or as large as an amphibian or a small mammal.
- **Herbivores** subsist primarily on vegetation and decayed organic material in the environment.

FIGURE 9-2 Digestive system of fish.

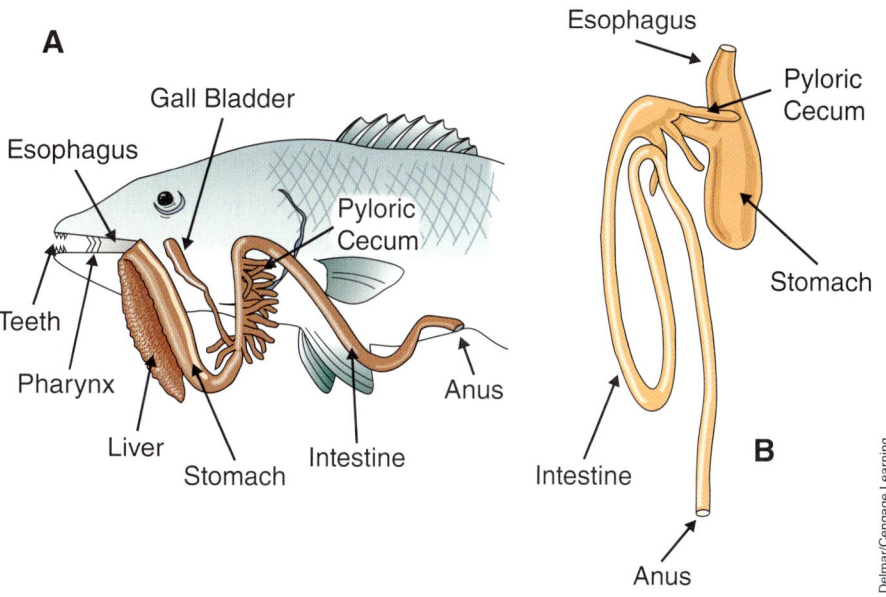

TABLE 9-2	DIGESTIVE SYSTEM STRUCTURES AND FUNCTIONS
Structure	**Functions**
Teeth	Grasping, holding, crushing, depending on species
Pharynx	Opening to the gills
Esophagus	Short, simple passage to stomach, lined with mucus-secreting cells
Stomach	Walls lined with cells secreting hydrochloric acid and pepsinogen for initial stages of protein digestion; holding compartment for feed
Pyloric cecum	Secretes enzymes for digestion; increased surface area for absorption of nutrients
Intestine	Secretes enzymes for digestion; increased surface area for absorption of nutrients
Gall bladder	Stores and releases bile for digestion and absorption of fats
Liver	Synthesis or storage from absorbed nutrients, production of bile, removal of some waste products from blood

➤ **Omnivores** consume almost any food source, either plant or animal in origin.

Certain anatomic changes in the mouth of fish occurred through evolutionary development. Fish can be classified according to their feeding habits into the following categories:

➤ **Predators**. Trout are an example of fish that feed on animals generally large enough to be seen with the naked eye. Teeth are well developed and act as a means of grasping and holding the prey. Some predators rely primarily on sight to hunt, whereas others rely on the senses of taste and touch or on lateral-line sense organs.

- **Grazers.** The mullet is an example of a fish that grazes in the same sense as mammalian grazers. Generally, mullets graze continuously on the bottom of the water habitat for either plants or small animal organisms. Food is taken in well-defined bites.
- **Strainers.** The menhaden is an example of a fish that selects food primarily by size rather than type. An adult menhaden can strain in excess of 6 gal (22.7 l) of water per minute through its gill rakers. Through this process of rapid straining, the menhaden is able to concentrate a relatively large mass of plankton and other organisms.
- **Suckers.** The buffalofish is an example of a fish that feed primarily on the bottom of its habitat, sucking in mud and filtering and extracting digestible material.
- **Parasites.** Some fish, like the lamprey, attach themselves to other animals and exist on host's body fluids.

Behavior

Fish also develop behavioral feeding patterns that are sensitive to environmental stimuli. By knowing the behavioral patterns of the particular species of fish, the producer can adapt a system of feeding that will best use labor and feed. Some environmental influences on feeding behavior include:

- Sensory use. Some fish depend largely on their sense of sight in hunting food, whereas others rely primarily on taste, touch, and smell.
- Season of the year. Some fish cease feeding activity during their spawning season. In temperate climatic regions, most fish start increasing feed intake in the spring, when the water temperature starts to rise. The peak growth period for most fish occurs in the spring and summer.
- Time of day. Some fish show peak feed activity in the dawn and dusk hours of the day.
- Physical contact with food. Quite often the texture of a potential food source is felt before the fish will consume it.

Even though the digestive anatomy, physiology, and feeding habits differ in fish compared to warm-blooded domestic animals, the nutritional requirements remain expressed in the same terms—energy, protein, vitamins, and minerals. Most of the research on these nutritional requirements centers around catfish, salmon, and trout.

ENERGY REQUIREMENTS

Energy is not a nutrient, but it is released during the **breakdown** (metabolic oxidation) of carbohydrates, amino acids, and fats. Because fish are cold-blooded, they expend no energy to maintain body temperature. This allows more energy for growth, activity, and reproduction. Several other factors affect energy use by fish:

- Age. As age increases, the **metabolic rate** of fish generally decreases.
- Composition of the diet. If a **diet** has a high protein or mineral content, metabolism increases, in order for the fish to eliminate waste products that could possibly build up and become toxic.
- Light exposure. Darkness decreases the energy requirement in some species. Fish grown in constant light do not grow as well as those of the same species having a rest period of darkness.
- Physiological activity. Salmon have high metabolic rates during the spawning season. Conversely, during winter rest, fish have extremely low metabolic rates.
- Size. In general, smaller fish have higher metabolic rates than larger fish.
- Species. Metabolic rates vary according to the characteristic behavioral patterns of the species. For example, **sedentary** fish have lower metabolic rates than do pelagic (open sea) fish.
- Temperature of the water. As water temperature increases, its ability to carry oxygen decreases. In response to the reduced oxygen-carrying capacity of the water, the respiration rates of fish increase, resulting in higher metabolic rates. For every 18°F increase in water temperature, the metabolic rate doubles. Figure 9-3 shows how increasing temperature increases the amount of feed recommended for trout.

Additionally, the metabolic rate decreases 5 percent for each degree F decrease from Standard Environmental Temperature (SET). Thus, when the water temperature is 20°F below Standard Environmental Temperature, growth—for all intents and purposes—ceases. However, the fish still eat to accommodate their metabolic needs for swimming, **osmoregulation**, and respiration.

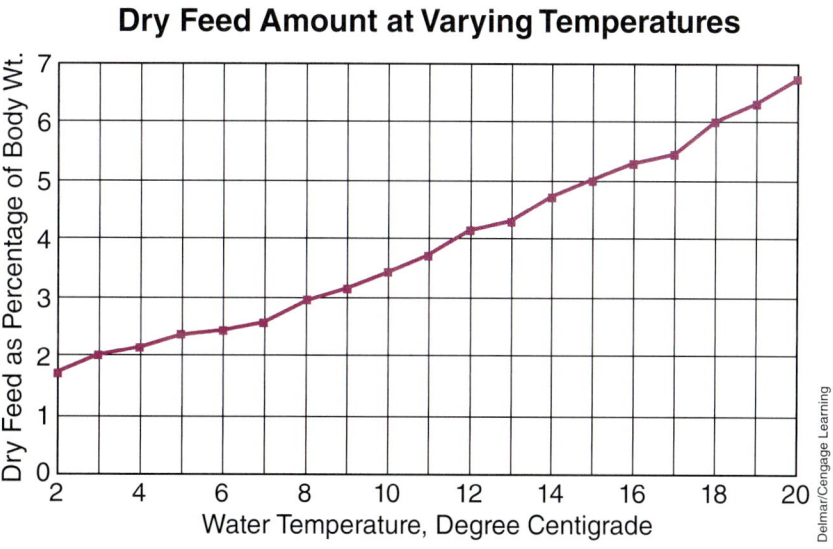

FIGURE 9-3 Increasing water temperature increases the recommended amount of feed for trout, 2 to 3 in.

Other environmental factors such as water flow rates, water composition, and pollution put certain stresses on fish and result in their metabolic rates being altered in keeping with the severity of the stress.

Energy Losses

Feeds and feedstuffs contain energy, but not all the energy goes toward growth and reproduction. Energy losses occur as feed is digested and metabolized. As feed moves through the digestive processes, energy is lost in the feces, urine, and gill excretions. Energy is also lost as heat. Figure 9-4 illustrates the loss of energy as intake energy (gross) loses fecal energy, becoming digestible energy. Then digestible energy loses energy to urine or gill excretions to become metabolizable energy. Metabolizable energy loses heat energy, becoming net energy. This is the energy available for maintenance, growth, and reproduction. Digestible energy (DE) and metabolizable energy (ME) are more exact measures of the energy required by fish.

Like traditional livestock, fish derive their energy from three sources—carbohydrates, fats, and proteins. Because carbohydrates are used rather inefficiently in most fish systems, the primary energy sources are fats and proteins.

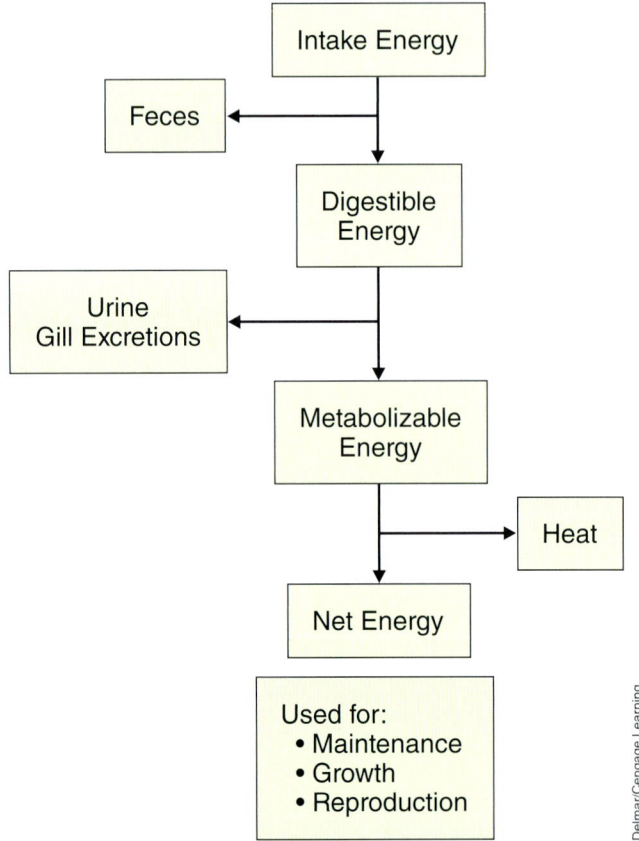

FIGURE 9-4 Energy contained in a feed is lost to digestive processes and heat, leaving net energy to be used for maintenance, growth, and reproduction.

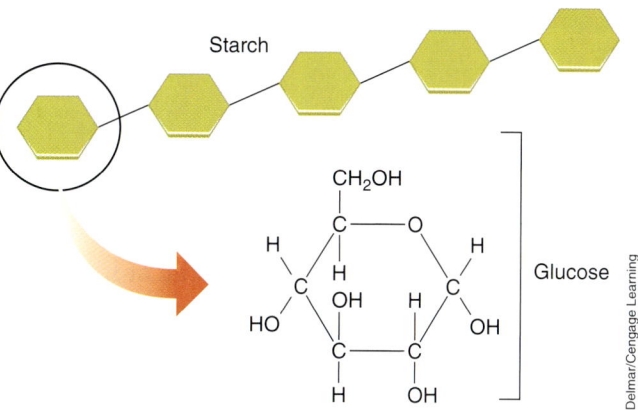

FIGURE 9-5 Starch is a long chain of glucose molecules. Digestion requires the starch to be broken down into simple sugars—glucose.

Carbohydrates

Fish digest simple sugars efficiently (Figure 9-5). As the sugar becomes larger and more complex, digestibility decreases rapidly. Warmwater fish digest dietary carbohydrates better than coldwater or marine fish. The ability to use carbohydrates as an energy source varies among the fish species. No specific NRC requirement is established for carbohydrates in the diet of fish. Some form of digestible carbohydrate should be included in the diet. Carbohydrates improve growth and provide precursors for some amino acids and nucleic acids. Also, carbohydrate is the least expensive source of dietary energy. In warmwater fish, cereal grains provide inexpensive sources of carbohydrates, but their use is limited in coldwater fish. Digestible carbohydrates in trout feed are generally lower than the levels in catfish feed. In nutrition, carbohydrates spare protein because less protein will be used for energy. An excess of dietary carbohydrates can cause livers to enlarge and glycogen to accumulate in the liver. A general recommendation is a diet of no more than 12 percent digestible carbohydrates. Fats and proteins supply most of the energy in fish diets.

Fat

Each gram of fat contains 2.5 times the energy in a gram of carbohydrates or proteins. The digestibility of fat varies, depending on:

- Amount in the diet
- Type of fat
- Water temperature
- Degree of unsaturation
- Length of carbon chain

Animal fats and fats that are highly **saturated** have a lower digestibility. On the other hand, in highly unsaturated fats—fats that fish can readily digest—there is danger of **oxidation** of the fats, resulting in feed spoilage. **Antioxidants** are routinely added to most fish diets to prevent fats from becoming rancid in storage.

Besides being an important source of energy for fish, dietary fats provide **essential fatty acids** (EFA) needed for normal growth and development. Fish cannot **synthesize** these fatty acids. Also, dietary fats

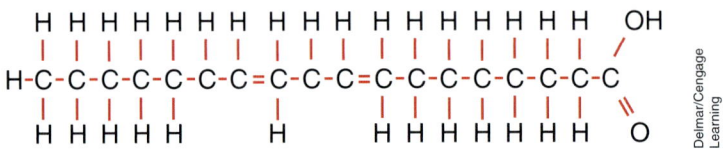

FIGURE 9-6 Diagram of linoleic acid, an essential fatty acid.

assist in the absorption of fat-soluble vitamins. Freshwater fish require a dietary source of linoleic acid and/or linolenic acid (Figure 9-6). These are both 18-carbon fatty acids. Marine fish, like the yellowtail or red sea bream, require a dietary source of eicosapentaenoic acid (EPA) and/or docosahexaenoic acid (DHA). These are 20- and 22-carbon fatty acids respectively.

Channel catfish, coho salmon, and rainbow trout require linolenic acid or EPA and/or DHA. Table 9-3 indicates the essential fatty acid requirements for several species of fish.

Essential fatty acid deficiency signs include skin lesions, shock syndrome, heart problems, reduced growth rate, reduced feed efficiency, reduced reproductive performance, and increased mortality. In the body essential fatty acids function as a part of cell membranes and the precursor for biochemicals that perform a variety of metabolic functions.

Fish diets are formulated to meet the optimum ratio of energy to protein for each species. Fats serve as an important source of energy, but no definite percentage of dietary fat can be given without considering the type of fat, as well as the protein and energy content of the diet. Table 9-4 lists some general guidelines for dietary fat for different situations in various species of fish. Too much dietary fat can result in an imbalance of the digestible energy to crude protein ratio and excessive deposition of fat in the body cavity and tissues.

TABLE 9-3 ESSENTIAL FATTY ACID REQUIREMENTS OF FISH[1]

Species	Requirement
Channel catfish	1.0 to 2.0% linolenic acid or 0.5 to 0.75% EPA[2] and DHA[2]
Chum salmon	1.0% linoleic acid and 1.0% linolenic acid
Coho salmon	1.0 to 2.5% linoleic acid
Common carp	1.0% linoleic acid and 1.0% linolenic acid
Rainbow trout	0.8 to 1.0% linolenic acid 20% of fat as linolenic acid or 10% of fat as EPA and DHA
Tilapia	0.5 to 1.0% linoleic acid
Red sea bream	0.5% EPA and DHA Yellowtail 2.0% EPA and DHA

[1]National Research Council (1993). Nutrient requirements of fish. Washington, DC: National Academy Press.
[2]EPA is eicosapentaenoic acid and DHA is docosahexaenoic acid.

TABLE 9-4 GUIDELINES FOR DIETARY FAT IN FISH

Species	Situation	Percentage of Fat
Trout	Starter diet	12 to 16
	Grower diet	8 to 10
	Production diet	6 to 8
Catfish	82°F (28°C)	12
	73°F (23°C)	5
Carp	82° to 73°F (28° to 23°C)	10 to 15
	<68°F (20°C)	10

PROTEIN REQUIREMENTS

Proteins are long chains of amino acids linked by bonds called **peptide bonds**. Figure 9-7 shows diagrams of four amino acids. All amino acids contain nitrogen, so all proteins contain nitrogen. In fact, measuring nitrogen content is a method of calculating protein content. Metabolism of protein for energy produces nitrogen end products. Fish eliminate these through the gills, feces, and urine. These nitrogen end products can cause problems in fish ponds.

Protein serves three purposes in the nutrition of fish:
1. Provide energy
2. Supply amino acids
3. Meet requirements for functional proteins—enzymes and hormones—and structural proteins.

The requirement for protein in fish diets is essentially a requirement for the amino acids in the dietary proteins. Some amino acids the fish cannot synthesize are called indispensable or **essential amino acids**. These include 10 amino acids:

- Arginine
- Histidine
- Isoleucine
- Leucine
- Lysine
- Methionine
- Phenylalanine
- Threonine
- Tryptophan
- Valine

Some of the dietary requirements for methionine and phenylalanine can be met by the amino acids cystine and tyrosine, respectively.

The amino acid requirement given by the NRC is shown in Table 9-5 for catfish, trout, salmon, carp, and tilapia.

Research evidence suggests that large differences exist among fish species in their requirements for amino acids. Some of these differences are probably caused by differences in growth rate, feed intake, and the source of the amino acid in the diet.

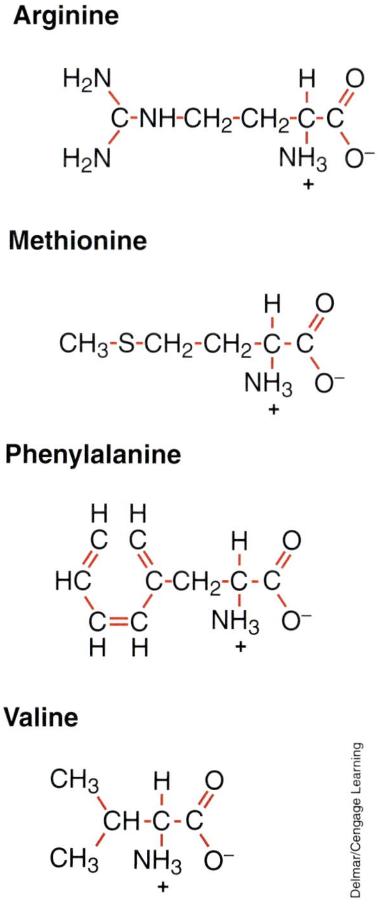

FIGURE 9-7 Diagram of some essential amino acids.

TABLE 9-5 AMINO ACID REQUIREMENT FOR FINFISH[1]

	Channel Catfish	Rainbow Trout	Pacific Salmon	Common Carp	Tilapia
Energy Base[b] (kcal DE/Kg diet)	3,000	3,600	3,600	3,200	3,000
Protein, crude (digestible),	32(28)	38(34)	38(34)	35(30.5)	32(28)
Amino acids %					
Arginine, %	1.20	1.5	2.04	1.31	1.18
Histidine, %	0.42	0.7	0.61	0.64	0.48
Isoleucine, %	0.73	0.9	0.75	0.76	0.87
Leucine, %	0.98	1.4	1.33	1.00	0.95
Lysine, %	1.43	1.8	1.70	1.74	1.43
Methionine + cystine, %	0.64	1.0	1.36	0.94	0.90
Phenylalanine + tyrosine, %	1.40	1.8	1.73	1.98	1.55
Threonine, %	0.56	0.8	0.75	1.19	1.05
Tryptophan, %	0.14	0.2	0.17	0.24	0.28
Valine, %	0.84	1.2	1.09	1.10	0.78

[1]National Research Council. (1993). Nutrient requirements of fish. Washington, DC: National Academy Press.

When proteins in most feedstuffs are properly processed, they are highly digestible. For a variety of protein-rich feedstuffs, the digestibility ranges from 75 to 95 percent. As dietary carbohydrate increases, the digestibility of protein tends to decline. Also, overheating during drying or processing reduces protein's nutritive value. But, insufficient heating of soybean meal decreases the availability of protein.

Protein requirements for fish are considerably higher than those for warm-blooded land animals. Protein requirements of fish decline with age.

Animal protein sources are generally considered to be of higher quality than plant sources, but animal protein costs more. In diets, a combination of protein sources yields better conversion rates than any single source.

Fish do not have the ability to use nonprotein nitrogen sources. Such nonprotein nitrogen sources as urea and diammonium citrate, which even non-ruminant animals can use to a limited extent, have no value as a feed source for fish. In fact, nonprotein nitrogen can be toxic at high levels.

Factors such as age, species, stocking density, water temperature, and production stage should govern the protein level incorporated in the diet. Table 9-6 indicates some recommended protein levels in practical fish diets.

TABLE 9-6 PROTEIN LEVELS IN PRACTICAL FISH DIETS

Species	Percentage As-Fed Basis		
	Starter	Grower	Production
Trout	40 to 55	35 to 40	30 to 40
Chinook Salmon	40	—	—
Catfish	35 to 40	25 to 36	28 to 32
Carp	43 to 47	37 to 42	—

A protein deficiency or indispensable amino acid deficiency is observed as a reduction in weight gain. But some specific amino acid deficiencies manifest as disease conditions. Cataracts form in salmonids, including rainbow trout, when given diets are deficient in methionine or tryptophan. A tryptophan deficiency also causes a lateral curvature of the spinal column, or scoliosis, in some salmonids. In trout, a tryptophan deficiency disrupts the metabolism of the minerals calcium, magnesium, sodium, and potassium.

In fish diets, protein and energy should be kept in balance. A deficiency or excess of energy reduces growth rates. When dietary energy is deficient, protein is used for energy. When dietary energy is in excess, feed consumption drops, and this lowers the intake of the necessary amounts of protein for growth.

VITAMIN REQUIREMENTS

Vitamins are organic compounds required in the diet for normal growth, reproduction, and health. They function in a variety of chemical reactions in the body. The simple digestive system of the fish establishes a definite need for the supplementation of vitamins in fish diets. Vitamin requirements for fish resemble those of nonruminant animals such as pigs and chickens. Fish and humans are among the few higher animals that require a dietary source of vitamin C. (See Figure 9-8 for an example of the effects of vitamin C deficiency in channel catfish.)

Vitamins are divided into two categories, water soluble and fat soluble. Water-soluble vitamins include—

- Thiamin
- Riboflavin
- Pyridoxine (Vitamin B_6)
- Pantothenic
- Niacin
- Biotin
- Folate
- Vitamin B_{12}
- Choline

FIGURE 9-8 Vitamin C deficiency in channel catfish causes a spinal deformity. The top catfish is normal, and the bottom catfish was deprived of dietary vitamin C for eight weeks.

- Myoinositol
- Vitamin C

Choline, myoinositol, and vitamin C serve a variety of functions. Choline functions as a:

- Component of membranes
- Precursor of acetylcholine, a chemical for nerve transmission
- Provider of methyl (CH_3) groups for chemical reactions

Myoinositol is also a component of membranes and is involved in sending signals during several body processes. Vitamin C is involved in the formation of **connective tissue**, bone matrix, and wound repair. It also facilitates the absorption of iron from the intestine and helps prevent the peroxidation of fats (lipids) in tissues.

Most water-soluble vitamins serve as **coenzymes** in the body's biochemical reactions. **Enzymes** are biological catalysts. Most enzymes are proteins, and they are unique for each biochemical reaction. Coenzymes then work with, or become part of, an enzyme.

The fat-soluble vitamins are:

- Vitamin A
- Vitamin D
- Vitamin E
- Vitamin K

Fat-soluble vitamins are absorbed in the intestine along with fats in the diet. Unlike water-soluble vitamins, fat-soluble vitamins can be stored in body tissues. Excessive amounts in the diet can cause a toxic condition called hypervitaminosis. Functions of the fat-soluble vitamins are quite specific. Vitamin A is necessary for sight, proper growth, reproduction,

SEARCHING FOR ALTERNATIVE PROTEIN SOURCES IN FISH DIETS

Scientific research is key to advancing aquaculture. Many areas need research; for example, scientists are searching for alternative protein sources or ways to replace animal (fish) protein with plant protein in fish diets. Finding alternative protein sources could improve profitability. The ARS funds some of this research and the results are reported on their website (http://www.ars.usda.gov/research/programs/programs.htm?NP_CODE=106.)

One project investigated two products that are potential substitutes for soybean meal in catfish diets—cottonseed meal and distillers dried grains with solubles (DDGS). Cottonseed meal is a local product that is generally priced competitively (on a protein basis) with soybean meal. Distillers dried grains with solubles may become abundant as ethanol plants come on line. Results show that about 50 percent of soybean meal can be replaced with cottonseed meal with supplemental lysine in channel catfish diets without negatively affecting fish performance. Further, DDGS can be used in channel catfish diets up to at least 30 percent of the diet when the diet is supplemented with lysine. Use of DDGS in the diet appears to improve feed efficiency.

Another research project looked into developing all-plant-protein diets for rainbow trout that are fish-meal free, Weight gain for fish that were fed diets containing plant proteins was equivalent to fish that were fed the fish-meal based diets. The study provided evidence that growth rates of trout that are fed fish-meal-free diets, using conventional and concentrated plant protein ingredients, can support growth up to about 80 percent of that observed when fish consumed meal-based feeds.

Continuing scientific research improves fish diets and our understanding of fish nutritional needs.

---

resistance to infection, and maintenance of body coverings. As in many land animals, fish can use betacarotene as a vitamin A precursor. Vitamin D helps the body mobilize, transport, absorb, and use calcium and phosphorus. It works with two hormones from an endocrine gland, the parathyroid. Vitamin E is the name given to all substances that act like alpha-tocopherol (chemical name for active form of vitamin E). Vitamin E, working with selenium, protects cells against adverse effects of oxidation. Vitamin K is required for the normal blood-clotting process. Many animals can synthesize vitamin K in their intestines. Fish, however, do not possess the bacteria to do this.

Table 9-7 summarizes the NRC recommendations and the effects of a vitamin deficiency for each of the fat-soluble and water-soluble vitamins.

MINERAL REQUIREMENTS

Fish can absorb a number of minerals directly from the water: calcium (Ca), magnesium (Mg), sodium (Na), potassium (K), iron (Fe), zinc (Zn), copper (Cu), and selenium (Se). This reduces the mineral requirement in

TABLE 9-7 VITAMIN RECOMMENDATIONS AND DEFICIENCIES IN FISH

Vitamin	NRC[1] Recommended Requirements (mg/kg as fed)	Vitamin Deficiency Symptoms
Fat soluble:		
A (IU/kg)	1,000 to 2,500	Hemorrhages; ascites, edema, visual dysfunctions; poor growth; high mortality; anemia; twisted gill opercula
D (IU/kg)	500 to 2,400	Poor growth; muscular disorders; increased fat content of liver
E	50	Anemia; red blood cell fragility; clubbed gills; low hematocrit; poor growth; edema; ascites; high mortality; ceroid in spleen, kidney, liver
K	R[2]	Anemia; prolonged blood coagulation time
Water soluble:		
B_{12}	0.01	Anorexia; anemia; depressed growth
Biotin	0.15	Lesions of colon; anemia; blue slime disease; spasms; muscle degeneration; anorexia; red blood cell fragility
Choline	400 to 1,000	Depressed growth and feed conversion; internal hemorrhaging; anemia; yellow-colored livers
Folate	1.0 to 1.5	Dark coloration; anemia; fragility of caudal fin; anorexia; lethargy; depressed growth; erythropenia; ascites
Myoinositol	300	Depressed growth; anemia; fin erosion; dark coloration; anorexia; stomach distention
Niacin	10 to 14	Lesions of colon; lethargy; edema of stomach and intestines; loss of coordination; photophobia; swollen gills; tetany; lesions of skin; anorexia; sunburn
Pantothenic acid	15 to 20	Clubbed gills; mumpy appearance; anorexia; lethargy; prostration; poor weight gain; flared opercles (nutritional gill disease)
Pyridoxine B_6	3	Irritability and fits; gasping; anorexia; ataxia; weight loss; ascites; anemia; rapid onset of rigor mortis; blue-green coloration on back
Riboflavin	4 to 9	Anorexia and poor growth; cloudy lens; hemorrhaging in eyes; photophobia; loss of coordination; xerophthalmia
Thiamin	1	Muscle and kidney degeneration; ataxia; anorexia; fatty liver; loss of coordination; depressed growth; internal hemorrhaging; anemia
Vitamin C (Ascorbic acid)	25 to 30	Impaired collagen production; lordosis and scoliosis; poor growth; anorexia; impairment of wound healing

[1] Values in the table are ranges for catfish, trout, salmon, carp, and tilapia. The National Research Council (NRC), Nutrient Requirements of Fish (1993), National Academy Press, Washington, DC, lists individual requirements.
[2] R: Required in diet, but quantity not determined.

the diet. But this also makes research on dietary mineral requirements difficult and inconclusive. Most researchers agree that fish require all of the minerals required by other animals.

Based on their requirement or use by an animal, minerals are divided into two groups: **macrominerals** and **microminerals**. Macrominerals are present in the body in relatively large quantities. The macrominerals include:

- Calcium (Ca)
- Chlorine (Cl)
- Magnesium (Mg)
- Phosphorus (P)
- Potassium (K)
- Sodium (Na)

Calcium and phosphorus are most directly involved in the development and growth of the skeleton, and they act in several other biochemical reactions. Fish absorb calcium directly from the water by the gills and skin. The requirement for calcium is determined by the water chemistry.

Dietary phosphorus is more critical. Phosphorus is derived from dietary phosphate. Phosphorus deficiency signs include poor growth, reduced feed efficiency, and bone deformities. The availability of phosphorus in feedstuffs varies widely. Feedstuffs from seeds contain phosphorus in a form known as **phytin**. The availability of phosphorus in phytin is low. Simple-stomach animals lack the enzyme to release the phosphorus.

Magnesium functions with many enzymes as a cofactor. The dietary requirement can be met from either the water or the feed. Deficiencies of magnesium cause anorexia, reduced growth, lethargy, vertebrae deformity, cell degeneration, and convulsions.

Sodium, potassium, and chlorine are electrolytes. Sodium and chlorine reside in the fluid outside the cells. Potassium resides inside the cells—an intracellular cation (positively charged ion). Because of the abundance of these elements in the environment, deficiency signs are difficult to produce.

Table 9-8 lists the NRC requirements for the macrominerals.

Microminerals are present in very small (micro) amounts in the bodies of fish, but they are still important to fish health.

The microminerals include:

- Copper (Cu)
- Iodine (I)
- Iron (Fe)
- Manganese (Mn)
- Selenium (Se)
- Zinc (Zn)

Copper is a part of many enzymes, and it is required for their activity. Although it is necessary for fish health, copper can be toxic at concentrations of 0.8 to 1.0 m per l in the water. Fish are more tolerant of copper in feed than in water.

TABLE 9-8 MACROMINERAL REQUIREMENTS FOR FINFISH[1]					
	Channel Catfish	Rainbow Trout	Pacific Salmon	Common Carp	Tilapia
Energy Base[b] (kcal DE/Kg diet)	3,000	3,600	3,600	3,200	3,000
Protein, crude (digestible), %	32(28)	38(34)	38(34)	35(30.5)	32(28)
Macrominerals					
Calcium, %	R[2]	1E[3]	NT[4]	NT	R
Chlorine, %	R	0.9E	NT	NT	NT
Magnesium, %	0.04	0.05	NT	0.05	0.06
Phosphorus, %	0.45	0.6	0.6	0.6	0.5
Potassium, %	R	0.7	0.8	NT	NT
Sodium, %	R	0.6E	NT	NT	NT

[1]National Research Council (1993). Nutrient requirements of fish. Washington, DC: National Academy Press.
[2]R: Required in diet, but quantity not determined.
[3]E: Estimated.
[4]NT: Not tested.

Iodine is necessary for the formation of hormones from the thyroid gland. Fish can obtain iodine from either water or feed. Similar to land animals, a deficiency causes the thyroid gland to grow, a condition similar to goiter.

Iron is necessary for the formation of **heme** compounds. These compounds carry oxygen. Because natural waters are low in iron, feed is considered the major source of iron. Iron deficiency causes a form of **anemia**. At high levels, iron can be toxic and cause reduced growth, diarrhea, liver damage, and death.

Manganese functions as a part of enzymes or as a cofactor. Although it can be absorbed from the water, it is more efficiently absorbed from the feed. A deficiency causes reduced growth and skeletal abnormalities.

Selenium protects cells and membranes against **peroxide** danger. Selenium deficiencies cause reduced growth. Both selenium and vitamin E are required to prevent muscular dystrophy in some species. When dietary selenium exceeds 13 to 15 mg per kg of **dry feed**, it becomes toxic, resulting in reduced growth, poor feed efficiency, and death.

Zinc is also a part of numerous enzymes. Dietary zinc is more efficiently absorbed than that dissolved in water. Dietary calcium and phosphorus, phytic acid protein type(a form of zinc), all affect zinc absorption and use. A zinc deficiency causes suppressed growth, cataracts, fin and skin erosion, dwarfism, or death.

FIGURE 9-9 Fish nutritionists examine fish in an experiment involving rainbow trout and new vitamin formulation.

Other trace minerals such as fluoride and chromium may be important, but evidence is limited.

OTHER DIETARY COMPONENTS

Many fish diets contain other ingredients that can affect them (Figure 9-9). Some of these ingredients are natural, others are added. These ingredients include substances such as water, fiber, hormones, antibiotics, antioxidants, pigments, **binders**, and feeding stimulants.

Water

All diets contain water. The water may be a part of the feedstuffs, come from the air, or be added. The less water in a diet, the easier the storage and handling. When moisture in a diet exceeds 12 percent, the feed is more susceptible to spoilage. Some commercial diets contain high moisture levels because fish seem to prefer moist feed.

Fiber

Fiber refers to plant material such as cellulose, hemicellulose, lignin, pentosums, and other complex carbohydrates. These are indigestible, and they do not play an important role in nutrition. Fiber adds bulk to a feed but increases the amount of fecal material produced. The goal in commercial aquaculture is to limit the diet's fiber content and use highly digestible feeds.

Hormones

Researchers have evaluated the use of various natural and synthetic hormones on fish. These hormones include growth hormone, thyroid hormones, gonadotropin (GnH), prolactin, insulin, and various steroids—androgens and estrogens. Hormones are used for two purposes: (1) induced or synchronized spawning and (2) sex reversal. Induced or synchronized spawning increases the availability and dependability of seed. Sex steroids reverse the sex of salmonids, carps, and tilapia, producing a monosex culture of sterile fish. This improves growth rate, prevents sexual maturation, and reduces flesh quality.

Antibiotics

With the arsenal of antibiotics available for humans and other livestock, only two have received FDA approval for use in fish—sulfadimethoxine/ormetoprim and oxytetracycline. When these antibiotics are used in feed, the quantity fed, the feeding rate, and the withdrawal time must be strictly controlled. Only licensed manufacturers can add antibiotics to feeds in the United States. Unlike land livestock, fish do not demonstrate any benefit from subtherapeutic levels of antibiotics in their feed.

Antioxidants

Fish feeds containing high levels of fats often use antioxidants. Oxidation of the fats affects the nutritional values of the fat and some vitamins. Synthetic vitamin E in diets usually has little antioxidant activity, so synthetic antioxidants like ethoxyquin, BHT, BHA, and propyl gallate are used. (See Figure 9-10.)

Pigments

Pigmentation of the skin and flesh in fish comes from carotenoids. Fish cannot make these carotenoids, so they must be present in the diet. In salmonids, the carotenoids astaxanthin and canthaxanthin are responsible for the red to orange color of their flesh. In the wild, these carotenoids come mainly from zooplankton. Some of the natural materials used to pigment the flesh of salmonids include crab, brill, shrimp, and yeast. Yellow pigmentation of the flesh of catfish is undesirable. It is caused by the carotenoids lutein and zeaxanthin from plant material in the diet.

Pellet Binders

Binders improve stability in the water, firmness, and reduce fines during processing and handling. Widely used binders are sodium and calcium bentonites, lignosulfates, carboxymethylcellulose, hemicellulose, guar gum alginate, and some new inert polymers.

Feeding Stimulants

Fish use sight and smell to find food, but the taste of the food determines its acceptance. Researchers and manufacturers continue to attempt to increase the palatability and acceptance of feed. This is especially important

FIGURE 9-10 An example feed tag.

NET WEIGHT 50 POUNDS OR 22.7 KILOS

MERMAID AQUAFEEDS

STURGEON GROWOUT PELLETS
4.0 MM SLOW SINKING PELLETS

GUARANTEED ANALYSIS
Crude protein, minimum.......... 36%
Crude fat, minimum............................ 7%
Crude fiber, maximum....................... 4%
Moisture, not more than 10%

INGREDIENTS
Plant protein products (including soybean meal, wheat gem), Marine protein products (including fish meal), Dried yeast culture, Fish oil, Dicalcium phosphate, Vitamin and Mineral premix containing a stabilized source of Ascorbic acid (Vitamin C), Ethoxyquin (antioxidant).

WARNING
Store in a dry, well ventilated area, free from rodents and insects.
Do not use moldy feed.
Use within 90 days of production date.

MANUFACTURED BY MERMAID AQUAFEEDS
Pierson-Florida
2356 State Rd 415
800.256.3478 - sales@mermaidfeeds.com

---

in starter and larval feeds. In general, carnivorous fish respond to alkaline and neutral substances. Herbivorous fish respond to acid substance. Besides increasing feed consumption, some compounds act as deterrents. This is the last thing an aquaculturalist wants.

Antinutrients and Toxins

Some naturally occurring substances can contaminate fish feeds and affect the performance of fish. Some **antinutrients** occur in plant and animal feedstuffs. These are described in Table 9-9.

Other feed contaminants occur by natural processes or find their way into the feed. These are listed and briefly described in Table 9-10.

TABLE 9-9 ANTINUTRIENTS IN FEEDSTUFFS

Name	Source	Effect	Prevention
Trypsin inhibitors	Raw soybeans	Inhibits digestive enzyme trypsin	Heat processing
Phytic acid (Phytate)	Soybean meal and other feedstuff from plants	Reduces availability of protein and minerals like Zn, Mn, Cu, Ca, Fe	Limit plant feedstuff; increase nutrient level
Gossypol	Glanded cottonseed meal	Depresses growth; damages organs and tissues; acts as a carcinogen	Limit use of glanded cottonseed meal
Cyclopropenoic fatty acids (CFAs)	Cottonseed meal Lesions, glycogen deposition;	elevated fatty acids; act as a carcinogen	Limit use of cottonseed meal
Glucosinolates	Rapeseed	Act as antithyroid agents upon enzymatic hydrolyses	Use low glucosinolate variety like canola; limit use of rapeseed or canola
Erucic acid	Rapeseed oil	Death; problem with skin, gills, kidneys, and heart	Avoid rapeseed oil
Alkaloids	Contamination of cottonseed or soybean meal	Growth depression and death	Quality cottonseed and soybean meal
Thiaminase	Some raw fish preparation	Destroys the vitamin thiamin	Heating or ensiling; feed thiamin in separate diet

TABLE 9-10 FEED CONTAMINANTS FROM NATURAL PROCESSES AND ENVIRONMENTAL CONTAMINATION

Name	Source	Effect	Prevention
Mycotoxins	Aflatoxins produced by the mold Aspergillus Flavus or fungus Fusarium tricintum	Carcinogenic; death; reduced growth; reduced feed consumption	Check for contamination; dry feed
Algal and marine toxins	Toxins of algae such as Ganyaulax spp. Gyrodiniun spp	Death	Identify toxic algae and eliminate
Oxidative rancidity	Autoxidation of unsaturated fats	Production of free radicals, peroxides, aldehydes, and ketones reduces nutritional value	Add synthetic or natural antioxidants to feed
Mercury	Industrial contamination	Gill problems; but may accumulate in muscle tissues and be a human health hazard when fish consumed	Selenium reduces toxicity and rate of accumulation
Cadmium	Industrial water	Liver necrosis; death	EDTA to chelate
Arsenic	Marine fishmeal	Potential toxicity in organic complex not known	—
Polychlorinated biphenyls (PCBs)	Industrial wastes; fish oil and fishmeal Accumulate in fat; widespread; can cause	liver enlargement, liver dysfunction, and decreased thyroid activity	Check feedstuffs and water quality
Pesticides	Accidental	Varies depending on pesticide; most accumulate in tissues; may affect human health or marketability of product	Careful application

SUMMARY

Though fish convert feed to human food very efficiently, the feeding cost of production needs to be controlled. Feeding fish requires an understanding of the process of digestion, the digestive system, and fish nutrition.

Fish consume feed for energy. They use this energy for growth, activity, and reproduction. In the fish diet, feeds containing protein, fats, and carbohydrates supply energy. These feeds enter the digestive system, where enzymes break down the protein, fats, and carbohydrates to simpler compounds that the fish uses for energy and to form tissue, enzymes, and bone. Protein in the diet also supplies 10 essential amino acids, and fat in the diet supplies essential fatty acids. Fat-soluble and water-soluble vitamins are also supplied by the diet. Minerals are supplied by the diet and by the water.

STUDY/REVIEW

Success in any career requires knowledge. Test your knowledge of this chapter by answering these questions or solving these problems.

True or False

1. Carbohydrates are made up of amino acids.
2. Antinutrients are necessary components in a fish diet.
3. Microminerals are just as important as macrominerals.
4. Protein contains nitrogen.
5. Vitamins A, D, E, and K are carried in the water portion of the diet.

Short Answer

1. Name two land animals that come close to being as efficient as fish in the conversion of feed to food.
2. Name five structures in the digestive tract of fish.
3. List five categories of feeding habits.
4. Give two examples of feeding behavior.
5. List four factors that increase the energy requirement for fish.
6. List three factors that decrease the energy requirement for fish.
7. As fish digest feeds, in what forms is energy lost?
8. List three factors that affect the digestibility of fat.
9. What chemical element is contained in all amino acids and becomes a waste product that can present problems?
10. Name 10 water-soluble vitamins.
11. Name all the fat-soluble vitamins.
12. Vitamin E acts with which micromineral?
13. Deficiency of which vitamin causes nutritional gill disease?
14. Name the six macrominerals.
15. Name six microminerals.
16. Name two reasons for using hormones.
17. What two antibiotics have FDA approval for use in fish?
18. Name six naturally occurring substances that could contaminate fish feed and adversely affect performance.

Essay

1. What is the function of the digestive tract?
2. Define nutrition.

3. Name two essential fatty acids and describe the effect of a deficiency.
4. Give two reasons for including dietary fat in fish diets.
5. Give three reasons for including protein in the diet.
6. Name eight essential amino acids.
6. Describe a protein deficiency.
7. Describe the effects of a vitamin C deficiency.
8. Why are mineral requirements difficult to determine in fish diets?
9. Why should fiber be avoided in fish diets?
10. Why are antioxidants used in fish diets?
11. Describe why pigments are important in fish diets.

KNOWLEDGE APPLIED

1. Obtain some freshly killed fish for dissection. Identify the external structures. Next, identify the major internal structures, especially those associated with the digestive system.
2. Compare the nutrition information contained on the side panel of a cereal box for humans to the information contained on a feed tag for fish feed. What did you learn about human nutrition?
3. Develop a short report on one of the nutritional deficiencies described in Table 7-7. If possible, give some type of presentation.
4. Produce a poster that describes how amino acids in the fish diet become proteins in the fish body. Start with amino acids in the feed, and end with amino acids in the protein of a fillet.
5. Search the Internet for recent cases of natural or environmental contamination that have affected aquaculture production in a region. Develop a report on one of these cases and indicate how the problem was solved. Use Table 7-10 for suggesting some areas of concern.

LEARNING/TEACHING AIDS

Books

Committee on Animal Nutrition, National Research Council. (1993). *Nutrient requirements of fish*. Washington, DC: National Academy Press.
Halver, J. E., and Hardy, R. W., Eds. (2002). *Fish nutrition, 3rd Ed*. New York, NY: Academic Press.
Hepner, B. (2009). *Nutrition of pond fishes*. Cambridge, MA: Cambridge University Press.
McClanahan, T. and Castilla, J. C. (2007). *Fisheries management: Progress toward sustainability*. Ames, IA: Blackwell Publishing Professional.
Lim, C. and Webster, C. D., Eds. (2006). *Tilapia: Biology, culture, and nutrition*. Binghamton, NY: The Haworth Press Inc.
Lovell, R. T. (1998). *Nutrition and feeding of fish*. New York, NY: Springer Publishing Co.

Stickney, R. R. (1991). *Culture of salmonid fishes*. Boca Raton, FL: CRC Press.

Tucker, C., and Robinson, E. H. (1996). *Channel catfish farming handbook*. Boston, MA: Kluwer Academic Publishers.

Webster. C. D. (2006). *Tilapia: Biology, culture, and nutrition*. Boca Raton, FL: CRC Press.

Webster. C.D. and C.E. Lim (2002). *Nutrient requirements and feeding of fish for aquaculture*, Oxfordshire, England: CABI

Internet

Internet sites represent a vast resource of information. The URLs (uniform resource locator) for the World Wide Web sites can change. Using a search engine such as Google, find more information by searching for these words or phrases: aquaculture nutrition, feed conversion, gastrointestinal tract, digestive system, absorption, metabolic rate, nutrients, carbohydrates, protein, fats, digestible energy, metabolizable energy, essential fatty acids, amino acids, vitamins, minerals, fat-soluble, water-soluble, enzymes, macrominerals, or microminerals.

For some specific Internet sites refer to Appendix Tables A-11, A-12 and A-14.

The goal of feeding fish is to provide a nutritionally balanced diet that supports maintenance, growth, reproduction, and health at a reasonable cost. Diet ingredients should facilitate manufacturing processes and produce a palatable feed with the desired physical properties (Figure 10-1). Finally,

CHAPTER 10

Feeds and Feeding

the diet that aquaculturalists feed to fish should have a minimal effect on water quality.

OBJECTIVES

After completing this chapter, the student should be able to:

- Identify methods for preparing feed and feeding fish
- Name potential ingredients for fish diets
- List the information needed to use a least-cost ration program
- Explain the importance of winter feeding of catfish
- Discuss the different feeding practices for different species of fish
- Calculate the amount of feed needed
- Calculate the feed conversion ration (FCR)
- Calculate feed cost
- Describe the protein and energy levels of typical feeds
- Discuss the effect of water temperature on feeding requirements and weight gain
- Explain the relationship of body weight to feed amount
- Discuss the relationship of the type of digestive system in larval fish to the time of first feeding

Understanding of this chapter will be enhanced if the following terms are known. Many are defined in the text, and others are defined in the glossary.

KEY TERMS

- Automatic feeder
- Coolwater species
- Demand feeder
- Extruded
- Feed conversion ratios
- Humectants
- Least-cost
- Microencapsulation
- Mycotoxins
- Natural foods
- Planktonic
- Premixes
- Proximate composition
- Yolk sac

DIET FORMULATION AND PROCESSING

Nutrient and energy recommendations made by the NRC or any other entity are just that—recommendations. Fish size, management, and environment can profoundly influence dietary requirements. Recommendations are adjusted up or down depending on the best judgment of the aquaculturalist or nutritionist.

Formulating Diets

When formulating fish diets, consider protein first, adjusting the energy to the optimal ration. Protein provides essential amino acids.

The amount of carbohydrate used depends on the fish species and the type of processing. Fat added to the diet provides energy and needs to meet the essential fatty acid requirement. Consider mineral supplements after determining the mineral content of the major feedstuffs. Vitamin **premixes** supply vitamins in diets because of the variation in vitamin content and availability of feedstuffs. Processing destroys some nutrients. To overcome this, these nutrients are supplied at higher rates.

Feeding represents a major share of the costs associated with production. **Least-cost** formulation is used to develop nutritionally complete diets at the minimum cost. Least-cost computer programs require the following information:

- Nutrient requirements of the fish
- Availability of the nutrients to the species of fish
- Energy content of the ingredients
- Minimum and maximum restrictions supplied in ingredients
- Costs of the ingredients

Table 10-1 shows a typical catfish feed.

Ingredients

Feedstuffs used to prepare fish diets can be classified as sources of protein, energy, essential fatty acids (EFS), minerals, and vitamins. Special ingredients include those to enhance palatability, to preserve, to increase growth, or to pigment the flesh.

Feed composition tables list the common feedstuffs used in fish diets and indicate the energy content, protein composition, amino acid content, vitamin content,

FIGURE 10-1 Large feed mills in the Mississippi delta produce complete catfish feeds and buy large quantities of the feed ingredients.

TABLE 10-1 FORMULA FOR A NUTRITIONALLY COMPLETE 32 PERCENT PROTEIN CATFISH FEED SUITABLE FOR PELLETING OR EXTRUDING

Ingredient	Lbs/Ton	Percent
Menhaden fish meal	160.0	8.00
Soybean meal, 48% protein	965.0	48.25
Corn	582.0	29.10
Rice bran or wheat shorts	100.0	10.00
Dicalcium phosphate	20.0	1.00
Pellet binder	40.0	2.00
Fat (sprayed on finished feed)	30.0	1.50
Trace mineral mix Manganese Iodine Copper Zinc Iron Cobalt	1.0 23.00 g/ton 5.00 g/ton 3.00 g/ton 136.00 g/ton 40.00 g/ton 0.05 g/ton	0.05
Vitamin mix Thiamine Riboflavin Pyridoxine Pantothenic acid Nicotinic acid Folic acid Vitamin B_{12} Choline chloride (70%) Ascorbic acid Vitamin A Vitamin D3 Vitamin E Vitamin K	2.5 10.00 g/ton 12.00 g/ton 10.00 g/ton 32.00 g/ton 80.00 g/ton 2.00 g/ton 0.008 g/ton 500.00 g/ton 340.50 g/ton (4,000,000 IU) (2,000,000 IU) 50.00 g/ton 10.00 g/ton	0.125
Coated ascorbic acid	0.75	0.0375

and mineral content of the feedstuff. These tables are used by aquaculturalists and fish nutritionists to prepare diets that economically meet the nutritional needs of fish.

The best source of protein for fish diets is generally an animal source. Oddly, the best source of a high-quality protein is fishmeal prepared from good-quality whole fish. It is highly digestible and palatable. Fishmeal from processing waste has less high-quality protein and a high mineral (ash) content. Meat and bonemeal from livestock and poultry by-products are fair but not as good as fishmeal. Dried blood meal is rich in protein but low in some essential amino acids.

Of the plant protein sources for fish diets, soybean meal is widely available and its amino acid content is the best of the plant proteins. Other plant protein sources such as cottonseed and peanut meal have been used in fish diets, but they lack some essential amino acids. Table 10-2 compares the composition of some protein sources.

TABLE 10-2 PROTEIN SOURCES FOR FISH DIETS

Feedstuff	Dry Matter (%)	Average Digestible Energy (Kcal/kg)	Crude Protein (%)	Fat (%)	Ash (%)	Problems
Blood meal	93	3188	89.2	0.7	2.3	Low in methionine and unbalanced
Canola meal	93	2549	38.0	3.8	11.1	Higher in fiber and tannins
Cottonseed meal	92	2792	41.7	1.8	6.4	Limiting in lysine and methionine
Fishmeal	92	3778	62.0	7.1	20.7	None
Fishmeal, by-product	92	—	50.8	9.6	10.4	Less high-quality protein
Meat and bonemeal	94	3058	50.9	9.7	29.2	High mineral (ash) content; use high-quality ration
Peanut meal	92	3370	49.0	1.3	5.9	Limiting in lysine and methionine
Poultry by-product	93	3546	59.7	13.6	14.5	Less high-quality protein
Poultry feather meal	93	3325	83.3	5.4	2.9	Digestibility low unless hydrolyzed
Soybean meal	90	3010	44.0	0.9	5.8	Heating destroys most anti-nutritional factors

Grains contribute carbohydrates to fish diets. Grains such as corn, wheat, and barley contain 60 to 70 percent starch. Besides supplying energy to a diet, starch acts as a binding agent in steam-pelleted and **extruded** fish feed.

Fats and oils used in fish diets supply energy and coat pellets to reduce abrasiveness and dustiness. Fats and oils also provide the essential fatty acids (EFAs). Marine fish oils represent one of the best sources of EFAs.

Feed composition tables provide useful guides to selecting and planning ingredients for fish diets. However, major ingredients of a fish diet should be subjected to a complete analysis to assure the producers of its quality. This analysis should include:

- **Proximate composition** (protein, fat, ash)
- Limiting amino acids
- Essential fatty acids
- Digestibility
- Test for **mycotoxins**
- Screening for pesticides and other contaminants

Some feedstuffs vary considerably in quality and composition. Established standards help producers evaluate an ingredient's value.

Processing

To maintain water quality and allow efficient consumption, feeds are processed into water-stable granules or pellets. Five processing methods prepare fish feeds:

1. Steam pelleting
2. Extrusion
3. Moist or semi-moist
4. Crumbling
5. Microencapsulation

In steam pelleting, moisture, heat, and pressure force ingredients into large, compact particles that sink in water. Steam helps gelatinize any starch, which helps bind the ingredients. All ingredients are finely ground before pelleting. During the process, fiber and fat content can prevent firm bonding. Sometimes special binding agents are used. After pelleting, the pellets are cooled and dried immediately.

The extrusion process uses the feed mixture in the form of a dough. To produce the feed, this dough is forced through a small opening at high pressures and temperatures. This process produces a feed that will float because of water vapor trapped in the feed. After extrusion, the feed is dried immediately. Overall, the extrusion process requires more elaborate equipment and more moisture, heat, and pressure than pelleting.

Aquaculturalists prefer extruded feeds in large ponds because they can see the fish feed.

Moist or semi-moist feeds are prepared by adding moisture and a binding agent such as carboxymethyl-cellulose or ground wet animal tissue to the dry feed ingredients. These soft pellets are more palatable, but they are more susceptible to spoilage. Moist or semi-moist feeds are frozen, or they contain **humectants**, like propylene glycol, that prevent bacterial growth. Also, they contain fungistats, like propionic or sorbic acid. For best quality, these diets are stored in hermetically sealed containers, under nitrogen, and kept at a low temperature until used.

Crumbling prepares diets for small fish. In this process the diet is first pelleted. Pellet size is reduced by crumbling. Next, screening separates the crumbles by particle size. Crumbling increases the surface area of the feed that is exposed to water, so crumbling requires some compaction to prevent leaching of water-soluble vitamins.

Microencapsulation is the coating of small feed particles with a substance that is insoluble in water but digestible by the enzymes in the fish's digestive tract. This process reduces disintegration and leaching in the water. Some compounds used include nylon cross-linked proteins, calcium alginate, and oils.

FEEDING AQUATIC ANIMALS

Proper feeding of crustaceans, mollusks, alligators, and frogs is important to commercial ventures. Chapters 7 and 8 contain limited but sufficient information about feeding these species. What follows are common feeding practices for prominent finfish: channel catfish, rainbow trout, striped bass, tilapia, and larval fish.

Catfish

Most catfish feed manufacturers now use the least-cost instead of fixed-feed method of feed formulation, where the formula varies, within limits, as ingredient prices change. The kind or amount of ingredients needed to provide essential nutrients for catfish is not secret. Feed manufacturers should be willing to reveal the type and amount of ingredients in their feed. These manufactured feeds are delivered to the catfish farms and stored in bulk tanks (Figure 10-2).

Form and Size

Not only must the feed contain all the essential nutrients, it must also be palatable to the catfish and of a size that can be eaten. If catfish do not or cannot eat the feed, maximum growth is not achieved, costing the producer money. The feed must be offered in such a way and at such a time that promotes total consumption. The forms and sizes of feed available include four types:

- Meal
- Crumbles
- Floating (expanded or extruded) pellets
- Sinking (hard or compacted) pellets

Feed size and form used depends on fish size, water temperature, and type of management. Meal and crumbles are used for fry and small fingerlings. Although more expensive, extruded or floating feed is

FIGURE 10-2 Large trucks deliver processed feeds to bulk storage tanks on catfish farms.

generally preferred when water temperatures are above 65°F because feeding behavior is much easier to monitor. Most producers feel that be able to observe fish when feeding is well worth the extra cost.

Sinking feed is used when the water temperature falls below 65°F because catfish reduce their feeding activity at colder temperatures and seldom come to the surface. Topping or multiple harvest is the most common production scheme used in intensive pond production of catfish. The size of catfish in a pond at any given time may vary from 0.02 to 2 lbs or larger. Because it is not practical to feed the catfish two or more sizes of feed every day, most farmers compromise by feeding one size pellet.

Feeders

Catfish may be fed by hand from the bank or a boat, or using some type of mechanical feeder. Hand feeding more than 10 acres of intensively cultured catfish ponds is too time-consuming and laborious. Larger farms use some type of mechanical feeder.

A blower-type feeder with a 1- to 3-ton hopper, mounted on a truck bed or pulled by a tractor, is best. A blower-type feeder can be calibrated to blow a known amount of feed per minute or equipped with a scale that allows the operator to measure the amount fed. The hopper of the blower is filled from a bulk storage tank. (See Figure 10-3.)

Two other types of mechanical feeders are the **demand feeder** and the **automatic feeder**. Neither has any place in the intensive pond production of catfish because frequent observation during feeding is impossible. The demand feeder is activated by the catfish, thus allowing a few large, aggressive "hogs" to consume most of the feed. When

FIGURE 10-3 Blower-type feeder mounted on a truck bed, receiving feed from a storage bin located between catfish ponds.

demand feeders are used in intensive production ponds, there is a large difference in the size of the fish produced. Waterfowl, particularly coots, quickly learn to use demand feeders and may consume more feed than the catfish. The automatic feeder is programmed to release specific amounts of feed at predetermined times during the day. It must be carefully monitored to avoid over or underfeeding the catfish. Unless large numbers of automatic feeders are used, many catfish will not get enough feed, whereas a small number of more aggressive fish will eat most of the feed.

Feeding Rates

Several factors affect the amount of feed a catfish will eat:
- Water temperature
- Water quality
- Size of the feed
- Palatability or taste of the feed
- Frequency of feeding
- Feeding method
- Location of feeding sites
- Type of pellet used
- Health of the fish

Table 10-3 gives the amount to feed daily, based on average expected gains, at stocking rates of 1,000 5-in. fingerlings per acre. To get the amount to feed per acre at higher stocking rates, divide the number of catfish stocked per acre by 1,000 and multiply the answer by the daily amount to feed per acre in Table 10-3.

Estimated feeding rates given in Table 10-4 calculate the amount to feed different size catfish at water temperatures above 70°F. A good estimate of the total weight of catfish in the pond uses this formula: Amount to feed daily = percentage body weight fed × total weight of fish in pond.

EXAMPLE About 40,000 lbs of catfish in ponds are being fed at a rate of 2.5 percent of their body weight daily. Amount to feed daily × 0.025 = 40,000 lbs. = 1,000 lbs of feed.

The amount fed daily must be adjusted at least every two weeks or the catfish soon will be underfed, causing a reduction in both growth and profits. This is best done by taking a sample of fish from the pond, usually by seine, and counting and weighing them. Then use the formula given to calculate the total weight of catfish in the pond at that time.

Total weight in pond = wt. fish in sample × no. fish in pond/no. fish in sample

TABLE 10-3 FEEDING GUIDE BASED ON AVERAGE EXPECTED GAINS WITH A FEED CONVERSION OF 1,000 5-INCH FINGERLINGS PER ACRE

		Column 1	Column 2	Column 3	Column 4
Week	Water Temp. °F	Wt. of 1,000 Fish at Beginning	% of Body Wt. Fed Daily	Wt. of Food Fed/acre/day 1,000 Fish	Feed Conversion
1	55–60	34.0	1.0	0.3	1.75
3	60–65	37.4	1.5	0.6	1.75
5	65–70	41.9	2.0	0.8	1.75
7	70–75	49.4	2.5	1.2	1.75
9	75–80	59.9	3.0	1.8	1.75
11	80–85	75.9	3.0	2.3	1.75
13	85–90	95.4	3.0	2.9	1.75
15	90–95	120.9	3.0	3.6	1.75
17	90–95	151.8	3.0	4.6	1.75
19	90–100	193.4	3.0	5.8	1.75
21	90–95	242.9	3.0	7.3	1.75
23	85–90	310.1	3.0	9.3	1.75
25	75–85	389.6	3.0	11.7	1.75
27	65–75	490.1	2.5	12.3	1.75
29	60–65	595.1	2.0	11.9	1.75

TABLE 10-4 ESTIMATED PERCENT OF BODY WEIGHT CONSUMED DAILY BY DIFFERENT SIZE CHANNEL CATFISH AT TEMPERATURES OF 70°F (21°C) AND ABOVE

Average Weight (Pounds)	Pounds per 1,000 Fish	Estimated Body Weight Consumed Daily
0.02	20	4.0
0.06	60	3.0
0.25	250	2.7
0.50	500	2.5
0.70	750	2.2
1.00	1,000	1.6
1.50	1,500	1.3

> **EXAMPLE** There are 45,000 catfish being fed at 3 percent of their body weight daily, stocked in a 10-acre pond. To adjust the amount to be fed daily for the next two weeks, a sample of fish is seined, counted, and weighed. The sample contains 200 fish weighing 80 lbs. The new amount to feed daily is calculated:
>
> $$\text{Total weight in pond} = \text{wt. of fish in sample} \times \frac{\text{no. fish in pond}}{\text{no. fish in sample}}$$
>
> $$\text{Total weight in pond} = 80 \text{ lbs.} \times \frac{45{,}000 \text{ fish}}{200 \text{ fish}} = 18{,}000 \text{ lbs of catfish in the pond}$$
>
> $$\text{Lbs. to feed daily for next 2 weeks} = 3\% \times 18{,}000 \text{ lbs} = 0.03 \times 18{,}000 \text{ lbs} = 540 \text{ lbs of feed}$$
>
> Total expected weight of fish = 795.4 lbs.   Total weight of food fed = 1,331.2 lbs.

Rather than seining a sample of fish and counting and weighing them every two weeks, culturalists can estimate fish growth. Some assumptions can be based on a pond's historical data or on industry averages about the percent of body weight fed daily and feed conversion factors. Table 10-3 is a feeding guide that shows how to make these calculations. It can be used with good results, but it is much better to use **feed conversion ratios** and percent of body weight to feed daily, which are valid for ponds. It is a good practice to remove a sample of fish occasionally for counting and weighing to calculate the weight of fish present and see how close growth estimates have been.

Another method of estimating the amount of feed to use daily when the water temperature is above 65°F is to feed the fish what they will eat in 10 to 15 minutes. If feed is still floating on the surface at the end of 15 minutes, the fish are being overfed and increasing costs as a result.

Feeding Practices

Manner and time of feeding, as well as the amount and type of feed, can have a profound effect on the growth and size variation and the quality of the catfish produced. A large variation in the size of catfish produced usually is the result of underfeeding or feeding in a small area of the pond. In underfeeding, the larger, more aggressive catfish eat a larger share of the feed and become bigger at the expense of the smaller catfish. This also happens when feed is offered in only a small area of the pond because the larger, more aggressive catfish quickly learn where the feed will be put in the pond and are there waiting for it. To produce catfish uniform in size, and to maximize profits, it is equally important that catfish be fed the proper amount of feed daily and that the food be distributed as evenly over the pond as possible (Figure 10-4).

Feeding twice daily, if possible, will usually improve feed consumption and feed conversion. This means that one-half of the daily allowance is fed in the early morning, and the other half in the late morning. If the

FIGURE 10-4 Tractor-pulled feeder and feed storage bins near catfish ponds in Mississippi.

catfish are fed only once a day, morning is the preferred time because feeding in the late afternoon increases the amount of fat deposited, and this can affect the quality of the processed fish.

Feed should not be offered until the oxygen level of the pond water is at least 4 parts per million (ppm) or milligrams per liter (mg/l) or higher because feed consumption drops dramatically at lower oxygen concentrations. Oxygen requirements for catfish increase greatly during feeding, so it is best not to feed in the late afternoon when oxygen concentrations in the water are decreasing.

Feeding seven days a week maximizes growth. Production time can be decreased by four weeks when compared to feeding only six days a week.

Winter Feeding

Winter feeding is an important management practice. It means more profit for the farmer, and the catfish will be in better condition during the winter and spring to withstand stresses that can cause disease outbreaks.

Two basic winter feeding programs are: (1) give sinking feed at 0.5 to 1 percent of body weight on alternate days when water temperature is above 49°F; or (2) give sinking feed at 0.5 to 1 percent of the body weight whenever the water temperature at a depth of 3 ft is 54°F or higher.

Trout (Salmonids)

Aside from the fish's final sale price, the amount and suitability of feed used for trout farming will be the primary factor determining the profitability of production. Digestive systems of trout and other salmonids are naturally equipped to process foods consisting primarily of protein (mostly from fish), and they can obtain a limited amount of energy from fat and carbohydrates. Although a farm could produce its own fish food, it is usually uneconomical to do so.

Form and Size

Diets for fry and fingerling trout require a higher protein and energy content than diets for larger fish. Fry and fingerling feed should contain approximately 50 percent protein and 15 percent fat. Feed for larger fish should contain about 40 percent protein and 10 percent fat. The switch to lower protein formulations usually occurs at transition from a crumble feed to a pelleted ration, called a grow-out diet or production diet.

Feeders

After selecting a quality feed and setting the amount of feed, the next consideration is how to feed the fish. Specific methods for feeding trout are somewhat dependent upon fish size. First-feeding fry should be fed a small amount by hand at least 10 times per day until all the fish are actively feeding. After this period, an automatic feeder is most practical, with two or three hand feedings daily to observe the fish.

As the fry grow, feeding frequency can be gradually decreased to about five times per day. Trout can hold roughly 1 percent of their body weight in dry feed at each feeding, so frequency should be adjusted accordingly. Fry gain weight rapidly, so they should be sample counted weekly for the first four to six weeks on feed, and the daily feed ration should be adjusted according to their weight. Feed should be distributed over at least two-thirds of the water surface when fry are less than 2 in. This assures easy access to the feed and will help to achieve size uniformity within the population.

Although the use of a published feeding chart is strongly recommended, charts are only guides, and individual judgment should be exercised based on observations. Overfeeding allows feed to settle to the tank bottom, where small trout will ignore it. Excess feed leads to deterioration of water quality and promotes disease. Remove excess feed promptly.

After fingerlings are moved out to tanks or earthen ponds, a variety of feeding alternatives are available. Hand feeding is generally not practical on a large commercial farm except in certain situations. Several types of automatic and mechanical feeders are available for trout farming, including electric, water-powered, and solar-powered feeders with variable timers. Some automatic feeders use compressed air to blow feed out over the water surface at set intervals, and truck or trailer mounted units that have hydraulically operated blower feeders.

One type of feeder commonly used on commercial trout farms is the demand feeder (Figure 10-5). This consists of a hopper for holding the feed pellets and a movable disc below the hopper opening that is attached to a pendulum extending to the water. Trout greater than 5 in can be readily trained to feed themselves, and, with careful adjustment of the feeders, rapid weight gain and efficient feed use can be reached. Demand feeders

FIGURE 10-5 Demand feeder for trout.

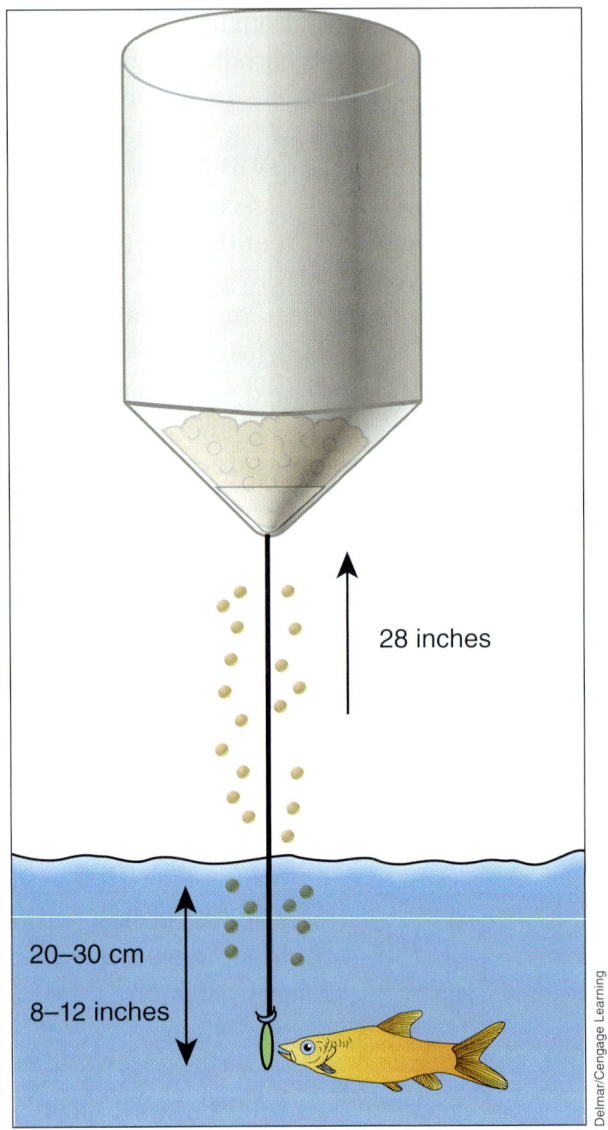

can eliminate the sharp oxygen decline that occurs when fish are fed by hand or machine a few times each day. Demand feeders also reduce labor costs associated with daily hand feeding. Enough feed for several days can be loaded. Disadvantages include the tendency to allow overfeeding due to improper feeder adjustment, and food release only in a small section of the pond or tank. Overfeeding with demand feeders can be a problem with larger trout.

Even if demand feeders are used, feeding according to a feed chart is recommended for best performance. When feeding by hand or with a mechanical distribution system (Figure 10-6), feed should be distributed throughout the pond and should not accumulate on the bottom. In concrete tanks, trout will feed on some pellets that fall to the bottom, but trout will rarely pick up pellets from the bottom of earthen ponds.

FIGURE 10-6 An automatic feeder used at a large trout facility in Idaho.

Feeding Rates

The primary goals in feeding trout are to grow the fish as fast and efficiently as possible, maintaining uniformity of growth with the least degradation of water quality. The amount of feed required by trout is dependent on water temperature and fish size; during normal production, trout should be fed seven days per week with a high-quality commercially prepared diet formulated for trout. Due to higher metabolic rates, smaller fish need more feed relative to their body weight than do larger fish, and fish in warmer water need more feed than fish in cooler water. Because fish are poikilothermic (cold-blooded) their body temperatures and metabolic rates vary with environmental temperatures.

In trout, the minimum temperature for growth is approximately 38°F. At this temperature and below, appetites may be suppressed and digestive systems operate very slowly. Trout will require only a maintenance diet of 0.5 percent to 1.8 percent body weight per day, depending upon fish size at these temperatures; more than this will result in poor food conversion and wasted feed. Above 38°F, the metabolism and growth rate of trout will increase with temperature until approximately 65°F, depending upon the genetic strain being cultured. Optimum temperatures for efficient growth are from 55° to 65°F. At these temperatures, feeding rates should be at maximum levels (1.5 percent to 6.0 percent body weight per day). Above 65°F, the metabolic rate will continue to increase until the temperature approaches lethal levels. Oxygen-carrying capacity of the water and respiratory requirements of the fish will limit the amount of food to be processed efficiently.

In very warm water (above 68°F), a trout's digestive system does not use nutrients well, and a larger proportion of the consumed feed is only partially digested before being eliminated. This nutrient loading of the

HOW LONG WILL FISH MEAL BE USED TO FEED FISH?

At the present time, feed for salmonids (trout, charr, salmon) contains high levels of fish meal. Other cultured fish—such as tilapia, catfish, and carp—can grow and feed without fish meal in their diet. Typically, about 35 to 55 percent of a salmonid diet consists of protein. In an average salmonid feed, 60 percent of that protein is likely to be fish meal. With increasing pressure on the world's wild-harvest fisheries, some individuals are questioning the logic and sustainability of continuing to harvest fish to feed other carnivorous fish. In addition, fish feed is typically the biggest variable production cost for farmers, and the price of fish meal is on the rise. Finding a practical protein alternative for expensive fish meal has been a top priority for researchers.

The use of fish meal in aquaculture causes other environmental concerns as well. It contains levels of phosphorus far beyond the requirement for optimal growth in fish. The excess phosphorus goes into the water, causing problems such as eutrophication or excess algae growth.

Researchers are using biotechnology to produce alternative plant-based protein sources. Plant protein has the potential to address the problem of phosphorus pollution because plants do not contain such high phosphorus levels. Moreover, the use of plant protein in aquaculture would help take the pressure off wild fish stocks.

Research at some institutions is focusing on the investigation of local crops as new sources for fish feed protein. Some of the potential fish meal replacements include distiller's by-products, legumes, and canola. Wheat, canola, and canola oil are already being used to some extent in feed for aquaculture. In order for crops to be used as the main protein source for fish, the crops must first be processed into a concentrate. Biotechnology is often used in this processing. Plant protein also requires processing because plants contain antinutritional compounds. These compounds must be destroyed during processing, or they could harm the fish.

There is hope. Researchers have been able to experimentally replace a large proportion of dietary fish meal at a reasonable cost while still obtaining optimal growth rates. The next challenge is to replace fish meal altogether, in a way that maintains production efficiency while promoting economic and environmental sustainability.

---

water, coupled with generally lower oxygen levels in warm water, can easily lead to respiratory distress and should be avoided. Under these conditions, feeding rates should be reduced enough to maintain good water quality and avoid wasting feed.

The best way to determine the correct amount and sizes of feed needed for trout production is to use a published feeding chart, usually provided by the feed manufacturer. The chart should be used as a guide, and it may need to be adjusted to fit specific conditions on individual farms. Overfeeding will cause the fish to use feed less efficiently and will not increase growth rates significantly. Knowing the number and size of the fish in a facility helps culturalists determine an appropriate amount of feed. At water temperatures above 55°F, a sample count of the fish, every week, can be used to adjust feeding percentages. In cooler waters, a sample count every two weeks usually is adequate. Good growth records help predict the seasonal growth rate.

At times, feeding should be restricted or stopped altogether, such as when water temperatures drop much below 40°F or rise much above 68°F. Feeding rates should also be reduced when fish are sick, as appetite will be depressed. Fish should always be kept off feed for a period before handling or transporting. For routine handling, such as grading or vaccinating, 24 hours without food is sufficient. If fish are to be transported a long distance or are to be processed, they should be kept off feed for a minimum of three to four days, longer if temperatures are low.

Special Purpose Feeding

Commonly used specialty feeds (Figure 10-7) for trout include those containing antibiotics or carotenoid pigments (canthaxanthin). Antibiotic treated feeds should be used only with diagnosis of a bacterial condition susceptible to treatment. Carotenoid pigmented feeds impart a pink or red coloration to the flesh of fish and do not affect their health or growth rate. Successful pigmentation can be achieved in approximately three months when fish are actively growing and in approximately six months during colder water conditions. Other specialty diets include an enriched diet for broodfish and a high-fat diet (16 to 24 percent fat) for producing an oilier fish used for smoking or for specialty markets.

FIGURE 10-7 Feed mill in Idaho producing trout feed.

Other Coolwater Fish

For many years, fish culture was classified into two major groups. Coldwater hatcheries cultured trout and salmon, and warmwater hatcheries cultured any fish not a salmonid. Muskellunge, northern pike, walleye, and yellow perch prefer temperatures warmer than those suited for trout but colder than those water temperatures most favorable for bass and catfish. The term **coolwater species** gained general acceptance in referring to this intermediate group.

Extensive pond culture of coolwater fish involves providing sufficient quantities of microorganisms and plankton as **natural foods** through pond fertilization programs. If larger fingerlings are to be reared, the fry are transferred, when they reach approximately 1.5 in. in length, to growing ponds where minnows are provided for food. A major problem in extensive pond culture is that the fish culturalist is unable to control the food supply, diseases, or other factors. Many times it is extremely difficult to determine the health and growth of pond-cultured fish.

Coldwater fishes can be successfully reared on dry feed. Starter feed is distributed in the trough by automatic feeders set to feed at five-minute intervals from dawn to dusk. Coolwater fish will not pick food pellets off the tank bottom, so it is necessary to continually present small amounts of feed with an automatic feeder. In some situations, coolwater fry are started on brine shrimp and then converted to dry feed.

Tiger muskie fry aggressively feed on dry feeds. Fry often follow a food particle through the entire water column before striking it. Hand-feeding or human presence at the trough does not disrupt feeding activity. When fish reach a length of 5 to 6 in, human presence next to a trough or tank can disrupt feeding activity completely. Cannibalism generally is a problem only during the first 10 to 12 days after initial feeding, when the fish are less than 2 to 3 in long. Removing weak and dying fry reduces cannibalism.

The methods developed for estimating feeding rates for salmonids can be adapted for use with coolwater species.

Tilapia

Pond culture is the most popular method of growing tilapia. One advantage is that the fish are able to use natural foods. Management of tilapia ponds range from extensive systems, using only organic or inorganic fertilizers, to intensive systems, using high-protein feed, aeration, and water exchange. The major drawback of pond culture is the high level of uncontrolled reproduction that may occur in grow-out ponds. Tilapia recruitment, the production of fry and fingerlings, may be so great that offspring compete for food with the adults. The original stock becomes stunted, yielding only a small percentage of marketable fish weighing 1 lb or more.

With optimal temperatures, feeding rates depend on fish size and density. Optimal daily feeding rates for tilapia are:

Tilapia Weight (g)	Feed as a Percent of Body Weight
30	3.5
50	3.0
100	2.5
175	2.0
450	1.5

If densities are high, suboptimal feeding rates may have to be used to maintain suitable water quality, increasing culture duration.

Feeding in Cage Culture

After proper stocking, the most important aspect of cage culture is providing good quality feed in the correct amounts to the caged fish. The diet should be nutritionally complete, containing vitamins and minerals. Commercial pellet diets for tilapia, catfish, or trout are best. Protein content should be 32 to 36 percent for 1- to 25-gm tilapia and 28 to 32 percent for larger fish.

Floating feeds allow observation of the feeding response and are effectively retained by a feeding ring on the cage. Because it takes about 24 hours for high-quality floating pellets to disintegrate, fish may be fed once daily in the proper amount, but twice-daily feedings are better.

Sinking pellets can be used, but extra care must be taken to ensure that they are not wasted. Sinking pellets disintegrate quickly in water and have a greater tendency to be swept through the cage sides. More than one feeding is needed each day. Tilapia cannot consume their daily requirement of feed for maximum growth in a single meal of short duration. Fish of less than 25 gm should be fed at least three times day.

Sinking pellets may be:
- Slowly fed by hand, allowing time for the fish to eat the feed before it sinks through or is swept out of the cage
- Placed in shallow, submerged trays
- Placed in demand feeders

Feeding slowly by hand is inefficient. Use of a tray allows quick placement of feed onto the tray, but multiple daily feedings are still required. The correct amount of feed must be weighed daily. Feeding rate tables or programs are required to make periodic increments in the daily ration. Feeding adjustments can be made daily, weekly, or every two weeks. The fish should be sampled every four to six weeks to determine their average weight and the correct feeding rate for calculating adjustments in the daily ration. Adjustments can be made between sampling periods by estimating fish growth based on an assumed feed conversion ratio (feed weight divided by fish weight gain).

EXAMPLE With a feed conversion ratio of 1.5, the fish would gain 10 gm for every 15 gm of feed. The correct feeding rate, expressed as percent of body weight, is multiplied by the estimated weight to determine the daily ration. Recommended feeding rates are listed in Tables 10-5 and 10-6.

TABLE 10-5 EXPECTED AVERAGE FINAL WEIGHTS FOR DIFFERENT CULTURE PERIODS AND INITIAL WEIGHTS OF TILAPIA

	Starting Weight 30 g	Starting Weight 60 g	Starting Weight 100 g
Length of Growing Season (Weeks)	Expected Average Final Weight (g)[1]		
12	200	270	350
16	250	340	440
20	310	410	520
24	370	480	600
28	420	550	690

[1] Values are for male populations.

TABLE 10-6 RECOMMENDED DAILY FEEDING RATES, EXPRESSED AS PERCENTAGE OF BODY WEIGHT, FOR TILAPIA OF DIFFERENT SIZES

Fish Weight (grams)	Feeding Rate (percent)	Fish Weight (grams)	Feeding Rate (percent)
1	11.0	30	3.6
2	9.0	60	3.0
5	6.5	100	2.5
10	5.2	175	2.5
15	4.6	300	2.1
20	4.2	400+	1.5

Feeding rate tables serve as guides for estimating the optimum daily ration, but they are not always accurate under a wide range of conditions, such as fluctuating temperatures or dissolved oxygen. Demand feeders can be used to eliminate the work (feed weighing, fish sampling, calculations) and uncertainty of feeding rate schedules by letting the fish feed themselves. Fish quickly learn that feed is released when they hit a rod that extends from the funnel into the water. Demand feeders and feeding rate schedules produce comparable growth and feed conversion, but demand feeders reduce labor by nearly 90 percent. Feeding rate schedules may still be used with demand feeders by adding a computed amount of feed daily instead of refilling the feeder whenever it is nearly empty.

With high-quality feeds, good growing conditions, and effective feeding practices, feed conversion ratios as low as 1:3 have been obtained. Generally, feed conversion ratios will range from 1.5 to 1.8 lbs of feed per 1 lb of fish.

Tank Culture of Tilapia

Fry are given a complete diet of powdered feed (40 percent protein) that is fed continuously throughout the day with automatic feeders. The initial feeding rate can be as high as 20 percent of body weight per day under ideal conditions—good water quality and temperature (86°F). It is gradually lowered to 15 percent by day 30. During this period, fry grow rapidly and will gain close to 50 percent in body weight every three days. Therefore, the daily feed ration is adjusted every three days by weighing a small sample of fish in water on a sensitive balance. If feeding vigor diminishes, the feeding rate is cut back immediately and water quality is checked.

Feed size can be increased to various grades of crumbles for fingerlings (1 to 50 gm), which also require continuous feeding for fast growth.

During the grow-out stages, the feed is changed to floating pellets to allow visual observation of the feeding response. Recommended protein levels are 32 to 36 percent in fingerling feed and 28 to 32 percent in feed for larger fish. Adjustments in the daily ration can be made less often (for example, weekly) because relative growth, expressed as a percentage of body weight, gradually decreases to 1 percent per day as tilapia reach 1 lb in weight. The daily ration for adult fish is divided into three to six feedings that are evenly spaced throughout the day.

If feed is not consumed rapidly (within 15 minutes), feeding levels are reduced. Dissolved oxygen (DO) concentrations decline suddenly in response to feeding activity. Although DO levels generally decline during the day in tanks, feeding intervals provide time for DO concentrations to increase somewhat before the next feeding. Continuous feeding of adult fish favors the more aggressive fish, which guard the feeding area, and causes the fish to be less uniform in size. With high-quality feeds and proper feeding techniques, the feed conversion ratio should average 1:5 for a 1-lb fish.

OTHER WARMWATER FISH

As long ago as 1924, fish culturalists attempted to increase yield and survival of smallmouth bass by providing a supplemental feed of zooplankton.

Ground fresh-fish flesh also was successfully used, but costs were prohibitive. These early attempts were discouraging, but culturalists have continued to rear bass fry to fingerling size on naturally occurring foods in fertilized earthen ponds. This method generally results in low yields and is unpredictable.

Interest in supplemental feeding of bass has been renewed in recent years due to successful experimental use of formulated pelleted feeds with largemouth bass fingerlings. The best success in supplemental feeding has been obtained by rearing bass fry in earthen ponds on natural feed to an average length of 2 in. before they are put on an intensive training program to accept formulated feed.

Striped Bass

Striped bass fingerlings often are fed supplemental diets in earthen ponds when zooplankton blooms have deteriorated or larger fish are desired. The fingerlings are fed a high-protein (40 to 50 percent) salmonid type of formulated feed at the rate of 5 lbs per acre per day. This is increased gradually to a maximum of 20.0 lbs per acre per day by the time of harvest. The fish are fed two to six times daily.

When striped bass fingerlings reach a length of approximately 1.5 in. they will accept salmonid-type feeds readily. Success can be anticipated when a training program is followed. Striped bass fingerlings can be grown to advanced sizes in ponds, cages, or raceways. Young striped bass and hybrid striped bass require the essential fatty acids eicosapentaenoic or docosahenaenic acid in the diet for normal growth.

TIME OF FIRST FEEDING

The time of first feeding depends on the type of digestive system in the larval form and the changes that occur during metamorphosis. Larval fish can be divided into three groups.

The first group includes salmonids and catfish. These fish have a functional stomach before changing from the food supplied by the **yolk sac** to external feed. (See Figure 10-8.)

The second group includes fish such as the striped bass and many marine fish. Larval stages of these fish possess rudimentary digestive tracts with no functional stomach or digestive glands. The digestive systems of these fish go through a complex metamorphosis.

Larvae of fish such as carp represent the third group. These larval fish develop a functional digestive system but remain stomachless throughout life. Those larval fish with immature digestive tracts represent greater

FIGURE 10-8 Sketch of fry with yolk sac.

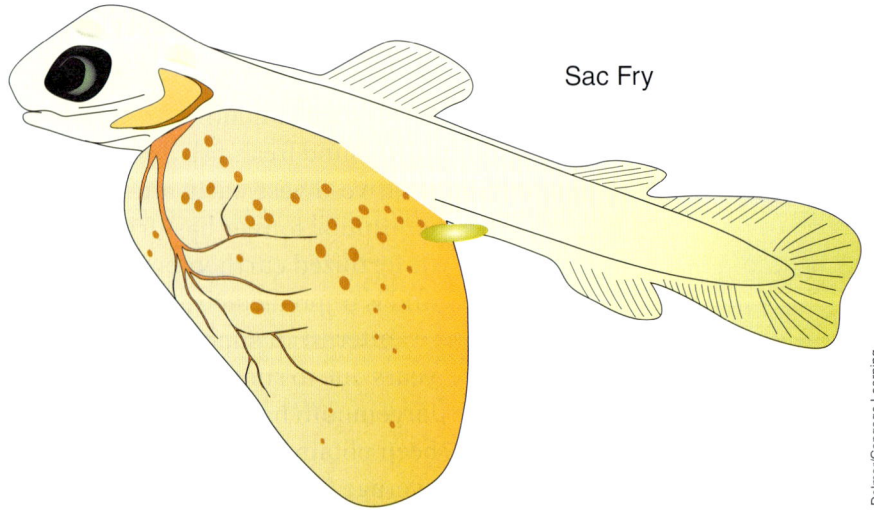

difficulty feeding. The most common practice is to offer food when the fry swim up. Swim up occurs when the fry have absorbed enough of their yolk sac to enable them to rise from the trough bottom and maintain a position in the water column. A considerable amount of work has been conducted to determine when various salmonid fry first take food. Brown trout begin feeding on food approximately 31 days after hatching in 52°F water, whereas food was first found in the stomachs of rainbow trout fry 21 days after hatching in 50°F water. The upper alimentary tract of rainbow trout fry remains closed by a tissue plug until several days before swim up. Thus, to feed rainbow trout fry before swim up would be useless. Yolk absorption is a useful visual guide to determine the initial feeding of most fish species. Most studies indicate that early feeding of fry during swim up does not provide them with any advantage over fry that are fed later, after the yolk sac has been absorbed. Many culturalists start feeding when 50 percent of the fry are swimming up because if fry are denied food much beyond yolk-sac absorption, some will refuse to feed. No doubt, starvation from lack of food will lead to a weakened fry that cannot feed even when food is abundant.

The initial feeding time for warmwater fishes is much more critical than for coldwater species because metabolic rates are much higher at warmer water temperatures. This leads to more rapid yolk absorption and the need for fish to be introduced to feed at an earlier date.

FEED CALCULATIONS

To manage fish nutrition, the aquaculturalist must make some calculations. Specifically, aquaculturalists need to calculate the amount of feed required, the amount of storage required and, finally, the feed conversion ratio. These calculations help determine an operation's profitability.

Feed Requirements

Feed requirements for fish change with age, size, health, and water conditions. Feeding charts are available for catfish and rainbow trout of different sizes and at different water temperatures. The following examples illustrate problems and solutions related to feed requirements.

EXAMPLE 1. A pond is stocked with 45,000 fish that weigh 50 lbs per 1,000 fish. The desired feeding rate is 3 percent of their weight daily. How much feed is needed for one day and for one week?

STEP 1.	$\dfrac{45,000 \text{ fish}}{1,000 \text{ fish} \times 50 \text{ lbs.}}$	= 2,250 lbs. of fish stocked
STEP 2.	2,250 lbs. × 0.03/day	= 67.5 lbs. of feed daily
STEP 3.	For 1 week: 67.5 lbs./day × 7 days/week	= 472.5 lbs.

EXAMPLE 2. A 12-acre pond contains 2,000 lbs of fish per acre. A bacterial disease is diagnosed and double-strength (2×) Terramycin medicated feed is needed for disease treatment. The daily recommended feeding rate is 1.5 percent body weight per day for a total of ten days. How much medicated feed should be ordered and fed to the sick fish?

STEP 1.	Total pounds of fish in pond: 2,000 lbs./acre × 12 acres	= 24,000 lbs. of fish
STEP 2.	Pounds of medicated feed required per day: 24,000 lbs. × 0.015/day	= 360 lbs./day
STEP 3.	Total pounds of medicated feed required for total treatment time of ten days: 360 lbs./day × 10 days	= 3,600 lbs.
STEP 4.	If the medicated feed comes in 50-lb. bags, how many are needed? $\frac{3,600 \text{ lbs.}}{50 \text{ lbs./bag}}$	= 72 bags

Storage Requirement

A producer needs to determine whether buying bulk feed is feasible based on farm size, storage time, and availability of feed sizes and loads. Table 10-7 illustrates how farm size and feeding ratios affect storage.

The following formula can be used to calculate feed storage time:

Storage time in days = Feed bin capacity in pounds
Maximum or average feeding rate in lbs/acre/day

TABLE 10-7 CAPACITY IN DAYS OF FEED FOR TWO SIZES OF BULK STORAGE FEED BINS FOR FIVE FARM SIZES AND THREE FEEDING RATES

Farm Size in Water Acres	Maximum Feeding Rates					
	50 lbs/acre/day Bin Size		75 lbs/acre/day Bin Size		100 lbs/acre/day Bin Size	
	10 Ton	23 Ton	10 Ton	23 Ton	10 Ton	23 Ton
15	26.7	61.3	17.8	40.9	13.3	30.7
40	10.0	23.0	6.7	15.3	5.0	11.5
70	5.7	13.1	3.8	8.8	2.9	6.6
100	4.0	9.2	2.7	6.1	2.0	4.6
140	2.9	5.8	1.9	4.4	1.4	3.3

EXAMPLE 3. The storage time or capacity in days of feed for a 5-acre pond fed at a maximum of 75 lbs/acre/day can be calculated as follows:

STEP 1. Determine the maximum feed requirement for 1 day:

5 acres × 75 lbs/acre/day = 375 lbs/day

STEP 2. If a person is considering bulk purchase and storage in a 10-ton bin, then the capacity in days for feed is: 10 tons × 2,000 lbs/ton = 20,000 lbs feed capacity of 10-ton bin

$$\frac{20,000 \text{ lbs}}{375 \text{ lbs/day}} = 53.3 \text{ days storage time or capacity}$$

This is a long storage time in the bin before new feed is purchased. Storage time should not exceed 30 to 45 days in the summer. For this small farm, bagged feed is recommended to reduce storage time and assure feed quality and freshness.

Feed Conversion Ratio

Feed conversion ratios (FCR) are calculated to determine the cost and efficiency of feeding the fish. FCRs are affected by the quality of feed, size and condition of fish, number of good feeding days related to temperature and water quality, and feeding practices. FCRs for catfish can vary from less than 1.5 to as high as 4 or more. FCRs much higher than 2 should be reduced. The FCR is a ratio of the pounds gained by fish after consuming a known amount of feed. In commercial fish ponds, fish get little nutrition from natural food organisms.

To determine the FCR requires records of the amount of feed that is fed to the fish in each pond, fish losses, and pounds of fish harvested. The FCR can be calculated monthly when fish are sampled and when fish are harvested.

Using this formula, the FCR for any period can be determined—

$$FCR = \frac{\text{Total pounds feed fed}}{\text{Final fish weight: Initial fish weight or weight gain between sampling periods}}$$

EXAMPLE 1. About 67,500 fingerlings weighing 50 lbs per 1,000 fish were stocked in a pond. Later, the fish were sampled and the average fish weight was 0.25 lb., or 250 lbs per 1,000 fish. During this time, 10 tons plus 1,600 lbs of feed were fed. No fish losses were observed. What is the feed conversion ratio?

$$\text{Feed Conversion Ratio (FCR)} = \frac{\text{Amount feed fed}}{\text{Weight gain of fish}}$$

STEP 1. Convert all feed weight to one unit (pounds) instead of having two units (tons and pounds).

One ton = 2,000 lbs
10 tons × 2,000 lbs/ton = 20,000 lbs
= 20,000 lbs + 1,600 lbs
= 21,600 lbs total feed

STEP 2. Determine the weight gain of fish for this period.
The final weight can be calculated using two methods:

Average fish weight × Number fish = Total weight

$$\frac{0.25 \text{ lb}}{\text{fish}} \times 67{,}500 \text{ fish} = 16{,}785 \text{ lbs}$$

$$\frac{67{,}500 \text{ fish}}{1{,}000 \text{ fish}} \times 250 \text{ lbs} = 16{,}785 \text{ lbs}$$

STEP 3. The initial weight of fish was:

$$\frac{67{,}500 \text{ fish}}{1{,}000 \text{ fish}} \times 50 \text{ lbs} = 3{,}375 \text{ lbs}$$

STEP 4. Determine the feed conversion ratio using the following formula:

$$\text{Feed Conversion Ratio} = \frac{\text{Amount feed fed}}{\text{Final weight} - \text{Initial weight}}$$

$$\text{FCR} = \frac{21{,}600 \text{ lbs feed}}{16{,}875 \text{ lbs} - 3{,}375 \text{ lbs}}$$

$$= \frac{21{,}600 \text{ lbs feed}}{13{,}500 \text{ lbs. weight gained}} = 1.6$$

The FCR means that during the grow-out time, the fish consumed an average of 1.6 lbs of feed to gain 1 lb in weight.

EXAMPLE 2. A pond had an estimated standing crop of 22,500 lbs of fish at the last sampling. A new sample estimated the total fish weight at 33,000 lbs. Between these two samplings, about 2,500 pounds of fish were lost. During this time, 11 tons plus 1,400 lbs of feed were fed. What is the feed conversion ratio?

STEP 1. Convert all feed weights to pounds.
(11 tons × 2,000 lbs/ton) + 1,400 lbs =
22,000 lbs + 1,400 lbs = 23,400 lbs feed

STEP 2. Final weight − Last weight = Weight gain
33,000 lbs − 22,500 lbs = 10,500 lbs gained

During this period, 2,500 lbs of fish were lost. These fish were part of the last weight sampling when fish weighed 22,500 lbs. Also, these fish did consume feed before they were lost. From the standpoint of the feed conversion ratio, these lost fish should be included. From the standpoint of economic return, both the weight gain and the feed that the fish consumed were lost. The bottom line is: How many pounds of feed were fed, and how many pounds of fish were produced and marketed? Examine the feed conversion ratio from both standpoints. First, determine the FCR including the fish that were lost.

$$\text{FCR (including lost fish)} = \frac{23{,}400 \text{ lbs feed}}{10{,}500 \text{ lbs} + 2{,}500 \text{ lbs}}$$

$$= \frac{23{,}400 \text{ lbs feed}}{13{,}000 \text{ lbs gain}} = 1.8$$

Next, determine the feed conversion ratio that does not include the lost fish that consumed feed but that you will not market.

$$\text{FCR} = \frac{23{,}400 \text{ lbs feed}}{10{,}500 \text{ lbs (alive)}} = 2.23$$

Fish losses increase the conversion ratio. Even though fish were converting well at 1.8 lbs, the production cost is 2.23 lbs because of the fish losses.

When fish are harvested, their weight should always be recorded and included in the total weight of fish produced to determine the feed conversion ratio. Record-keeping forms should be used to record and use the information that is needed to determine the feed conversion ratio.

EXAMPLE 3. A fish farm has 45 acres of water. The expected annual average production per acre is 3,500 lbs of fish. Approximately how much feed will need to be purchased for the year, and what will be the total feed cost if feed is expected to cost $240/ton? From past experience, the producer expects an FCR of 1.8.

STEP 1. Determine the total pounds of fish expected to be produced on the farm for the year.

3,500 lbs per acre × 45 acres = 57,500 lbs fish

STEP 2. Determine the amount of feed required to produce 157,500 lbs of fish, assuming that this weight represents weight gained and not the total weight of fish produced. Remember that the initial weight of the fish when they were stocked is not taken into account in this example.

$$\text{FCR} = \frac{\text{Pounds of feed fed}}{\text{Weight gain of fish}}$$

$$\text{FCR} = \frac{\text{Pounds of feed fed}}{157{,}500 \text{ lbs weight gain}} = 1.8$$

Pounds of feed = 1.8 × 157,500 lbs = 283,500 lbs

$$\frac{283{,}500 \text{ lbs}}{2{,}000 \text{ lbs}} \times 1 \text{ ton} = 141.75 \text{ tons of feed}$$

STEP 3. Determine the approximate cost of feeding the fish for the year.

141.75 tons × $240/ton = $34,020

TABLE 10-8 COST OF FEED IN CENTS TO PRODUCE A 1-LB FISH AT DIFFERENT FEED CONVERSION RATES AND FEED PRICES

	Feed Cost per Ton				
FCR	250	275	300	325	350
1.5	18.8	20.6	22.5	24.4	26.3
1.6	20.0	22.0	24.0	26.0	28.0
1.7	21.3	23.4	25.5	27.6	29.8
1.8	22.5	24.8	27.0	29.3	31.5
1.9	23.8	26.1	28.5	30.9	33.3
2.0	25.0	27.5	30.0	32.5	35.0

Feed Costs

The production of food fish in the United States involves the use of high-protein feeds. The feed bill is a major production cost. The cost of feeding fish is determined by the feed conversion efficiency of fish and cost of feed. Fish farmers should understand how to estimate their feed requirements over time for planning purposes, and they should know their feed costs.

Table 10-8 illustrates how the feed conversion ratio (FCR) and feed price affect the cost of producing fish. With this information, the per acre feed cost can be estimated.

EXAMPLE. If 4,000 lbs of fish were produced per acre, the feed was $250/ton, and the FCR was 1.8, then what is the total cost of feeding these fish per acre?

From Table 10-8, the cost of feed in cents to produce a 1-lb. fish with feed at $250/ton and FCR at 1.8 is 22.5 cents or $.225.

4,000 lbs/acre × $.225/lb. = $900/acre

If the same fish had an FCR of 2.0 instead of 1.8 with feed at $250/ton, then:

4,000 lbs/acre × $.25/lb. = $1,000/acre or $100/acre higher.

For a 15-acre pond, that means
$100/acre difference × 15 acres = $1,500 either saved or spent.

AQUATIC PLANTS

In the presence of sunlight and chlorophyll, plants convert carbon dioxide and water to sugar and then to starch. This process is photosynthesis.

To release the energy stored in starches during photosynthesis, plants use the process of respiration. The energy released is used for growth and reproduction.

Aquatic plants obtain most of the nutrients they need from the water. Depending on the plants' water temperature, salinity and soil (substrate) varies. Water temperature varies from 50°F to 85°F, depending on the plant. Salinity varies from 35 to 0 ppt. Some plants actually attach themselves to the ground or bottom of the water facility. These are called benthic. Other plants float free. These are called **planktonic**.

SUMMARY

Some anatomical and behavioral traits of fish species influence how they are fed. Digestion and feeding are affected by factors such as age, diet, light, activity, size, species, and water temperature.

Feeding practices vary, depending on the species of fish being fed and the facility in which they are being raised. For all producers, keeping records and calculating the feed conversion ratio (FCR) are essential. The FCR indicates how efficiently feed is being converted to human food.

STUDY/REVIEW

Success in any career requires knowledge. Test your knowledge of this chapter by answering these questions or solving these problems.

True or False

1. Mycotoxins are a good source of plant proteins.
2. Floating feed is produced by the extrusion process.
3. Water temperature influences the amount of feed a fish will eat.
4. Excess feeds can cause the water quality to deteriorate.
5. Floating feeds are fed so that the fish can see the food.
6. Fry swim up when they get their first feed.
7. Cost and efficiency of feeding fish can be determined by calculating the storage requirement.
8. Fish are fed high-carbohydrate diets.

Short Answer

1. Feed conversion ratios for catfish should be around _____.
2. As catfish grow, does the percentage of body weight consumed daily in feed increase or decrease?
3. What causes a large variation in the size of catfish in a pond?
4. Fingerling trout require a diet higher in _____ than do larger fish.
5. For trout, feeding should be restricted or stopped when the water temperature drops below _____ °F or rises above _____ °F.
6. _____ can be fed to trout to produce a pink or red flesh coloration.
7. What is the first consideration when formulating a fish diet?
8. What is the best source of high-quality protein for fish diets?
9. Name five methods of processing fish feed.
10. What is the advantage of feeding fish sinking pellets?
11. What is the common method of feeding intensively cultured catfish?
12. Give three disadvantages of using a demand feeder.
13. As catfish grow, what happens to daily feed consumption?
14. What is the most common method of feeding trout?
15. Some fish are able to use natural foods in the pond. Name two general categories of natural foods.
16. Name two factors that can effect the time of first feeding.
17. A pond is stocked with 50,000 fish that weigh 75 lbs per 1,000 fish. The desired feeding rate is 2.2 percent of their daily weight. How much feed is needed for one day and for one week?
18. Fifty thousand fish weighing 50 lbs per thousand were stocked in a pond. Later, the fish were sampled and the average weight of the fish was 250 lbs per thousand fish. During this time, 14 tons of feed were fed. No fish losses were observed. What is the feed conversion ratio?

19. A pond has an estimated standing crop of 30,000 lbs of fish. A new sampling estimated the total fish weight at 40,000 lbs. Between the two samplings, about 3,000 lbs of fish were lost. During this time, 16 tons of feed were fed. What is the feed conversion ratio?
20. A fish farm has 50 acres of water. The expected annual average production per acre is 2,800 lbs of fish. Approximately how much feed should be purchased for the year, and what will be the total cost if the feed is priced at $275 per ton? The producer expects a feed conversion ratio of 1.6.
21. If 4,500 lbs of fish were produced per acre, with the feed cost at $300 per ton and the feed conversion ratio at 1.9, what would be the total cost of feeding these fish per acre?

Essay

1. Compare the tilapia feeding in an extensive system to tilapia feeding in an intensive system.
2. Describe three methods used to distribute feed to fish.
3. Describe how fish growth is determined.
4. Discuss the relationship between feeding and dissolved oxygen (DO).
5. Computer programs that compute least-cost rations require what type of information?
6. Why do producers use floating feed?
7. What effect does warm water have on the trout digestive system?
8. Why is it important that fish rapidly consume all of the feed?
9. Commonly, fish are offered food when the fry swim up. Describe the swim up.

KNOWLEDGE APPLIED

1. Obtain five samples of high-protein feeds common to your area. Take these samples to a laboratory for protein analysis. Compare their values to those obtained from feed composition tables.
2. Fecal material is a reality of any livestock production operation. Research the differences and similarities in the composition of trout (salmonid), sheep, beef, and dairy fecal waste.
3. Obtain samples of protein feeds—for example, soybean meal, cottonseed meal, meat meal, and fish meal. Use published composition tables and compare each feed. Observe differences in the smell, texture, and other characteristics. Compare cost.
4. Use a computer program to balance a catfish or trout diet. Figure the cost of the diet.
5. Obtain feed labels from trout, catfish, pig, and chicken feed. Compare the contents of each and compare the price.
6. Contact suppliers of aquaculture equipment and obtain information on demand feeders and compare the different types.
7. Obtain samples of floating feed and sinking feed. Place these in water and observe their behavior.
8. Collect samples of fish feed. Develop a display in small bottles. Label the type, size, protein, and energy content.

LEARNING/TEACHING AIDS

Books

Committee on Animal Nutrition, National Research Council. (1993). *Nutrient requirements of fish*. Washington, DC: National Academy Press.

Halver, J. E., and Hardy, R. W., Eds. (2002). *Fish nutrition, 3rd Ed*. New York, NY: Academic Press.

Hepner, B. (2009). *Nutrition of pond fishes*. Cambridge, MA: Cambridge University Press.

Lim, C. and Webster, C. D., Eds. (2006). *Tilapia: Biology, culture, and nutrition*. Binghamton, NY: The Haworth Press Inc.

Lovell, R. T. (1998). *Nutrition and feeding of fish*. New York, NY: Springer Publishing Co.

McClanahan, T. and Castilla, J. C. (2007). *Fisheries management : Progress toward sustainability*. Ames, IA: Blackwell Publishing Professional.

Webster. C. D. (2006). *Tilapia: Biology, culture, and nutrition*. Boca Raton, FL: CRC Press.

Webster. C. D. and C. E. Lim (2002). *Nutrient requirements and feeding of fish for aquaculture*, Oxfordshire, England: CABI

Internet

Internet sites represent a vast resource of information. The URLs (uniform resource locator) for the World Wide Web sites can change. Using a search engine such as Google, you may find more information by searching for these words or phrases: diet formulation, nutrient requirements, essential fatty acids, feed conversion ratio, microencapsulation, proximate composition, feed composition tables, feed processing, or humectants.

For some specific Internet sites refer to Appendix Tables A-11, A-12, and A-14.

Disease is rarely a simple association between a pathogen (disease-carrying organism) and a host fish. Usually other circumstances must be present for active disease to develop. Management practices limiting stress help prevent disease outbreaks.

This chapter makes no attempt to cover all diseases of aquatic animals. Rather, this

CHAPTER 11

Health of Aquatic Animals

chapter presents the importance of good management practices and their interaction with some representative diseases. Chapters 4, 6, 7, and 8 provide some additional details for specific species.

OBJECTIVES

After completing this chapter, the student should be able to:

- Outline fish health management
- List behavioral signs of sick fish
- List common physical signs of sick fish
- List common stressors of fish
- Outline general management measures for preventing disease outbreaks
- List and compare treatment methods
- Complete a list of general guidelines for treatment of fish diseases
- Calculate treatments for fish ponds
- Define terms associated with disease conditions
- Discuss disease resistance
- Define terms associated with severity of disease or condition
- Discuss the role of stress in fish diseases

- Describe the immunization of fish
- List signs of stress and disease
- Discuss common diseases caused by pathogenic viruses
- Discuss common diseases caused by pathogenic bacteria
- Describe a fungal infection
- Name and describe a common pathogenic protozoan parasite
- Name and describe a common pathogenic crustacean parasite
- Describe a grub or fluke infection
- Name and describe a common pathogenic worm parasite
- List noninfectious diseases and give examples
- Describe an infestation of lice

Understanding of this chapter will be enhanced if the following terms are known. Many are defined in the text, and others are defined in the glossary.

KEY TERMS

Abnormalities	Hydrate
Acidosis	Immune
Alkalosis	Infiltration
Antibiotic	Injections
Antibodies	Inoculation
Antigen	Lethargic
Antigenicity	Mortality
Bacteria	Mucus
Bath	Prognosis
Cellular	Secondary
Chronic	Susceptible
Diagnosis	Therapeutic
Dip	Transmission
Flush	Vaccine
Fungus	Virus
Hemoglobin	Visceral
Hemorrhagic	

HEALTH MANAGEMENT

Disease is any condition of an aquatic animal that impairs normal physiological functions. Fish disease outbreaks increase production costs because of the investment lost in dead fish, the cost of treatment, and decreased growth during convalescence. In nature, we are less aware of fish disease problems because sick animals are quickly removed from the population by predators. In addition, fish are much less crowded in natural systems than in captivity. Parasites and **bacteria** may be of minimal significance under natural conditions, but they can contribute to substantial problems when animals are crowded and stressed under culture conditions.

Fish health management describes management practices that are designed to prevent fish disease. Once fish get sick, salvage is difficult.

Successful fish health management begins with disease prevention rather than treatment. Good water-quality management, nutrition, and sanitation prevent fish diseases. Without this foundation, outbreaks of opportunistic diseases are impossible to prevent. As Figure 11-1 illustrates, disease results from the interaction of the host fish with the environment and the pathogen.

The fish is constantly bathed in potential pathogens, including bacteria, fungi, and parasites. Poor water quality, poor nutrition, or **immune** system suppression generally associated with stressful conditions allow these potential pathogens to cause disease. Medications used to treat diseases buy time for fish and enable them to overcome opportunistic infections, but they are no substitute for proper animal husbandry.

Daily observation of fish behavior and feeding activity allows early detection of problems when they do occur so that a **diagnosis** can be made before the majority of the population becomes sick (see Figure 11-2). Successful treatment starts early in the course of the disease, while the fish are still in good shape.

STRESS AND DISEASE

Stress is a condition in which an animal is unable to maintain a normal physiologic state because of various factors adversely affecting its well-being.

FIGURE 11-1 Disease rarely results from a simple contact between the host fish and the disease-causing organism (the pathogen).

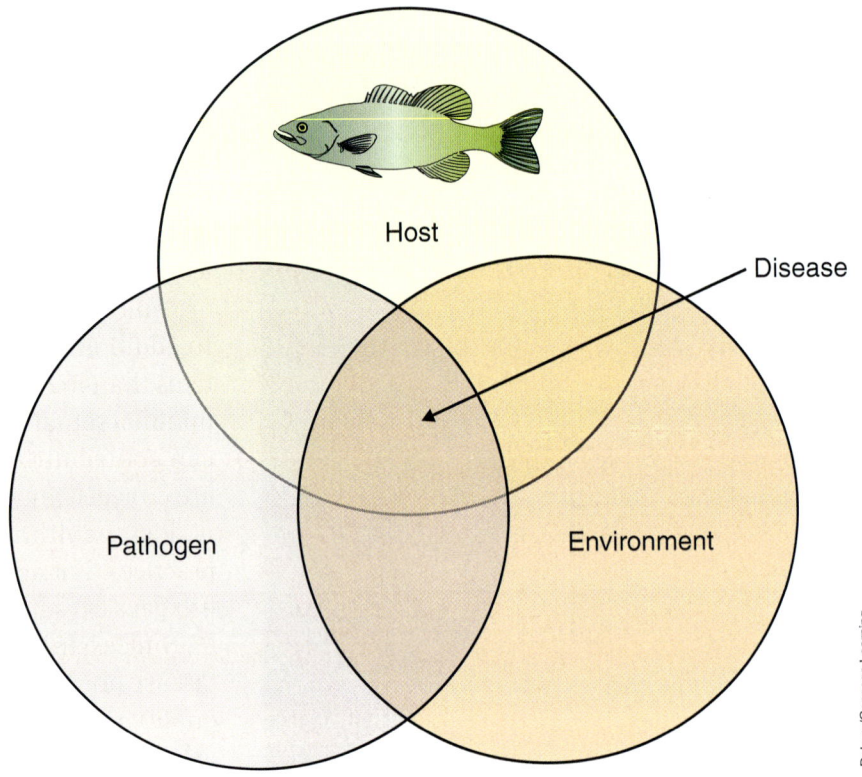

FIGURE 11-2 Checking golden shiner minnow baitfish for disease.

Stress is caused by placing a fish in a situation that is beyond its normal level of tolerance. Specific examples of things that can cause stress (stressors) are listed in Table 11-1.

Stress increases blood glucose levels. This is caused by a secretion of hormones from the adrenal gland (see Figure 11-3). Stored sugars, such as glycogen in the liver, are metabolized. This creates an energy reserve that

TABLE 11-1 STRESSORS IN FISH		
Types of Stressor	Example	Causes
Chemical stressors	Poor water quality Pollution Diet composition Metabolic wastes	Low dissolved oxygen, improper pH Intentional or accidental like chemical treatments and insecticides Type of protein and amino acids Nitrogenous—accumulation of ammonia or nitrite
Biological stressors	Population density Other species of fish Microorganisms Macroorganisms	Crowding Aggression, territoriality, lateral swimming space requirements Pathogenic and nonpathogenic Internal and external parasites
Physical stressors	Temperature Light Sounds Dissolved Gases	Important influence on immune system of fish
Procedural stressors	Handling Shipping Disease Treatments	

FIGURE 11-3 Stress triggers a chain of events that result in an "alarm reaction" fight or flight. If the stress is not removed the energy reserves become depleted and the fish becomes exhausted. Its ability to resist disease is compromised and the fish may become sick or die.

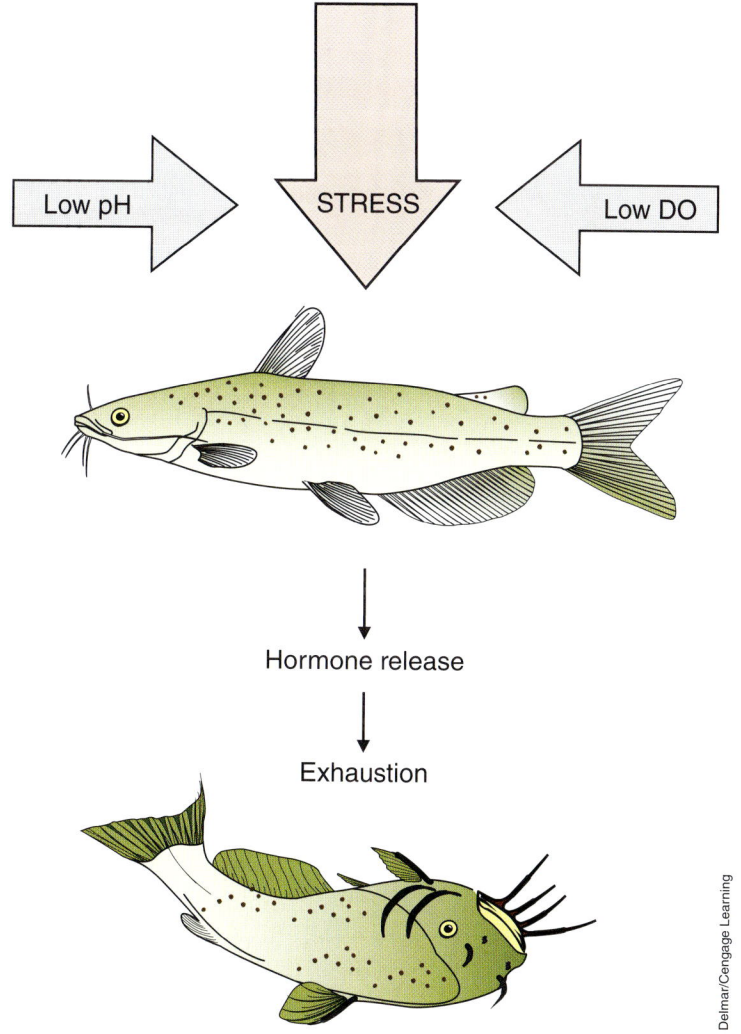

prepares the animal for an emergency action. Osmoregulation is disrupted because of changes in mineral metabolism. Under these circumstances, a freshwater fish tends to absorb excess water from the environment (over-**hydrate**). A saltwater fish will tend to lose too much water to the environment (dehydrate). This disruption requires that extra energy be used to maintain osmoregulation. Respiration increases, blood pressure increases, and reserve red blood cells are released into the circulation. The inflammatory response is suppressed by hormones released from the adrenal gland.

DISEASE RESISTANCE

All fish do not get sick and die each time a disease outbreak occurs. Many factors affect how an individual responds to a potential pathogen. The pathogen—bacteria, parasite, or **virus**—must be capable of causing disease. The host (fish) must be in a **susceptible** state, and certain environmental conditions must be present for a disease outbreak to occur, such as one or more of the stressors in Table 11-1.

PROTECTIVE BARRIERS AGAINST INFECTION

Fish possess four protective barriers to protect against infection:
1. Mucus
2. Scales and skin
3. Inflammation
4. Antibodies

Mucus (slime coat) is a physical barrier that inhibits entry of disease organisms from the environment into the fish. It is also a chemical barrier because it contains enzymes (lysozymes) and **antibodies** (immunoglobulins), which can kill invading organisms. Mucus also lubricates the fish, which aids movement through the water, and it is also important for osmoregulation.

Scales and skin function as a physical barrier protecting the fish against injury. When these are damaged, a window is opened for bacteria and other organisms to start an infection.

Inflammation is a **cellular**, nonspecific response to an invading protein. An invading protein can be a bacterium, a virus, a parasite, a **fungus**, or a toxin. Inflammation is characterized by pain, swelling, redness, heat, and loss of function. It is a protective response and an attempt by the body to wall off and destroy the invader.

Antibodies, a specific cellular response, are molecules formed to fight invading proteins or organisms. The first time the fish is exposed to an invader—an **antigen**—antibodies form that protect the fish from future infection by the same organism. Exposure to sublethal concentrations of pathogens is extremely important for a fish to develop a competent immune system. An animal raised in a sterile environment will have little protection from disease. Young animals do not have an immune response that works as efficiently as the immune response in older animals. Figure 11-4 illustrates the formation of antibodies.

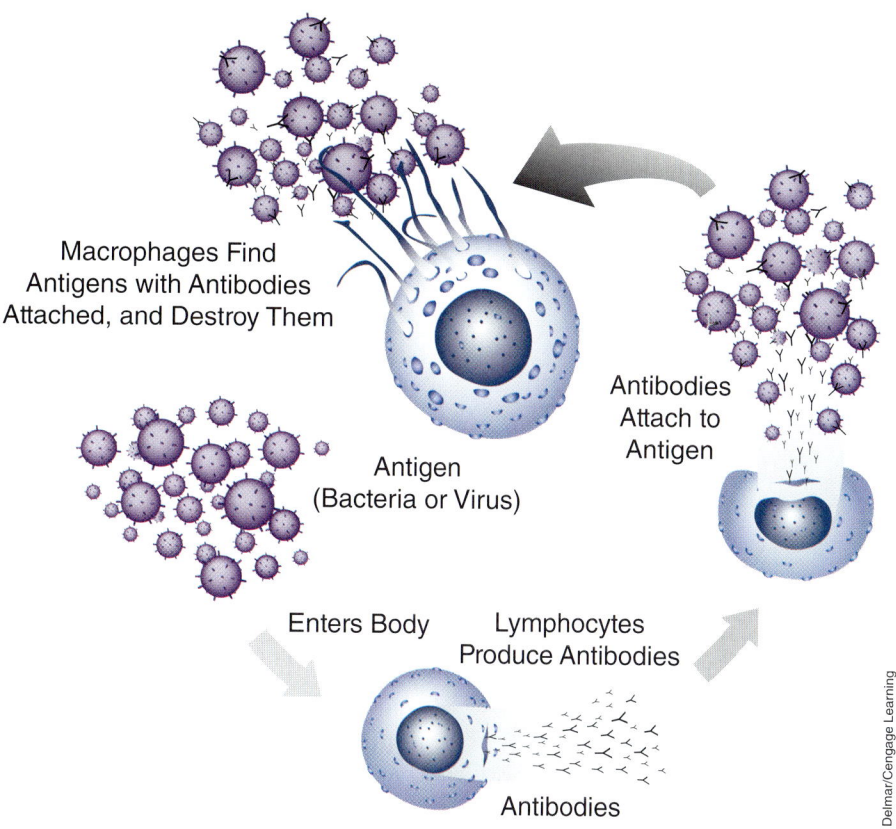

FIGURE 11-4 The formation of antibodies—the immune response.

Effect of Stress on Protective Barriers

Any stress causes chemical changes in mucus that decreases its effectiveness as a chemical barrier against invading organisms. Stress upsets the normal electrolyte—sodium, potassium, and chloride—balance, which results in excessive uptake of water by freshwater fish and dehydration in saltwater fish. The need for effective osmoregulatory (electrolyte balance) support from mucus components is increased.

Handling physically removes mucus from the fish. This decreases chemical protection, osmoregulatory function, and lubrication. The fish uses more energy to swim at a time when its energy reserves are already being used up metabolically. Handling also disrupts the physical barrier against invading organisms.

Chemical stress, like disease treatment, often damages mucus, resulting in loss of protective chemical barrier, loss of osmoregulatory function, loss of lubrication, and damage to the physical barrier created by mucus.

Scales and skin are most commonly damaged by handling stress. Any break in the skin or removed scale creates an opening for invasion by pathogenic organisms. Trauma caused by fighting (reproductive stress or behavioral stress) can potentially cause skin breaks or scale loss.

Parasitic infestations can result in damage to gills, skin, fins, and loss of scales, which can create breaks in the skin where bacteria may enter. Many times, fish that are heavily parasitized actually die from bacterial infections. The parasite problem, associated physical damage, and stress response allow the bacteria in the water to invade the fish, causing a lethal disease.

Temperature stress, particularly cold temperatures, can completely halt the activity of antibodies of the immune system, eliminating an important first defense against invading organisms. Excessively hot temperatures are also very detrimental to fish.

A sharp decrease in temperature severely impairs the fish's ability to quickly release antibodies against an invading organism. The time lapse required to mount an antibody response gives the invader time to reproduce and build up its numbers, giving it an advantage that may allow it to overwhelm the fish.

Prolonged stress severely limits the effectiveness of the immune system. This increases the opportunities for an invader to cause disease. Also, any stress causes hormonal changes that decrease the effectiveness of the inflammatory response.

Prevention of Stress and Disease

Good management prevents stress. This means maintaining good water quality, good nutrition, and sanitation.

Good water quality involves preventing accumulation of organic debris and nitrogenous wastes, maintaining appropriate pH and temperature for the species, and maintaining dissolved oxygen levels of at least 5 ppm. Poor water quality is a common and important stressor of cultured fish and precedes many disease outbreaks.

Good nutrition means feeding a high-quality diet that meets the nutritional requirements of the fish. Each species is unique, and the nutritional requirements of different species vary. Chapter 9, Fundamentals of Nutrition in Aquaculture, provides information on nutrition and digestion.

Proper sanitation implies routine removal of debris from fish tanks and disinfection of containers, nets, and other equipment between groups of fish. Organic debris that accumulates on the bottom of tanks or vats is an excellent medium for reproduction of fungal, bacterial, and protozoal agents. Prompt removal of this material from the environment decreases the number of agents the fish is exposed to. Disinfection of containers and equipment between groups of fish minimizes **transmission** of disease from one population to another.

DISEASE TYPES

Two broad categories of disease affect fish: infectious and noninfectious diseases. Infectious diseases are caused by pathogenic organisms present in the environment or carried by other fish. They are contagious, and some type of treatment may be necessary to control the disease outbreak. In contrast, noninfectious diseases are caused by environmental problems, nutritional deficiencies, or genetic defects. Noninfectious diseases are not contagious and usually cannot be cured by medications.

Infectious Diseases

Infectious diseases are broadly categorized as parasitic, bacterial, viral, or fungal.

Parasitic

Parasitic diseases of fish are most frequently caused by microscopic organisms called "protozoa" that live in the aquatic environment. A variety of protozoans that infest the gills and skin of fish cause irritation, weight loss, and eventually death. Most protozoan infections are relatively easy to control using standard fishery chemicals such as copper sulfate, formalin, or potassium permanganate.

Bacterial

Bacterial diseases are often internal infections and require treatment with medicated feeds containing antibiotics approved for use in fish by the Food and Drug Administration. Typically, fish infected with a bacterial disease will have **hemorrhagic** spots or ulcers along the body wall and around the eyes and mouth. They may also have an enlarged, fluid-filled abdomen and protruding eyes. Bacterial diseases can also be external, resulting in erosion of skin and ulceration. Columnaris is an example of an external bacterial infection that may be caused by rough handling.

Viral

Viral diseases are impossible to distinguish from bacterial diseases without special laboratory tests. They are difficult to diagnose, and no specific medications are available to cure viral infections of fish. Immunization can protect fish from a viral disease, but vaccines do not exist for all viral diseases.

Fungal

Fungal diseases are the fourth type of infectious disease. Fungal spores are common in the aquatic environment but are not normally a problem in healthy fish. When fish are infected with an external parasite, bacterial infection, or injured by handling, the fungi can colonize diseased tissue on the fish's exterior. These areas appear to have a cottony growth or may appear as brown matted areas when the fish are removed from the water. Potassium permanganate is effective against most fungal infections. Because fungi are usually a **secondary** problem, diagnosis and correction of the original problem are important.

Noninfectious Diseases

Noninfectious diseases can be broadly categorized as environmental, nutritional, or genetic. Environmental diseases are the most important in commercial aquaculture. Environmental diseases include low dissolved oxygen, high ammonia, high nitrite, or natural or man-made toxins in the aquatic environment. Proper techniques for managing water quality enable producers to prevent most environmental diseases.

PARASITIC DISEASES

Parasites obtain their food from the host. This reduces the performance of the host and leads to other diseases. Parasites produce large numbers of eggs, and many spend different phases of their life cycle in different hosts.

Tapeworms

Causative organisms include the Asian tapeworm (*Bothriocephalus opsarichthydis*), the bass tapeworm (*Proteocephalus ambloplitis*), the catfish tapeworm (*Corallobothrium fimbriatum* and *Ligula intestinalis*), and others. The species of tapeworm (Figure 11-5) varies with the species of host fish.

All fish are susceptible, especially black basses, Chinese carps, catfishes, sunfishes, and golden shiners.

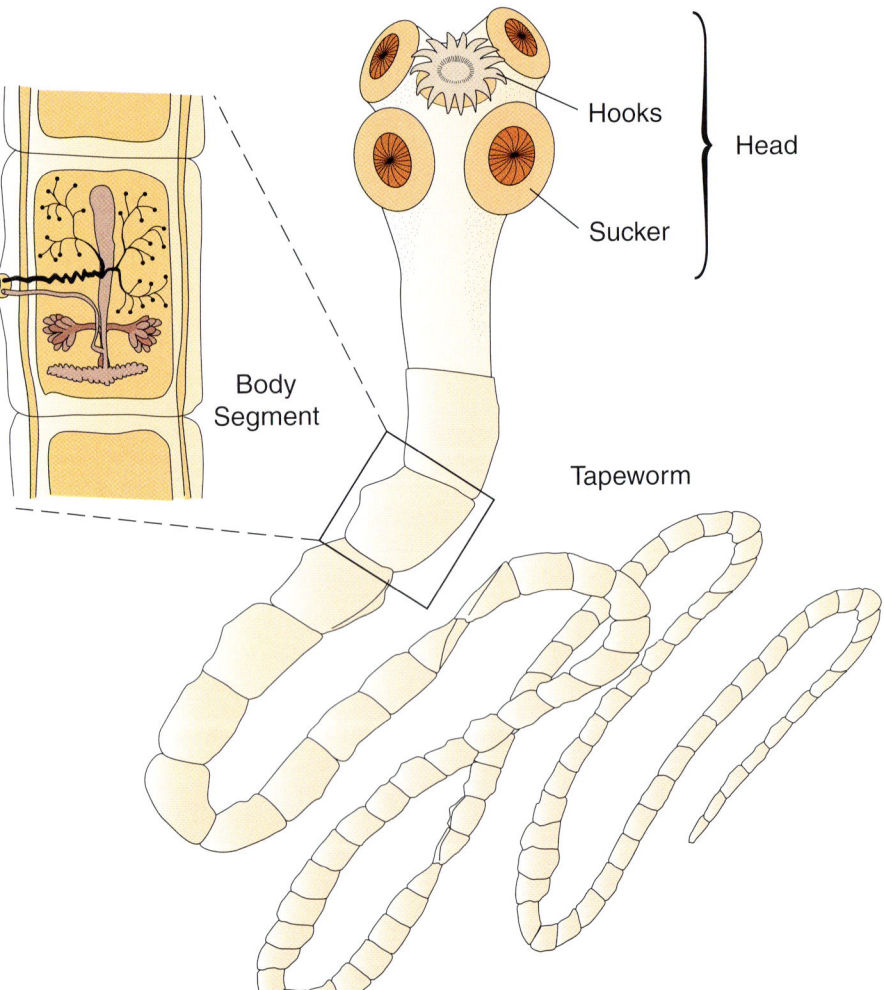

FIGURE 11-5 A tapeworm and its parts.

Signs

Fish often show no outward indication, but fish may be listless, lose weight, or become sterile. In some cases, such as severe infections of Asian tapeworms in grass carp, the abdomen becomes distended and the intestine blocked. Microscopic verification determines presence of tapeworms.

Contributing Factors

Use of broodfish infected with tapeworms, purchase of contaminated fry and fingerlings, or the use of surface water containing tapeworm-infested hosts contribute to an infection. Droppings of fish-eating birds in and near the pond can introduce tapeworms.

Prevention and Treatment

Aquaculturalists avoid maintaining or purchasing infected fry, fingerlings, and broodfish. They drain, dry, and disinfect ponds between fish crops and eliminate or reduce exposure to intermediate hosts. Currently, no **therapeutic** agents are available.

Life Cycle of the Asian Tapeworm (*Bothriocephalus opsarichthydis*)

During their life cycle, many parasitic worms require a number of animals as intermediate hosts. The Asian tapeworm, for example, requires a copepoda (a small, nearly microscopic free-swimming organism) as an intermediate host. (See Figure 11-6.) The Asian tapeworm probably develops in any fish species that eats the infected copepoda. The tapeworm often causes problems in grass carp, golden shiner, fathead minnow, Colorado squawfish, and mosquitofish.

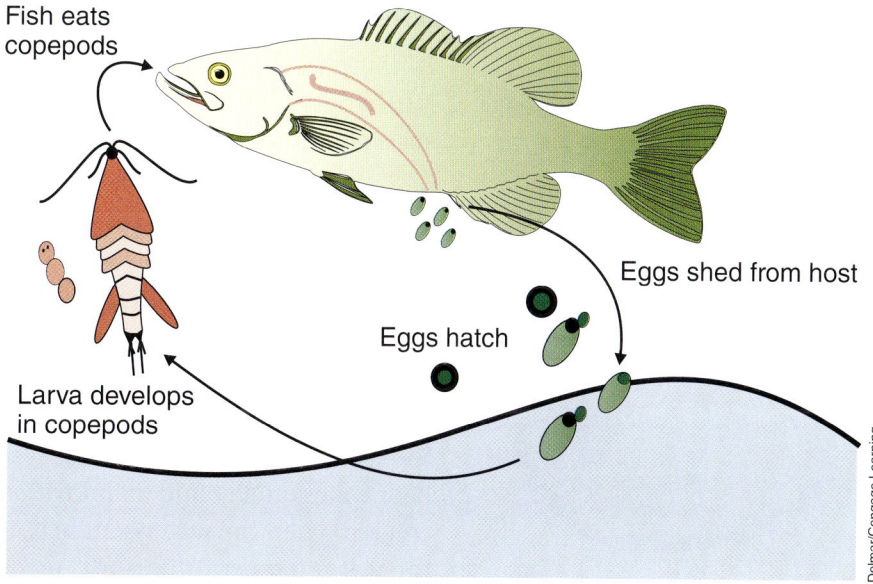

FIGURE 11-6 Life cycle of Asian tapeworms.

After introduction of a parasite into a pond or reservoir, only the removal of all fish and disinfection of the pond bottom to kill the copepoda or other intermediate hosts can ensure its eradication. The ponds should be filled with copepoda-free water—usually well water—and stocked with only parasite-free broodfish.

Trichodiniasis

Protozoa of the genus *Trichodina* cause trichodiniasis. (See Figure 11-7.) All freshwater fish are susceptible.

Signs

Affected fish are listless and **lethargic**. Fish also eat less and may have frayed fins.

Contributing Factors

Reduced water quality, including low dissolved oxygen and high concentrations of organic material, contribute to the disease. Fluctuating temperatures during fall and spring and malnutrition can debilitate the fish and accelerate the buildup of the parasite. Microscopic examination verifies the presence of Trichodina.

Prevention and Treatment

Maintain good water quality, including dissolved oxygen concentrations of 4 ppm or above. Offer the fish an adequate amount of a nutritionally complete feed, and avoid overcrowding fish, especially fingerlings. Possible therapeutic agents include formalin, potassium permanganate, and copper sulfate.

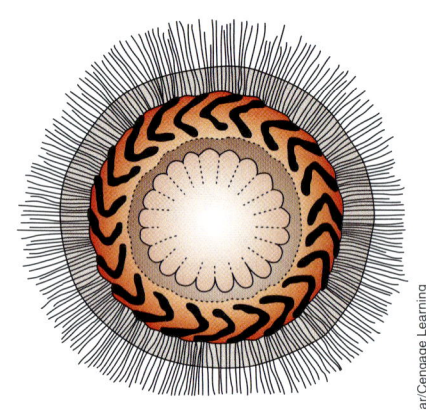

Tricodiniasis
FIGURE 11-7 Protozoan causing trichodiniasis.

Ichthyophthiriasis

Ichthyophthiriasis multifiliis, a ciliated protozoan (Figure 11-8), causes ichthyophtiriasis, commonly called Ich, and all freshwater fish are susceptible.

Signs

Parasites appear as small raised spots, resembling sprinkled table salt, over the entire body surface and fins. (See Figure 11-9.) Affected fish may flash against the bottom or sides of a tank. Heavily infected fish often congregate at the intake or outlet of the pond or tank.

Contributing Factors

Poor water quality and malnutrition contribute to Ich. Susceptibility is increased when the water source is contaminated with wild fish and when the water temperature is 60° to 75°F. Microscopic examination verifies the presence of the Ich protozoa.

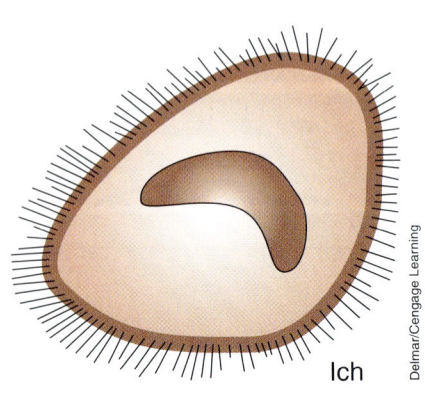

Ich
FIGURE 11-8 Ciliated Ich protozoan

FIGURE 11-9 Tiny white spots on the fins and skin are noticeable after Ich reaches the mature.

Prevention and Treatment

Avoid contaminated water supply, nets, and other equipment. Provide water of good quality, and offer nutritionally adequate feeds. Formalin, table salt, copper sulfate, and potassium permanganate can be used as therapeutic agents.

Life Cycle of Ich Parasite

The life cycle of the Ich parasite involves several stages. (See Figure 11-10.) Because treatments for one stage often do not affect the other stages, the disease frequently recurs. The adult stage of this protozoan emerges from the skin and mucus of the infected fish and drops to the bottom of the pond, where it forms a cyst. The organism then divides many times and produces many tiny juvenile forms. Under suitable conditions, these juveniles, called "theronts," emerge and swim about, seeking a fish of any species to penetrate. When they find one, they burrow into it and develop to maturity. Only the free-swimming stages of this parasite are vulnerable to treatment.

Monogenetic Flukes

Monogenetic flukes (Figure 11-11) include flukes of the genera *Gyrodactylus*, *Dactylogyrus*, and *Cleidodiscus*. Representatives of these genera are similar in size and general appearance. All warmwater fish are susceptible.

Signs

Fish show discomfort, sometimes rubbing or flashing against the tank or pond walls and bottom or becoming listless and staying near the edge of the pond. Gills may be flared on small fish. When flukes are abundant, their primary damage, combined with secondary bacterial infection, may cause death.

FIGURE 11-10 Life cycle of Ich parasite.

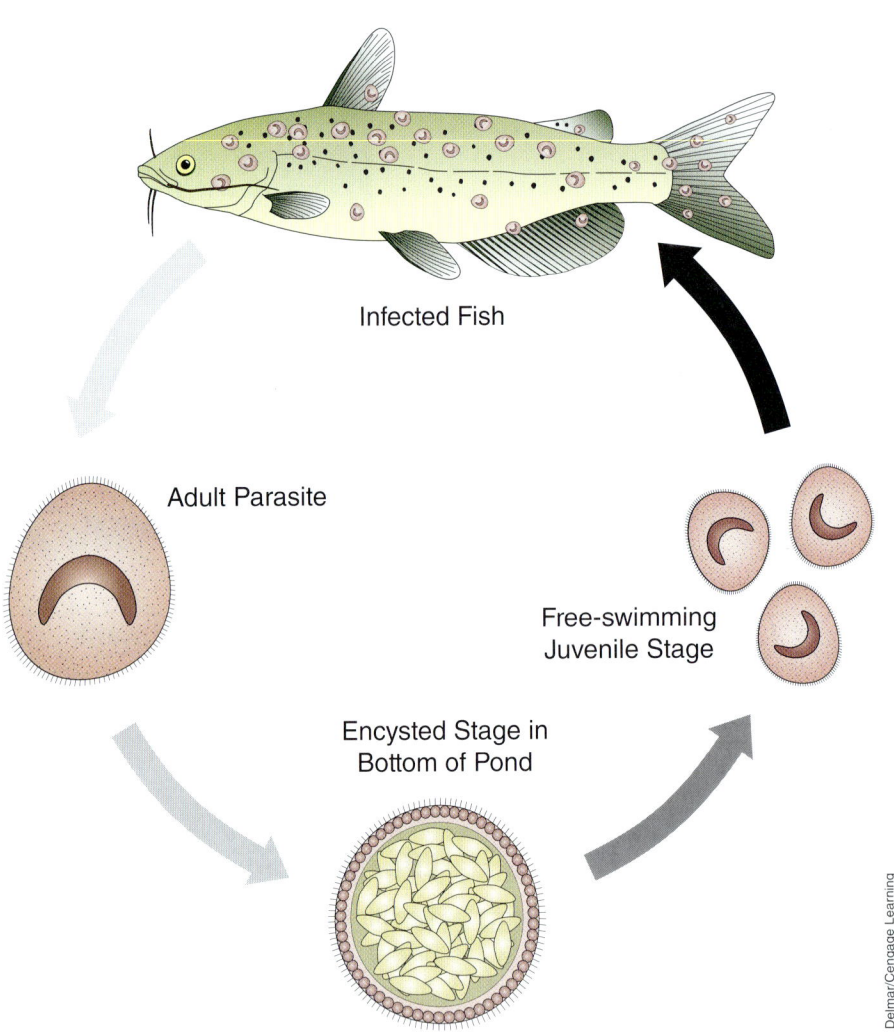

Contributing Factors

Poor water quality, accompanied by inadequate nutrition, crowding, and fluctuating water temperatures all can contribute. Microscopic examination verifies the presence of flukes.

Prevention and Treatment

Maintain good water quality, offer adequate feeds, and avoid overcrowding. Masoten (Dylox) formalin and potassium permanganate may be used as therapeutic agents.

Anchor Parasites

This disease is caused by a parasitic crustacean (copepoda) *Lernaea cyprinacea*. All freshwater fish, but especially baitfish, catfish, and carp, are susceptible.

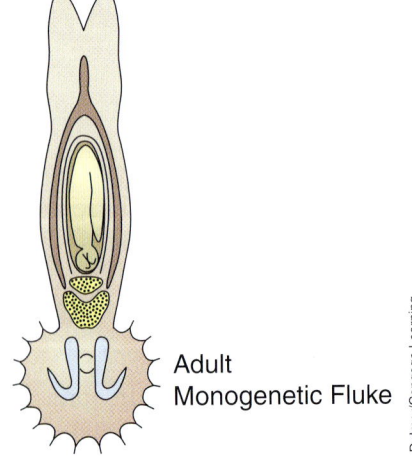

FIGURE 11-11 Adult monogenetic fluke.

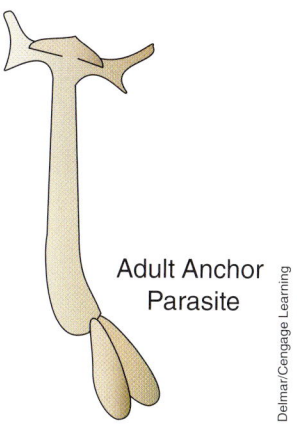

FIGURE 11-12 Adult anchor parasite.

Signs

Signs include small reddish lesions on the external surface, often surrounded by fungus. The parasite resembles a shaft of a small barb (similar to a broom straw) inserted into the fish's flesh. The anchor tends to prevent detachment of the parasite. Figure 11-12 illustrates an anchor parasite.

Contributing Factors

Disease-free fish become infected by the stocking of fish contaminated with the anchor parasite into parasite-free populations. Movements of wildlife such as ducks or muskrats from pond to pond can spread the parasite.

The presence of anchor parasites can be verified by visual examination. *Prevention and Treatment.* The best prevention is to avoid stocking parasite-infected stocks of fish. Masoten (Dylox) can be used as a therapeutic agent.

Costiasis Disease

Protozoa of the genus *Ichtyobodo* cause Costiasis disease. (See Figure 11-13.) It is sometimes called Costia. All freshwater fish are susceptible.

Signs

Blue-gray film sometimes appears over the body surface. Fish do not feed and are often listless and lethargic. In gross examinations (visible to the unassisted eyes), gill filaments may appear ragged.

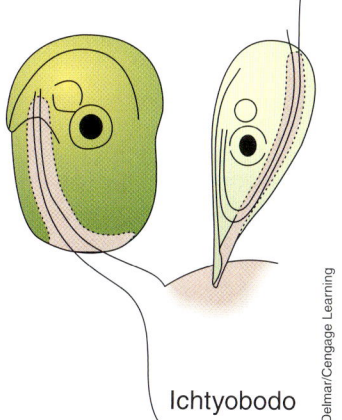

FIGURE 11-13 Protozoan parasite Ichtyobodo causes costiasis.

Contributing Factors

Overcrowded conditions aggravated by fluctuating water temperatures that are common in fall and spring contribute to susceptibility. Malnutrition can also contribute to the susceptibility to Costia. Microscopic examination verifies the presence of the protozoa.

Prevention and Treatment

Good water quality and nutritionally complete feeds help prevent infections of Costia. Some therapeutic agents include table salt, formalin, or acetic acid, and copper sulfate followed by potassium permanganate.

Fish Lice

Fish lice (Figure 11-14) are parasitic branchiurans of the genus *Angulus*. These are related to the anchor parasite. All freshwater fish are susceptible.

Signs

Infected fish will flash or rub against the tank or pond bottoms or sides. Also, they will be listless and show red spots. When infections are heavy, fish start dying.

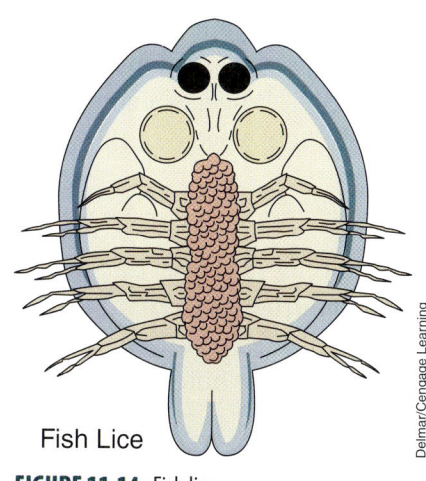

FIGURE 11-14 Fish lice.

Contributing Factors

Stocking of lice-contaminated fish into parasite-free populations allows the lice to spread. Depending on size, examination by eye or microscope verifies the presence of fish lice.

Prevention and Treatment

Stocking of parasite-free fish is the best prevention. Masoten (Dylox) can be used as a therapeutic agent.

Fish Grubs (Larval Flukes)

Grubs of the genera *Crassiphiala* (black spot), *Clinostomum* (yellow grub), and *Posthodiplostomum* (white grub) cause the disease signs. Bluegills and other sunfishes, black basses, and most minnows are susceptible.

Signs

Signs include small pigmented nodules—either black, cream-colored, or white— in the flesh and **visceral** cavity, and sometimes in the gills. Fish may show a loss of equilibrium and become deformed.

Contributing Factors

Infected fish and the presence of intermediate hosts—fish-eating birds—and snails contribute to the spread of fish grubs. Microscopic examination verifies the presence of grubs.

Prevention and Treatment

Eradicating snails in ponds and removing bird roosts in the vicinity eliminate the intermediate hosts. No therapeutic agents are available.

Life Cycle of Black Spot Grub

The spots typical of black spot disease result from a reaction of the fish's body to the presence of larval parasitic worms in its flesh. The worms reach the adult stage only when an infected fish is eaten by a susceptible bird host. As shown in Figure 11-15, the life cycles have at least two free-swimming stages and involve three kinds of hosts: fish, snails, and birds. When an animal other than a bird eats an infected fish, no danger of infection exists.

Whirling Disease

Whirling disease is a protozoan parasite, *Myxobolus cerebralis*, that affects the nervous systems of trout species. It was first introduced into the United States from Europe in the 1950s, probably in infected trout. The disease spread as these infected trout were transferred to other hatcheries and/or stocked into open waters and is now found in the wild in other states. So far, whirling disease has not had a major impact in eastern states. In some western states, however, it is severe.

FIGURE 11-15 Life cycle of the black spot grub.

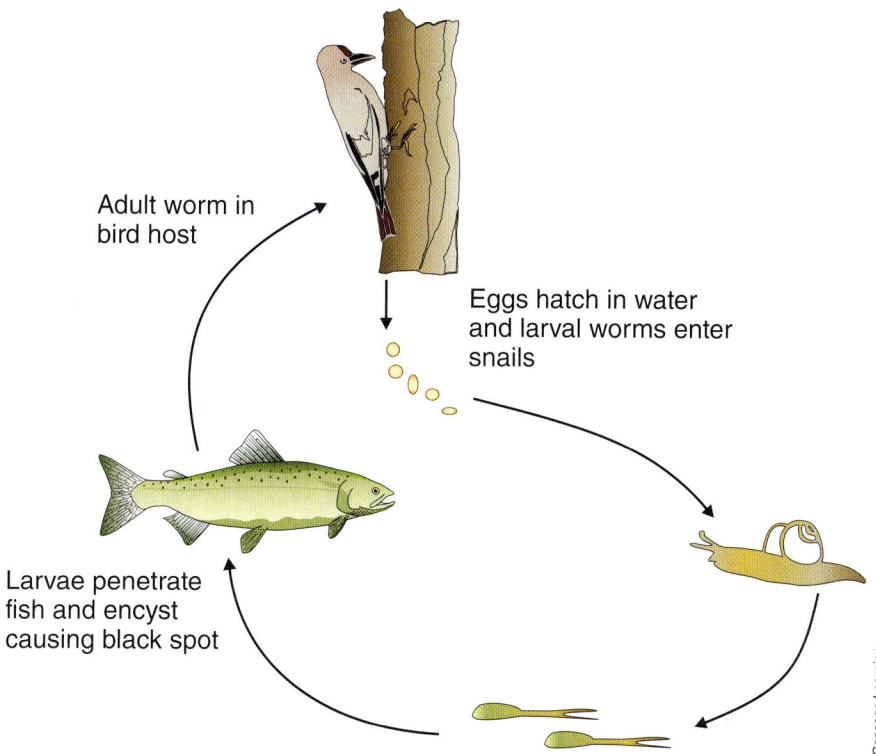

This parasitic infection attacks juvenile trout and salmon, but does not infect warmwater species. All species of trout and salmon may be susceptible to whirling disease. Other members of the trout and salmon family, such as mountain whitefish, are also at risk. Brown trout become infected with the parasite, but they appear to have an immunity to the infection.

Whirling disease has no known human health effects.

Signs

This parasite has a free-swimming stage that enters young trout, attacking the cartilage. In severe infections, inflammation around the damaged cartilage places pressure on the nervous system, causing the fish to "whirl" when startled. Seriously infected fish have a reduced ability to feed or escape from predators, and mortality is high.

Contributing Factors

Spores formed by the parasite while inside the fish are released upon death. These spores are then ingested by a common aquatic worm (*Tubifex tubifex*). After a few months inside the worm, the parasite changes into the free-swimming infective stage and is released into the water where it infects a fish host to complete its life cycle.

It may also be transmitted by birds, and it is possible that fishermen could carry the disease on fishing equipment.

FIGURE 11-16 Fish with fungus have a cotton furlike appearance.

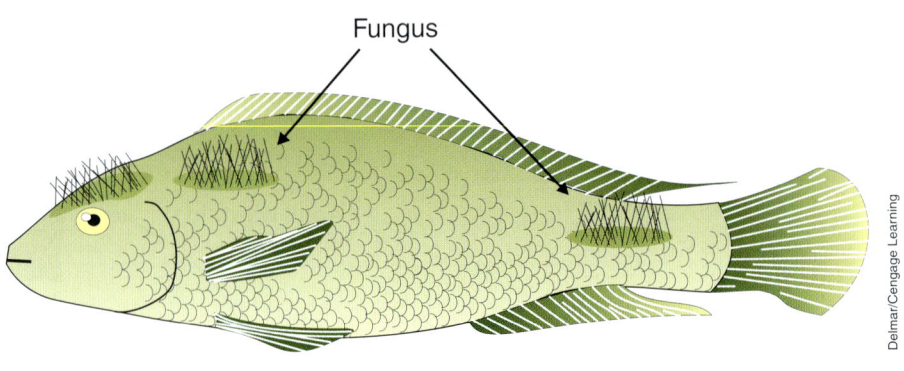

Prevention and Treatment

Major research efforts are being aimed at preventing whirling disease. Several promising areas of investigation include the identification of genetically resistant species, interruption of the parasite's life cycle, and identification of the environmental and fish life-history factors that allow the disease to prevail. No apparent cure exists.

FUNGUS

Fungi, usually of the genera *Saprolegnia* and *Achlya*, cause the disease signs. All freshwater fish can be affected.

Signs

Fish have a general cottonlike or furlike appearance (Figure 11-16), usually associated with localized discolored areas or lesions. Fungus assumes the color of materials suspended in the water.

Contributing Factors

Fungal infections are generally secondary and indicate other adverse conditions. Fungal infections seldom become established on healthy fish unless they have been subjected to stress or injury. Stress conditions include prolonged periods of very low temperatures, malnutrition, and possibly low dissolved oxygen. Microscopic examination verifies the presence of fungi.

Prevention and Treatment

Fish farmers maintain good water quality and feed nutritionally adequate feeds throughout the year. Feeding just before winter and in early spring is especially important. Copper sulfate and potassium permanganate can be used as therapeutic agents.

BACTERIAL DISEASES

Bacteria are microscopic one-celled organisms having various shapes. They often cause disease, though some bacteria are beneficial.

Bacteremia (*Hemorrhagic Septicemia*)

Bacteria like *Aeromonas hydrophila* or *Pseudomonas fluorescens*, and possibly other bacteria, cause bacteremia, or bacteria in the blood. All fish can be affected.

Signs

Infected fish are listless and lethargic and eat less feed. They show shallow, irregular-margined reddish sores or ulcers on the sides; popeye; enlarged (swollen), fluid-filled belly; raised scales; red streaks in the fin rays and at the bases of the fins; and a reddened area around the anus.

Contributing Factors

Outbreaks may occur in spring when the water warms, particularly when the fish spawn, are handled, moved, or are overcrowded. Also, outbreaks occur when dissolved oxygen content of the water is low, and possibly when other conditions such as malnutrition and disease weaken the fish. Laboratory culture confirms bacteremia.

Prevention and Treatment

Precautions include avoiding rough handling and overcrowding, especially during summer, maintaining good water quality, and providing a well-fortified feed containing greater than recommended levels of ascorbic acid (vitamin C).

Possible therapeutic agents include **antibiotics** like oxytetracycline (Terramycin) in the diet. The addition of oxytetracycline to fish transport water may retard the transfer of the bacterium but will not cure infected fish.

Columnaris Disease

The bacterium *Flexibacter columnaris*, sometimes called *Cytophaga columnaris* or *Chondrococcus columnaris*, is responsible for columnaris disease. All fish are susceptible.

Signs

Affected fish show discolored patches on the body with little or no hemorrhaging or sloughing of scales. Discolored patches and scale loss superficially resemble damage caused by fungus infections. Other signs include mouth and barbel erosion, fin erosion, tail loss, and decayed areas in gills.

Contributing Factors

Mechanical injury caused by rough handling, especially when water temperature exceeds 68°F, can contribute to an infection. Overcrowding in holding and transport facilities, poor water quality (low dissolved oxygen), fluctuating water temperatures, and malnutrition can also contribute to a columnaris infection. Microscopic examination and laboratory culture verify the presence of columnaris.

Prevention and Treatment

Preventive measures include avoiding rough handling and overcrowding, especially during summer, maintaining good water quality, and providing a well-fortified feed containing greater than recommended levels of ascorbic acid (vitamin C).

Possible therapeutic agents include water treatments with potassium permanganate or Diquat. The medication of feed with oxytetracycline may be helpful if the infection is systemic (internal).

Enteric Redmouth

This infection in trout is caused by the bacterium *Yersinia ruckeri*. Sometimes the disease is called redmouth or Hagerman redmouth. Enteric redmouth (ERM) disease occurs in salmonids throughout Canada and much of the United States. The disease has been reported in rainbow trout, steelhead, cutthroat trout, coho, chinook, and Atlantic salmon. The bacteria was first isolated in 1950 from rainbow trout in the Hagerman Valley, Idaho.

Signs

ERM is characterized by inflammation and erosion of the jaws and palate. Trout typically become sluggish, dark in color, and show inflammation of the mouth, opercula, isthmus, and base of the fins. Reddening occurs in the body fat and the posterior part of the intestine. The stomach may fill with a colorless, watery liquid, and the intestine fills with a yellow fluid. ERM often produces sustained, low-level **mortality**, but it can cause large losses.

Contributing Factors

Disease spread can be linked with fish movement. Large-scale outbreaks occur if chronically infected fish are stressed during hauling or exposed to low dissolved oxygen or other poor environmental conditions. Diagnosis of infections can be determined only by isolation and identification of the bacterium.

Prevention and Treatment

The best control is to avoid the pathogen. Fish and eggs should be obtained from sources known to be free of ERM. Vaccination is available and can be administered efficiently to fry for long-term protection. Some drugs, such as sulfamerazine, may be required to stop mortality during an outbreak.

Enteric Septicemia of Catfish (ESC)

The bacterium *Edwardsiella ictaluri* causes enteric septicemia (ESC). Because of one of the signs of the disease, it is sometimes called "hole-in-the-head disease." It affects channel catfish.

Signs

Fish may have a "hole-in-the-head" lesion between the eyes. This may appear as a white or reddish raised area before the hole appears. Fish will also have pimple-like lesions over the general body surface, a typical dropsy appearance, bloody-looking internal organs, or yellowish or reddish fluid in the body cavity. Fish may cease feeding, become listless, hang tail-down in the water, and spasmodically swim rapidly in circles.

Contributing Factors

Low dissolved oxygen, high ammonia and nitrite concentrations, and water temperatures between 70° and 82°F appear to be the conditions usually associated with the onset of the disease. Laboratory culture verifies the existence of the columnaris bacteria.

Prevention and Treatment

Aquaculturalists should maintain good-quality water, keep dissolved oxygen level above 4 ppm, and provide good-quality feed containing supplemental ascorbic acid (vitamin C). Also, the aquaculturalist should avoid using broodfish that have a history of the disease. Possible therapeutic agents include antibiotics such as oxytetracycline in the feed.

Ulcer Disease of Goldfish

The warmwater strain of *Aeromonas salmonicida* causes ulcer disease of goldfish, especially the broodfish. Sometimes ulcer disease is called "goldfish furunculosis."

Signs

Fish show ulcers or lesions with irregular margins on the sides of the fish. The lesions start as small white spots and progress to large hemorrhagic (bloody) sores. Scales are lost at the site of the ulcer.

Contributing Factors

Stress associated with handling, transportation, and spawning contributes to the development of the disease. The condition is aggravated by water temperatures of 55° to 75°F, and poor nutrition. Laboratory culture verifies the presence of the bacteria.

Prevention

Fish farmers should collect, handle, and transport the broodfish in winter when the water temperature is less than 55°F, use young broodfish, and offer a nutritionally complete feed that is fortified with ascorbic acid (vitamin C). No therapeutic agents are available.

Chilodonelliasis

In cold water, the bacterium *Chilodonella cyprini*, and, in warm water, the bacterium *C. hexasticha* can cause chilodonelliasis. (See Figure 11-17.) Most freshwater fish are susceptible.

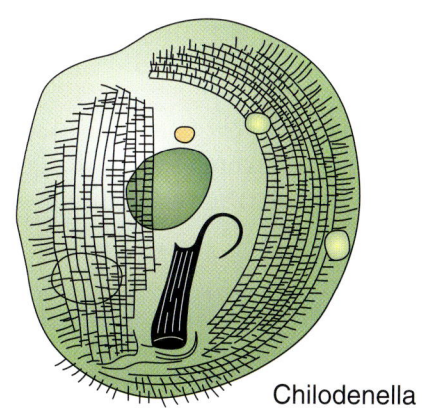

FIGURE 11-17 The bacteria that cause chilodonelliasis.

BACTERIUM OR VIRUS—WHAT'S THE DIFFERENCE?

A bacterium is a microscopic, single-celled organism, very different from a virus. The plural of bacterium is bacteria. Bacteria occur everywhere life exists. They possess a tough, rigid outer cell wall through which they absorb their food. Some bacteria have a slimy outer capsule, and some may have a whiplike flagella to propel them through liquids. If flagella are positioned all around the bacterium, it is called "peritrichous," but if flagella are at each end, it is called "lophotrichous." Some bacteria may simply drift in air or water currents.

Bacteria generally reproduce by splitting into two. This is called "binary fission," and it may occur once every 15 to 30 minutes. Under favorable conditions, one bacterium could conceivably form over 150 trillion bacteria in 24 hours. This usually does not happen, however. Bacteria are very numerous and quite tough. A pinch of soil contains millions, and some bacteria can survive freezing, intense heat, drying, and some disinfectants. To survive adverse conditions, bacteria form spores, which can remain active for years.

Bacteria can be classified by their shapes. Bacteria shaped like a sphere are called cocci. Those shaped like rods are called bacilli. Spirillum are in a spiral shape. Vibrio are comma shaped. Mycobacteria are very small rods. Flexibacter form long thin rods. The figure illustrates some of the common shapes.

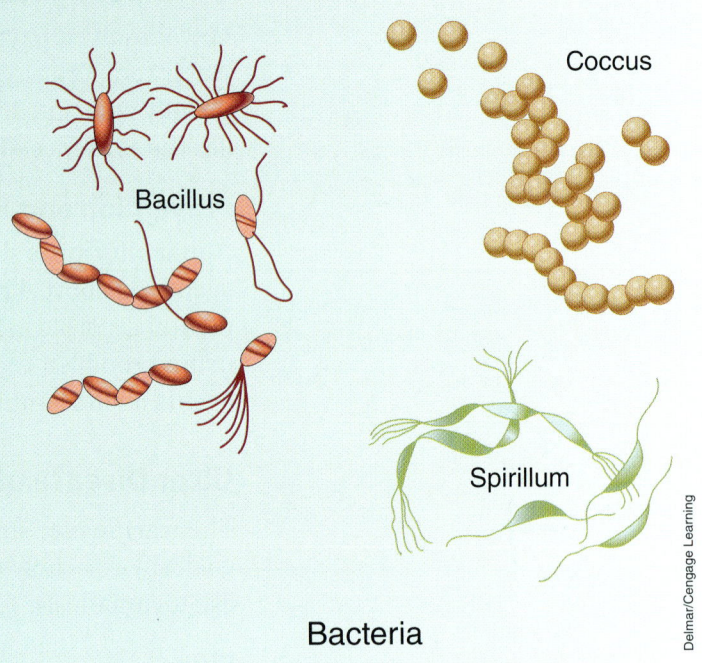

Bacteria

Signs
Fish with chilodonelliasis have bright red gills that sometimes bleed when touched.

Contributing Factors
Water of reduced quality containing especially high levels of organic matter contributes to infections. The condition is aggravated by crowding, malnutrition, and by water temperatures of 40° to 70°F. Verification requires microscopic examination.

Prevention and Treatment
Maintain good water quality, and offer the fish adequate amounts of good-quality feed, especially in the late winter and early spring.

Formalin and potassium permanganate can be used as therapeutic agents.

Prevention and Treatment

The only prevention is to avoid infected eggs, fish, and feed. No therapeutic agents are available. When an infection occurs, infected fish must be destroyed and the facility disinfected.

Infectious Pancreatic Necrosis

All salmonids and goldfish are susceptible to the virus that causes infectious pancreatic necrosis (IPN).

Signs

Affected fish may show an overall darkening of the body, popeyes, abdominal swelling, and hemorrhages at the base of the fins. Fish may be observed spiraling along the long axis of the body. Also, the death rate increases.

Contributing Factors

IPN virus spreads through feces, eggs, and seminal and ovarian fluids from parent to progeny. Fish surviving the disease become carriers. Susceptibility decreases with age; first-feeding fry are most susceptible.

Prevention and Treatment

Fish farmers do not use fish infected with the IPN virus. No treatment is known for IPN. Infected fish must be destroyed and the facility disinfected.

NONINFECTIOUS DISEASES

Infectious diseases are transmitted from fish to fish by an identifiable pathogenic agent. One fish can give the disease to another. Noninfectious diseases result from an animal's reaction to environmental changes. One affected fish cannot pass the condition on to others.

Oxygen Starvation

Reduced dissolved oxygen levels cause oxygen starvation.

Signs

Affected fish gather at the water inflow or outlet. Also, fish will be observed gasping at the water surface. Oxygen starvation may be noted as sudden mortality.

Prevention and Treatment

Producers monitor dissolved oxygen levels and try to predict any sudden drops in dissolved oxygen. Treatment is some form of aeration.

Alkalosis

Water that becomes too basic for the species causes **alkalosis**. In other words, the pH increases to a level higher than the species can tolerate.

Signs

When the pH is high for an extended period, fish die. Alkalosis can also cause corroding of the skin and gills or a milky turbidity of the skin.

Prevention and Treatment

Aquaculturalists must monitor pH level and maintain the pH in a range optimal for the species being cultured. When the problem continues to reoccur, the farmer must determine the cause and correct it. In some cases, the pH level can be reduced by adding alum or agricultural gypsum.

Acidosis

Acidosis is caused by a drop in the pH to a level too low for the species.

Signs

Affected fish shoot through the water with sudden rapid fin movements. Fish gasping for air sometimes jump out of the water. The skin may have a milky turbidity or be red and inflamed. Also, brown deposits will be noted on the gills. Death of fish can occur very quickly, or it can take a slow course.

Prevention and Treatment

Monitoring the pH and maintaining the pH at an optimal range is the best prevention. Liming raises the pH and total hardness of the water. The farmer must determine the cause of the imbalance and prevent its reoccurrence.

Nutritional Deficiency

A nutritional deficiency can be caused by unsuitable food, too much food, too little food, or a vitamin deficiency. Nutritional diseases can be very difficult to diagnose. A classic example of a nutritional disease of catfish is "broken back disease," caused by vitamin C deficiency. The lack of dietary vitamin C contributes to improper bone development, resulting in deformation of the spinal column. (See Figure 11-19.) Another important nutritional disease of catfish is "no blood disease," which may be related to a folic acid deficiency. Affected fish become anemic and may die. The condition seems to disappear when the deficient feed is discarded and a new feed provided.

Signs

The signs of nutritional deficiency vary depending on the deficiency type. Some common signs include body deformities; slow, weak movements; loss of appetite; slow growth; and a hollow-bellied profile. Chapter 9, Fundamentals of Nutrition in Aquaculture, Table 9-7, lists the effects of various deficiencies.

FIGURE 11-19 Nutritional deficiency of vitamin C causes "broken back" disease or spinal curvature in catfish.

Prevention and Treatment

Fish farmers must feed a nutritionally balanced ration designed for the species being cultured. Also, a farmer must be careful not to overfeed or underfeed fish. Calculating the feed conversion ratio provides information for adjusting feed levels. Some conditions caused by nutritional deficiency cannot be reversed. For those conditions that can be reversed, the type, amount, or time of feeding should be changed.

Pfiesteria

Pfiesteria is a dinoflagellate. It is not a bacterium or a virus. Dinoflagellates are a natural part of marine ecosystems and are generally referred to under the broad heading of "algae." Using their flagella, whiplike tails, they propel themselves about, acting very much like animals, though half contain chloroplasts and perform photosynthesis like other algae. Actually, neither plant nor animal but "protists," these tiny organisms cause red tides, or mahogany tides. This single celled dinoflagellate is approximately 7 microns wide at its girth (.007 millimeter wide).

Historically, dinoflagellates have not caused toxic blooms in the Chesapeake Bay as they have in other parts of the world (in Florida, Maine, and the Pacific Northwest, for example). Some fish kills, however, raise concerns that harmful algal blooms may be on the rise.

Although it is a dinoflagellate, Pfiesteria exhibits behavior different from normal "red tide" organisms. Pfiesteria appears to use its toxin to stun fish and then feed on them.

Signs of Pfiesteria include fish kills and fish with characteristic Pfiesteria lesions. Fish die when they can no longer maintain homeostasis within their environment because of the lesions.

Well-documented human health effects linked to Pfiesteria have occurred in laboratory conditions, where researchers were working with the organism in close proximity and in high concentrations. Others, including fishers, a water-skier, and those monitoring fish kills, have also complained of skin lesions and other health effects, such as headaches and light-headedness. Specifically, these human health problems have been linked to Pfiesteria.

Some individuals who fish in affected waters have reported sores and less specific ailments, such as fatigue and light-headedness. Making a clear connection between generalized symptoms and Pfiesteria can prove difficult.

Poisoning

Any toxic substance or toxic levels of a substance in water can poison fish.

Signs

Predictably, the signs vary with the poison. The most noted sign may be sudden mortality.

Prevention and Treatment

Before building ponds, aquaculturalists should test the soil and water for pesticide contamination or other toxic substances. Water should be monitored for ammonia, nitrate, iron, and other potential hazards. Treatment varies depending on the toxin present. Emergency measures require dilution of the toxin with fresh, clean water or a total water change. Some conditions cannot be reversed.

Brown Blood Disease

High nitrite levels in the water oxidize **hemoglobin** in the blood to methemoglobin.

Signs

Signs of brown blood disease include loss of appetite, topping, and brown-colored blood. Fish may die suddenly.

Prevention and Treatment

Brown blood disease can be prevented by monitoring nitrite levels in water and anticipating high nitrite levels when water temperature and pH rise. Sodium chloride, or common table salt, when applied at a rate of 5 ppm per acre-ft effectively reverses the effects of nitrite.

Anemia

Poor nutrition or a chronic disease can cause anemia in fish. Anemia is the inability of the blood to carry oxygen.

FIGURE 11-20 Genetic abnormalities such as no tail are rare and not very significant to the industry.

Signs

Anemic fish are lethargic, lose their appetite and color, and have pale gills. Eventually, they die.

Prevention and Treatment

The best prevention and treatment is feeding a nutritionally complete diet and controlling disease.

Gas Bubble Disease

Water that is supersaturated with oxygen or nitrogen causes gas bubble disease. This situation is found naturally in well and spring water when ice melts or when air is introduced into water lines or pumps.

Signs

Affected fish show bubbles under the skin and in the gill tissues. Also, fish rustle when they are taken out of the water.

Prevention and Treatment

Fish farmers monitor dissolved oxygen levels and maintain these levels at optimum for the species. Also, farmers control algae growth and avoid algae blooms. The situation can be corrected by mechanical aeration.

Genetic

Genetic abnormalities include conformational oddities such as lack of a tail (Figure 11-20) or presence of an extra tail. Most of these are of minimal significance. Using unrelated fish as broodstock every few years minimizes inbreeding and the expression of genetic abnormalities.

DETERMINING THE PRESENCE OF DISEASE

The most obvious sign of sick fish is the presence of dead or dying animals. The careful observer can usually tell that fish are sick before they start dying because sick fish often stop feeding and may appear lethargic. Healthy fish should eat aggressively if fed at regularly scheduled times. Pond fish should not be visible except at feeding time. Fish that are observed hanging listlessly in shallow water, gasping at the surface, or rubbing against objects indicate that something may be wrong. These behavioral abnormalities indicate that the fish are not feeling well or that something is irritating them.

In addition to behavioral changes, some physical signs should alert producers to potential disease problems in their fish. These include the presence of sores such as ulcers or hemorrhages, ragged fins, abnormal body conformation like a distended abdomen or dropsy, and exophthalmia, or popeye. When these abnormalities are observed, the fish should be evaluated for parasitic or bacterial infections.

When fish are suspected of getting sick, the first thing to do is check the water quality. Low oxygen is a frequent cause of fish mortality in ponds, especially in the summer. High levels of ammonia are also commonly associated with disease outbreaks when fish are crowded in vats or tanks. In general, it is appropriate to check dissolved oxygen, ammonia, nitrite, and pH during a water-quality screen associated with a fish disease outbreak.

Ideally, daily records should also be available for immediate reference when a fish disease outbreak occurs. These should include the dates fish were stocked, size of fish at stocking, source of fish, feeding rate, growth rate, daily mortality, and water quality. (See Figure 11-21.) This information is needed by an aquaculture specialist to solve a fish disease problem.

FIGURE 11-21 Fish records are important to determining disease.

AquaFood Farms Health Records

Week of _____

Observation	Date	Time	Loc#1	Loc#2	Loc#3
DO level					
pH					
Total alkalinity					
N					
Total ammonia N					
Unionized ammonia					
Chloride					
Total hardness					
Temperature					
Observed signs of disease					
External parasites					
Feeding behavior					
Number of mortalities					

FIGURE 11-22 Tissue samples from rainbow trout and others in the Mat-Su Valley were collected and analyzed as part of the National Wild Fish Health Survey, Alaska Department of Fish and Game Pathology lab, Anchorage, AK. Photo by Doug Palmer, Kenai Fish and Wildlife Office.

Good records, a description of behavioral and physical signs exhibited by sick fish, and results of water-quality tests provide a complete case history for the diagnostician.

Fish submitted to a diagnostic laboratory should be collected live, placed in a freezer bag without water, and shipped on ice to the nearest facility (Figure 11-22). Small fish can be shipped alive by placing them in plastic bags that are partially filled—30 percent— with water. Oxygen gas can be injected into the bag prior to sealing it. Insulated containers for shipping live, bagged fish minimize temperature fluctuations during transit. In addition to fish samples, a water sample collected in a clean jar should also be submitted.

DISEASE TREATMENT

The following questions help determine whether or not treatment is warranted:

1. What is the **prognosis**? Is the disease treatable, and what is the possibility of a successful treatment?
2. Is it feasible to treat the fish where they are, considering the cost, handling, prognosis, and other factors?
3. Is it worthwhile to treat, or will the cost of treating exceed the value of the fish?
4. Can the fish withstand the treatment considering their condition?
5. Do the loss rate and the disease warrant treatment?

Before any treatment is started, the following four factors must be known:

1. The volume of water in the holding or rearing unit to be treated.
2. Fish species and ages, because different species and ages react differently to the same drug or chemical.
3. The toxicity of the chemical to the particular species of fish to be treated and the effect of water chemistry on the toxicity of the chemical should also be known.
4. The disease being treated. Although this factor appears to be self-evident, it is often forgotten.

Methods of Treatment

Various methods of treatment and drug application control fish diseases.

Dip

In the **dip** method, a strong solution of a chemical is used for a relatively short time. This method can be dangerous because the solutions used are concentrated. The difference between an effective dose and a killing one is usually very slight. Fish are usually placed in a net and dipped into a strong solution of the chemical for a short time, usually 15 to 45 seconds, depending on the type of chemical, the concentration, and the species of fish being treated.

Flush

This method is fairly simple and consists of adding a stock solution of a chemical to the upper end of the unit to be treated, then allowing it to **flush** through the unit. An adequate water flow must be available so that the chemical can be flushed through the unit or system in a short time period. This method cannot be used in ponds.

Prolonged

There are two types of prolonged treatments: a short term, or **bath**, treatment and an indefinite prolonged treatment.

Bath

The required amount of chemical or drug is added directly to the rearing or holding unit and left for a specified time, usually one hour. The chemical or drug is then quickly flushed with fresh water. Several precautions must be observed with this treatment to prevent serious losses. Although a treatment time of one hour may be recommended, fish should be observed during the treatment period. At the first sign of distress, fresh water is added quickly. Use of this method requires extreme caution to ensure that the chemical is evenly distributed throughout the unit to prevent the occurrence of a hot spot of the chemical.

Indefinite

Usually this method is used for treating ponds or hauling tanks. A low concentration of a chemical is applied and allowed to dissipate naturally. This is generally one of the safest methods of treatment. One major drawback is the large amounts of chemicals required, which can be prohibitively expensive. As in the bath treatment, the chemical must be distributed evenly throughout the unit to prevent hot spots.

Feeding

In the treatment of some diseases, the drug or medication must be fed or in some way introduced into the stomach of the sick fish. This can be done by either incorporating the medication in the food or by weighing out the correct amount of drug, putting it in a gelatin capsule, and then using a balling gun to insert it into the fish's stomach. This type of treatment is based on body weight.

Injections

Large and valuable fish, particularly when only small numbers are involved, can at times be treated best by injecting the medication into the body cavity—intraperitoneal (IP)—or in the muscle tissue—intramuscular (IM). Most drugs work more rapidly when injected IP than IM. IP injections require caution to ensure that no internal organs are damaged. The easiest location for IP **injections** is the base of one of the pelvic fins. For IM injections, the best location is usually the area immediately next to the dorsal fin.

CALCULATING TREATMENTS

Calculation of treatment levels, units of measure, terminology, and treatment levels used in prescribing treatment rates are often confusing, not only to the fish farmer, but also to many biologists. Even though most people are familiar with pounds, ounces, gallons, acres, and feet, it can be confusing to make the conversions necessary for a prescribed treatment.

Most aquaculture systems, including ponds, tanks, and raceways, eventually require some type of chemical treatment to combat a disease or aquatic plant problem or improve water quality. Some treatments, like fertilizers or lime, are based on the surface area of water, but most treatments include the total volume of water in the production or holding unit.

All commercial aquaculturists should know how to calculate treatment rates, determine the amount of chemical or material needed, and apply the treatment. Before any treatment is applied, the water, fish, chemical, condition of the problem, and method of treatment should be known and understood. Producers can experience high economic losses when treatment rates are not properly calculated.

Before any calculation is made, the units of measurements should be determined. The unit of measurement selected should be familiar and convenient for the specific situation. For example, the large volume of water in ponds is usually expressed as acre-feet, whereas the volume of a small tank may be expressed in gallons or cubic feet. Another decision is whether to use the English or metric system of measurement. A working knowledge of both systems is important because reports or publications may use either one. Metric units are easiest to work with when small volumes or weights are involved. To make calculations easy, refer to Tables A-1 and A-2 in the Appendix for conversion factors and parts per million equivalents.

Basic Formula

Most treatments can be calculated using this basic formula:

$$\text{Amount of chemical needed} = V \times C.F. \times \text{ppm desired} \times \frac{100}{\%AI}$$

Where

V = The volume of water in the unit to be treated.

$C.F.$ = A conversion factor that represents the weight of chemical that must be used to equal one part per million (1 ppm) in one unit of the volume (V) of water to be treated. The unit of measurement for the results is the same as the unit used for the C.F. (pounds, grams, or other unit). Table A-2 in the Appendix contains a list of these conversion factors for various units.

ppm = The desired concentration of the chemical in the volume (V) of water to be treated expressed in parts per million.

100 = 100 divided by the percent active ingredient (AI) contained in %AI the chemical to be used.

Most chemicals are 100 percent AI unless otherwise specified, so this value is usually 1. The percent AI is found on the label of most fishery chemicals. Chapter 10, Water Requirements for Aquaculture, provides example problems using the basic formula.

IMMUNIZATION

Vaccines given to fish cause them to produce antibodies to a specific disease, giving them immunity to the disease. Technology in fish vaccination has advanced rapidly in the past few years for four reasons:

1. The list of antibacterial drugs that could legally be used is extremely small, and the prospects for enlarging the list were dim.
2. The effectiveness of the few available antibacterial drugs was rapidly being diminished because of the development of antibiotic resistance among the bacterial fish pathogens.
3. The danger that this antibiotic resistance might be transmissible to microorganisms of public health concern, and the real possibility that drugs now approved for use in fish culture would have their approval revoked.
4. The viral infections in fish could not be treated with any of the antibiotics available.

The advantages of immunizations include:
- Does not produce antibiotic resistant bacteria
- Can be applied to control viral and bacterial diseases
- Vaccination is convenient and economical
- Requirements for licensing vaccines are less stringent than those for antibiotics.

Vaccination Methods

Unfortunately, the biggest single factor working against the widespread use of fish vaccination was the lack of a safe, economical, and convenient technique for vaccinating large numbers of fish. Attempts at oral vaccination have been unsuccessful, and alternative procedures devised include mass **inoculation**, **infiltration**, and spray vaccination. The mass inoculation method works well with fish in the 5 to 25 gm range, and individual operators are able to vaccinate 500 to 1,000 fish per hour. Cost of the technique seems reasonable, but the number of fish that can be treated is limited by the human resources available for short-term employment and by the size of the inoculating tables.

The infiltration method or hyperosmotic immersion allows vaccination of up to 9,000 fish (1,000 per lb.) quickly and safely in approximately four minutes. The method uses a specifically prepared buffered hyperosmotic solution. Through osmosis, fluid is drawn from the fish body during its immersion in the buffered prevaccination solution. The fish are then placed into a commercially prepared vaccine that replenishes the body fluids and simultaneously diffuses the vaccine or bacterin into the fish.

TABLE 11-2. USDA LICENSED BIOLOGICS (VACCINES) FOR FISH

Product Name/Trade Name	Species	Disease
Aeromonas Salmonicida Bacterin Biojec 1500	Salmonids	Furunculosis
Aeromonas Salmonicida-Vibrio	Salmonids	Furunculosis, vibriosis
Autogenous Bacterin Autogenous Bacterin	Fish	Bacterial diseases
Vibrio Anguillarum-Ordalii Bacterin	Salmonids	Vibriosis
Vibrio Anguillarum-Ordalii-Yersinia Ruckeri Bacterin	Salmonids	Vibriosis, yersiniosis (enteric red-mouth disease)
Yersinia Ruckeri Bacterin	Salmonids	Yersiniosis (enteric red-mouth disease)
Vibrio Salmonicida Bacterin	Salmonids	Vibriosis
Vibrio Salmonicida Bacterin	Salmonids	Vibriosis
Vibrio Anguillarum-Salmonicida Bacterin	Salmonids	Vibriosis
Aeromonas Salmonicida Bacterin	Salmonids	Furunculosis
Autogenous Bacterin	Fish	Bacterial diseases
Edwardsiella Ictaluri Bacterin	Catfish	Enteric septicemia
Vibrio Anguillarum-Ordalii Bacterin	Salmonids	Vibriosis
Vibrio Anguillarum-Ordalii Bacterin	Salmonids	Vibriosis
Yersinia Ruckeri Bacterin	Salmonids	Yersiniosis (enteric red-mouth disease)

Fish are spray vaccinated by removing them briefly from the water and spraying them with a vaccine from a sand-blasting spray gun. **Antigenicity**—ability to produce antibodies—of the preparations is markedly enhanced by the addition of bentonite, an absorbent.

Vaccines approved by the USDA and licensed for fish are listed in Table 11-2.

SUMMARY

Disease is any condition of an aquatic animal that impairs normal physiological function. Prevention of disease is better than treatment. Prevention of fish diseases is enhanced by good management of the water environment and the feed and by observing fish frequently for signs of disease.

Management practices must minimize stress on fish to decrease the occurrence of disease outbreaks from parasites, bacterial, viral, or fungal pathogens. When disease outbreaks occur, the underlying cause of mortality should be identified as well as underlying stress factors that may be compromising the natural survival mechanisms of the fish. Correction of stressors, such as poor water quality and crowding, should precede or accompany disease treatments.

Stress compromises the fish's natural defenses so that it cannot effectively protect itself from invading pathogens. A disease treatment is an artificial way of slowing down the invading pathogen so that the fish has time to defend itself with an immune response. Any stress that adversely effects the ability of the fish to protect itself will result in an ongoing disease problem. As soon as the treatment wears off, the pathogen can build up its numbers and attack again. Rarely would a treatment result in total annihilation of an invading organism. Disease control depends upon the ability of the fish to overcome infection as well as the efficacy of the chemical or antibiotic used. Effects of the disease and treatment can be very costly to the producer.

For some diseases, vaccinations immunize the fish. The continued development of vaccines is important. Few treatments are approved for fish diseases. Vaccines are effective for bacterial and viral infections and are easier to approve.

STUDY/REVIEW

Success in any career requires knowledge. Test your knowledge of this chapter by answering these questions or solving these problems.

True or False

1. Disease is any condition of an animal that impairs normal function.
2. All bacteria cause disease.
3. Fish can get lice.
4. Fish can be vaccinated.
5. Only a pathogen is necessary to cause disease.

Short Answers

1. In aquaculture, why is fish health management more important than the treatment of disease?
2. List two general practices that help prevent fish diseases.
3. Give an example of a chemical stressor.
4. Give an example of a biological stressor.
5. Give an example of a physical stressor.
6. Which of the following is not one of the four protective barriers protecting fish against an infection?
    a. antibodies
    b. dorsal fin
    c. mucus
    d. scales and skin
7. Handling, crowding, and poor water quality are examples of _____ that decrease the fish's ability to resist disease.
8. Name the two broad categories of diseases that affect fish.
9. An example of an infectious disease is:
    a. acidosis
    b. alkalosis
    c. Ich
    d. brown blood
10. What three things interact to produce a disease in fish?
11. Name four protective barriers that normally protect fish against infection.
12. Give three examples of stress affecting the normal protective barriers.
13. List the four broad categories of infectious diseases.
14. Name a disease caused by a ciliated protozoan.
15. What disease gives fish a cottonlike or furlike appearance?

16. What is another name for "hole-in-the-head" disease?
17. When fish receive a treatment, in what ways can the treatment be administered?
18. Name three methods of vaccinating fish.

Essay

1. Define health management.
2. Name four types of stressors and give an example of each.
3. What controls osmoregulation?
4. List three good management practices that prevent stress.
5. Distinguish between bacterial, viral, and fungal diseases.
6. List three broad categories for noninfectious diseases and give an example of each.
7. Describe the life cycle of a fluke.
8. When describing a disease in animals, why should the disease be described in terms of signs rather than symptoms?
9. Explain why treatment of viral diseases is difficult.
10. What causes acidosis?
11. Describe four behavioral patterns that indicate that fish are healthy.
12. Name five physical abnormalities that indicate the presence of disease.
13. Why are records important to fish health management?

KNOWLEDGE APPLIED

1. Based on the information in this chapter, develop a checklist of things to observe when checking fish for signs of disease. Make the checklist complete enough that it could be given to a new employee.
2. Visit or contact a disease diagnostic laboratory. Find out which species the lab will perform diagnostic procedures on. Determine some of the diseases the lab dealt with in the past year. Ask what the requirements are for submitting an animal for diagnosis and how much the diagnosis costs. Also, ask what diseases they have to report.
3. Using the formula for calculating the amount of chemical needed, develop a series of problems for calculating the treatment of fish in ponds, raceways, or tanks. For example, how much potassium permanagate is needed to treat a pond 700 ft. long × 700 ft. wide × 4 ft. deep with a concentration of 2 ppm? Potassium permanagate is 100 percent active ingredient. Or, for example, how much formalin is needed to treat a circular tank that is 4 ft. in diameter and has a water depth of 3 ft. with 250 ppm? Formalin liquid is 100 percent active. (More sample problems are available in the resources listed at the end of the chapter.)
4. Visit a veterinary-supply store. Ask the veterinarian or sales representative to discuss the types of vaccines sold. Also, find out what types of restrictions are placed on the sale of vaccines, antibiotics, and other

medicines. As an alternative, ask a veterinarian or sales representative to visit class and discuss vaccines, antibiotics, and other medicines.

5. From a biological supply house or the biology department at a college or high school, obtain prepared microscope slides of bacteria. Observe the different shapes—rods, spheres, or spirals. Also, if available, obtain prepared slides of microscopic parasites, such as some of the protozoa that infect fish.

6. Research and report the role of bacteria in human health and commerce. Some bacteria cause disease, whereas other bacteria have a role in human welfare.

LEARNING/TEACHING AIDS

Books

Adams, S. (1990). *Biological indicators of stress in fish*. Bethesda, MD: American Fisheries Society.
Andrews, C., Exell, A., and Carrington, N. (2002). *The manual of fish health. 2nd Ed*. Surrey, England: Interpet Publishing.
Jepson, L. (2004). *The super simple guide to common fish diseases*. Neptune City, NJ: TFH Publications, Inc.
McClanahan, T. and Castilla, J. C. (2007). *Fisheries management: Progress toward sustainability*. Ames, IA: Blackwell Publishing Professional.
Noga, E. J. (2000). *Fish disease: Diagnosis and treatment*. Ames, IA: Iowa State Press.
Pillay, T. V. R., and Kutty, M. N. (2005). *Aquaculture principles and practices*. Ames, IA: Blackwell Publishing Ltd.
Untergasser, D. and Axelrod, H.R. (1989). *Handbook of Fish Diseases*. Neptune City, NJ: TFH Publications, Inc
Webster, C. D. and Lim C. (2001). *Nutrition and Fish Health*. Boca Raton, FL:CRC Press

Internet

Internet sites represent a vast resource of information. The URLs (uniform resource locator) for the World Wide Web sites can change. Using a search engine such as Google, find more information by searching for these words or phrases: fish health, pathogenic virus, pathogenic bacteria, fungal infection, fish disease resistance, protozoan parasite, crustacean parasite, immune systems, stressors (chemical, biological, physical, procedural), antibodies, transmission of disease, noninfectious fish diseases, parasitic fish diseases, bacterial fish diseases, viral diseases, infectious fish diseases, fish vaccines, or fish immunizations.

For some specific Internet sites refer to Appendix Tables A-11, A-12 and A-14.

High-quality water and plenty of it (Figure 12-1, page 350) are the primary considerations for any aquaculture facility. This is true for finfish, shellfish, and crustaceans. Water provides oxygen and food, serves as an excretory site, helps regulate body temperature, and may harbor disease-causing organisms. Gaining an understanding of water helps aquaculturalists become more productive.

CHAPTER 12

Water Requirements for Aquaculture

OBJECTIVES

After completing this chapter, the student should be able to:

- Explain why water is important in aquaculture
- Explain the quality features of water for aquaculture
- Define terms related to water-quality management
- Calculate water needs and filling time
- Calculate treatments for volumes of water
- List causes of dissolved oxygen loss
- List signs of dissolved oxygen efficiency
- Describe the prevention of oxygen depletion
- List methods of correcting dissolved oxygen deficiency
- Know what causes turbidity
- List the purposes of liming
- Discuss aquatic plant control methods
- List ways to dispose of wastewater
- Describe the properties of water
- List cations and anions found in water
- Describe why and how aquatic solutions change
- Explain how changes in water affect aquatic life

- Match compounds and elements with their chemical formulas and symbols
- Discuss the importance of oxygen in water-quality management
- Discuss the role of temperature in oxygen management
- List chemicals, compounds, and elements that are detrimental to water quality
- Understand the importance of nitrogen compounds in water-quality management
- Complete statements about pH and water quality
- Select from a list methods of managing the pH cycle
- Select general guidelines for water-chemistry management

Understanding of this chapter will be enhanced if the following terms are known. Many are defined in the text, and others are defined in the glossary.

KEY TERMS

Acid	Hydroponics
Aeration	Ions
Alkaline	Liming
Anaerobic	Solubility
Anions	Thermal stress
Beer's law	Organic
Buffers	pH
Cations	Salinity
Colorimetric	Salts
Conductance	Saturation
Dissolved oxygen	Settling pond
Effluent	Titrant
Fertilization	Titrimetric
Heat capacity	Turbidity
Heavy metals	Water hardness

WATER QUALITIES, MEASUREMENTS, AND ALTERATIONS

Hydrogen and oxygen associate to form water, H_2O. In nature, water is not a pure substance. It consists, to some degree, of a number of dissolved and suspended substances. Unless the water is particularly muddy, these substances consist predominately of dissolved ions. **Ions** are elemental forms or groups that carry an electrical charge in solution. Positive ions are called **cations**, and examples are calcium, potassium, magnesium, and sodium. Negative ions or **anions** are substances like carbonate, bicarbonate, sulfate, and chloride. (See Figure 12-2.)

Substances that dissolve readily in water to form simple ions are called **salts**. Because many of the dissolved ions in water would combine upon evaporation to form salt compounds, the content of simple ions in solution is frequently referred to as "salt content."

A sample of natural surface water also contains **organic** matter. Organic matter consists of living material, its excretions, and decomposing material. This is usually only a small percentage of a residue. Organic matter can affect water chemistry by requiring oxygen for living processes as the complex organic matter is built up from and torn down into simple compounds.

Gases are also present in water, either dissolving at underground pressures in the case of groundwater or under normal pressure at the surface from the gas mixture of the air. Some gases enter the water as they are formed by living agents in water or by chemical release from muds in the bottom.

Units of Measure

The most common unit of measure of trace elements, gases, ions, and pesticides is parts per million (ppm). A part per million is the same as a milligram per liter (mg/l). Smaller concentrations are sometimes reported as parts per billion (ppb), which is the same as a microgram per liter (mcg/l). Larger concentrations are sometimes reported as parts per thousand (ppt) or grams per liter (g/l).

Sometimes, and particularly with ionized components of salts, a measure called equivalents per million (epm) is used. Equivalents per million is the same as the expression milliequivalents per liter (meq/l). Standard multiplication factors convert epm to mg/l. Rarely, the measure grains per gallon is used. One grain per gallon equals 17.1 parts per million.

FIGURE 12-1 Water, lots of it, for aquaculture.

FIGURE 12-2 Cations or positive ions and anions or negative ions in water.

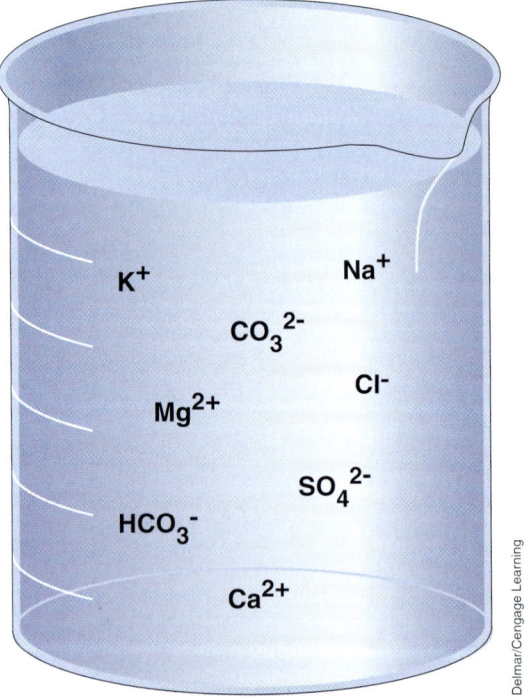

pH

The **pH**—one of the most common water tests—is a measure of hydrogen ions in the water. The pH scale spans a number range of 0 to 14 with the number 7 being neutral (Figure 12-3). The pH scale is logarithmic, so every one-unit change in pH represents a 10-fold change in acidity. Measurements above 7 are basic and below 7 are acidic. The farther a measurement is from 7, the more basic or acidic is the water. **Acid** and **alkaline** (basic) death points for fish are approximately pH 4 and 11. Growth and reproduction can be affected between pH 4 and 6 and pH 9 and 10 for some fishes (Figure 12-4). Also, pH affects the toxicity of other substances, such as ammonia and nitrite.

The pH of some ponds may change during the course of a day and is often between 9 and 10 for short periods of afternoons. Fish can usually tolerate such rises that result when carbon dioxide, an acidic substance, is used up by plants in photosynthesis. The most common pH problem for pond fish is when water is constantly acidic. The nature of the bottom and watershed soils is usually responsible. Water with a stable and low pH is only correctable with **liming**.

Temperature

Water temperature helps determine which species may or may not be present in the system. Temperature affects feeding, reproduction, immunity, and the metabolism of aquatic animals. Drastic temperature changes can be fatal to aquatic animals. Not only do different species have different requirements, but optimum temperatures can change or have a narrower range for each stage of life.

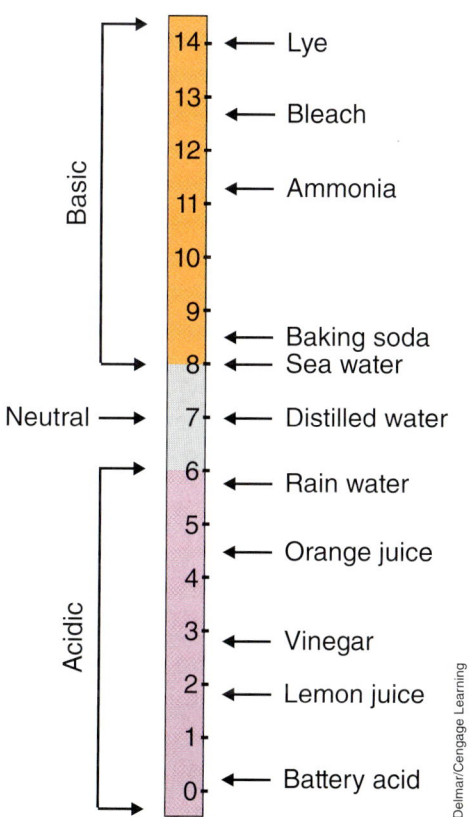

FIGURE 12-3 The approximate pH of some common substances.

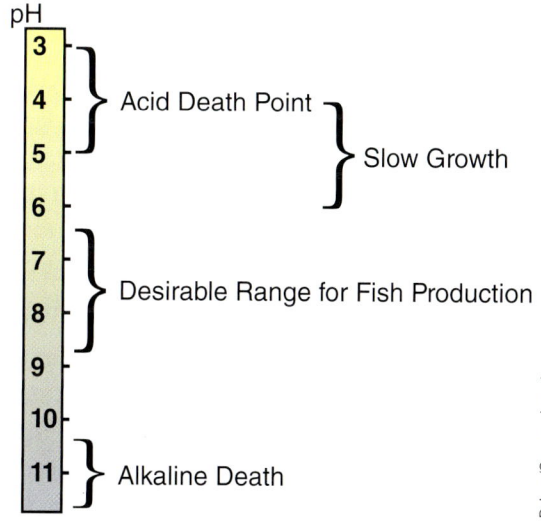

FIGURE 12-4 The effect of pH on fish survival.

HOH—WATER!

Water—the stuff of life—covers three-fourths of the earth's surface and represents a major component of the bodies of plants and animals. A 200-lb human would weigh only about 90 pounds (41 kg) with all the water removed. This miracle liquid forms from two gases—hydrogen and oxygen. Two atoms of hydrogen (H) and one atom of oxygen (O) combine to form water, H_2O or HOH. The water molecule is bipolar, having charged poles like a magnet, giving water unique properties.

Depending on the temperature, water exists in three forms. It is a liquid, between 32° and 212°F (0° and 100°C). It is a gas or vapor at temperatures above 212°F (100°C), and water becomes a solid (ice) at temperatures below 32°F (0°C).

Of all the naturally occurring substances, water has the highest specific heat. This makes it a good coolant in biological systems and makes it resist rapid temperature changes. Specific heat is the amount of heat required to raise the temperature of a substance 1°C.

Water is the universal solvent, dissolving almost everything. It is powerful enough to dissolve rocks, yet gentle enough to hold an enzyme in a fragile cell. As a solvent, it acts as a medium for biochemical reactions and carries waste products and nutrients.

This miracle liquid supports life.

All species tolerate slow seasonal changes better than rapid changes. **Thermal stress** or shock can occur when temperatures change more than 2° to 3°F in 24 hours.

The **heat capacity** of water is very high, making it resistant to changes in temperature. This moderates the daily and seasonal climatic changes in temperature. But cooling is often impractical, and heating is possible, but costly.

Water temperature is an important variable in many chemical tests and electronic measures for water quality. Many determinations require adjustment for the temperature or noting the temperature at the time of sampling.

Water temperature affects many biological chemical processes. Spawning is triggered by temperature. Temperature differences between the surface and bottom waters help produce vertical currents moving nutrients and oxygen throughout the water column. Temperature influences the **solubility** of oxygen and the percentage of un-ionized ammonia in water.

Salinity

Salinity is the measure of the total concentration of all dissolved ions in water. Sodium chloride (NaCl) is the principal ionic compound in seawater, but most inland ponds contain substantial concentrations of other ionic compounds (salts), such as compounds of sulfate and carbonate. Salinity also can be measured according to the density the salts produce in the water, the refraction they cause to light, or by electrical **conductance**. The result in all cases is reported in parts per thousand (ppt) salinity.

Standard seawater is 35 ppt or more. Well waters sometimes accumulate high amounts of dissolved ions due to ionization of compounds of underground minerals or as a result of leaching from the high salt content of arid land surfaces.

Dissolved ionic substances can be measured by electrical conductance. On laboratory reports this may be shown as specific conductivity. Conductivity is reported as microohms per cm or $EC \times 10^6$. From such an electrical measure, tables can be used to derive tons per acre-ft, parts per million, grains per gallon, and other measures. Conductivities in natural surface water measure from 50 to 1,500 microohms per cm.

Freshwater fish such as catfish, carp, tilapia, and trout tolerate some salinity and still grow and survive. Testing for water quality is an important part of ensuring that the fish are in the best environment possible (Figure 12-5).

Total Dissolved Solids

Another measure of the presence of dissolved ionic constituents is total dissolved solids. This measurement is made by weighing the residue of an evaporated sample after passing it through filter paper. If the sample is not filtered, the reported value is total solids (TS) instead of total dissolved solids (TDS). Because ionic compounds dominate the content of dissolved substance in most water samples, TDS reflects closely the numerical quantity derived in the measure of salinity.

TDS is reported in mg/l of whole sample. In both surface and ground waters of inland areas where TDS exceeds 2,000 mg/l the principal anions are sulfate and chloride. The measure of these ions in mg/l when added

FIGURE 12-5 Test kits provide useful, quick information about water quality.

together usually approximate one-half the measure of TDS. Although not the same as salinity, TDS sometimes serves as the only measure on which to make decisions regarding dissolved salts. As mentioned for salinity, 2,000 mg/l (2 ppt) is known to adversely affect sensitive species or younger stages of some species. Catfish handle 6,000 to 11,000 mg/l salinity quite well, depending on acclimation. Besides seawater, water that flows through limestone and gypsum dissolves calcium, carbonate, and sulfate, with resulting high levels of TDS.

Major Dissolved Ionic Components

Helpful tests determine various ionic constituents of water. Calcium (Ca) and magnesium (Mg) compounds are preferred as major ionic components. Solubility of liming compounds is affected by the presence of sodium. Aquatic species are more affected by total ionic presence than they are by individual ion concentrations.

Carbonate (CO_3^{2-}) and bicarbonate (HCO_3^-) are present in both surface and ground water supplies at levels consistent with their solubilities. Measurements normally range below 300 mg/l and are considered harmless to fish life. Carbonate and bicarbonate act as pH **buffers**, resisting changes in pH.

Sulfate (SO_4^{2-}) and chloride (Cl^-) are expected to range higher than carbonate/bicarbonate in well and surface waters where the water comes in contact with appropriate rocks, such as those composed of aluminum sulfate. Normal surface waters contain less than 50 mg/l of sulfate and chloride, but some well waters far from coastal areas reach 1,000 to 2,000 mg/l sulfate and several times that amount in chlorides.

Nitrate (NO_3^-) is generally nontoxic to fishes and can be expected to occur at less than 2 mg/l in natural surface water. Fish can tolerate several hundred mg/l. In some recycled waters or where feeding causes enrichment, nitrate could climb to several mg/l. Phosphate (PO_4^{3-}), fluoride, and silicate are minor constituent anions. Phosphate, like nitrate, is usually present in slight amounts (less than 0.1 mg/l) in natural surface and well water. Aside from promoting unwanted growth of algae in ponds, phosphate is considered harmless.

Fluoride concentrations in surface water would be considered normal at less than 0.5 mg/l, high at 1 to 2 mg/l, and rare at over 10 mg/l. Fish react differently to fluoride according to overall water conditions and species. In some cases, 3 mg/l causes major losses, but normal populations have been recorded in lakes where concentrations reach over 13 mg/l.

Silica is rather unreactive and harmless and is normally present in pond waters at less than 10 mg/l.

Potassium (K^+) usually represents a very low percentage of surface water cations. Calcium (Ca^{2+}) and magnesium (Mg^{2+}) are typically greater and vary from site to site in proportion to sodium. Sodium (Na^+) is much more soluble than the other cations and can range into the thousands of mg/l where the others are limited to hundreds of mg/l. Water with 5 or less mg/l calcium is considered very low and 10 mg/l or less low in calcium. Problems to fish life from maximum cation amounts are typically associated with intolerance by fish to extremely high total dissolved solids (salts).

Some waters fed by wells have calcium and magnesium ions dissolved to perhaps twice that which is considered possible for natural surface water and yet will maintain living populations of fishes and other aquatic animals. Excess sodium has a detrimental influence on liming efforts and can cause liming to have little positive effect. Often, it is difficult to obtain water samples that reflect the conditions of various waters and conditions (Figure 12-6).

Sodium Absorption Ratio

This test is used to check the alkali hazard of irrigation water. It compares sodium concentration with calcium and magnesium to assess the potential for sodium buildup in cropland.

Sodium Percentage

This is another irrigation water test, also known as sodium hazard, that compares the amount of sodium to all cations present. The effect of sodium on aquatic microplant production is not well understood. For crops, the sodium percentage should be less than 60.

Total Alkalinity and Total Hardness

This is a measure of water's basic substances. Because in natural water these substances are usually carbonates and bicarbonates, the measurement is expressed as mg/l of equivalent calcium carbonate. In some cases,

FIGURE 12-6 Obtaining a water sample that reflects the condition of all the water is difficult.

TABLE 12-1 WATER HARDNESS LEVELS

Water Hardness	As ppm CaCO$_3$
Soft	0 to 20
Moderately soft	20 to 60
Moderately hard	61 to 120
Hard	121 to 180
Very hard	> 180

such as many groundwater and western ponds, sodium carbonate is the predominant basic substance. These basic substances resist change in pH (buffering), and where an abundance of calcium and magnesium bicarbonate is dissolved, the pH will stabilize between 8 and 9. If an abundance of sodium carbonate is present, the pH may exceed 9 or 10. Some laboratory forms report carbonate (CO_3^{2-}) and bicarbonate (HCO_3^-) in addition to total alkalinity. These are typically derived from alkalinity measurements by multiplication by standard conversion factors.

Total hardness is the measure of the total concentration of primarily calcium and magnesium expressed in milligrams per liter (mg/l) of equivalent calcium carbonate ($CaCO_3$). Calcium and magnesium are usually present in association with carbonate as calcium carbonate or magnesium carbonate. Total hardness relates to total alkalinity and indicates the water's potential for stabilizing pH. Waters can be high in alkalinity and low in hardness if sodium and potassium are dominant. Table 12-1 lists water hardness classifications as ppm of $CaCO_3$.

Fish do best when measurement of hardness or alkalinity measures between 20 and 300 mg/l. Below 20 mg/l can result in poor production. Cases where total hardness is considerably below the measure for total alkalinity is also not desirable. Liming increases total hardness and total alkalinity.

Lime Requirement

This test determines the amount of liming needed to neutralize acid because of leached-out cropland cations. Sediment (mud) is removed and analyzed to determine the amount of ground limestone needed to bring the pH of soils to an acceptable level and hold it there for two to five years. For aquaculture use, the test determines the amount of lime necessary to raise the pH of the bottom mud to 5.8 and raise the water hardness to acceptable levels. The lime requirement of the bottom mud must be satisfied before lasting effects can be expected in the water column. The lime requirement is reported in lbs per acre and may amount to one to several tons of lime per acre. Liming efforts may be futile where sodium presence is great or where muds are extremely acid.

Suspended Solids and Turbidity

Suspended solids (unfiltered residue) measure less than 2,000 mg/l in muddy pond waters (Figure 12-7). Many times this amount is needed to directly affect fingerlings and adult fishes. Muddiness can affect natural food production at 250 mg/l suspended solids by shutting out sunlight. Also, **turbidity** interferes with reproduction of some fish at less than 500 mg/l. Sometimes turbidity is used as a measure of suspended solids and is given in turbidity units. Turbidity units are derived by light transparency. Turbidity in fertile surface waters is largely due to organic material, particularly algae. An expected measurement for water is 2 units, whereas algae-rich waters measure 200 units.

FIGURE 12-7 A Secchi disk can be used to determine turbidity.

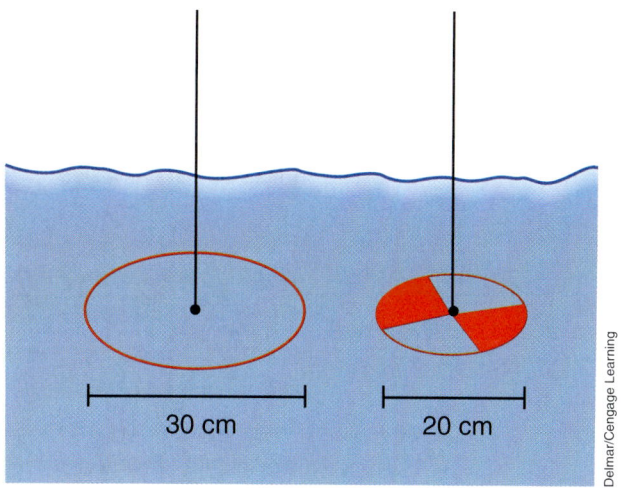

Trace Elements

Trace elements include ionic constituents of water that dissolve to a small extent. Some elements are routinely included in reports of irrigation water and soil tests because of the impact on plants. Boron is most common in this category. More generalized water reports include certain metals that could have toxic effects, and some laboratories offer metal analyses as special order items.

Aquatic organisms of all types are sensitive to metal poisoning when the concentration of these reaches a certain level in the water column. Certain fish groups tend to be more sensitive than others to particular metals. Copper, for example, is more toxic to rainbow trout than to channel catfish. Exact levels of tolerable metals in solution that are considered safe for aquatic life are the subject of much discussion and disagreement. The toxic nature of metals is influenced by water hardness. A metal may poison fish in very soft water at a rate that would have to be increased 10-fold to produce the same effect in hard water. Because the concentration of a metal required to produce toxicity may differ according to the overall water chemistry, a clear statement on the toxicity of the various metals is difficult to find.

Trace elements normally present in unpolluted surface waters at concentrations of less than one mg/l include:

Aluminum (Al)	Lead (Pb)
Arsenic (As)	Manganese (Mn)
Barium (Ba)	Mercury (Hg)
Beryllium (Be)	Molybdenum (Mo)
Cadmium (Cd)	Nickel (Ni)
Chromium (Cr)	Selenium (Se)
Cobalt (Co)	Silver (Ag)
Copper (Cu)	Zinc (Zn)
Iron (Fe)	

Availability of trace elements at toxic levels is affected by water hardness, dissolved organics, and suspended clays. Toxic action is also influenced by the form, for example, free ion or bound in organic compound. Whereas trace elements can be toxic, some are essential for the health of aquatic life. Except for aluminum, arsenic, barium, and iron, these elements should be considered potentially harmful when present in concentrations above 0.1 mg/l.

Dissolved Gases

Dissolved gases determine the basic suitability of water for fish survival. These gases include oxygen (O_2), carbon dioxide (CO_2), nitrogen (N_2), ammonia (NH_4^+ and NH_3), hydrogen sulfide (H_2S), chlorine (Cl_2), and methane (CH_4). Dissolved gases are usually not found on water analysis report forms because the manner by which samples are collected and shipped can cause gas measurements to be inaccurate. Ammonia is the most stable of the group, and if a sample is processed within a day after collection, it should measure fairly accurately. Other measures are best taken at the water site, using appropriate meters or chemical test procedures.

Dissolved Oxygen

Aquatic life requires **dissolved oxygen** (DO). It varies greatly in natural surface water and is characteristically absent in groundwaters. Most aquatic animals need more than a 1 ppm concentration for survival. Depending on culture circumstances, aquatic animals need 4 to 5 ppm to avoid stress. Concentrations considered typical for surface water are influenced by temperature but usually exceed 7 to 8 mg/l (ppm). In ponds, dissolved oxygen fluctuates greatly due to photosynthetic oxygen production by algae during the day and the continuous consumption of oxygen due to respiration. Dissolved oxygen typically reaches a maximum during the late afternoon and a minimum around sunrise. Cloudy weather, rain, plankton die-offs, and heavy stocking and feeding rates result in low levels of dissolved oxygen, which can stress or kill fish. As Table 12-2 indicates, the gains and losses of the dissolved oxygen in ponds are very close.

Oxygen is only slightly soluble in water. Water may be frequently supersaturated with oxygen in ponds with algae blooms. For example, at sea level at a temperature of 77°F, pure water contains about 8 ppm of

TABLE 12-2 DISSOLVED OXYGEN LEDGER FOR PONDS		
Source	Gain (ppm)	Loss (ppm)
Photosynthesis by phytoplankton	6–20	—
Diffusion	1–5	—
Plankton respiration	—	5–15
Fish respiration	—	2–6
Diffusion	—	1–5
Respiration by other organisms	—	1–3
Total	7–25	9–29

oxygen when 100 percent saturated, but during the afternoon hours, levels of 10 to 14 ppm in ponds with healthy algae blooms are not uncommon (Figure 12-8).

As water warms, is raised to higher altitudes, or becomes more saline, its oxygen holding capacity declines. Water saturated with oxygen at 59°F contains about 9.8 ppm, whereas water at 86°F is saturated at about 7.5 ppm.

Aquaculturalists measure dissolved oxygen with oxygen meters or chemical test kits, which give results in mg/l. Guidelines for oxygen management usually report that oxygen levels should be maintained above 4 mg/l (ppm) to avoid stress. Most warmwater fish experience significant oxygen stress at levels of 2 mg/l, and that levels of less than 1 mg/l (ppm) may result in fish kills (Figure 12-9). While these guidelines are accurate,

FIGURE 12-8 Increasing the water temperature or the salinity reduces its oxygen-holding capacity.

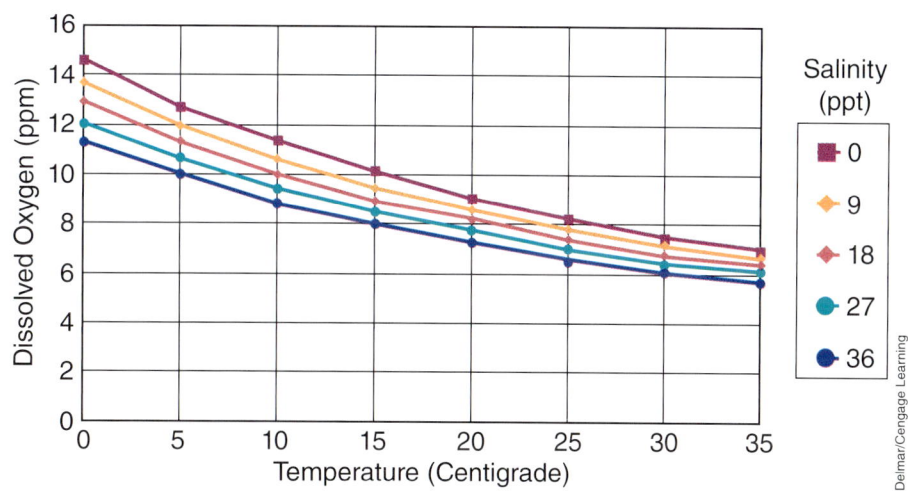

FIGURE 12-9 Fish need dissolved oxygen at 5 ppm. Less stresses fish and can be lethal.

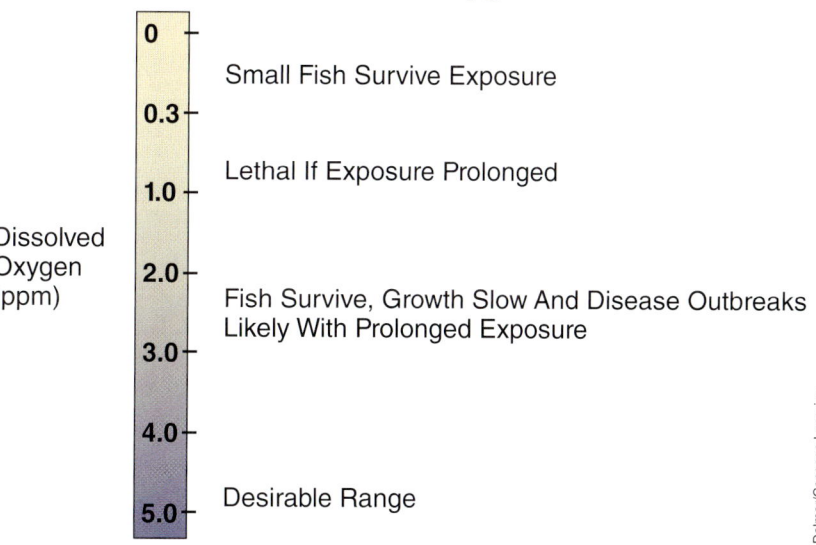

fish actually respond to the percent **saturation** of oxygen rather than the oxygen content in water. A reading of 1 mg/l at 30°C (13.3 percent saturated) is a higher concentration than 1 mg/l at 15°C (10.2 percent saturated) and represents more available oxygen.

If dissolved oxygen reaches low levels, fish will show signs, including:
- Not eating and acting sluggish
- Gasping for air at the surface
- Grouped near water inflow pipe
- Slow growth
- Outbreaks of disease and parasites.

Proper water management prevents the problems from depletion of dissolved oxygen. Management techniques include:
- Monitoring dissolved oxygen at critical times
- Avoiding overfeeding
- Proper stocking level
- Avoiding over-fertilization
- Controlled plant growth
- Some form of aeration (Figure 12-10)
- Keeping water circulating

Carbon Dioxide

Carbon dioxide (CO_2), a minor component of the atmosphere, is highly soluble in water. Most carbon dioxide in pond water occurs as a result of respiration. Levels usually fluctuate inversely to dissolved oxygen, being low during the day and increasing at night, or whenever respiration occurs at a greater rate than photosynthesis (Figure 12-11). Carbon dioxide is present in surface water at less than 5 mg/l (5 ppm) concentrations but may exceed 60 mg/l (60 ppm) in many well waters and 10 mg/l where fish are maintained in large numbers. Some aquatic

FIGURE 12-10 Pond aeration increases the dissolved oxygen.

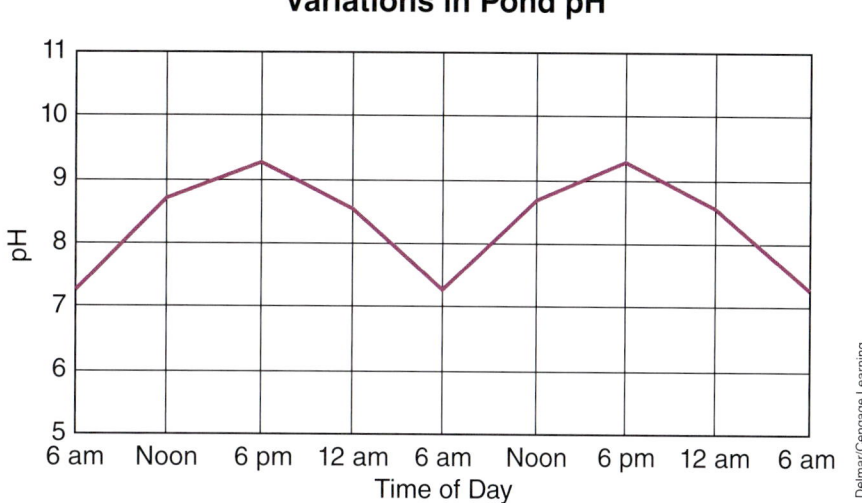

FIGURE 12-11 Carbon dioxide (CO_2) content of the water affects the pH of water in a pond during the course of a day.

animals, including fish, can endure stress and survive at up to 60 mg/l. If oxygen is lowered into its stress-causing range, carbon dioxide limitation is reduced to 20 mg/l.

Carbon dioxide interferes with aquatic animals' ability to extract oxygen from water, contributing to the stress of fish during periods of low oxygen. Aerating water to improve its oxygen content drives off excess carbon dioxide.

Adding quick lime ($Ca(OH)_2$) to water rapidly removes carbon dioxide without affecting oxygen content. This improves the ability of fish to use available oxygen. Carbon dioxide acts as an acid in water, lowering pH as it increases in concentration. Carbonate buffers in water neutralize carbon dioxide and stabilize pH fluctuations within the range tolerated by fish. Waters low in alkalinity and hardness may experience extremes of pH due to its poor buffering against changes in carbon dioxide concentrations.

Hydrogen Sulfide

Hydrogen sulfide (H_2S), rotten-egg gas, is present in some well waters but is so easily oxidizable that exposure to oxygen readily converts it to harmless form. Its toxicity depends on temperature, pH, and dissolved oxygen. Any measurable amount after providing reasonable **aeration** could be considered to have potential to harm fish life.

Hydrogen sulfide occurs in ponds as a result of the **anaerobic** decomposition of organic matter by bacteria in mud. Hydrogen sulfide is toxic to fish and interferes with normal respiration. Toxicity is increased at higher temperatures and a pH less than 8 when the largest percentage of hydrogen sulfide is in the toxic un-ionized form. Vigorous aeration or splashing is usually sufficient to remove hydrogen sulfide from well water.

In ponds, hydrogen sulfide can be released from anaerobic mud when the bottom is disturbed by seining and harvest activities. Liming ponds raises mud pH and reduces the potential for the formation of hydrogen sulfide. Potassium permangenate at 2 to 6 mg/l (2 to 6 ppm) removes hydrogen sulfide from water and reverses the effects of its toxicity to fish.

Ammonia

Ammonia is present in slight amounts in some well and pond waters. As fishes become more intensively cultured or confined, ammonia can reach harmful levels. Any amount is considered undesirable, but stress and some death loss occurs at more than 2 mg/l (2 ppm), and at more than 7 mg/l (7 ppm) fish loss increases sharply. Ammonia is a waste product of protein metabolism by aquatic animals (Figure 12-12).

In water, ammonia occurs either in the ionized (NH^+) or un-ionized (NH_3) form, depending upon pH. Un-ionized ammonia is considerably more toxic to fish and occurs in greater proportion at high pH and warmer temperatures (Figure 12-13). For example, at 82.4°F and pH 8, 6.55 percent of the total ammonia is present in unionized form. At pH 9, 41.23 percent of the ammonia is unionized. Unionized ammonia is stressful to warmwater fish at concentrations greater than 0.1 mg/l and lethal at concentrations approaching 0.5 mg/l. Concentrations of 0.0125 ppm cause reduced growth and gill damage in trout.

Test kits for determining ammonia in water measure total ammonia. To determine if a large percentage of the ammonia is in un-ionized form, pH is also measured. A pH above 8 in the presence of ammonia concentrations

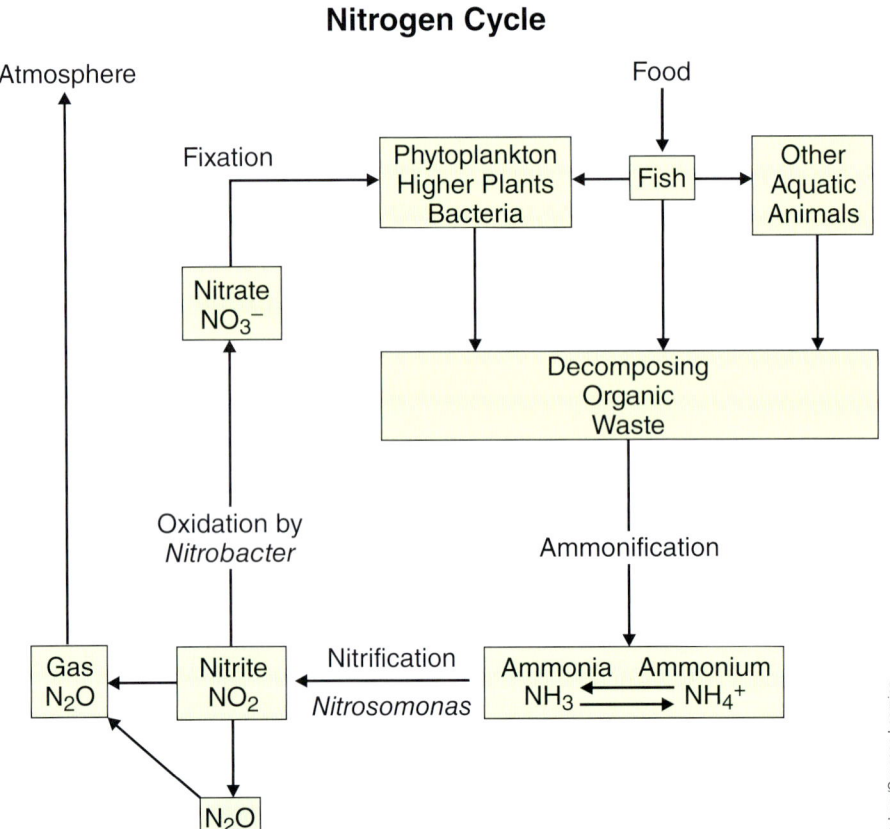

FIGURE 12-12 Nitrogen cycle. Feed and living tissue all contain nitrogen. Eventually, the nitrogen is released and converted to several forms.

FIGURE 12-13 As temperature and pH increase, so does un-ionized ammonia.

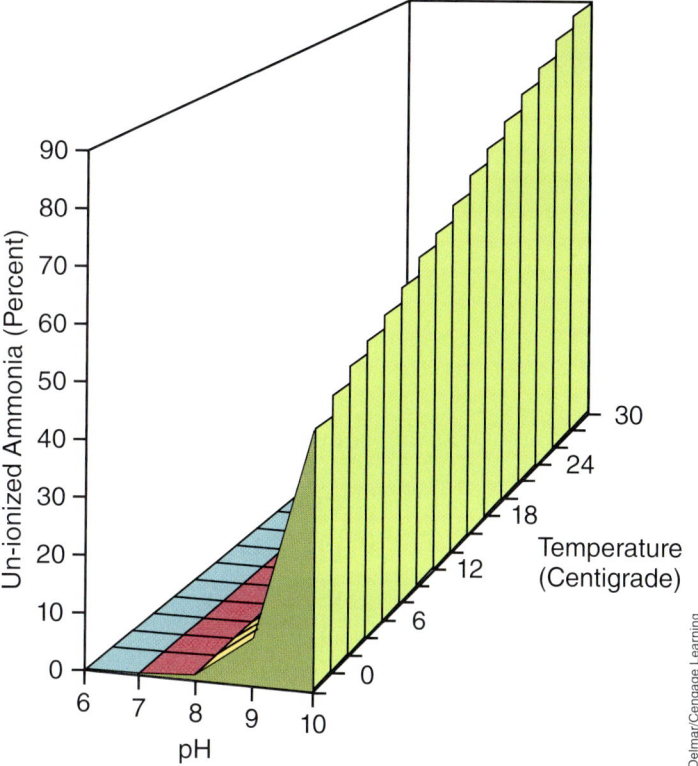

above 0.5 mg/l is cause for concern. Algae use ammonia as a nitrogen source for making proteins. Concentrations usually remain low in ponds with phytoplankton blooms. The greatest concentration of ammonia often occurs after plankton die-offs, at which time pH is low due to high levels of carbon dioxide, and the majority of ammonia is present in the relatively nontoxic ionized form.

Management is the key to preventing problems from ammonia. The management techniques are similar to those for preventing oxygen depletion:

- Avoid overcrowding
- Avoid overfeeding
- Add freshwater
- Control plant growth
- Monitor the pH
- Remove fecal material

Nitrite

Nitrite (NO_2^-) is one of the basic products of organic matter decomposition. It acts as an intermediate stage in the conversion of ammonia to nitrate. These reactions occur in soils, mud, and water. Nitrite changes quickly to nitrate if oxygen is present. In culture ponds that are rich from feeding, a temporary accumulation of this chemical in harmful amounts sometimes occurs. Nitrite measurements of more than 1 mg/l (1 ppm) should be suspect in causing fish deaths.

The binding of nitrite with the hemoglobin molecule gives blood a chocolate brown color. Fish farmers call this condition brown blood disease. In medical terms it is known as "methemoglobinemia." The toxicity of nitrite to fish is lessened by the presence of chlorides in water. The addition of salts, either calcium chloride or sodium chloride at rates of 20 mg/l for each 1 mg/l of nitrite, is a standard treatment for preventing nitrite poisoning in freshwater ponds. Most warmwater fish can tolerate at least 0.4 mg/l of nitrite in freshwater without treatment if oxygen levels remain above 4 mg/l.

Nitrite should be monitored frequently if a problem is suspected because its concentration may increase rapidly in pond water, especially during spring and fall, or when algae blooms suddenly die.

Chlorine and Chloride

Chlorine is usually present at approximately 1 mg/l in municipal water supplies as a result of chlorination. Fish succumb quite easily at these levels.

Some aquaculturalists may mistake the difference between chlorine and chloride. Chlorine (Cl_2), in gaseous form or as hypochlorites, is widely used for disinfection of water supplies and in aquaculture for sterilization of equipment, tanks, and standing water in drained ponds. Chlorine acts as a powerful oxidizing agent, and it is toxic to fish at concentrations of less than 0.05 mg/l. Residual chlorine in municipal water supplies is normally between 0.5 and 2.0 mg/l. Water used for fish culture should not contain any residual chlorine to be considered safe. Chlorine can be removed from water by extended periods of aeration before use, or more rapidly by the addition of sodium thiosulfate at a rate of 7.0 mg/l for each 1 mg/l of chlorine. Sodium thiosulfate is not considered toxic to fish at concentrations required to remove chlorine from water.

Chloride is a by-product of chlorine dissociation in water, but it is also widely associated with numerous other compounds that are highly water soluble. Chloride occurs in a range of concentrations in water and is often used as an indicator to characterize aquatic environments as saline or fresh. Fish, too, are classified as freshwater or marine as determined by their physiological adaptation to salinity. Chloride is important to fish in osmoregulation and other physiological processes and is regarded as nontoxic within the tolerance range of each species. Tests to determine the chlorine or chloride content of water require completely different chemical procedures. The distinction between the two compounds and their significance in water is clear, but occasionally confused.

Nitrogen and Methane

Nitrogen (N_2) and methane (CH_4) are considered to play a critical role only at abnormally high levels. For example, nitrogen may be driven to high levels in waters that are plunged deeply at dam outfalls or through pump cavitation. As total gas concentrations exceed 115 percent of normal, fish are affected by bubble formation in the blood, called gas bubble disease. Total gases is a type of measure that is sometimes used in aquatic analyses, but it is not often seen in laboratory reports.

Organic Material and Its Breakdown Products

In water, organic material consists of living organisms and various dissolved organic chemicals of their excretion and dead matter and with its associated decomposition products. Organic chemicals or tissues consist mostly of the elements carbon, oxygen, hydrogen, nitrogen, and sulfur. Organic material decomposes constantly into its ultimate breakdown products of carbon dioxide, sulfide, ammonia, nitrate, hydrogen ion, and water.

Carbon is always a component of organic chemicals. In the analysis of organics, the characteristics of decomposition potential and carbon presence are used for measurement. Measurement of organics can give the pond owner an idea of overall enrichment and because decomposition requires oxygen, a general idea of what will be demanded of a pond's oxygen content.

Some organic materials, such as pesticides, will act as toxins to fish at very low levels. For low-level detection of these specific organic constituents, sophisticated instrumentation techniques such as chromatography are used.

Total Organic Carbon

This test is commonly reported on laboratory forms and is given in mg/l. Natural surface water would be expected to contain 10 mg/l, and water that receives regular feeding could build up to over 30 mg/l. Water with decomposing plant life could also be expected to have high organic carbon. The measure of total organic carbon includes dissolved organic carbon and suspended material, commonly called "particulate organic carbon."

COD

Chemical oxygen demand (COD) is a speedy and reliable estimate of organic load that is reported in mg/l. A normal measure would read less than 10. A measure of 60 would be considered rich.

BOD

Biochemical oxygen demand (BOD) is a standard test for organic material. It is reported in mg/l per hour or per total test time. On a per-hour basis, 0.5 would be considered rich. BOD is usually measured over a five-day test period.

Chlorophyll

The measure of this photosynthetic pigment gives an estimation of plant life suspended in the water column. Unfertile ponds range up to 20 micrograms per liter (mcg/l), and fertile ponds with rich phytoplankton blooms range from 20 to 150 mcg/l.

Oil and Grease

Most oils and grease float at the surface. Aquatic animals suffer at rates above 0.1 mg/l where the oily products are not those of the natural release of plants or other living agents of the pond. Roughly, sudden die-off of fishes would be expected to occur when such concentrations exceed 100 mg/l. An oil slick is usually apparent in cases of pollution, but a sudden die-off of an algae population will cause a slick from the oil material in the plants.

Pesticides

Pesticides are not a natural part of the environment. Their presence is determined by past human activities. A knowledge of the historical application of pesticides in the watershed gives a clue to which chemicals should be tested for. Test laboratories offer scans of pesticides known to retain their integrity months and years after application. New pesticides disappear from detection soon after application. Insecticides are more toxic to crustaceans than fish because of their closer kinship to insects. Pesticide groups include:

- Organochlorides like DDT and toxaphene at 0.01 to 0.5 mg/l
- Organophosphates like methyl parathion and malathion at 1 to 10 mg/l
- Carbamates like carbaryl and methomyl at 0.1 to 10 mg/l
- Pyrethroids like permethrin at 0.1 to 10 mg/l
- Fungicides like benomyl and captan at 0.05 to 5 mg/l
- Herbicides at 0.05 to 500 mg/l (aquatic and terrestrial toxicity varies greatly)

OTHER FACTORS

Several factors relate to the water requirements for aquaculture. These include the soil type, the effect of rainfall, the rate of evaporation, the amount of light, water odor, and water color.

Soil

The soil must hold water, so clay-type soils are desirable. Soil core samples at various places around the site determine if adequate clay is present to prevent excess seepage. The local office of the Natural Resources Conservation Service (NRCS; http://www.nrcs.usda.gov/) can provide assistance. Table 12-3 shows seepage in various soil types. Ponds can be built in soils that have high percolation rates if they are lined with a layer of packed clay or plastic.

Once water is added to a pond, the soil in the pond bottom becomes an important part of the system. Nutrient and oxygen fluxes between the pond bottom (mud) play an important role in determining water quality. The mud receives much of the organic matter entering the pond. Organisms produced in and on pond bottoms become food for crustaceans and fish. Some fish lay eggs in the pond bottom.

TABLE 12-3 SOIL TYPES AND SEEPAGE LOSSES

Soil Type	Seepage Losses (inches/day)
Sand	1–10
Sandy loam	0.52–3
Loam	0.32–0.8
Clayey loam	0.1–0.6
Loamy clay	0.01–0.2
Clay	0.05–0.4

Rainfall

Rainfall varies from place to place, year to year, and season to season. Rainfall influences the amount of water in a pond. Also, precipitation falling on the land may become runoff. As runoff, it dissolves various substances that may affect the water used for aquaculture. Rain gauges are inexpensive and provide one more piece of helpful information for making management decisions. Rainfall affects runoff pollution from land, the temperature, pH, and the total dissolved solids of surface water.

Evaporation

Evaporation tends to follow the amount of sunlight (solar radiation) and the air temperature. As the solar radiation and air temperature increase, evaporation increases. Humidity and wind speed also affect evaporation. Low humidity and high wind speed increase evaporation. Ponds need compensation for evaporative losses.

Light

The amount of light striking a pond influences the amount of photosynthesis. As the light increases, photosynthesis increases. Clouds, roughness of the water surface, and turbidity decrease the amount of light available for photosynthesis.

Odor

Water's odor affects its recreational value and the taste of fish and other aquatic foods. Odor comes from sources like municipal and industrial waste discharges, decomposing vegetable matter, and microbial activity. The human nose is the best instrument for detecting odors in water. Table 12-4 can be used to classify and record odor from a water sample.

Color

Water color results from dissolved substances and suspended material. It can provide useful information about the water's source and content. Transparent water with low amounts of dissolved or suspended substances indicates low productivity and appears blue. Dissolved organic materials add a yellow or brown color. Some algae and dinoflagellates produce reddish or deep yellow colors in water. Water rich in phytoplankton appears green. Soil runoff produces various red, brown, yellow, and gray colors.

Methods like the Forel-Ule Color Scale and the Borger Color System attempt to standardize the descriptions of the apparent color of water. This method for determining the color of water depends on the accuracy of the eye comparing water samples against a standardized color

TABLE 12-4 ODOR CLASSIFICATIONS

Description	Nature of Odor	Examples of Odor
Aromatic (spicy)	Camphor, cloves, lavender, lemon	Same
Balsamic (flowery)	Geranium, violet, vanilla	Same
Chemical	Industrial wastes or treatments	
	Chlorinous	Chlorine
	Hydrocarbon	Oil refinery wastes
	Medicinal	Phenol and iodine
	Sulfur	Hydrogen sulfide (rotten egg)
Disagreeable (pronounced, unpleasant)	Fishy	Uroglenopsis, Dinobryon (dead algae)
	Pigpen	Anabaena algae (visit a pig farm to sample this distinctive odor)
	Septic	Stale sewage
Earthy	Damp earth	
	Peaty	Peat
Grassy	Crushed grass	Same
Musty	Decomposing straw	
	Moldy	Damp cellar
Vegetable	Root vegetables	Same

scale or system. When using the color scale, determinations should be made against a white background. The white quadrants of a Secchi disk can be used.

OBTAINING WATER

The prime consideration for any aquaculture facility is a plentiful supply of high-quality water. This is true whether the organisms to be grown are fish or shellfish. Also, it is an economic necessity that the water be relatively inexpensive. A secondary consideration is whether the soil in the area will hold water. Suitable alternatives cost proportionately more.

For crawfish and shrimp, the water depth required is usually 1.5 to 3 ft, whereas, for finfish, recommended depths are from 3.5 to 5 ft. Each acre-ft of water contains about 326,000 gal. A 1-acre pond, 5 ft deep, requires over 1.6 million gallons for a single filling. Some flushing of the pond is usually desirable and is practiced, which means an additional 0.4 million gal per acre per year. Planning any type of aquaculture facility requires estimating the water needs (Figure 12-14).

Calculating Water Needs

If water supply is inadequate, then a facility may not meet its production goals. A suitable water supply in both volume and quality is essential for the intensive, commercial production of fish.

FIGURE 12-14 The effect of pumping rate (gallons per minute) becomes very evident in larger ponds.

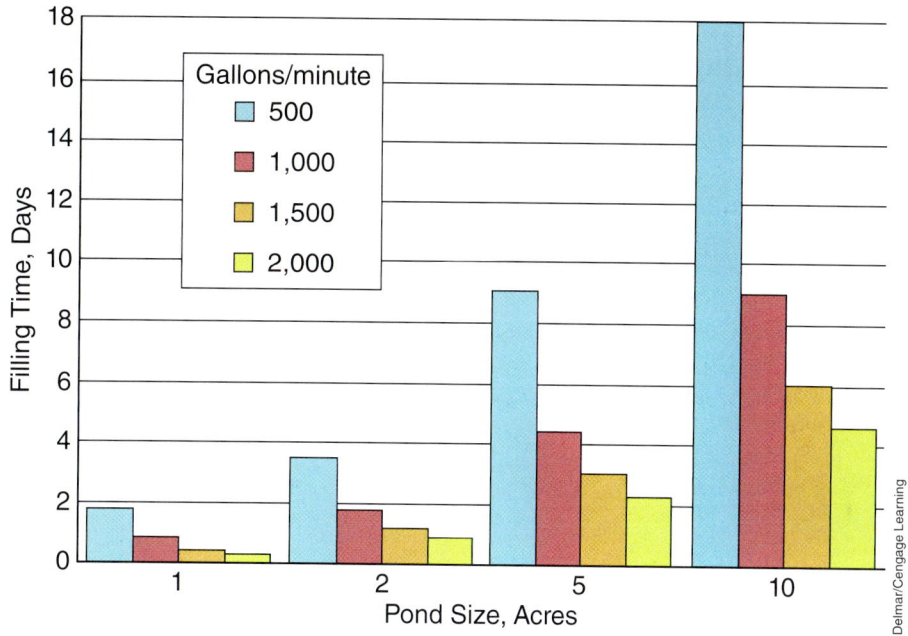

The following examples will illustrate the steps and calculations required to estimate the water requirements for various production units.

EXAMPLE 1. A producer wants to construct four ponds in the same area and service all with one water well. The ponds vary in size, and the owner wants to be able to fill any pond within seven days. The pond sizes are 6 acres, 4 acres, 5.5 acres, and 3.5 acres. The average water depth in each pond is 5 ft. What flow rate in gallons per minute (gpm) is required from the service well to fill any pond in at least seven days?

STEP 1.	Determine the volume of water in the largest pond. If the largest pond can be filled in seven days, then any smaller ponds will fill in seven days or less. Volume in acre-ft = 6 acres × 5 ft average depth	= 30 acre-ft
STEP 2.	Convert 30 acre-ft to gallons: 30 acre-ft × 325,850 gal/acre-ft	= 9,775,500 gal
STEP 3.	Determine the minimum flow rate needed to fill the pond in seven days. In this case, we are not including any adjustment for seepage, evaporation, or rainfall. $$\frac{9{,}775{,}500 \text{ gal}}{7 \text{ days} \times 24 \text{ hours/day} \times 60 \text{ minutes per hour}}$$ To be on the conservative side, a 1,000-gpm well should be adequate.	= 970 gpm

EXAMPLE 2. With the 1,000-gpm well from Example 1, what would be the filling time in days for the smallest pond of 3.5 acres and 5 ft. average water depth?

STEP 1.	Determine the volume of water in the pond: Volume in acre-ft = 3.5 acres × 5 ft. average depth	= 17.5 acre-ft
STEP 2.	Convert 17.5 acre-ft to gallons: 17.5 acre-ft × 325,850 gal./acre-ft	= 5,702,375 gal.
STEP 3.	With a flow rate of 1,000 gpm, the filling time in minutes is: $\dfrac{5{,}702{,}375 \text{ gal.}}{1{,}000 \text{ gal./min.}}$	= 5,702 minutes
STEP 4.	Convert 5,702 minutes to days: 5,702 minutes × 60 minutes/hr. 95 hours × 24 hours/day	= 95 hours = 3.96 or 4 days to fill pond

EXAMPLE 3. A fish hatchery facility is being planned that will include six fish-holding tanks, each 4 ft wide and 30 ft long. The average water depth in each is 3 ft. The water supply for these tanks has to provide at least two complete water exchanges per hour in all tanks simultaneously. The facility will also have 20 troughs, each 2 ft wide and 10 ft long with an average depth of 1 ft. A water supply of 5 gpm is required for each trough, and all may need water at the same time. What is the minimum water requirement in gallons per minute (gpm) for this facility?

STEP 1.	Determine the water requirement for the six holding tanks. Determine the volume of water that the six holding tanks contain. Volume per tank = 4 ft (width) × 30 ft (length) × 3 ft (depth) Convert 360 ft³ to gallons: 360 ft 3 × 7.48 gal/ft³ 2,693 gal/tank × 6 tanks	= 360 ft³/tank = 2,693 gal/tank = 16,158 gal capacity for tanks
STEP 2.	The desired filling time for all tanks is 1 hour or 60 minutes. The gpm required to fill tanks is calculated as follows: $\dfrac{16{,}158 \text{ gal capacity of all tanks}}{60 \text{ minutes}}$ = 270 gpm required to fill all tanks in 60 minutes	

STEP 3. Determine how much water is required to supply two complete water exchanges per hour to all tanks. Two exchanges per hour means that the total volume of water in all tanks needs to be completely replaced every 30 minutes.

$$\frac{16{,}158 \text{ gal in tanks}}{30 \text{ minutes}} = 540 \text{ gpm for two water exchanges per hour in all tanks}$$

The water requirement for the six holding tanks is 540 gpm. It is not 540 gpm plus 270 gpm because the maximum requirement of 540 gpm for continuous flow use can be reduced to 270 gpm to fill all tanks within 60 minutes. If fewer tanks are filled at one time or the flow rate is increased, then the filling time will decrease.

STEP 4. Now, determine the water requirements for the 20 troughs. The water requirement for these troughs needs to be added to the water requirement for the holding tanks because all facilities may be in use at the same time. First, determine the total volume of water in the troughs:

Volume per trough = 2 ft (width) × 10 ft (length) × 1 ft (depth) = 20 ft^3/trough

Convert 20 ft^3 to gallons:
20 ft^3 × 7.48 gal/ft.^{3} = 150 gal/trough

Total water volume in troughs:
150 gal/trough × 20 troughs = 3,000 gal

The desired filling time for all troughs is a maximum of 60 minutes. To determine the water requirement, do the following calculations:

$$\frac{3{,}000 \text{ gal. capacity all troughs}}{60 \text{ minutes}} = 50 \text{ gpm required to fill all troughs in 1 hour}$$

STEP 5. Determine the flow rate required for all troughs once they are full of water. The required continuous flow rate for each tank is 5 gpm.

5 gpm/trough × 20 troughs = 100 gpm

For the troughs, the water requirement is 100 gpm because only 50 gpm is needed to fill the tanks within the desired time of 1 hour.

STEP 6. To determine the total water requirement for the facility, the water requirements for tanks and troughs are added. If more water is needed for other purposes that would occur when all other facilities are operating, then these water requirements are also calculated and added to the total requirement. The total water requirement in gpm for facility is 540 gpm for six holding tanks plus 100 gpm for 20 troughs equals 640 gpm.

Sources

Water for aquaculture comes as surface water or ground water. Either source may require some pumping.

Surface Water

If water from a stream or reservoir is used to fill the pond, the water should be screened or filtered. A series of screened boxes or a gravel filter bed are useful at the intake source. Unfiltered water at the pond inlet should run through a fine mesh screen box or saran filter sock. The primary purpose of the screens and filters is to prevent undesirable wild fish, fish eggs, and predaceous invertebrates from entering the pond.

Ground Water

Water pumped from wells is often low in oxygen and may contain high levels of undesirable gases, such as hydrogen sulfide and carbon dioxide. It is beneficial to aerate this water prior to its entering a pond or tank, especially if the inflowing water rapidly replaces the pond or tank water. A splashboard or other device that splashes, sprays, or agitates the water as it enters the pond is normally used.

Pumping

The final deciding factor in any aquaculture production facility is operating cost. Pumping costs contribute significantly to operating costs. The costs to pump an acre-ft of water include, using a diesel-powered pump, diesel fuel, depreciation, interest, and maintenance costs of the well and pumping equipment. The cost of wells is extremely variable and impossible to relate to pumping lift or the amount of fuel required to pump a given quantity of water (Figure 12-15). Well costs must be estimated on an individual basis.

FIGURE 12-15 Pumping water contributes significantly to the cost of an operation.

MANAGING WATER

Managing water requires maintaining the quantity, testing the quality, and maintaining the quality. Other factors include aquatic life, evaporation, and seepage.

Quality Measurements

A practical understanding of water quality is essential to the aquaculturalist to allow an evaluation of environmental conditions and implementation of effective management strategies. Often, aquaculturalists do not have extensive training in water chemistry and, as a result, may misinterpret or misapply information about water quality and its management. Though the chemistry of water is a complex subject, most aspects of general importance to fish farmers can be simplified to allow for easier understanding and practical approaches to management.

Pond water is tested for many reasons. The selection of water tests would be different for each. Reasons for testing include:

- A catastrophe with sudden die-off of fish and no clue to a cause
- A catastrophe with sudden die-off of fish and a good idea of cause
- Little or no die-off but with knowledge or suspicion of some accident or malicious action in the vicinity of the pond
- Evaluation of suitability of pond water or water source

If a probable cause is known, the test requirements can be easily narrowed down. If a pesticide is suspected, it is very helpful for the pond owner to do an initial investigation to determine what type of pesticide was used. Otherwise the testing may be no more than searching for a needle in a haystack because of the great variety of manufactured pesticides.

When the cause of the catastrophe is a mystery, or where water is merely being evaluated, a general spectrum of tests that evaluate parameters known to commonly affect aquatic life can be conducted. Table 12-5 lists common tests, sample selection, and methods.

TABLE 12-5 SUMMARY OF WATER MANAGEMENT METHODS

Parameter	Procedure	Method*
Air Temperature: See Temperature, Air		
Alkalinity	Collect samples in plastic or glass bottles. Fill completely and cap tightly. Avoid excessive agitation and prolonged exposure to air. Keep samples cool in refrigeration unit or ice chest, and analyze within 24 hours. Warm to room temperature before analyzing.	Test kit; titrimetric
Ammonia	Collect samples in clean glass or plastic bottles. Samples not analyzed immediately may be preserved colorimeter by reducing the pH to 2 or less with sulfuric acid. Refrigerate samples and analyze within 24 hours. Warm to room temperature and neutralize before analysis.	Test kit; electronic
Biochemical Oxygen Demand (BOD)—see Oxygen, Biochemical Demand (BOD)		
Carbon Dioxide	Collect samples in clean glass or plastic bottles. Fill completely and cap tightly. Avoid excessive agitation or prolonged exposure to air. Analyze as soon as possible but samples can be stored at least 24 hours by cooling to a temperature lower than the source.	Test kits; titrimetric

TABLE 12-5 SUMMARY OF WATER MANAGEMENT METHODS (CONTINUED)

Parameter	Procedure	Method
Chemical Oxygen Demand (COD): See Oxygen, Chemical Demand		
Chloride: See Salinity		
Depth, Pond	Install staff gauge in each pond and read at same time each day, before restoring to specified depth.	Read and record
Dissolved Oxygen: See Oxygen, Dissolved		
Evaporation and Inflow	Surface inflow/outflow and evaporation should be determined.	Read and record
Hardness	Collect samples in plastic or glass bottles that have been washed with detergent, rinsed with tap water, and rinsed with 1:1 nitric acid solution and deionized water. Store only if prompt analysis is not possible.	Test kits; titrimetric and colorimetric, electronic
Nitrate	Collect samples in clean plastic or glass bottles. Store at 39°F (4°C) if samples analyzed in 24 to 48 hours. Warm to room temperature before running test.	Test kits; colorimetric, visual and electronic; Electronic meter
Nitrogen, Total Kjeldahl	Collect samples in clean glass or plastic bottles. Samples should be refrigerated and analyzed within 24 hours. Samples preserved by adjusting the pH to 2 or less with sulfuric acid can be stored in a refrigerator for up to 28 days.	Digestion required; colorimetric, electronic
Oxygen, Biochemical Demand (BOD)	Determined by measuring the dissolved oxygen in a freshly collected sample and comparing it to the dissolved oxygen level in a sample collected at the same time, but incubated at 20°C for five days. The difference between the two oxygen levels is the BOD.	Test kits
Oxygen, Chemical Demand (COD)	Collect samples in glass bottles. Use plastic bottles only if they are known to be free of organic contamination. Test as soon as possible. To store samples, bring pH to 2 or less with sulfuric acid.	Test kits; titrimetric or colorimetric
Oxygen, Dissolved	Measure directly in water source or collect samples. Sampling and sample handling important for meaningful results. Dissolved oxygen changes due to many variables.	Test kit; Winkler titration Electronic meter
pH, Water	Take readings at 10 in. (25 cm) below the water surface. Collect samples in clean plastic or glass bottles. Fill bottles completely and cap tightly. If a probe is used, calibrate using a precision thermometer. Calibrate meter with standard buffers at pH 7 and pH 10.	Test kit; colorimetric; visual method Calibrated electronic meter
Phosphorus, Total	Collect one sample by pooling three samples. For most reliable results, analyze immediately. Samples could be refrigerated and analyzed within 24 hours.	Test kit; colorimetric, electronic
Pond Soil Characteristics—see Soil, Characteristics		
Pond Temperature—see Temperature, Water		
Rainfall	Install three rain gauges at site. Read and empty at 24-hour intervals, or more frequently to prevent gauge overflow. Report average of three readings.	Read rain gauge
Salinity	Collect samples pooled from three levels if collected from a pond. Use clean glass or plastic bottles. Samples can be stored for up to 28 days in sealed containers.	Test kit; titration Test kit; electronic colorimeter Meter; conductivity Electronic hydrometer refractometer
Secchi Disk Visibility—see Visibility, Secchi Disk		
Seepage	Determine seepage from a 24-hour water balance, preferably when there is no rainfall, inflow, or outflow.	Read and record

Parameter	Procedure	Method
TABLE 12-5 SUMMARY OF WATER MANAGEMENT METHODS *(CONTINUED)*		
Soil, Characteristics	Determine soil type and particle size.	Soil triangle and International system of particle sizes
Solar Radiation	Install Solar Monitor and Quantum Sensor and read the cumulative radiation each day and at end of each time interval.	
Temperature, Air	Install three maximum-minimum thermometers in the shade near ponds; read at 24-hour intervals and report average maximum and average.	Thermometers
Temperature, Water	Take readings at 10 in (25 cm) below the water surface. Ideally, in a pond take readings also at mid-water and 10 in (25 cm) above the bottom. If probe is used, calibrate using a precision thermometer.	Certified thermometer Electronic meter
Total Dissolved Solids (TDS)	Collect sample, measure volume, dry, and weigh dried sample. A convenient alternative is to test the conductivity that can be used to estimate the TDS.	Gravimetric Electronic meter
Total Kjeldahl Nitrogen—see Nitrogen, Total Kjeldahl		
Total Phosphorus—see Phosphorus, Total		
Wind Speed	Install totalizing anemometer, read at 24-hour intervals.	Anemometer
Visibility, Secchi Disk	At two locations in each pond, calculate Secchi Disk Visibility.	Secchi disk

*Many of the colorimetric, titrimetric and other methods have some type of electronic/sensor equivalent.

Test Methods

Whether the water is tested in a laboratory or in the field by an aquaculturalist, general test methods include **titrimetric**, **colorimetric**, and electronic meters. For many water quality tests, companies sell test kits complete with all of the chemicals and standards.

Titrimetric

Titrimetric analyses use a solution of known strength—the **titrant**—which is added to a known or specific volume of sample in the presence of an indicator. The indicator produces a color change, indicating that the titration is complete. To calculate the results, the amount of titrant used is measured. A microburette or precision pipet adds the titrant.

Colorimetric

Beer's law states that the higher the concentration of a substance, the darker the color produced in a test reaction. This law provides the basis for determining the concentration of many substances in water samples. Known chemical reactions produce typical colors. The concentration that these colors represent is determined visually by comparing the color that is obtained from a sample to a set of standards. Because the human eye's interpretation can be quite subjective, electronic colorimeters provide a more accurate indication of color intensity. Colorimeters consist of a light source passing through a sample that is measured on a photo

Electronic Meters

Modern electronics provides the aquaculturalist with a variety of electronic meters designed to measure specific water-quality factors, including pH, total dissolved solids, conductivity, dissolved oxygen, temperature, and turbidity. Like the chemical methods, standard solutions are important. They are used to calibrate the electronic meter. Using test kits or electronic meters, aquaculturalists regularly check the oxygen, pH, carbon dioxide, and ammonia of water in production.

Flesh Analysis

Fish flesh is often taken in large-scale fish kills to analyze for possible toxicants. Pesticides and **heavy metals** are the most common items investigated. Some metals will normally be found in much greater concentration in the flesh than in the water in which that fish swims. Some pesticides characteristically also accumulate in quantities much greater than are present in the water.

Sediments

Sediments or muds from pond bottoms are usually checked for heavy metals and pesticides. Measurements of heavy metals in sediments are typically much higher than those measured in the water column. Usual measures of lake sediments in mg/l include:

- Cadmium (Cd), 0.1 to 1.5
- Cobalt (Co), 4 to 40
- Chromium (Cr), 50 to 250
- Copper (Cu), 20 to 90
- Iron (Fe), 11,000 to 70,000
- Lead (Pb), 10 to 100
- Lithium (Li), 15 to 200
- Manganese (Mn), 100 to 1,800
- Mercury (Hg), 0.1 to 1.5
- Nickel (Ni), 30 to 250
- Zinc (Zn), 50 to 250

Sediment pesticides are often checked when establishing a particular site as suitable for fish culture. The very persistent pesticides are the object of such checks, and organochlorines are usually tested as "scans." Sediments are also checked for pesticides when they are suspect in a catastrophe situation. The testing laboratory should be advised as to which type of pesticide has been used. Otherwise, many useless tests may be conducted that result in a very high expense to the pond owner.

Other Management Factors

Other water-management factors include pond **fertilization**, insect control, vegetation control, evaporation control, and seepage.

Pond Fertilization

Ponds stocked with fish eggs, fish fry, or fingerlings benefit from fertilization. The nutrients provided stimulate the productivity of the natural food chain, resulting in abundant zooplankton, the preferred natural food of most young fish. A mix of organic and inorganic fertilizers is used most often. Quantities of organic fertilizers used are dependent on quality and nutrient content. Cottonseed meal or chicken manure at 300 to 500 lbs per acre will help promote development of the zooplankton bloom. Chapters 4, 7, and 8 provide examples of using animal manure to fertilize ponds. Phytoplankton growth is stimulated by adding 8 to 10 lbs per acre of phosphorus in a liquid or granular fertilizer mix. The most commonly used inorganic fertilizers are ammonium phosphate or liquid polyphosphates. When using a liquid fertilizer, application rates are usually about 1 to 1.5 gal per acre.

In any case, natural water fertility determines the need for fertilization. Older ponds with nutrient rich sediments may respond without fertilization. Fertilization should take place when or shortly after the pond is filled. At the time that the fish are fed a prepared diet, the need for fertilization is eliminated.

Aquatic Insect Control

The larvae and adults of many insects prey on small fish, particularly their eggs and fry. Newly filled ponds attract insects as sites for egg laying. As a result, ponds that are newly prepared for stocking young fish also grow a crop of predaceous insects. These insects reduce the likelihood of fish surviving to grow to larger sizes. Insects and their larvae can be eliminated by applying chemicals to the water that are specifically approved for this use. State Departments of Agriculture list legal chemicals.

When selecting a chemical for use in fish ponds, it must be labeled for use in ponds, and labeled specifically for the type of fish that is being grown, such as foodfish, baitfish, or goldfish. Chemicals that are toxic to insects may also be toxic to some fish. Treatments are generally more effective at pH below 9 and at water temperatures between 65° and 85°F. The appropriate chemical is mixed with water and sprayed or otherwise well dispersed around the pond edge. The product label contains a great deal of information about the product and should be read thoroughly and carefully before each use.

A major drawback of using chemicals is that many are also toxic to desirable zooplankton, an important food for young fish. To benefit from insect control and zooplankton production, timing of application is critical. Ponds should be treated for insect control five to seven days before stocking fish fry. If eggs are stocked, application should occur five to seven days before eggs are expected to hatch. This treatment will control insects but allow zooplankton to develop because their recovery time is shorter. Following initial control, applications should not be repeated until approximately three weeks following the initial treatment or until fish have depleted the zooplankton supply and are feeding on prepared diets. In ponds used for producing small fish or fingerlings, regular control of insects at two- or three-week intervals will enhance survival if ponds receive supplementary feeding to compensate for depletion of zooplankton populations. Fish longer than 1 in. (2.5 cm) are much less susceptible to predation by insects.

In foodfish ponds where these chemicals cannot be used, a mixture of 4 gal of diesel fuel and 1 qt of motor oil per surface acre can be applied to kill air-breathing predaceous insects. This treatment clogs the breathing tubes of the insects when they rise to the surface for air. To minimize damage to zooplankton populations, the diesel fuel should be aerated for 24 hours prior to application. It is approved for use on foodfish.

Aquatic Vegetation Control

Proper preparation, filling, and fertilizing of ponds help prevent aquatic weed problems. A pre-filling application of herbicide can prevent weed and algae growth. It also reduces phytoplankton bloom development. A phytoplankton bloom established in a pond out-competes aquatic weeds. If weeds become established, they prevent development of a satisfactory bloom. For small weed infestations, manual removal is best. Once weeds are removed, development of the phytoplankton bloom will prevent their reestablishment. If extreme weed problems develop, aquatic herbicides may be required and the advice of a weed specialist should be considered as to the type and amount of herbicide to use.

In aquaculture, the grass carp is often used for biological weed control at densities dependent upon the extent of weed infestation. For rapid control, grass carp of about 8 to 12 in. long should be stocked at about 100 per acre. Larger carp can be stocked at lower densities. As the grass carp control the weeds, they should be removed because they compete with other fish for feed added to the pond. At harvest time, due to their large size, they can physically damage small fish if confined in the same net. A large-mesh seine to remove the grass carp solves this problem. Grass carp can be used at densities as low as 1 to 2 per acre for maintenance after weed control has been achieved. In states where grass carp are legal, triploid (sterile) grass carp may be required.

Evaporation and Rainfall

In some areas, evaporation from a pond surface exceeds the amount of rainfall that enters the pond directly. And as most aquaculture ponds are not constructed to receive water from an outside drainage area, additional water is required for maintenance of the desired depth of water in the aquaculture ponds.

Seepage

Ponds should have sufficiently high clay content to hold water well. Proper pond construction, in addition to a careful selection of the pond site, is imperative.

CALCULATING TREATMENTS

Most aquaculture systems, including ponds, tanks, and raceways, eventually require some type of chemical treatment to combat a disease or aquatic plant problem or improve water-quality conditions. Some treatments, like fertilizers or lime, are based on the surface area of water, but most treatments include the total volume of water in the production or holding unit.

Aquaculturalists should know how to calculate treatment rates, determine the amount of chemical or material needed, and apply the treatment. Before any treatment is applied, the water, fish, chemical, condition of the problem, and method of treatment should be known and understood. Producers can experience high economic losses when treatment rates are not properly calculated.

Before any calculation is made, the units of measurements should be determined. The unit of measurement selected should be familiar and convenient for the specific situation. For example, the large volume of water in ponds is usually expressed as acre-ft, whereas the volume of a small tank may be expressed in gallons or ft.^{3} Another decision is whether to use the English or metric system of measurement. A working knowledge of both systems is important because reports or publications may use either one. Metric units are easier to work with when small volumes or weights are involved.

To make calculations easy, refer to Tables A-2, A-5 and A-6 in the Appendix for conversion factors and parts per million equivalents.

Basic Formula

Most treatments can be calculated using this basic formula:

$$\text{Amount of chemical needed} = V \times C.F. \times \text{ppm desired} \times \frac{100}{\%AI}$$

Where: V = The volume of water in the unit to be treated.

C.F. = A conversion factor that represents the weight of chemical that must be used to equal one part per million (1 ppm) in one unit of the volume (V) of water to be treated. The unit of measurement for the results is the same as the unit used for the C.F. (pounds, grams, or other unit). Table A-2 in the Appendix contains a list of these conversion factors for various units.

ppm = The desired concentration of the chemical in the volume (V) of water to be treated expressed in parts per million.

$\frac{100}{\%AI}$ = 100 divided by the percent active ingredient (AI) contained in the chemical to be used.

Most chemicals are 100 percent AI unless otherwise specified, so this value is usually 1. The percent AI is found on the label of most fishery chemicals.

To understand how the basic formula is used to calculate treatments, review the following examples that represent a variety of practical situations.

EXAMPLE 1. How much copper sulfate is needed to treat a 10-acre pond with an average water depth of 4 ft with a 1.3 ppm treatment?

Select the unit of measurement and then determine the volume of water in the pond. Volume (V) of water in the pond can be expressed as acre-ft.

10 acre × 4 ft. = 40 acre-ft

The conversion factor (C.F.) for acre-ft is 2.7 lbs. This weight is required to give 1 ppm in 1 acre-ft of water.
The parts per million (ppm) or concentration of copper sulfate desired is 1.3.

Copper sulfate is 100 percent active (AI), so when 100 is divided by 100, the result is 1. The amount of copper sulfate needed is determined by using the correct numbers in the basic formula as follows:

$$\text{Weight of chemical needed} = V \times C.F. \times \text{ppm desired} \times \frac{100}{\%AI}$$

lbs. of copper sulfate = 40 acre-ft × 2.7 lbs. × 1.3 ppm × 1 = 140.4 lbs.

EXAMPLE 2. How much formalin is needed to treat a holding tank 20 ft long by 4 ft wide that has a water depth of 2.5 ft with 250 ppm?

Determine volume (V) of water in tank by multiplying length × width × depth in ft.
Volume in ft.³ = 20 ft. × 4 ft. × 2.5 ft. × 200 ft.³

The conversion factor (C.F.) for ft³ is 0.0283 gm, or the weight of chemical needed to give 1 ppm in 1 ft³ of water.

The parts per million (ppm) desired is 250.

Formalin is regarded as 100 percent active (AI) for treatment purposes, so 100 divided by 100 equals 1.

Formalin needed = 200 ft³ × 0.0283 gm × 250 ppm × 1 = 1,415 gm

Formalin is a liquid, so the weight in grams must be converted to a volume unit. This is done by dividing the weight, 1,415 gm, by 1.08, the specific gravity of formalin.

$$\frac{1,415 \text{ gm}}{1.08 \text{ gm/cm}^3} = 1,310 \text{ cm}^3$$

To convert cubic centimeters to another unit of volume for measuring in fluid ounces, refer to Appendix Table A-3. Multiply the conversion factor of 0.0338 × 1,310 cm³ to obtain volume in fluid ounces, 44.3 fl oz.

EXAMPLE 3. How much potassium permanganate is needed to treat a circular tank that is 6 ft in diameter and has a water depth of 4 ft with 10 ppm?

Volume of circular tank is: $V = 3.14 \times r^2 \times d$
Volume = 3.14 × (3 ft × 3 ft) × 4 ft. = 113 ft.³

Conversion factor (C.F.) for ft³ is 0.0283 gm.

The desired ppm is 10.

Potassium permanganate is considered to be 100% AI.

The values are substituted into the basic formula to obtain grams of potassium permanganate needed:
113 ft.³ × 0.0283 gm × 10 ppm × 1 = 32 gm

DISPOSING OF WATER

Each year, agricultural practices come under closer scrutiny for evidence that they may deteriorate the environment. Some claims against agriculture are valid, some are not. Agriculturalists and aquaculturalists respond to valid claims by changing practices. Clean, plentiful water is the life blood of successful aquaculture. Wastewater is the major pollutant from aquaculture.

Water used in aquaculture carries with it any uneaten feed, fish wastes or excrement, chemicals, dead fish, sediments, algae, and escaped fish. This wastewater is known as **effluent**, and it is produced by hatcheries, rearing tanks, processing plants, ponds, raceways, flushing, harvesting drawdown, and haul tanks. Aquaculturalists recognize that wastewater harms the environment and must be creatively dealt with. Improper disposition of wastewater can cause damage to the aquatic species being cultured. Finally, wastewater disposal is regulated by laws, regulations, and agencies.

Pollution

Wastewater pollution can cause a number of problems for the environment. Dead fish look unsightly and smell. Nutrients in the wastewater cause the aging or eutrophication of a body of water. The wastewater increases sedimentation and may contain weed seeds. Wastewater smells because of decaying materials and the gases released. Finally, wastewater may contain a species of fish that could genetically damage wild fish or become a trash fish in the wild.

Treating and Disposing

Along with plans for adequate clean water, disposal of wastewater receives considerable attention. Basically, six methods are used to treat wastewater:

1. Settling ponds or vats
2. Irrigation water
3. Percolation ponds
4. Filtering systems
5. Hydroponics
6. Chemical additives

Settling Ponds or Vats

Wastewater moves into the settling area, where the solids settle to the bottom and the top water is released into the environment or a stream. Solid material needs to be removed occasionally. (See Figure 12-16.)

Irrigation Water

As long as the wastewater contains no chemicals harmful to a field crop, it can be used for irrigation. Some of the substances in the wastewater serve as fertilizer for the crop. (See Figure 12-17.)

FIGURE 12-16 Settling ponds allow solid material to settle to the bottom. Eventually, they have to be cleaned.

FIGURE 12-17 Wastewater from aquaculture production facilities can be disposed of by irrigation.

Percolation Ponds

These ponds are similar to **settling ponds**, but the bottoms of the ponds are porous. Water absorbs into the ground. This method has the potential of polluting groundwater supplies.

Filtering Systems

Elaborate filtering systems prepare wastewater for release into the environment. These filtering systems are often expensive and require a high level of maintenance.

Hydroponics

Wastewater from aquaculture provides the nutrified water for raising plants hydroponically. **Hydroponics** is the growing of land (field) crops with their roots dangling in nutrified water. Often, the wastewater from aquaculture contains nutrients below the level required by the plants.

Chemical Additives

Chemicals or biochemicals added to wastewater can remove pollutants. This method requires care because the improper addition of chemicals to the water could also create pollution.

Aquaculturalists deal with wastewater by using a modification or a combination of the six basic methods. What they choose to do could be influenced by the regulations that apply to wastewater in their area.

Regulations and Agencies

Federal, state, and local governments all establish regulations for wastewater disposal. The federal Environmental Protection Agency (EPA) requires any aquaculture facility that produces more than 100,000 lbs or discharges wastewater for more than 30 days to obtain a National Pollution Discharge Elimination System (NPDES; http://cfpub.epa.gov/npdes/) permit. The EPA sets minimum standards to be met and delegates specific regulations to the states. Another federal agency, the U.S. Army Corps of Engineers, has jurisdiction in some areas over streams and rivers and the water in them.

The specific responsibilities of inspection and regulation often fall to state and local agencies. Aquaculturalists should contact these agencies for complete details. Agencies to contact include the Natural Resources Conservation Service (NRCS) and the National Institute of Food and Agriculture (NIFA [formerly Cooperative State Research Education and Extension Service or CSREES]). Ignorance of the regulations is no defense, and failing to follow regulations can result in fines.

SUMMARY

Successful aquaculture depends on a reliable source of high-quality water. Water provides oxygen and food, serves as an excretory site, regulates body temperature, and may harbor disease-causing organisms. Water quality must be monitored and maintained. Water contains cations and anions. Their amounts and types affect the water quality. One major change influenced by ions is the pH. Water too acidic or too alkaline stresses aquatic life.

Water contains dissolved gases. The most important of these is oxygen. Low oxygen levels also stress fish and can result in a rapid fish kill. Increased water temperature decreases the level of oxygen in water. When oxygen levels threaten fish, supplemental aeration is used. Carbon dioxide in water comes from the respiration of plants and animals. High levels of carbon dioxide interfere with the fish's ability to obtain oxygen and also decreases the pH of the water.

Nitrogen in water comes primarily from the metabolic wastes of the animals. Nitrogen compounds in the water include ammonia, nitrates, and nitrites. Levels of ammonia are especially detrimental to the fish culture even at very low levels in the un-ionized form. Un-ionized ammonia increases as the pH increases.

Water for aquaculture supports a very complex system. Successful aquaculture requires a thorough understanding of water chemistry, constant observation, and testing. Common tests to monitor water quality use colorimetric, titrimetric, and electronic methods. Water management for aquaculture includes consideration of soil types used in ponds, rainfall, evaporation, light, odor, and color. Water management also includes math skills to determine volumes and treatment levels.

After the water is used in aquaculture, the final problem is disposing of the wastewater. Aquaculturalists choose the best method of disposal for their needs.

STUDY/REVIEW

Success in any career requires knowledge. Test your knowledge of this chapter by answering these questions or solving these problems.

True or False

1. An increase in pH means that the water is more acidic.
2. A Secchi disk is used to measure dissolved oxygen content.
3. Phytoplankton increases the oxygen content of the water in a pond.
4. Cations are negatively charged ions.
5. Plants produce most of the nitrogen in a pond.
6. Electronic meters can determine the dissolved oxygen content of water.

Short Answer

1. Which of the following is not a cation?
    a. chloride
    b. calcium
    c. sodium
    d. potassium
2. Fifteen parts per million (ppm) is the same as:
    a. 15 grams per pound
    b. 10 ounces per quart
    c. 15 milligrams per liter
    d. 15 grams per liter
3. A pH of 7 would be:
    a. basic
    b. neutral
    c. acidic
    d. 7 ppm
4. Respiration by fish adds carbon dioxide to the water. This _____ the pH of the water.
    a. increases
    b. decreases
    c. does not effect
    d. neutralizes
5. Name three processes affected by water temperature.
6. Name three cations and three anions found in water.

7. What is the most common measurement used to describe the amount of gases, ions, pesticides, and elements in water?
8. How does pH affect un-ionized ammonia?
9. What effect does increased water temperature have on oxygen content?
10. List three ways to measure salinity.
11. What is the relationship between water hardness—carbonate and bicarbonate—and pH changes?
12. TDS closely reflects what other measure?
13. What simple instrument is used to measure turbidity?
14. What is the purpose of liming and what is used to lime a pond?
15. Name four potentially harmful trace elements that may be found in water.
16. List five gases that could be found in a water sample.
17. What is the optimal level of dissolved oxygen for fish?
18. Name four events that reduce the dissolved oxygen in water.
19. What processes in water produce carbon dioxide and what process uses it?
20. What is the optimal pH range for aquatic animals?
21. How can hydrogen sulfide be removed from water?
22. Name four forms in which nitrogen may exist in water.
23. What chemical element is always a part of organic compounds?
24. An earthen pond is 10 acres with an average depth of 5 ft. How many gallons and how many acre-ft of water will be required to fill this pond? Also, if the well used to fill this pond delivers 1,500 gal per minute, how many days and hours will it take to fill the pond?
25. A concrete raceway 5 ft wide and 45 ft long with an average depth of 3 ft holds how many gallons?
26. List four heavy metals that might be found in the bottom soil of ponds.
27. A tank needs a potassium permanganate treatment of 10 ppm. How much potassium permanganate is needed to treat a rectangular tank 10 ft long and 2 ft wide with an average water depth of 1.5 ft? Assume the potassium permanganate to be 100 percent AI.
28. List three reasons why wastewater from aquaculture facilities is considered a source of pollution.
29. Name four methods used to treat wastewater from aquaculture facilities.

Essay

1. Why does photosynthesis in a catfish pond cause the pH to increase?
2. Does increased carbon dioxide in water raise or lower the pH? Why?
3. List three management practices that prevent or help overcome low dissolved oxygen levels in water.
4. What form of nitrogen is the most harmful to fish? Why?
5. What poor management practices cause high levels of ammonia in water?

6. Explain the difference between chlorine and chloride.
7. Why is soil type an important consideration in pond water management?
8. How would pesticides get into water used for aquaculture?
9. What is the best instrument for detecting odor? Name six odor descriptions commonly used.
10. Other than routine testing, give three other reasons that may cause an aquaculturalist to test water quality.
11. Explain the difference between a titrimetric test and a colorimetric test.
12. Why are ponds fertilized?

KNOWLEDGE APPLIED

1. Take a field trip to a local municipal wastewater facility to observe its procedures. Find out what water-quality measurements/tests they use and compare these to aquaculture.
2. Plans for simple hydroponic projects can be easily obtained. Construct a simple hydroponic demonstration project. Determine what nutrients the plants require in the water. Compare this to the nutrients expected in aquaculture wastewater.
3. Construct a Secchi disk. Use a 12-in by 12-in piece of sheet metal. Mark an 8-in (20 cm) circle on the sheet metal. Make a small hole in the center. Paint the top of the disk with flat white paint. Divide the circle into equal quadrants. Paint two opposite quadrants flat black. Insert an eye bolt through the hole in the center and place a lead weight on the bolt before attaching the nut. Attach a calibrated line to the eye bolt. Use the Secchi disk to take readings on a pond. Lower the disk into the water until it just disappears. Record the measurement on the calibrated line. Lower the disk until it disappears. Then, raise it until it reappears. Record this measurement and use the average of the two.
4. From one of the suppliers listed in Learning/Teaching Aids, obtain test kits for dissolved oxygen, ammonia, and pH. Practice using these test kits on various water sources. If possible, obtain an electronic pH meter and a dissolved oxygen meter and compare their readings to those obtained from the kits.
5. Take measurements of various shapes of containers that hold water. Use the measurements to determine the volume of water they will hold. Convert English volume measurements to metric volumes. For raceways or ponds, practice stepping off the measurements and using this to estimate the volume.
6. Obtain soil samples of clay, loam, and sand. Examine the properties of each in dry, damp, and wet states. On a field trip, practice identifying soil types.
7. In a fairly large body of water such as a pond, determine the temperature at various depths. For example, record the temperature at 1-ft intervals. Explain any differences.
8. Because ponds and other water-holding facilities are found in many shapes, estimating the surface area and volume provides an excellent chance for applied geometry. For example, ponds can be a combination of a rectangle and a triangle. Some tanks are round.

9. Track and record the weather for several months. Record the wind speed, hours of daylight, amount of cloud cover, amount of precipitation, and the maximum and minimum temperatures.
10. Visit the Web sites for "Equipment and Supplies." Develop a report on two of the methods used to monitor water quality.

LEARNING/TEACHING AIDS

Books

American Public Health Association, American Water Works Association, and Water Pollution Control Federation. (2009). *Standard methods for the examination of water and wastewater*. Washington, DC: APHA.

Barnabe, G., and Barnabe-Quet, R. (2002). *Ecology and management of coastal waters: The aquatic environment*. New York, NY: Springer Publishing.

Creswell, R.L. (2000). *Aquaculture Desk Reference*, Dade City, FL: Florida Aqua Farms Inc.

Johnson, R.L., Holmquist, D. and Redding, K. (2007). *Beaverton, OR: Vernier Software & Technology*.

Hach Company. (2010). *Hach water analysis handbook procedures*. Loveland, CO. http://www.hach.com/download-resources McLarney, W. (1998). *Freshwater aquaculture: A handbook for small scale fish culture in North America*. Point Roberts, WA: Hartley & Marks Publishers, Inc.

U.S. Department of Agriculture. *Water: The yearbook of agriculture 1955 (House Document No. 32)*. 84th Congress, 1st Session.

Equipment and Supplies

Carolina Biological Supply Company: http://www.carolina.com/
Hach Company: http://www.hach.com/ LaMotte Company: http://www.lamotte.com/aquarium_fish_farming.html
NASCO: http://www.enasco.com/ (search "aquaculture") Vernier: http://www.vernier.com/ (search "water")
YSI, Inc.: http://www.ysi.com/

Internet

Internet sites represent a vast resource of information. The URLs (uniform resource locator) for the World Wide Web sites can change. Using a search engine such as Google, find more information by searching for these words or phrases: water quality management, dissolved oxygen, oxygen management in fish, pH (hydrogen ions in water), heat capacity, total dissolved solids, sodium absorption ratio, water hardness, secchi disk, oxygen holding capacity, nitrogen cycle, chemical oxygen demand, biochemical oxygen demand, titrimetric, colorimetric, effluent water, settling, or ponds.

For some specific Internet sites refer to Appendix Tables A-11, A-12 and A-14.

Ponds and raceways are the most common structures for raising fish. With water supplies reaching maximum and new species being seriously cultured, many other structures for culture are being used. These include tanks, silos, cages, and recirculating or closed systems.

CHAPTER 13

Aquatic Structures and Equipment

Aquaculture has developed some unique equipment and some unique uses for equipment. This chapter cannot cover all aquaculture structures and equipment, but some additional equipment is discussed in related chapters. For example, Chapter 10, Water Requirements for Aquaculture, discusses testing equipment, and Chapters 4, 5, and 6 discuss equipment used for harvesting.

OBJECTIVES

After completing this chapter, the student should be able to:

- Distinguish between four types of ponds
- Identify factors in pond site selection
- Explain important pond construction requirements
- Define tank and raceway culture
- List advantages and disadvantages of tank and raceway culture
- Define cage culture
- List advantages and disadvantages of cage culture
- Describe cage design requirements

- Identify site-specific factors that determine costs
- List factors to consider when planning pond size
- Identify layout and design considerations
- List advantages of small versus large ponds
- Identify four types of aerators
- Describe the seines used on ponds
- Describe a live-car or sock
- Identify the need for boats, tractors, trucks, and pumps
- Identify species for pond, cage, raceway, tank, or silo culture
- List steps in determining a site's water quality
- Determine whether soil is suitable for pond construction
- Compare some of the biological concerns with cages and closed systems

Understanding of this chapter will be enhanced if the following terms are known. Many are defined in the text, and others are defined in the glossary.

KEY TERMS

Borings	Percolation
Cage	
Feeding ring	Relift pump
Flow index	Seine
Freeboard	Silos
Impoundment	Slopes
Levee	Sock
Live-cars	Stratification
Mud line	Topography
Net pens	Watershed
Paddlewheel	

PONDS

With the large production of catfish in the United States, ponds are the most common type of structure for raising fish. Four types of ponds are natural, **impoundment**, excavated (Figure 13-1), and **levee** (Figure 13-2). Each type can be used for fish production. Natural ponds are located in low areas that fill with ground water or act as catch basins for runoff and springs. An impoundment pond is built by constructing a dam across a ravine or streambed, retaining the water behind the dam. The simplest and most common type of constructed pond is the excavated pond, which is made by digging a pit. Levee-type ponds are constructed by pushing up dikes on flat land and filling them with water from wells.

Pond harvesting techniques depend on the size range of the animals, the intensity of production, and the species cultured. Technology exists for harvesting most cultured aquatic animals. Some harvest systems, such as those used in intensive catfish pond culture, are well developed and modern. For other species, harvesting equipment and techniques need to be improved. Animals reared for recreational fishing have special handling and transporting requirements that are not readily adaptable to mechanized harvesting methods and remain labor-intensive and expensive. Table 13-1, page 393 lists characteristics of each pond type.

Design Standards and Construction

Well-designed ponds save money and time in construction and operation. The requirements of the fish and the methods of harvest are factors that will determine the size and depth of the pond. Another consideration in design is whether the pond will be used for spawning, fish-out, and single or multiyear production.

When more than one pond is planned, the ponds should be located so that each pond can be drained and filled separately from the others. This will avoid the spread of disease and allow more timely harvests of various ponds independently.

Size

Fish can be grown in ponds of any size. Fish ponds range in size from 0.1 acre to 20 acres. Site conditions, limitations of construction equipment, and harvest

FIGURE 13-1 Small earthen excavated ponds for trout.

FIGURE 13-2 Levee-type ponds constructed side-by-side are popular all over the Mississippi Delta. Tractor is driving on the levee pulling a blower feeder.

methods will determine pond size. Ponds of less than 1 acre are easier to manage and harvest. The pond should not be so large as to require stocking more fish than the aquaculturist is prepared to purchase. Fingerlings are expensive, and they can make up a large part of the investment in fish farming.

TABLE 13-1 FISH POND TYPES

Pond Type	Recommended Soils	Recommended Fish Species	Advantages	Limitations
Natural	—	Sunfish Crappie Bullhead Bass	No construction cost; presence of natural foods.	Not drainable; no control of pond design or location; difficult to harvest; difficult to manage or renovate; possible wild fish or weed problems.
Impoundment	Clay	All species	Drainable, allowing control of water level; easy to manage, renovate, and harvest; pond bottom can be dried for removal of weeds, wastes, disease, and wild fish.	High construction cost; often relies on runoff; possible siltation or pollution problems.
Excavated	Clay Sand Peat Gravel	Sunfish Crappie Bullhead Bass	Moderate construction cost; applicable to many types of soils and topography.	Usually not drainable; difficult to harvest; size limitation; renovation difficult; bottom uneven making seining ineffective; maintenance difficult.
Levee	Clay	All species	Bottom even; drainable; harvesting easy with correct equipment.	Expensive construction; flat land required; high level of management required.

Pond site and design may be the most important factors controlling profitability. Ponds that leak, have irregular bottoms, or routinely suffer from a shortage of water will not produce a consistent crop of fish.

Ideally, levee ponds built on flat land and filled with groundwater or surface water are more suitable for commercial catfish production. Some terrain is rolling and not conducive to this kind of construction. In hilly terrain, pond builders must take advantage of the natural formations by constructing dams across valleys between hillsides so that runoff from rainfall on the **watershed** will be stored behind the dam.

Water Supplies

Water to fill and maintain watershed ponds usually comes entirely from runoff, though groundwater (wells) and surface water (springs, streams, and reservoirs) can also be used as supplemental sources. The ratio of watershed to water surface acreage should be large enough so that ponds fill and sometimes overflow during rainy months but drop no more than 2 ft. during drier months. The best ratio of watershed to water surface varies according to the type of land on which the pond is built. For a watershed of heavy clay soil on open land, the best ratio is 5 acres of land for each surface acre of pond. For a sandy watershed in a wooded area, the best ratio is 30 acres or more of land for each surface acre of pond.

When a watershed is too small and unable to supply enough water to the pond, or when an outside source of water is needed for filling during dry periods, water from wells, streams, or rivers can be pumped into the pond. Water containing wild fish should be filtered to avoid introducing these fish into the pond.

When ponds are built in series in a valley, less watershed is needed to maintain an acre of water. Before harvest, water can be pumped or drained from one pond to another for storage. This procedure not only allows a producer to refill using the stored water immediately after harvest, but it also eliminates the possibility of draining nutrients into nearby natural waters.

Soil Characteristics

Good-quality soil that is at least 20 percent clay is necessary for building the cores of dams. This includes clay, silty clay, and sandy clay soils. Soil should be sampled by frequent **borings** along a proposed dam site to determine if the clay foundation is large enough to build the dam.

Borings for soil samples should also be taken from the proposed dam site and the shoreline to be sure that there is enough clay to build the dam. Usually, a good source of clay can be found in the hillside near the dam site. If such a source is available, using it to build the dam can add to the pond's size. If removing the clay will uncover rock formations, sand, or gravel in the pond bottom, it is best to leave the clay in place.

Pond construction in limestone areas can be especially risky because of the possibility of underlying cracks and sinks that may cause the pond to leak. In areas where the soil of the proposed pond bottom could leak, soils should be bored to check for quality. Approximately four borings per acre are sufficient, unless there are variations in soil type in the pond bottom. Figure 13-3 shows the parts of a dam, including the core and drainage system.

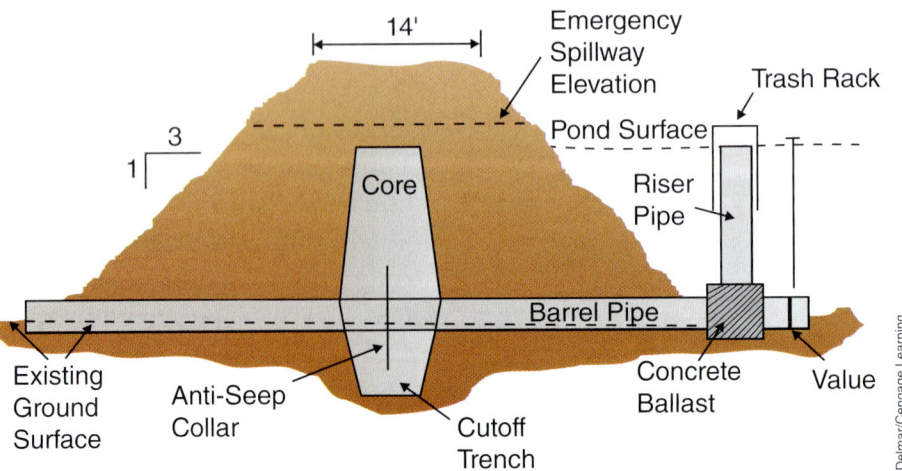

FIGURE 13-3 Cross-section of a typical dam at the drainpipe.

Topography

Topography affects the size and shape of a watershed pond. Generally, steep **slopes** in V-shaped valleys require dams of larger volume per water surface acre than sites with gently sloping hills and wide, flat valleys. So, ponds built in steep terrain usually cost more per pond acre than those built in gently rolling terrain.

Ideally, watershed ponds should be less than 10 ft deep at the drain. This depth allows the producer to harvest the pond without draining it. Deep ponds must be drained of much of their water before they can be **seined** for a complete harvest.

Some sites with gentle slopes and large flood plains allow for the construction of two-sided and three-sided watershed ponds (Figure 13-4). These ponds are usually constructed parallel to hills bordering a creek. Runoff is used as a water source, but the dam does not cross a hollow or draw. The great advantage of this kind of pond is that it does not have to be drained for harvest.

Other Considerations

A watershed pond site should be selected so that pipes and valves can be installed to drain the pond completely. The proposed shoreline should be excavated to provide a depth of at least 3 ft around the edge of the pond.

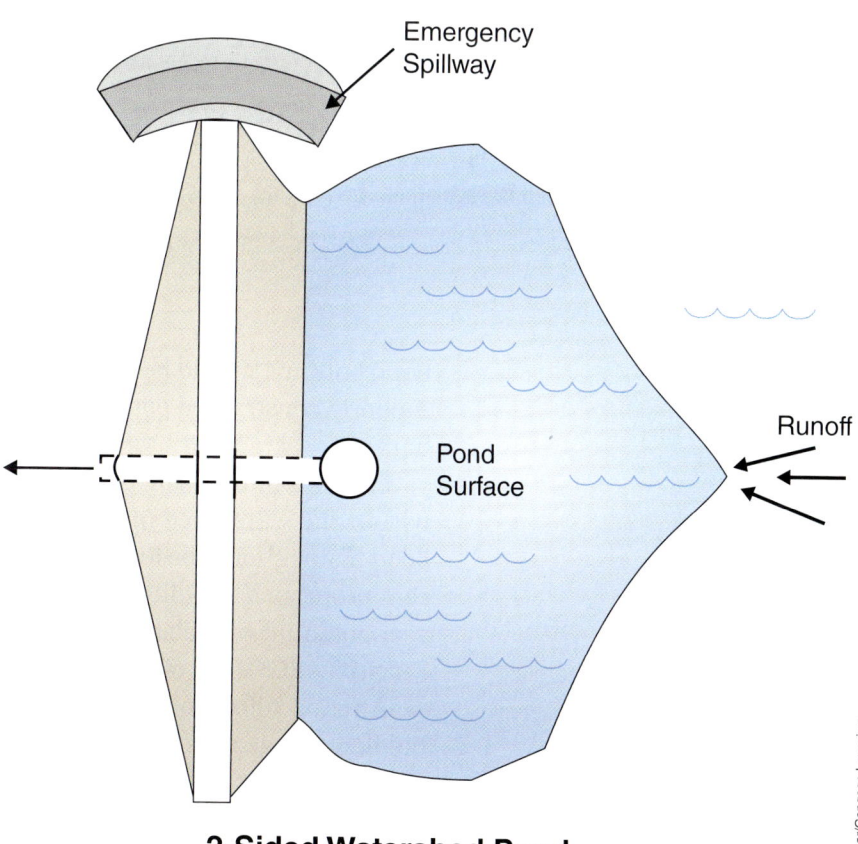

FIGURE 13-4 Diagram of conventional hill ponds.

CLAY

Gravel, sand—very coarse to fine—silt, and clay make up the soil. Clay is the smallest of these soil particles. An international system of soil classification compares clay with the other soil particles.

Name of Particle	Diameter limits (mm)
Gravel	above 2.00
Coarse sand	2.00 to 0.20
Fine sand	0.20 to 0.02
Silt	0.02 to 0.002
Clay	below 0.002

Clay particles are plate-shaped, and they are often composed of complex compounds such as kaolinite, illite, and montmorillonite. Clay is sticky and capable of being molded—highly plastic—when wet and very hard and cloddy when dry. Because of clay's small particle size and because all particles have the same negative charge, clay possesses some unique characteristics.

Clay in water forms a colloid. Because all clay particles have the same kind of electrical charge, they repel one another. Because the clay particles are so closely packed, they cannot move freely but maintain the same relative position. If a force is exerted upon the clay, then the particles slip by each other. This permits clay to take on a shape as in pottery or brick making. When the force is removed, the particles retain their new position, because the same electrical forces as before act upon them. Clays become permanently hard when baked or fired.

In ponds, clay can cause the water to remain muddy. The negatively charged clay particles repel each other and will not settle out. The addition of some positively charged particles is necessary to cause coagulation and precipitation of the clay particles. Compounds used to do this include limestone, alum, and gypsum.

Clay is important in pond construction. It keeps water in ponds. Without clay in the construction of a pond, water losses by seepage are excessive and expensive. A pond made of fine sandy loam or sandy clay loam can lose over 14 in of water a day. A pond made of clay or clay loam but not properly packed can lose over one in of water a day. Ponds constructed of properly compacted heavy clay lose only about 0.002 in of water in one day.

When clay is not available for pond construction, bentonite is incorporated into the pond. Bentonite is a clay product that swells when wet and fills minute holes in the soil. Also, organic matter, such as manure, or a plastic liner can be used to help prevent seepage.

Pond bottoms should be smooth and should slope gently to the drainpipe. A poorly constructed pond with an uneven bottom will cause incomplete harvests.

Floods from nearby rivers must not flow over the dam, and floods within the watershed must not weaken the dam. Ponds constructed in flood plains should be located so that they will not cause damage to adjacent property if flooding does occur. Information on floods and their 100-year potential is available from the USDA Natural Resources Conservation Service (NRCS) field office in each county.

After deciding on a dam site, the permanent waterline and the potential flood-stage waterline of the proposed pond should be marked off to make sure that water will not encroach on other property. Also, if the pond site contains 1 acre or more of wetland, the U.S. Army Corps of Engineers (http://www.usace.army.mil/) must grant a permit before the pond can be constructed.

Cost of Construction

Cost estimates include clearing, earthfill, excavation, pipe and drain, concrete, seeding, and road gravel. Each pond site is unique. In general, a large, shallow, one-sided watershed pond is relatively inexpensive to construct. A three-sided pond may cost about twice as much as a one-sided pond.

Site Selection of Levee-Type Production Ponds

Considerable thought and planning should go into selecting sites for commercial fish production ponds. Construction costs, ease and cost of operation, and productivity can be greatly affected by the site selected.

Water Availability

Water for filling levee-type ponds must come from a well, spring, reservoir, or stream because there is no watershed for runoff water to enter the pond. One of the first considerations in selecting a site for commercial fish ponds is to make sure that an adequate supply of suitable quality water is available for the size of the farm planned.

Usually, one well with a capacity of 2,000 to 3,000 gallons per minute (gpm) is adequate for four 20-acre ponds, or a minimum of 25 gpm per acre of pond surface. (See Figure 13-5.) Enough water must be available to fill the pond completely within 10 days, otherwise problems with vegetation and water-quality management will occur.

A local well-drilling company, a groundwater geologist, or the closest office of the U.S. Geological Survey should be able to tell if a well of the desired capacity can be developed at the site and provide information on the quality of groundwater.

Water Sources

Water for levee ponds for commercial fish production can come from a well, spring, stream, or reservoir. Of the four choices, a well is usually the best for several reasons. A spring is almost as good. Streams and reservoirs should be used only as last resorts because they usually contain various species of wild fish that can get into the pond and cause severe management problems. Streams and reservoirs can also become contaminated with pesticides or industrial chemicals that could kill fish. During droughts, water levels become so low in streams and reservoirs that they can no longer be used as sources of needed pond water. Wild fish in streams and reservoirs serve as constant sources of reinfection with various infectious diseases. The quality of water in streams and reservoirs also can change enough during droughts or floods as to be unusable.

Soil Characteristics

The soil must hold water, so clay-type soils are desirable. Soil cores taken at various places around the site ensure that adequate clay is present to prevent excess seepage. The local Soil Conservation Service Office can provide assistance. Although more costly, ponds can be built in soils that have high **percolation** rates if they are lined with a layer of packed clay or plastic.

FIGURE 13-5 Ponds require wells that can supply about 25 gal per minute per acre of pond surface need.

Topography

The topography or lay-of-the-land determines the amount of dirt that has to be moved during pond construction. Pond construction on flat land requires less dirt moving than building on rolling or hilly land. Levee-type ponds built on flat land usually require about 1,100 to 1,200 yd^3 of soil to be moved per acre.

Wetlands

If the site is classified as a wetland, a permit is required from the U.S. Army Corps of Engineers before clearing or building on it. In some states, permits are needed from one or more state agencies before any clearing or building can take place on wetlands.

Draining

For management purposes and for harvesting, ponds must be drained (Figure 13-6). The site should be selected with this in mind. Further, ponds need to drain by gravity flow. Pond draining should not cause flooding on a neighbor's land or block its drainage.

Flooding

The site under consideration should not be subject to periodic flooding. This can be checked at the local U.S. Geological Survey office or NRCS office.

Utility Right-of-Ways

Before building ponds over pipelines or under power lines, producers first should check locations of right-of-ways with the utility company to avoid possible legal problems later.

FIGURE 13-6 For management purposes ponds must be drained and dried. Here a small pond is being drained.

Pesticides

If row crops were ever grown on or adjacent to the site being considered, the site should be checked for pesticide residues. Three areas within a site that must be checked for pesticide-residue levels are:

- Low areas where runoff collects
- Any area where spray equipment, either aerial or ground, was filled with pesticides
- Any area in the site where pesticides were disposed of or were stored

Analysis of soils for pesticide residues can be done by commercial analytical laboratories or, in some states, by the state chemical laboratory.

CONSTRUCTION OF LEVEE-TYPE PONDS

Pond construction for the commercial production of aquatic animals is one of the most expensive and important aspects of developing a fish farm. Unless careful consideration is given to the design and the cost of pond construction, the producer may find that the layout is not suitable for the species of fish desired and the cost of building makes it impossible to receive a profit. As discussed, site selection is the first step in construction.

Size

Most commercial catfish ponds in the Mississippi Delta are built on a 20-acre land unit, which gives a surface area of about 17.5 acres of water, depending on the slope of the inside (wet) levee, the top width of the levee, and the height of the levee above the normal water level, often referred to as **freeboard**. The 20-acre size is a compromise between ease of management and cost of construction. A larger pond is much more difficult to manage, and ponds smaller than this are more expensive to construct because more valves and inflow and drainpipes must be used.

In addition to costing more to construct, smaller ponds decrease the amount of water area available for fish production. If a 20-acre land unit is made into two 10-acre pond units with 16-ft tops, 3:1 slopes and 1.5 ft of freeboard, about 0.34 surface acre of water will be lost due to the increased amount of levee needed. Decisions on actual pond size depend on what the producer wants to raise, the topography of the land, and the amount of land available for pond construction. Before deciding on pond size, the aquaculturalist should look at the cost of building several different size ponds and go with the size that is most economical and consistent with production goals and degree of management planned. A typical layout of levee-type catfish ponds is shown in Figure 13-7.

Shape

Pond shape is largely determined by the topography and by property lines. Most commercial levee-type fish ponds are rectangular because of the greater ease and economics in harvesting and feeding, although

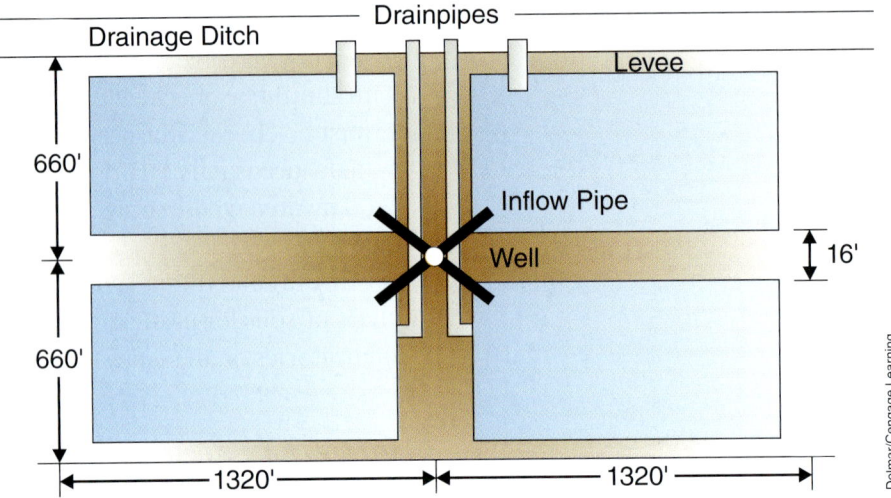

FIGURE 13-7 Layout of a typical levee-type catfish pond. Courtesy of Thomas L. Wellborn, "Construction of Levee-type Ponds for Fish Production," Southern Regional Aquaculture Center, Publication No. 101.

square ponds are cheaper to build. A square 20-acre pond requires 3,596 ft of levee, but a rectangular 20-acre pond that is 660 ft by 1,320 ft requires 3,822 linear ft of levee, a difference of 226 ft. A rectangular 10-acre pond (467 ft × 933 ft) requires 2,729 linear ft of levee, whereas a square 10-acre pond (660 ft × 660 ft) requires 2,569 ft of levee, a difference of 160 ft.

Ponds with a curving, irregular shape are difficult to harvest unless drained. They are also extremely difficult to manage with respect to water quality. Irregularly shaped ponds should not be constructed before carefully weighing the advantages and disadvantages of cost and management.

Levee Width

Levees should be at least 16 ft wide at the top. The main levees where the wells are located should have a 20-ft top width to allow an easier flow of traffic for feeding, harvesting, and water-quality management. Levees with top widths of less than 16 ft require more maintenance than do those with top widths of 16 ft or greater. Narrow levees also are a greater hazard to employees and equipment in wet weather.

If the soil type is such that the levees become impassable in wet weather except to four-wheel-drive vehicles, at least two sides of each pond should be graveled to permit all-weather access for feeding, harvesting, disease treatment, and water-quality management.

Freeboard and Depth

Freeboard is the height of the levee above the normal water level. The amount of freeboard should not exceed 2 ft, nor be less than 1 ft. Levees with a freeboard in excess of 2 ft are expensive to build. Excess freeboard makes it difficult to get equipment into and out of the pond. Levees with a freeboard of less than 1 ft are subject to erosion, thus increasing maintenance costs.

Pond depth is important because of management implications. The pond depth at the toe of the slope at the shallow end should never be less than 2.5 ft, nor greater than 3.5 ft. At the deep end of the pond, the maximum depth at the toe of the slope should not exceed 6 ft, with a 5-ft depth preferred. The pond bottom must be flat, free of all roots, stumps, and debris that might interfere with seining, and it should slope from the shallow end to the deep end at a rate of 0.1 to 0.2 percent along the long axis of the pond.

Slope of Levees

Slope is expressed as a ratio of the horizontal distance in feet for each 1 ft of height. For example, a 3:1 slope extends out 3 ft horizontally for each ft of height and a 6:1 slope extends 6 ft horizontally for each ft of height.

The actual slope of the levees will depend on the type of soil at the pond site, but for most soil types a 3:1 slope is satisfactory if properly compacted. Increasing the slope to 4:1 or 5:1 substantially increases the amount of soil required for the levees. This increases the construction. For example, a levee 6 ft high, with a 16-ft top width and 3:1 slope, contains 7.6 yds^3 of soil per linear ft. A levee with the same dimensions except for a 4:1 slope requires 8.9 yds^3 of soil, and a levee with 5:1 slope needs 10 yds^3 of soil per linear ft. The cross-section of a typical levee for a catfish pond is shown in Figure 13-8.

Orientation

The direction in which the pond's long axis is oriented depends mostly on the topography of the site and the property lines. Some say ponds should be oriented with the long axis parallel or at right angles to the prevailing winds. Levees of ponds with the long axis parallel to the prevailing winds are subject to erosion because of increased wave action, but the ponds are better aerated because of this increased wave action. Ponds oriented at right angles are subject to less levee erosion and are not as well aerated.

Site Preparation and Construction

All existing vegetation, roots, stumps, and topsoil must be removed from the site before starting levee construction in order to allow a good bond between the foundation soil and the fill material. After completion of the levee, several inches of topsoil are put back on the top and outside slopes so that a vegetative cover can be established quickly.

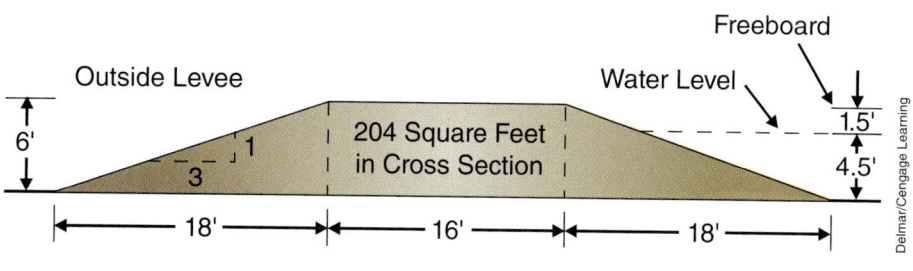

FIGURE 13-8 Cross-section of a typical levee for a commercial catfish pond. A levee with these dimensions contains 7.6 ft.^{3} fill material in each linear ft. Courtesy of Thomas L. Wellborn, "Construction of Levee-type Ponds for Fish Production," Southern Regional Aquaculture Center, Publication No. 101.

FIGURE 13-9 Dirt pan used to build catfish ponds.

Because of their speed, self-loading pans are the most efficient equipment to build pond levees (Figure 13-9). They also give the best compaction of fill material when complete wheel-track coverage is made over each layer of fill placed in the levee. For proper compaction, the soil must have at least 12 to 15 percent moisture. If the soil is dry during construction, each layer of fill dirt placed on the levee must be wetted before compaction. Laser-controlled pans increase the accuracy and speed of cuts and fills.

Bulldozers can be used to build pond levees, but they do not give good compaction. If bulldozers are used, using a sheepsfoot roller gives complete coverage over each fill layer. The area in the levee where the drain is located is left open during construction for drainage of any storm water during construction. As soon as the pond bottom and drain site are excavated to grade, a drainpipe of appropriate diameter is installed. A small, sloped ditch, about one-third of the diameter of the drainpipe, is dug to give uniform support for the pipe. Fill material is placed around the side and over the top of the drain and is hand compacted. This fill material must be moist during hand compaction to ensure a watertight seal around the drainpipe. The hand-compacted fill over the drain gives protection from heavy equipment during the completion of levee construction.

Drains

Ponds must be sited and constructed so that they can be drained by gravity flow. As you might expect, the lowest part of the pond must be higher than the ditch into which it is to be drained. The pond bottom must be flat and slope from the shallow to the deep end with a slope of 0.1 to 0.2 percent. A flat, sloping bottom is necessary for harvesting and draining.

A harvesting basin inside or outside the pond is not a part of levee ponds. Harvest basins do have a place in the production of certain types of fish, although they are expensive to construct. In the early years of commercial catfish production, harvest basins were routinely built, but they were quickly discarded because of problems and expense associated with them.

Levee-type ponds use several drain structures. The best type of drain structure to use is the one that most closely meets the producer's needs based on the pond design selected and estimated cost. The so-called "inside swivel drain," or modified Canfield outlet is the most common drain used in levee-type ponds (Figures 13-10 and 13-11). It is located at the lowest part of the pond. The pond's water depth is determined by the

FIGURE 13-10 The modified Canfield outlet or inside swivel drain controls the water level by pivoting the drainpipe up or down on its swivel joint.

FIGURE 13-11 Swivel drain in a newly constructed pond. Pipe is lowered to drain the pond.

length of the upright drainpipe, and the water level is adjusted by pivoting the upright drain up or down on its swivel joint. It must be held securely in position to prevent unplanned drainage. This can be done with a chain from the end of the upright drain to a post on the bank.

Swivel joints require grease. Maintenance of the swivel joints can be a problem because work has to be done under water or when the pond is drained.

Many farmers who raise catfish in levee-type ponds now use outside drains. The drainpipe is laid through the levee at the lowest point in the pond with the inside end extending at least 5 to 10 ft out from the toe of the slope to prevent clogging by dirt sloughing from the levee. This inside end is screened to prevent the loss of fish.

The outside end of the drainpipe should extend 5 ft past the toe of the slope to prevent excess erosion of the levee during draining. The outside end of the pipe is fitted with a T and a standpipe of the height necessary to maintain the desired normal water level in the pond. The end of the T is fitted with a valve for water level manipulation and complete draining if needed.

Another method is to have the outside standpipe 24 in. high, rather than the height of the normal water level in the pond, and fitted with a valve. The end of the T is capped. During rain, the valve is partially opened to remove excess water. This system permits rapid draining of 3 to 4 ft of water with only slight danger of wild fish entering the pond through the drainpipe. The pond can be completely drained by removing the cap at the end of the T.

The discharge end of the drainpipe needs to be at least 24 in above the surface of the water in the drainage ditch. This helps ensure that wild fish can be prevented from entering the pond through the drain.

Size of Drain

The drainpipe size needed is dictated by the pond size and how rapidly the farmer wants to drain it. Regardless of the pond size, a drain should allow the pond to be completely drained in no more than seven days, and preferably in five days.

Design Assistance

The local NRCS provides assistance in planning the design of fish ponds. In some areas of the United States, this agency is experienced in pond design and can help prevent severe design flaws. Using a computer program, the NRCS can also help calculate the cubic yards of dirt that must be moved during construction.

RACEWAYS AND TANKS

Most U.S. raceways are made of concrete, but a few are constructed of other materials, such as earth, stone, metal, concrete, or plastic, including fiberglass. Sites suitable for production in raceways are limited primarily by the water source. Most hatcheries in which raceways are used are

FIGURE 13-12 Raceways at the world's largest commercial trout production facility, Clear Springs in Buhl, Idaho.

located in the Snake River Valley of Idaho, where high-quality flowing water is available (Figure 13-12). The number of raceways in the effluents of electric power plants could increase as aquaculturists seek new sources of high-quality, heated water and the economics of production justify the cost of development and operation.

Raceways receiving water from power plant effluents have been used to culture marine and estuarine organisms, channel catfish, striped bass, rainbow trout, American eel, and other species.

The density or weight of fish that can be stocked per raceway depends on fish size, raceway size, and the available flow of quality water. These relations can be expressed as follows:

$$F = \frac{W}{L \times I}$$

where F = the **flow index**, W = weight of fish in pounds, L = length of fish in inches, and I = water inflow in gallons per minute.

The flow index is derived practically by the maximum weight of fish that can be raised in a raceway. Once the flow index has been established, the formula can be rewritten and used as follows:

$$W = F \times L \times I$$

If all measures are metric, then flow index and maximum weight can be figured in metric. This form of the equation is used to calculate the weight of fish that can be safely maintained as fish length increases or water flow changes. Most hatcheries of the U.S. Fish and Wildlife Service (http://www.fws.gov/) are operated with a flow index equivalent to 10 to 25, when F is calculated in metric units (1.0 to 2.5 when F is calculated in English units). Hatchery efficiency declines at flow indexes below 10, and water quality deteriorates at indexes above 25. Most raceways for salmonids have been designed for fish loads of 2 to 3 lb./ft.^{3} (32 to 48 kg/m^3).

Many who are interested in commercial fish farming strike upon the idea of raising catfish or other warmwater species in raceways. This idea seems appealing because yields per unit of growing area may be theoretically increased and the need for large ponds eliminated or reduced. Nevertheless, some factors make the raceway culture of warmwater fish rather risky. Some unique situations exist where raceway systems may be practical and profitable.

Raceways are generally constructed in a ratio of 5:1 (or greater) length to width, and with a depth of 3 to 5 ft. (See Figure 13-13.) Water should flow evenly through the system to eliminate areas of poor water circulation where waste materials or sediment may accumulate. Raceways may be constructed above ground or in the ground from cement or fiberglass, and even wood has been used. Fish cultured in raceways require a large quantity of good-quality water, preferably supplied by gravity flow from artesian wells or higher elevations. If pumping is required, operating costs may be high and risks increased due to possible failure of pumps or power supply.

Raceways should be considered only if an abundance of good-quality water is available. On the average, 1 to 3 gal/min of flow should be available for each ft^3 of raceway volume at densities of 3 lbs of fish per ft.^{3} If

FIGURE 13-13 Small rectangular tank used for fry and fingerlings.

supplemental aeration is used, the water requirement may be somewhat reduced. Water flow should be sufficient to keep solid waste material from accumulating in the raceway and to dilute liquid waste, primarily ammonia, excreted by fish.

For many intensively cultured species, water quality, not physical space, limits the number of animals that can be reared in a raceway or similar system. Water must be continuously added to flush away wastes and to maintain a quality environment.

In some raceways, effluents from downstream sections are pumped back upstream and reused. The effluent may first be improved by aeration, filtration, sedimentation, ozonation, or a combination of these and other processes before recycling. In single-pass raceways, where water is used only once before being discharged, incoming water quality depends on the source.

Effluent quality from intensive systems is a function of the number and size of fish or other organisms, the feeding rate, and the water-flow rate.

Silos

Silos are deep, cylindrical tanks similar in operation to horizontal raceways (Figure 13-14). Water is exchanged rapidly to maintain adequate oxygen levels and to remove waste products. The water quality in silos is usually no better than that of the incoming water, but dissolved oxygen may be increased by aeration and agitation, or in some units by the addition of gaseous or liquid oxygen. Silos may be constructed with sediment basins to remove solid waste before water is discharged.

Since rectangular raceways typically have length:width:depth ratios of 30:3:1, they occupy considerably more surface area than do circular tanks of similar capacity with a diameter:depth ratio of about 1:2. These

FIGURE 13-14 Concrete silos used for trout production.

deep, circular tanks or silos require a relatively large flow of water. Silos were initially developed in Pennsylvania for the culture of trout in areas where there was an adequate flow of high-quality water to provide a gravitational head sufficient to flush solid wastes from the tank.

Silos have been incorporated into water recirculation culture systems. Such systems require biological, mechanical, or chemical filters, or a combination of these, to remove waste metabolites and allow filtered effluent water to be recycled through the culture tanks. These systems have been used to culture striped bass, channel catfish, and salmonids.

Circular Tanks

Circular tanks constructed of plastic, concrete, and steel are widely used throughout the United States to culture aquatic species. Small 25- to 500-gal. tanks are used to spawn fish, to maintain fry and fingerlings, and to culture brine shrimp and other forage organisms. Circular tanks are used as flow-through units and also in water recirculation systems (Figure 13-15). Tank size correlates with growth, food conversion efficiency, and survival of rainbow trout.

CAGES AND PENS

Cage culture of fish uses existing water resources but encloses the fish in a cage or basket that allows water to pass freely between the fish and the pond. Modern cage culture began in the 1950s with the invention of synthetic materials for cage construction. In the United States, universities did not begin conducting research on cage rearing of fish until the 1960s. Today, cage culture receives more attention by both researchers and commercial producers. Factors such as increasing consumption of fish,

FIGURE 13-15 Circular fiberglass tank being used in a recirculating system at a high school.

declining wild fish stocks, and a poor farm economy produced a strong interest in fish production in cages. Many of the United States's small or limited-resource farmers look for alternatives to traditional agricultural crops. Aquaculture appears to be a rapidly expanding industry—one that may offer opportunities even on a small scale. Cage culture also offers the farmer a chance to use existing water resources that, in most cases, have only limited use for other purposes.

Cage culture of fish is not foolproof or simple. Cage production can be more intensive in many ways than pond culture and should probably be considered as a commercial alternative only where open pond culture is not practical.

Like any production scheme, cage culture of fish has advantages and disadvantages.

Advantages

Some distinct advantages of cage culture include:
- Many types of water resources can be used, including lakes, reservoirs (Figure 13-16), ponds, strip pits, streams, and rivers, which could otherwise not be harvested. Specific state laws may restrict the use of public waters for fish production.
- A relatively low initial investment is all that is required in an existing body of water.
- Catching is simplified.
- Observation and sampling of fish are simplified.
- Allows the use of the pond for sport fishing or the culture of other species.

These advantages are appealing. A potential fish farmer can produce fish in an existing pond without destroying its sport fishing and can try fish culture with reasonable risks.

FIGURE 13-16 Ponds often represent an unused resource for cage culture.

Disadvantages

Cage culture also has disadvantages, including:
- Feed must be nutritionally complete and kept fresh.
- Low dissolved oxygen syndrome (LODOS) is an ever-present problem and may require mechanical aeration.
- The incidence of disease can be high, and diseases may spread rapidly.
- Vandalism or poaching is a potential problem.

Good management of cage-cultured species prevents the disadvantages causing the failure of this production option.

Aquatic organisms may be confined at very high densities in cages or pens placed in ponds, lakes, rivers, and estuaries. Mollusk cages may be suspended from piers, long lines, or rafts. Water quality within cages and pens depends primarily upon both the quality of water in the surrounding area and the rate of exchange by circulation through these aquaculture units. Cages and other devices placed in estuarine areas with tidal flows are flushed by tidal action. In lakes and ponds, the swimming and feeding activity of fish cultured in the cages, as well as of wild fish around the cages, can generate sufficient water movement to permit water exchange in cages. Wind-generated waves also help to circulate and exchange water around cages and pens placed in ponds and lakes with little or no natural current.

Cages placed in rivers or industrial effluents take advantage of the flowing water to remove wastes. Cages, rafts, and other off-bottom systems may be moved from one site to another to take advantage of better environmental conditions. For example, caged fish cultured in the intake canal of a power plant during the spring and summer may be moved to the discharge canal to take advantage of the thermal effluent during the fall and winter.

Many factors may seasonally alter water quality, for example, nitrogen gas supersaturation in thermal effluents, agricultural runoff, industrial discharges, and reservoir drawdown for irrigation or flood control. These must be known and understood by the aquaculturist before placing cages or pens in a body of water.

Harvesting is a problem in cage and pen culture. Cages or pens are usually located away from land, and access is provided by boat or barge. Pond and raceway harvesting techniques are unworkable in this setting. Most cage and pen-reared fish must be dipped by hand and placed in containers, a labor-intensive operation. **Net pens** or cages do not permit partial or selective harvest.

Cage Construction and Placement

Cages for fish culture have been constructed from a variety of materials and in practically every shape and size imaginable. Basic cage construction requires that cage materials be strong, durable, and nontoxic. The cage must retain the fish yet allow maximum circulation of water

FIGURE 13-17 Cages positioned in a pond.

through the cage. Adequate water circulation is critical to the health of the fish, bringing oxygen into the cage and removing metabolic wastes from the cage. The location of the cage (Figure 13-17) may be critical to proper circulation. Mechanical circulation and aeration through the cage may be necessary if stocking densities are high, cages are large, or water quality deteriorates during production.

Cage Materials

Cage components consist of a frame, mesh or netting, **feeding ring**, lid, and flotation. Cage shape may be round, square, or rectangular. Shape does not appear to affect production. Cage size depends on the size of the pond, the availability of aeration, and the method of harvest. Most fish farming supply companies sell manufactured cages, cage kits, or materials for constructing cages. Common cage sizes are:

- Cylindrical—4×4 ft.
- Square—$4 \times 4 \times 4$ ft. and $8 \times 8 \times 4$ ft.
- Rectangular—$3 \times 4 \times 3$ ft. and $8 \times 4 \times 4$ ft.

The cage frame can be constructed from wood—preferably redwood or cypress—iron, steel, aluminum, fiberglass, or PVC. Frames of wood, iron, and steel—unless galvanized—should be coated with a water-resistant substance, such as epoxy, or with an asphalt-based or swimming-pool paint. Bolts or other fasteners used to construct the cage should be of rust-resistant materials.

Mesh or netting materials that can be used include galvanized wire, plastic-coated welded wire, solid plastic mesh, and nylon netting (knotted or knotless). Solid plastic (polyethylene) mesh is most commonly used for small cages because of its durability. Mesh size should be no smaller than 1/2 in. to assure good water circulation through the cage. A larger mesh size can be used if large fingerlings are stocked. The feeding ring or collar can be made of 1/8- or 3/16-in mesh and should be 12 to 15 in wide. The feeding ring keeps the floating feed from washing through the cage sides.

All cages should have lids to prevent fish from escaping and to keep predators—including people—out of the cages. Lids may be made from the same material as the rest of the cage, or they can be from other materials, such as plywood, masonite, aluminum, or steel. Plywood, masonite, and steel will need to be painted with exterior or epoxy paint. The cage manager needs to be able to observe feeding behavior and have easy access to the cage to remove uneaten feed and any dead fish.

Cage flotation can be provided by inner tubes, styrofoam, waterproofed foam rubber, capped PVC pipe, or plastic bottles. Cages can also be suspended from docks. Plastic bottles should be made of sturdy plastic, such as antifreeze or bleach bottles, and should have their caps waterproofed with silicon sealer. Floats should be placed around the cage so that it floats evenly with the lid about 6 in out of the water.

Cage Construction

Figure 13-18 shows the construction of a simple cage. The simplest cage design to construct is a 4×4 ft $\times$ cylindrical cage fashioned from 1/2-in plastic mesh. The mesh comes in a roll 4 ft wide by 50 ft long. A total of 21 ft of plastic mesh is used per cage. Thirteen feet of mesh is used for the cylinder, with two 4-ft panels for the bottom and lid. The plastic mesh is easily cut with tin snips. The cylinder is formed around two metal, PVC, or fiberglass hoops at the top and bottom of the cage. A third hoop is used to form the lid.

The cage can be laced together with 18-gauge bell wire—plastic coated solid copper wire—stainless steel wire, hog rings, or black plastic cable ties. White cable ties deteriorate in sunlight.

The basic design and construction principles apply to all cages (see Figure 13-19, page 414):
- Attach the netting or mesh securely to the frame
- Lace carefully leaving no gaps or holes
- Lids and feeding rings are essential

Cage Placement

Location of the cage in the pond may be critical to its success. Two factors to consider in cage placement are: (1) access to the cage and (2) maintenance of water quality. Daily feeding and cage management require easy access under almost any weather condition. Access may be by pier or by boat. Critical factors for locating cages in an area to maximize water quality are:
- Windswept location. It is important that the cage be in an area where it will receive maximum natural circulation of water through the cage. Usually, this is in an area that is swept by the prevailing winds.
- At least 6 ft of water depth. A minimum of 2 ft of water is needed under the cage to keep cage wastes away from the fish.

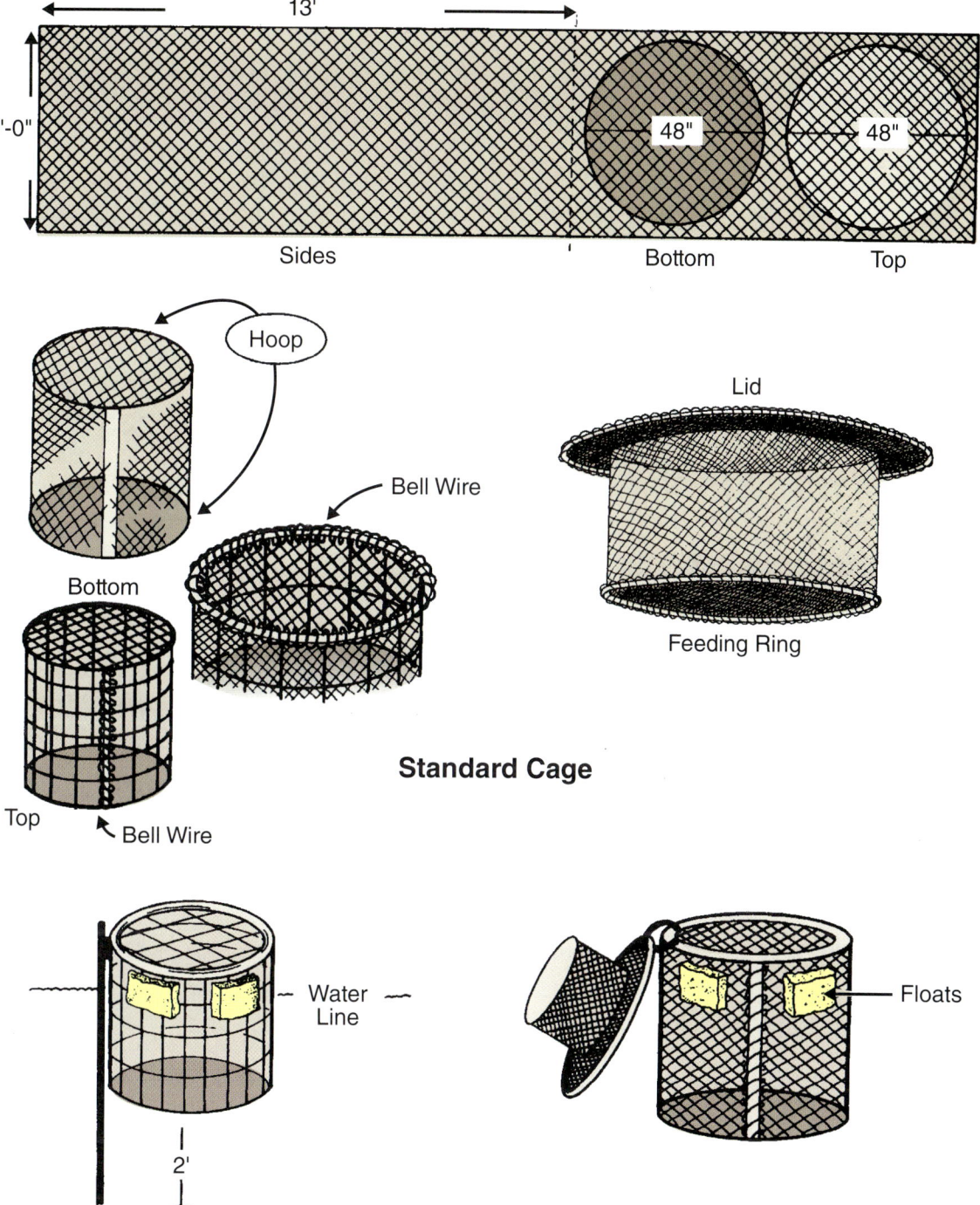

FIGURE 13-18 Construction of a 4 ft. × 4 ft. cylindrical cages.

- Away from coves and weed beds. Coves, weed beds, and overhanging trees can reduce wind circulation and potentially cause problems.
- Away from frequent disturbances from people and from animals like dogs and ducks. Disturbances from people frequently walking on the dock, fishing, or swimming near the cage, or from

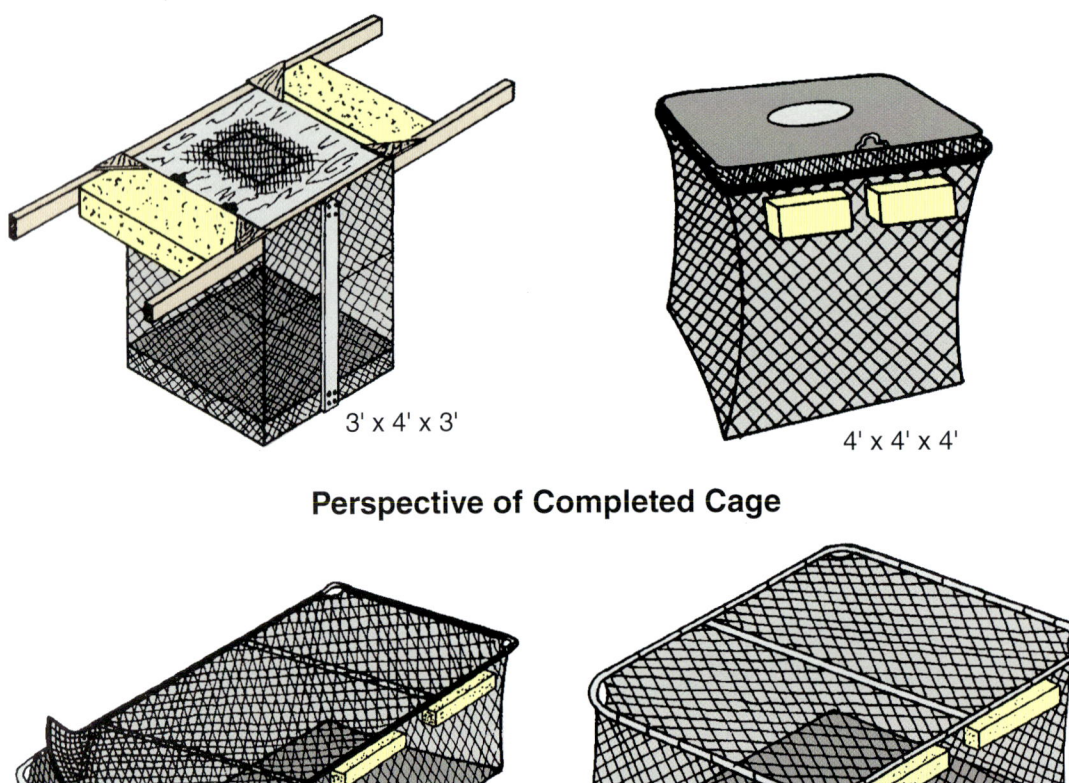

FIGURE 13-19 Some other common cage designs.

animals that frequent that area of the pond will excite the fish and can cause stress, injury, reduced feeding, and secondary disease.
➤ At least 10 ft from other cages. Cages should not be too close together. Close proximity to other cages may increase the likelihood of low dissolved oxygen syndrome (LODOS).

Access to electricity or to a location where a tractor-driven paddlewheel, irrigation pump, or other aeration device can deliver aerated water should be considered before locating cages. Aeration devices should move oxygenated water through the cages.

Large Cages

Larger cages have been built and used, particularly in large lakes, reservoirs, rivers, bays, and estuaries. Many times, large cages are called "net pens" because they are constructed from nylon netting. Some research and demonstrations include large cages for ponds that have built-in aeration devices. Large cages can be designed that maximize the number of fish sustainable by the pond and actually support increased densities.

OTHER MAJOR EQUIPMENT

Aquaculture uses a wide range of devices, such as aeration equipment, nets and seines, boats, pumps, tractors, and trucks.

Aeration Equipment

Aerators work by increasing the area of contact between air and water. Aerators also circulate water so that fish can find areas with higher oxygen concentrations. Circulation reduces water layering—**stratification**—and increases oxygen transfer efficiency by moving oxygenated water away from the aerator.

Fish farmers have used emergency aerators powered by tractor power takeoffs (PTOs) for many years. With intense production, the need for PTO aerators can be quite expensive because each aerator requires a tractor. More electric aerators are being used. Large tractor-powered aerators are still used as backups during severe oxygen depletions, equipment failure, or power outages.

Each producer decides which aeration device should be purchased or built. This decision is important and should be made with the specific application and associated costs of energy and equipment in mind. Most aerators are in one of the following categories: surface spray or vertical pump, pump sprayer, paddlewheel, diffused air, and propeller aspirator pump.

Surface Spray or Vertical Pump

Surface spray aerators have a submersible motor that rotates an impeller to pump surface water into the air as a spray. (See Figure 13-20.) They float, are lightweight, portable, and electrically powered. Units of 1 to 5 hp with pumping rates of 500 to 2,000 gpm are available.

FIGURE 13-20 Surface spray or vertical pump aerator.

Designed to be operated continuously during nighttime, cloudy weather, or when low dissolved oxygen concentrations are expected, surface spray aerators have prevented fish kills when used at 1.5 to 2 hp/acre. They are usually of little use in large ponds because of relatively low oxygen transfer rates and their inability to create an adequately large area of oxygenated water.

Pump Sprayer

Pump sprayer aerators are found on many fish farms. Most are powered by tractor power takeoffs or electricity. Some units are engine driven and require mounting on a trailer frame for transport. Pump sprayer aerators are equipped with either an impeller suction pump, an impeller lift pump, or a turbine pump. Some have a capped sprayer pipe or "bonnet" with outlet slits attached to the pump discharge. Others discharge directly through a manifold that has discharge slits on top and outlets at each end. Water is sprayed vertically through the discharge slits from each end of the manifold. This type is commonly referred to as a "T-pump" or "bankwasher" and directs oxygenated water along a pond bank where distressed fish often go. Pump sprayers typically have no gear reduction, which reduces mechanical failure and maintenance. These units do not erode the pond bottom, and minimum operating depth is reached when the intake is covered with water.

Paddlewheel Aerators

Paddlewheel aerators have been used on catfish farms for many years. Farm-made paddlewheels are usually made from ¾-ton truck differentials and vary with drum size and configuration, shape, number, and length of paddles (Figure 13-21). Units are powered by PTOs or driven by self-

FIGURE 13-21 Portable paddlewheel aerators can be powered by a tractor PTO, mounted engine, or an electric motor.

contained diesel engines. The self-contained units are usually on floats and attached to the pond bank or held in place by steel bars secured in the bank or pond bottom.

Increasing either the speed of the drum rotation (rpm) or paddle depth generally increases aeration capacity. Paddle depth affects oxygen transfer rates more than does the speed of rotation. This increase in capacity is not cost-free because horsepower requirements increase and oxygen transfer efficiency may decrease. The maximum rotational speed of a tractor-powered paddlewheel aerator for extended operation is limited by the tractor and the gear reduction of the paddlewheel.

The shape of the paddles is also important. For example, U, V, or cup shapes are more efficient than flat paddles. Paddlewheels create vibrations that can be reduced when paddles are arranged in a spiral pattern (Figure 13-22).

The oxygen transfer rate and power requirement increase with paddle immersion depth and the diameter of the paddlewheel drum. The size of the spray pattern likewise increases. The power required to operate a paddlewheel aerator at any given speed and paddle depth is constant. Fuel consumption and operating costs depend on the power source. Most producers do not have enough paddlewheel aerators for all ponds and move these units from pond to pond. When fish are stressed with low dissolved oxygen, they often go to shallow areas of the pond near the banks. A paddlewheel, though mobile, can be difficult to situate in the pond properly so that it is effective without damaging itself or the tractor.

Electric Paddlewheel

Electric paddlewheel units (Figure 13-23) are 4 to 12 ft long with paddles of triangular cross section and a total drum diameter of about 28 to 36 in. Paddlewheel speed is usually 80 to 90 rpm with a paddle depth of about 4 in, enough to load the motor. Motor sizes range from 0.5 to 19 hp and larger.

Methods used to reduce the motor speed to the desired aerator shaft speed include V-belts and pulleys, chain drive and gears, and gearboxes. Shafts of most electric motors run at 1,750 rpm, and most units are mounted on floats.

Diffused Air Systems

Diffuser aerators operated by low-pressure air blowers or compressors forcing air through weighted aeration lines or diffuser stones release air bubbles at the pond bottom or several feet below the water surface. (See Figure 13-24.) Efficiency of oxygen transfer is related to the size of air bubbles released and water depth. The smaller the bubble and the deeper it is released, the more efficient this type aerator becomes. When tested at normal catfish pond depths, these aerators were found to be inefficient compared to other devices.

Limited studies in commercial catfish ponds showed no improvement in fish production when a diffused aeration system was used. One of the biggest problems with diffused-air systems is clogging of the air lines and

FIGURE 13-22 Side wheel style paddlewheel aerator. This aerator requires the PTO of a 45-hp tractor.

FIGURE 13-23 Scientist taking an oxygen reading on a pond with brush aerators anchored in the background.

FIGURE 13-24 Small diffused air system being used to maintain catfish in a small laboratory tank.

diffusers so that periodic cleaning is required. Also, the air lines interfere with harvesting. Diffused air systems are used for supplemental oxygen in trout facilities.

Propeller-aspirator Pump

These aerators consist of a rotating, hollow shaft attached to a motor shaft. The submerged end of the shaft is fitted with an impeller that accelerates the water to a velocity high enough to cause a drop in pressure over the diffusing surface, which pulls air down the hollow shaft. Air passes through a diffuser and enters the water as fine bubbles that are mixed into the pond water by the turbulence created by the propeller. These are electrically powered, and models range from 0.125 to 25 hp.

Seines

For every 2 ft of pond width to be seined, 3 ft of seine length is required. The same ratio applies to pond depth. Floats can be made of styrofoam or plastic attached on 18-in centers.

Most catfish seines have a **mud line** on the bottom of the net. A mud line is made of many strands of rope or a roll of menhaden netting bound together. As the seine is drawn across the pond bottom, the mud line stays on top of the mud, eliminating the digging effect of lead-weighted lines.

Seines should be made of polyethylene or nylon. Catfish spines will not catch in polyethylene material. Nylon netting requires a net treatment to prevent spines from entangling. Also, seines need to be treated so they do not rot (Figure 13-25).

The mesh size to be used varies according to the minimum size of the fish to be captured. Buying the proper mesh seine for an operation allows capture of only fish that are large enough for the market. The size of the fish caught varies somewhat with the mesh width and fish condition and activity. Fish do not grade as well when water temperatures are cold.

FIGURE 13-25 After seines are made, they must be treated so that they do not rot in water. This "ferris wheel" is used to wrap the treated seine around while it dries.

Live-Cars or Socks

Holding live fish is sometimes necessary if the market cannot take all the catch in one day, or because of a delay between capture and hauling fish to market. Often, catfish producers may want to sell fish directly to consumers. In these cases, catfish can be held in **live-cars** or **socks**. Live-cars are net enclosures that can be placed in a pond to temporarily hold the fish. They are made of the same material as seines.

Holding fish in live-cars requires caution. Diseases, oxygen stress, weight loss, and poaching are common problems. Aeration may be necessary near the live-car at night, particularly in warm weather. Producers limit the time the fish are held to only a few days to reduce weight loss and prevent disease. Disease can occur in holding devices during any season but is much more prevalent when water temperatures are highest. Poachers can easily steal fish from unguarded holding facilities.

Typical live-cars are 20 to 40 ft long, 10 ft wide, and 4.5 ft deep, and are constructed of a mesh size that retains harvestable fish and allows the escape of the smaller fish. The escape of market-sized fish is prevented by attaching a line of floats to the upper edge of the live-car, or sometimes by sewing an apron of netting, 1 to 2 ft wide, with supporting floats, inside the top of the live-car. Stakes anchor the unit in the pond. (See Figures 13-26 and 13-27.)

Live-cars can be used advantageously by the catfish producer. They can be coupled with a harvesting seine to serve as a temporary holding container and grader. Hauling trucks can then be scheduled accurately because the fish quantity is known, and the truck can be loaded without delay. A filled live-car can be moved from the seine landing area to an area more accessible to the hauling truck (See Figure 13-27). Hauling efficiency is also

FIGURE 13-26 Workers prepare to haul another basket of catfish from the sock or live car, where fish are first gathered and held overnight before being sent to market.

FIGURE 13-27 Boom mounted brailer lowered to live haul truck during loading of catfish.

improved because the fish essentially empty their digestive tracts overnight, and can be pre-cooled by positioning the live-car near the well outlet. Live-cars can be used to hold small quantities of fish for processing or as repositories for selected fish during sorting.

Loading is simplified because fish cannot escape by swimming under the seine. Boom-mounted loading baskets can be lowered into the live-car and filled by crowding fish into them, eliminating time-consuming dip-netting. Workers stand outside the live-car to load fish and are much less likely to be injured by spines of catfish fins.

The fish capacity of a live-car depends on the unit size, the pond depth and temperature, the length of time fish are to be held, and whether a well discharge or a circulating pump is available.

Other Equipment

For harvesting fish, producers may also need:
- A seine reel for hauling in and storing the seine
- Seine stakes
- Tractors
- Sturdy dip nets
- Baskets
- Boots or chest waders
- Scales
- A boom
- A boat and motor

Transporting

Transporting live fish requires maximum care to avoid losses. In transport, fish are crowded into a relatively small amount of water. Agitators, blowers, compressed oxygen, compressed air, or liquid oxygen can be used individually or in combination to keep the fish alive. Transport containers are usually made of wood, fiberglass, or aluminum. Many types are commercially available.

Generally, the dissolved oxygen content in the water is the factor that determines whether the fish live or die. Fish should not be fed for at least 24 hours before transport so that excessive fish wastes do not accumulate during transport. Fish wastes and regurgitated feed consume large quantities of oxygen and can produce ammonia and carbon dioxide problems.

Transporting fish in cool weather or in cool well water increases fish survival. Cool water holds more oxygen than warm water, and fish consume less oxygen at lower temperatures. Also, large fish consume less oxygen by weight than small fish. When transporting fish, an oxygen probe is often in the hauling tank with the meter in the cab of the truck so the driver can monitor oxygen concentrations during transport.

Fish health and survival depend on the producer's ability to limit stress. Stress from netting, loading, hauling, and stocking weakens the fish and makes them more susceptible to disease and water-quality problems.

Successful hauling begins with the use of good pond management practices and careful harvesting. Fish stressed during production or harvest cannot be transported successfully. Fish are not fed for two to three days before harvesting and hauling because they can be handled much better if the intestinal tract is empty of feed. After harvesting, a common procedure is to haul the fish in cool water and then temper the hauling water to the receiving water before the fish are unloaded. The loading level in the transport unit varies with the size of fish to be transported, the length of the haul, and the temperature of the hauling water. For example, catfish of 1 to 2 lbs can be hauled at the rate of 4 lbs of fish per gallon of transport water if the trip requires fewer than 12 hours.

Tanks for hauling fish from the harvest area to the holding facilities must be fitted with water-aerating devices suitable for the size and species. Very small fish are sometimes injured by compressed air and agitators, but they tolerate much better the small bubbles of (bottled) oxygen coming from carbide aerator stones. Other fish are damaged little by agitators, compressed air, or sprayed water. About 1 lb of fish per gallon of water can be moved in cold weather, but this amount should be reduced to about 0.67 lb in warm weather.

Oxygen Testing Equipment

Intensive fish culture requires periodic checks of dissolved oxygen levels. During certain times of the year, these dissolved oxygen checks must be made several times in each 24-hour period. For small operations, a chemical test kit will suffice. Chemical test kits for oxygen determinations can be purchased. Many operations use electronic oxygen meters to save time and labor during all the dissolved oxygen checks required. (See Figure 13-28.) At least one backup oxygen meter is needed because these can easily be damaged. A chemical oxygen test kit may be needed to check the accuracy of oxygen meters.

FIGURE 13-28 The convenience of electronic oxygen testing meters taken one more step. The meter probe is on a long pole, and the technician quickly takes readings from many ponds.

Tractors

Depending on the operation, a tractor—90 to 100 hp—maybe needed for such things as pulling a feeder, providing power for a paddlewheel aeration device, a **relift pump**, and pulling a seine (Figure 13-29). For commercial catfish production, at least two tractors are needed for four 17.5- to 20-acre ponds.

PTO Relift Pump

Catfish producers need at least one relift pump (Figure 13-30) for every four 17.5- to 20-acre ponds. When the discharge end is capped and slotted, this pump is the second-most efficient emergency aeration device. Enough pipe should be available to pump water from one pond to an adjacent pond. Pumping good water from one pond into an adjacent pond with low oxygen levels can be a good way to keep fish alive until the problem can be corrected. Also, at times it may be necessary to remove water rapidly from a pond to correct certain water-quality problems.

FIGURE 13-29 Tractors are used to pull the seine through the pond.

FIGURE 13-30 PTO relift pump for aeration and quick exchange of water in ponds.

Boats

Boats are used for dispensing certain chemicals directly into the water for the control of diseases, aquatic weeds, and water-quality problems. (See Figure 13-31.) The boat should be powered by an outboard motor of about 10 hp. A trailer may be used for transporting the boat and motor from one pond to the next.

Crawfish producers use a boat for setting traps (Figure 13-32). Other aquaculturalists use boats to capture wild broodstock to produce seed.

Trucks

One or more ½- to ¾-ton pickup trucks are necessary for routine work around the farm. (See Figure 13-33.) The number needed depends on the size of the farm and its type, the number of ponds, and the number of employees.

FIGURE 13-31 Boats are used in commercial catfish production during harvest to push the seine.

FIGURE 13-32 Typical boat used to harvest crawfish. The wheel serves to hold the boat straight in the wind. This is a so-called "go-devil" rig.

FIGURE 13-33 Workers at the Creston NFH, net trout into a tank truck for stocking.

SUMMARY

Although pond culture still predominates, the use of raceways, tanks, silos, cages, and recirculating systems has increased. As aquaculture production increases and technology improves, these additional culture systems can be expected to become more widely used. Regardless of the culture system, planning is essential for successful aquaculture. Next, the aquaculturalist must completely understand the type of production facility being used.

Aquaculture has created new uses for traditional equipment and new equipment for new jobs. Each aquaculturalist needs to select carefully the best equipment for his or her facility and be knowledgeable in the use of the equipment.

STUDY/REVIEW

Success in any career requires knowledge. Test your knowledge of this chapter by answering these questions or solving these problems.

True or False

1. In the United States, raceways are the most common type of structure for raising fish.
2. For a potential pond site, the Natural Resources Conservation Service can provide information about soil types.
3. Most commercial catfish ponds on the Mississippi Delta are built on 5-acre units.
4. Paddlewheels are used to aerate raceways.
5. A live-car or sock sits on the back of a semi-truck and hauls fish to market.

Short Answer

1. Name three types of ponds that could be used for aquaculture.
2. Levee-type ponds are constructed by:
    a. digging a hole
    b. building a dam across a ravine
    c. pushing up dikes on flat land
    d. pouring concrete walls above the ground
3. What size of pond can fish be grown in?
    a. 1 acre
    b. 5 acre
    c. 0.1 acre
    d. any size
4. Two general sources for water to fill ponds or raceways include _____ and _____ water.
5. _____ is an essential soil type for pond construction and the core of dams.
6. Why is it important that the bottom of a pond be level?
7. When drilling a well, who should be able to tell if the well will produce the desired capacity of the facility?
    a. U.S. Geological Survey
    b. U.S. Fish and Wildlife Service
    c. U.S. Department of Agriculture
    d. U.S. Food and Drug Administration
8. Why is a well usually the best choice for a levee pond for commercial production?

9. Levee-type ponds should be constructed:
    a. on hilly land
    b. on flat land
    c. between mountains
    d. along the coastline
10. If crops were grown on, or adjacent to, a potential site, what should be checked before proceeding?
    a. pesticide residues
    b. crop residues
    c. organic matter
    d. depth of cultivation
11. List three sources of water for ponds.
12. Name three areas within a pond site that need to be checked for pesticide residues.
13. What is the best shape for a levee-type pond for commercial catfish production?
14. Give two reasons for making main levees 20 ft wide at the top.
15. How does the slope of a levee affect the amount of soil to be moved?
16. Name or describe two types of drains used in levee-type ponds.
17. For a raceway, what is the ratio of the length to the depth and the width?
18. What species can be used for cage culture?
19. Give three basic design principles for all cages.
20. Name four types of aerators.
21. A pond is 300 ft wide. What length must the seine be?
22. List two guidelines for transportation of live fish.
23. Name three uses for a tractor in aquaculture.

Essay

1. Describe why clay is essential for some pond construction.
2. Define topography and indicate how it affects pond construction.
3. Define freeboard.
4. Compare raceway construction to levee pond construction.
5. What is a silo?
6. Why would cage culture be beneficial for the novice aquaculturist?
7. What are three disadvantages of cage culture?
8. Why do cages need lids?
9. Why must water 6 ft deep be used for a cylindrical cage that is 4 ft. × 4 ft.?

10. Why should cages be placed 10 ft. apart?
11. What is a live-car and how do catfish producers use them?
12. How are boats used in aquaculture?

KNOWLEDGE APPLIED

1. Visit a construction company that specializes in earth moving. Observe equipment used and ask someone to explain how it is used. Also, ask what the cost per yard is for moving soil.
2. With the help of information from the NRCS and a good soils manual, determine the type of soil available in your area. Also, using a sample of clay and a sample of sand, mix each with water and observe the different behaviors. If possible, invite a NRCS official to visit the class and discuss and demonstrate soil types.
3. If water resources permit, use information resources at the end of this chapter and construct a small cage. If a cage cannot be built, use the information resources to design a cage and develop a list of materials. Next, calculate the cost to purchase the materials.
4. Obtain a test kit or electronic meter for measuring dissolved oxygen. Design some simple aeration methods for water samples. Using the test kits or an electronic meter, determine the effectiveness of these aeration methods.
5. With paper or a computerized drafting program, design a levee-type pond. Determine the slope of the levee, size of the pond, size of the levee, and calculate how many yards of soil will need to be moved to make the levees.
6. Develop a class demonstration to show how clay prevents water seepage from ponds.
7. Using a small hand-held GPS unit, mark the corners of a 5-, 10-, and 20-acre rectangular pond. Or create a class competition to see which team can come the closest to laying out and marking the corners of a 5-acre rectangular pond.

LEARNING/TEACHING AIDS

Books

Beveridge, M. (2002). *Cage aquaculture*. Hoboken, NJ: Wiley-Blackwell.

Lekang, O-I. (2007). *Aquaculture engineering*. New York, NY: Wiley-Blackwell.

McLarney, W. (1998). *Freshwater aquaculture: A handbook for small scale fish culture in North America*. Point Roberts, WA: Hartley & Marks Publishers.

Swann, L., Brown, J., Katz, S. and Merzdorf, R. (1998). *Getting started in freshwater aquaculture*. West LaFayette, IN: Purdue University.

Note: Many fact sheets are available from the Regional Aquaculture Centers and others listed in Appendix Table A-12. For example:

Aquaculture engineering, FAO Fisheries and Aquaculture Department: http://www.fao.org/fishery/topic/13265/en

Pond design, construction and repair, ALEARN, Auburn University: http://www.aces.edu/dept/fisheries/aquaculture/ponddesign.php

International water harvesting and aquaculture for rural development, Auburn University: http://www.ag.auburn.edu/fish/international/cage.htm

Cage culture of fish in the north central region, AquaNIC: http://www.ag.auburn.edu/fish/international/cage.htm

Raceway publications, AquaNIC: http://www.aquanic.org/systems/raceways/pubs.php

Cage publications (freshwater and marine), AquaNIC: http://www.aquanic.org/systems/cages/pubs.php

Internet

Internet sites represent a vast resource of information. The URLs (uniform resource locator) for the World Wide Web sites can change. Using a search engine such as Google, find more information by searching for these words or phrases: aquatic structures, ponds (levee-type, impoundment, excavated, natural), feeding ring, flow index, biofilters, aquaculture raceways, watershed, silo culture, soil characteristics, systems, cage fish culture, or aeration equipment.

For some specific Internet sites, refer to Appendix Tables A-11 and A-14.

An aquarium functions like a small pond, and, for this reason, it can teach some principles of aquaculture. In a small aquarium, you can observe in action the same set of rules and laws that control a big fish farm. Aquarium supplies can be found almost anywhere in the United States. Tropical

CHAPTER 14

Aquariums

aquarium fish come in over 1,000 varieties, so one can find a fish to suit almost any need.

OBJECTIVES

After completing this chapter, the student should be able to:

- Identify fish for a freshwater aquarium
- Describe the major groupings of aquarium fish based on their reproductive habits
- List plants that can be used in an aquarium
- Identify the supplies necessary to create an aquarium
- Explain the role of light, nutrients, and plants in an aquarium
- Describe how to set up a beginner's aquarium
- Discuss the management and maintenance of an aquarium

Understanding of this chapter will be enhanced if the following terms are known. Many are defined in the text, and others are defined in the glossary.

KEY TERMS

Algae
Bubble-Nest Builders
Cichlids
Dechlorinator
Diapause
Egg-Scatterers
Hypersaline
Phytoplankton

BACKGROUND

The object when establishing an aquarium is to create a small pond. A pond is not a sterile place, and the aquarium should not be sterile, either. Many aquarium manuals will tell the reader to bleach their gravel, tank, and everything else, sometimes on a regular schedule, making it a sterile place. Though this will kill many of the bacteria that could cause problems in an aquarium, it will also kill the helpful ones. Aquaculturalists learn about many interactions in ponds and water. If most of the ecosystem is killed with bleach or something else, it cannot be examined thoroughly. A good, healthy aquarium with nutrients, plants, and bacteria allows the study of the principles that control aquaculture.

FISH FOR THE AQUARIUM

Tropical aquarium fish come in over 1,000 varieties. Chapter 6, Raising Ornamental Fish, provides additional details. Different kinds of fish will require different care, different conditions, different spaces, and different equipment. Getting the equipment before knowing the type of fish could lead to the purchase of inappropriate or unusable equipment or to insufficient space.

The care and compatibility of the fish also needs to be researched. This research will determine what conditions the fish will need, what equipment will be needed, and how to set up the tank. Research of the fish determines the size of a tank. As a rule of thumb, you will need 1 in. of healthy mature small fish per gallon of water, and 1 in. of of fish per 3 gal of water for large or messy fish.

Aquarium fish can be divided into several major groups, depending on the way they reproduce (Figure 14-1).

Live-Bearers

Probably the most common fish in the pet store belong to this group. Live-bearers include guppies, swordtails, mollies, and platies. Most of them are extremely domesticated, available in many varieties and colors. They are called live-bearers because the young are born already swimming and eating. This makes them excellent fish for beginners.

FIGURE 14-1 Molly, guppy, platty, and swordtail — good for beginner projects in an aquarium.

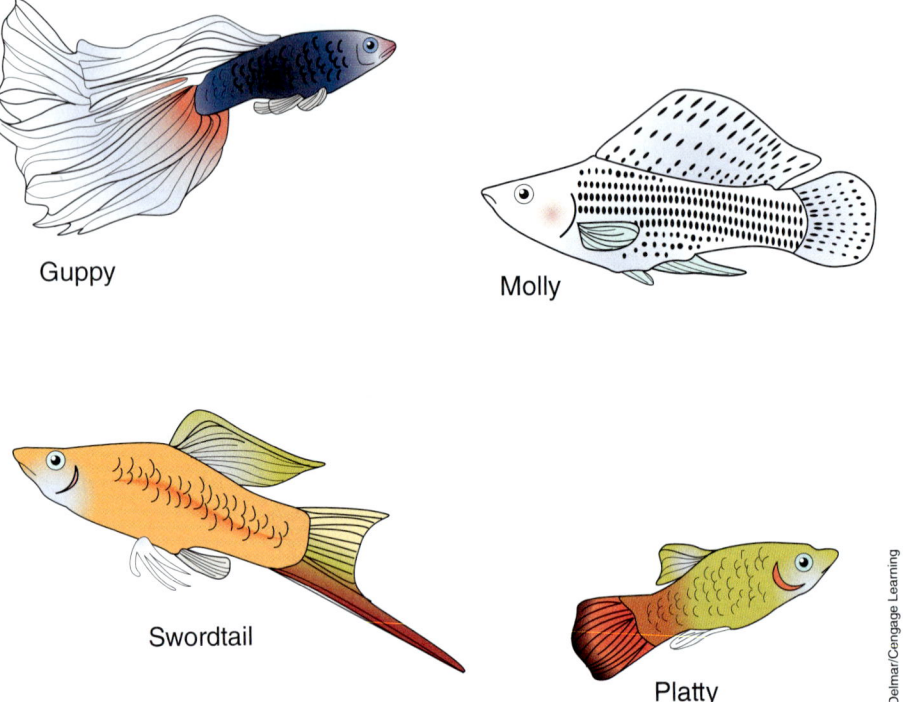

Mouth-Brooders

This is a unique group that developed an interesting way to help their young survive. One fish, usually the male, will make a depression in the sand or gravel where the eggs will be laid. As soon as the eggs are laid, one of the parents, usually the female, will pick up the eggs in its mouth. The eggs are then kept in the fish's mouth until they hatch. Even then, the young stay in the mouth of the parent until they are able to swim fast enough to take care of themselves. Fish from this group are also a good choice for the beginner.

Bubble-Nest Builders

The two most common fish in the **bubble-nest builders** group are the kissing gourami and the Siamese fighting fish (bettas). One parent, usually the male, makes a nest of bubbles that float on the surface. By using a sticky mucous, these bubbles will stay together and not pop. Some of this group will also mix floating plant material into the nest. The pair will lay their eggs under the nest. Some fish in this group have eggs that will then float up to the surface, but most of them place their eggs in the nest themselves (usually done by the male). Typically, the male guards the eggs. After hatching, the young are protected by the parent(s) until they are old enough to swim away.

Egg-Layers

Egg-layer fish lay eggs that stick to the side of the tank, a rock, or a plant. Alternatively, they make a nest. Most of them will guard their eggs until they hatch and for a short time afterward. The largest group of fish in this category is the **Cichlids** (pronounced si-klids), which includes the popular angelfish. Because they will guard their eggs and even raise the young for a while, they are easier than the egg-scatterers (description follows) to raise.

Egg-Scatterers

The fish in the **egg-scatterers** group lay their eggs in the middle of the water. The eggs then drift to the bottom, or they are scattered over plants. Tetras, barbs, and danios are the most common fish in this group. The eggs and fry for this group are usually very small, making it hard to feed the newly hatched fish. Only the advanced student should try to work with them.

Healthy fish must be used to start an aquarium. Fish should be eating well, brightly colored, and active. A reputable fish store sells only healthy fish. Frequent observation of the tank is necessary to maintain fish health. Fish should be checked for signs of disease, such as sores on their bodies, tattered fins, swimming on their sides, and other abnormalities.

CHOOSING AND ESTABLISHING AN AQUARIUM

Aquariums are sold in standard sizes ranging from 2 to 125 gal. Manufactured aquariums are available, with many supplies designed to fit them, such as tops, lights, pumps, and other equipment. After the aquarium, several other supplies are necessary. (See Figure 14-1.)

Supplies for the Aquarium

A general list of the basic supplies for an aquarium includes the following:
- Dip net at least 6 in wide.
- Heater large enough for tank. Heaters come in different sizes and watts. Dealers know what size is needed for an aquarium.
- A stand or a place to put the aquarium. Water weighs about 8 lb per gal. This requires a sturdy stand.
- A light.
- A top for the tank to keep some fish from jumping out.
- A thermometer to measure the temperature.
- Gravel or sand. Very fine gravel or coarse sand is better for plants. Natural colors are better to create a little pond.
- Rocks or driftwood for decoration and hiding places.
- A filter.
- Fish food. The type of food used depends on the type of fish in the aquarium.
- **Dechlorinator**. Tap water may contain chlorine, which will kill fish. Water can set for twenty-four hours to let the chlorine evaporate, or the chlorine can be removed by a dechlorinator.
- Gravel vacuum to clean the gravel.

434 AQUACULTURE SCIENCE

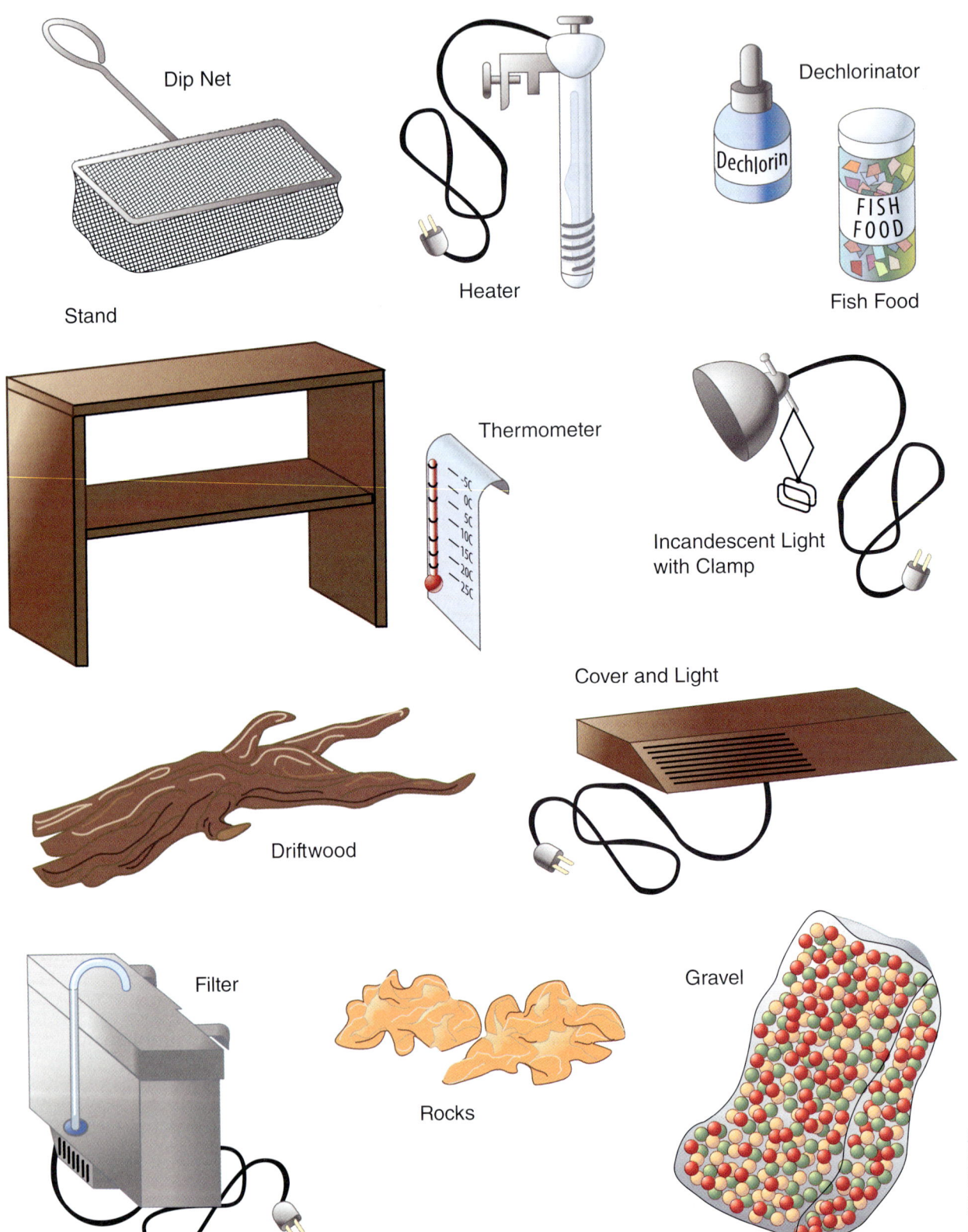

FIGURE 14-2 Materials and supplies necessary for an aquarium.

Why Light, Nutrients, and Plants

Most of the world's oxygen is produced in the surface layers of the oceans by tiny plants called **phytoplankton**. Phytoplankton comprise all the one-celled and small colonies of **algae** that drift in water. These plants do not have any roots and get their nutrients from the water. When a pond's water turns green, it is because of phytoplankton.

In aquaculture, nutrients are added to the pond or tank when the aquatic animals are fed. The nutrient level is usually higher than it would be in a natural system, and more plants grow because of this. In most aquaculture ponds, the plants are phytoplankton, and the water is green.

In the aquarium, phytoplankton is undesirable, because the fish could not be seen if the water were green. For this reason, plants are included in the aquarium. They will use the light and nutrients in the same way as phytoplankton. With just enough plants and not too much light or too many nutrients, plants will get rid of the phytoplankton. The plants will serve as the waste disposal system for the aquarium, just as they would in a pond.

Light

Sunlight or artificial light can provide light for an aquarium. Sunlight is cheap and natural, but it has a few drawbacks in the aquarium. It is hard to control the amount of sunlight that the tank will receive. Too much or too little will create problems. With sunlight, the aquarium will be dark when the sun goes down, and evening is when most people have time to spend with their aquariums. The best recommendation is to use some kind of artificial lighting.

Fluorescent or incandescent lights will reproduce the light necessary for plants to grow. The best situation combines fluorescent, incandescent, and sunlight in the aquarium. Some lights are specially made for aquariums, or lights can be made from fixtures bought at hardware and discount stores.

Different-sized aquariums will use different-sized lights. For example, a 5-gal aquarium can use one 60-watt incandescent bulb, but a 55-gal aquarium would do better with two 40-watt fluorescent bulbs. The aquarium shop can recommend the best lighting for a specific aquarium with growing plants.

Aquarium lights should be left on for at least 12 hours each day. You can leave them on longer, but anything less than 12 hours is not enough. Too much light will start producing green algae on the glass and rocks in the tank. If this happens, the lighting time should be shortened.

Nutrients

Nutrients in the aquarium will come from the fish feeds. Many companies sell aquarium-plant fertilizers. In a well-fed tank, these should not be needed. This is the major advantage over the sterile aquarium, where everything is constantly being cleaned.

Aquarium fish eat the food given to them, and their waste, in turn, becomes the food for the plants. Without the plants, these nutrients would build up and the tank would require constant cleaning. Even with a healthy plant community in the aquarium, the fish can be fed too much, and all of the nutrients would not be used. This eventually ends up as a fouled tank.

Laying a thin layer (1 in.) of peat on the tank bottom before adding sand or gravel provides another source of nutrients. Eventually, the plants with roots will use the peat to obtain some nutrients.

Plants

Hundreds of aquatic plants are suited for an aquarium. Most aquarium shops will carry a few different varieties and can order others from their suppliers. The following is a list of some of the major groups of plants that are commonly seen in the aquarium industry.

Sagittaria

A group of grass-like plants native to the Americas, sagittaria reproduce by sending off runners. A few plants eventually provide a nice, thick growth. The varieties available include a dwarf type that is excellent for planting in the front of the tank.

Valisneria

Valisneria is another grass-like plant that reproduces easily by runners. One variety has a beautifully spiraled leaf (corkscrew valisneria).

Cabomba

Grown and sold mostly as cuttings, cabomba is an ideal plant to use for filling in areas, as it grows quite thick. It grows best in bright light and when peat is added below the sand or gravel.

Sword Plants

A large group of plants from the Amazon region, sword plants are members of the genus Echinodorus. They come in many sizes and leaf shapes. Sword plants have a heavy root base and leaves that can reach over a foot long. They grow best with peat around the roots and a lot of light.

Cryptocorynes

Cryptocorynes comprise a large, diverse group of plants from the East, mostly Thailand and India. Cryptocorynes reproduce in the aquarium by sending out runners from the parent plant. Once established, they are extremely hardy and will live for years. Cryptocorynes sometimes lose all their leaves when transplanted, but they send up new ones from the roots if left alone. They like peat in the tank bottom.

Aponogetons

Usually sold as bulbs with a few leaves coming out, aponogetons are a good plant for the center of a tank. The bulbs will eventually die off, however, and they are difficult to resprout. The plant will last longer if the bulb is placed in peat. The unusual Madagascar lace plant is in this group.

FIGURE 14-3 Some typical aquarium plants.

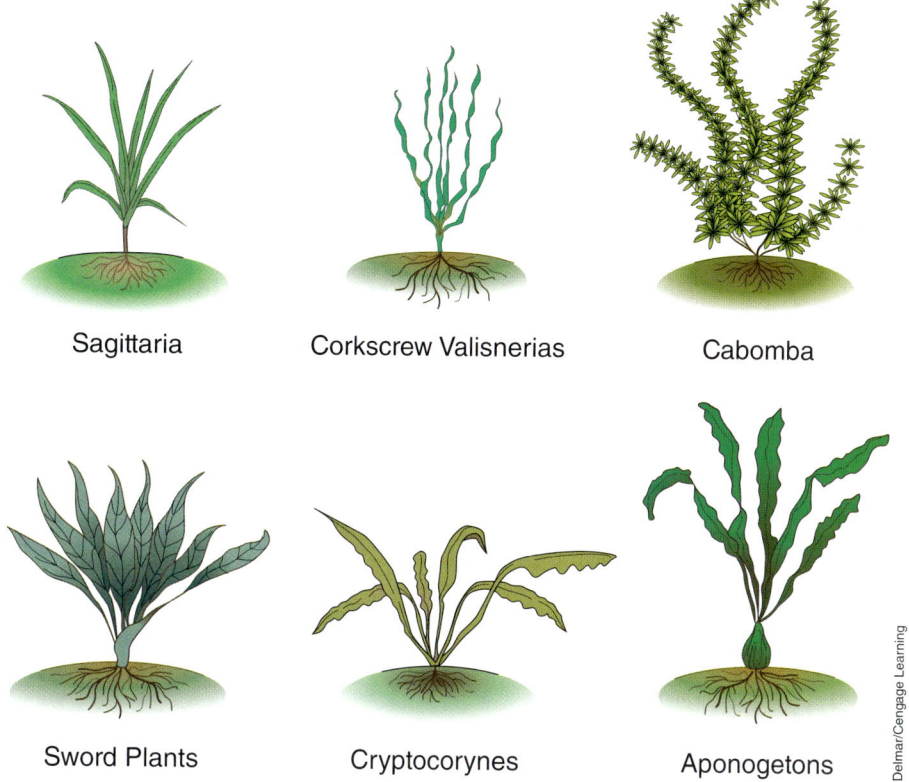

Figure 14-3 illustrates some of the aquarium plants.
When buying plants, look for:
1. Plants that are green and healthy with good roots. A mistreated plant is difficult to revive.
2. Inexpensive types for the beginner that will serve the purpose and help develop aquatic skills.

MANAGING THE AQUARIUM

Learning to manage an aquarium requires both the practice of aquaculture principles and the power of observation.

Feeding the tank

Most of the mistakes in aquaculture are made in feeding, both by underfeeding and overfeeding. Underfed fish will become sickly and more apt to catch diseases, and they certainly will not reproduce well. Overfeeding is a more common problem. This causes poor water quality, putting the fish at risk of disease.

Feeding fish is easy to do, using a simple rule of thumb that fish farmers follow. Feed fish once or twice a day, giving them the amount they eat in about 10 minutes. If they are still hungry, they need more food. If the fish do not eat all the food, then they received too much food. Following this rule of thumb will eliminate most of the problems in aquaculture of aquarium fish. When fish are overfed, any food lying on the bottom of the tank should be removed. This is done by using a 5- to 6-ft long piece of hose to siphon the uneaten food into a bucket.

There wide varieties and types of fish food on the market from which to choose. Some fish need to be fed more often, others need to be fed less often. Different fish have different dietary requirements, so selecting a variety of good, healthy foods that will meet a particular fish's nutritional needs is important.

Once the amount of food eaten is established, it should be recorded. This is important information because if fish ever stop eating this amount, it may mean problems. It is a good indication that something is wrong with either the fish or the water.

BRINE SHRIMP

Brine shrimp are important to aquaculture because newly hatched brine shrimp nauplii (larvae) provide a food source for many fish fry. Culture and harvesting of brine shrimp eggs represents another aspect of the aquaculture industry. The Great Salt Lake Brine Shrimp Cooperative, Inc., in Utah is a fully vertically integrated cooperative that harvests, processes, packages, and markets brine shrimp (*Artemia franciscana*) eggs on behalf of its members. The cooperative supplies highly nutritional, on demand, live feed at the hatchery level to commercial shrimp and fish farms in 55 countries around the world. The Utah Artemia Association is the brine shrimp industry's trade association. Members of this trade association have certificates of registration to harvest brine shrimp eggs from the Great Salt Lake. The industry harvests only the excess brine shrimp eggs from the Great Salt Lake. The association advocates policies that protect the Great Salt Lake ecosystem and the brine shrimp resource.

Brine shrimp live in **hypersaline** lakes in which the salt content may be 25%, predators and competitors are few, and algal production is high. The life cycle of Artemia begins from a dormant cyst that contains an embryo in a suspended state of metabolism (known as **diapause**). The cysts are very hardy and may remain viable for many years if kept dry. Water-temperature and salinity changes in Great Salt Lake occur in about February and cause the cysts to rehydrate and open to release the first growth stage, known as a nauplius larva.

Depending on the water temperature, the larvae remain in this stage for about 12 hours, subsisting on yolk reserves before molting to the second nauplius stage, which feeds on small algal cells and detritus (organic debris formed by the decomposition of plants or animals) using hair-like structures on the antennae, known as "setae." The nauplii molt about 15 times before reaching adult size of about 10 millimeters in length. Adult male shrimp are easily identified by the large pair of "graspers" on the head end of the animal. These are modified antennae and are used to hold on to the female during mating. The population of brine shrimp in Great Salt Lake includes both males and females and reproduces sexually.

Although the cysts are very small (about 200 micrometers in diameter), at times they become so numerous that they form large red-brown streaks on the surface of the lake.

Cleaning the aquarium

Even an aquarium that is well balanced with fish, plants, and light requires changing some of the water. The easiest way to do this is by using a siphon hose, vacuuming any wastes off the bottom at the same time. When adding new water:

1. Remove any chlorine, either by letting the water stand for 24 hours before adding to the aquarium or by using a dechlorinator. Most cities put enough chlorine into the drinking water to kill fish. In addition, some cities have ammonia in their tap water.
2. Add water that is the same temperature as the water in the aquarium. If the water is colder, add it slowly, allowing the fish to become accustomed to it.

If the fish are not overfed, the plants and the filter should take care of the aquarium wastes. If the water turns yellowish or green, changing one-quarter to one-half of the water will usually fix the problem. A well-balanced aquarium should never need to be totally emptied.

A certain amount of algae will grow on the aquarium glass. Algae should be cleaned off the insides of the front and sides of the aquarium, but it can be left on the back of the tank. Many commercial scrapers are specifically made for this, but a plastic scrub pad will work. A scrub pad used around the house to clean dishes or anything else should not be used.

For additional algae removal, an algae-eating catfish and/or other scavengers can be added to the aquarium to help.

To clean the aquarium gravel, a special piece of aquarium equipment called a gravel vacuum can be used. It is a rigid, plastic tube, generally about 2 in in diameter, which attaches to one end of a siphon tube to allow cleaning of the debris from the aquarium gravel. This helps maintain the water quality and appearance of the aquarium.

The filter also requires some maintenance. Outside power filters are convenient to clean. Most are sold with cartridges that just slip out for cleaning or replacement. To clean the cartridge, rinse it gently under a stream of cool water. The bacteria that help to break down the wastes in the aquarium live in this cartridge. A hot stream of water will destroy these bacteria. If the cartridge is too dirty to clean, replacement is necessary.

USING A BEGINNER AQUARIUM

Operating an aquarium can teach many of the principles of aquaculture (Figure 14-4). The following project is an example, with the goal to produce some live-bearing fish and to keep some records similar to those of a commercial fish farmer. Live-bearers are extremely domesticated and do well in an aquarium. Because the young come are born ready to swim and eat, they are easier to raise for the beginner. With enough cover for the young to hide in, only one tank is needed for both parents and young.

FIGURE 14-4 A completely set up aquarium.

The male of the live-bearers has a specially shaped fin on his underside that he uses to fertilize the eggs inside the female. This fin is one way to tell the males from the females.

The male chases the females around the tank. Then, getting next to them, he will bend this special fin around so that he can mate. After this, the eggs will start developing inside the female, and it will take another few weeks before she has her young. As the young mature, a dark spot develops at the bottom of the female's belly. In some fish, even the eyes of the young can be seen as they grow.

As soon as the young hatch, they can swim and eat small pieces of food. Flake food crumbled into a powder makes a good feed for the young. At each feeding, the adults should be fed first with regular flake food. With good plant cover for the young to hide in and regular feedings, many should survive. The parents may eat some of the young, and this is quite natural.

Broodfish in a properly maintained 10-gal aquarium will start filling the aquarium with young. At this point the operator of this sample aquarium can:

1. Start selling some of the young or give them to a friend.
2. Set up another aquarium to raise the offspring.

In an aquaculture operation, it is normal to have at least two separate tanks, one for the breeders and one for raising the young.

To get a good idea of the problems and solutions to breeding fish, a project like this needs to be continued for at least six months. Whenever anything is done with the aquarium, the operator needs to record it by writing down in a log or using a computer log to track the day-to-day management decisions, routine maintenance, and observations. Figure 14-5 is a log that could be used for the duration of a project.

FIGURE 14-5 A sample log for tracking an aquarium project.

Beginning date ____				Ending date ____			
Type of fish used ____							
Number of males ____				Number of females ____			
Size of aquarium ____ gallons							

Production data			
Date	No. young	Deaths	Feeding notes

Costs							
Aqua	Fish	Plants	Filter	Light	Food	Stand	Other

WATER QUALITY

Establishing an aquarium also provides an excellent opportunity to monitor water quality using the tests to monitor pH, dissolved oxygen (DO), carbon dioxide (CO_2), Ammonia (NH_3), Nitrite(NO_2), and hardness. Electronic or chemical methods (test kits or test strips) can be used to measure these water quality parameters. A format like that shown in Figure 14-6 can be used to record the measurement of these parameters in an aquarium over a period of time. Of course, maintaining the correct water temperature for the aquarium fish is very important, and this can easily be monitored with a thermometer. Chapter 12 provides more information on water quality measures.

FIGURE 14-6 Sample log for tracking water quality in an aquarium.

Date	pH	Dissolved oxygen	Carbon dioxide	Ammonia	Hardness

AQUARIUM CHECKLIST

To ensure the safe and successful operation of an aquarium, operators should build the following actions into a routine:

- Check that all pumps and filters are operating at least once a day.
- Check that the temperature is correct in the tank at least once a day.
- Check that all lights are working at least once a day.
- Check all electrical connections and cords at least once a week.
- Replace any carbon in your filter regularly. In an under gravel filter system, replace the carbon at least once a week, or remove it for better filtration. In other filter systems, the carbon should be changed once every three to five weeks.
- Clean other filter media regularly, or replace as necessary. For power filters and corner filters, the media and water flow should be checked every couple of weeks. For other external filters, check the water flow weekly, but you can probably check the media monthly. Of course, if the filter manufacturer recommends checking more often, follow their instructions.
- Clean the impeller assembly on any water pump at least once a year, and make sure to oil external water pumps as necessary based on the manufacturer's instructions.
- Check the tank for leaks at least once a month. though rare in modern aquariums, a leak could be very costly if not caught early.
- Check the stand regularly for any signs of cracks or other damage.

SUMMARY

An aquarium functions like a small pond, and it can teach some principles of aquaculture. The same set of rules and laws that control a big fish farm can be seen in a small aquarium. Aquarium supplies for them can be found almost anywhere in the United States. Setting up an aquarium involves selecting the fish and understanding their biology, in addition to selecting plants and other equipment and supplies. Once operational, a successful aquarium requires careful monitoring and observation in order to manage it properly.

STUDY/REVIEW

Success in any career requires knowledge. Test your knowledge of this chapter by answering these questions or solving these problems.

True or False
1. Bacteria in an aquarium is not good for the fish.
2. As a rule of thumb, 1 in of healthy mature fish need about 6 gal of water.
3. Some fish are live-bearers, whose young are born already swimming and eating.
4. Sagittaria is a group of grass-like plants native to the Americas that grow underwater.
5. Overfeeding aquarium fish is a common problem.

Short Answer
1. Identify types of fish by their common name that can be used in a freshwater aquarium.
2. Name the major groupings of aquarium fish based on their reproductive habits.
3. Identify three general signs of diseased fish.
4. List six supplies necessary, besides the tank fish, water, and plants, to create an aquarium.
5. What two means can be used to eliminate the chlorine from water before it is used in an aquarium?

Essay
1. Explain the role of light, nutrients and plants in an aquarium.
2. Describe how to set up a beginner's aquarium.
3. Discuss the management and maintenance of an aquarium.
4. Explain "feeding the tank." What in the tank is being fed?
5. Describe the parameters that measure water quality.

KNOWLEDGE APPLIED

1. Keep records on an aquarium for three months, and then use these records to write a report on the successes and failures of the project (see Figure 14-6). Include the following items: total expenses, types and amounts of food, list problems and solutions, sources of information, and any income from fish sales.
2. Collect data on the water quality of an aquarium (see Figure 14-7). On graph paper or with a computer spreadsheet, plot the pH, dissolved oxygen, carbon dioxide, ammonia, and hardness by date. Note any trends or patterns.
3. Hatch some brine shrimp and observe the nauplii (free-swimming first stage larvae) under a dissecting microscope or with a magnifying glass. Brine shrimp eggs can be obtained at most aquarium supply stores and the instructions for hatching are on the container. Brine shrimp nauplii are used to feed the fry of many species.

4. Grow aquatic plants in a small aquarium and track the dissolved oxygen (DO) in the water for two months before adding a few goldfish. Continue tracking the DO for another month. Plot your measurements on a graph.
5. Go online or go to an aquarium/pet shop and find 10 different aquariums for sale. Describe their sizes, features, and prices in a report.

LEARNING /TEACHING AIDS

Books

Boruchowitz, D. (2009). *The simple guide to freshwater aquariums*. Neptune City, NJ: TFH Publications, Inc.

Harper, D. (2006). *Aquarium fish*. New York, NY: HarperCollins Publishers.

Hargreaves, V. (2007). *The complete book of the freshwater aquarium: A comprehensive reference guide to more than 600 freshwater fish and plants*. San Diego, CA: Thunder Bay Press, Inc.

Hargrove, M. and M. Hargrove. (2006). *Freshwater aquariums for dummies*. Indianapolis, IN: Wiley Publishing, Inc.

Innes, W. T. (1994). *Exotic aquarium fish*. Surrey, England: T.F.H. Publications.

Jennings, G. (2006). *500 Freshwater aquarium fish: A visual reference to the most popular species*. Buffalo, NY: Firefly Books, Inc.

Ward, A. (2007). *Questions and answers on freshwater aquarium fishes*. Neptune City, NJ: TFH Publications, Inc.

Internet

Internet sites represent a vast resource of information. The URLs (uniform resource locator) for the World Wide Web sites can change. Using a search engine such as Google find more information by searching for these words or phrases: brine shrimp, freshwater aquarium, aquarium, aquarium plants, managing an aquarium. Or enter the names of any of the fish that could be used in an aquarium.

For some specific Internet sites refer to Appendix Table A-14.

A recirculating aquaculture system is an enclosed system where the only water replacement is the water lost to evaporation and cleaning. Recirculating systems in aquaculture create a great deal of interest in the aquaculture community in the United States and other parts of the world.

CHAPTER 15

Recirculating Systems

Given enough of the right resources and management skills, most species of fish grown in ponds, floating net pens, or raceways could be reared in commercial scale recirculating systems.

This chapter provides the concepts of recirculating aquaculture systems and guidelines on the management of a system.

OBJECTIVES

After completing this chapter, the student should be able to:

- List four advantages and four disadvantages of recirculating aquaculture systems
- Identify the four categories of waste solids
- Describe the foam fraction and its removal
- Describe the purpose of a biofilter and how to tell when it is working in a system
- Write the general chemical reactions for nitrification
- Describe the importance of pH in a recirculating aquaculture system
- Compare aeration to oxygenation
- List fish that can be grown in a recirculating aquaculture system

- Identify water quality parameters to monitor in a recirculating aquaculture system
- Explain the relationship of stress to disease control in fish
- Connect possible management changes to a recirculating aquaculture system, based on observations of the fish, water quality or feeding

Understanding of this chapter will be enhanced if the following terms are known. Many are defined in the text and others are defined in the glossary.

KEY TERMS

Ammonia-nitrogen	Nitrite-nitrogen
Aeration	Oxygenation
Biofilter	Settleable solids
Denitrification	Substrate
Foam fractionator	Suspended solids
Nitrification	

BACKGROUND

Recirculation systems have generally been expensive to build and operate—increasing the cost of producing fish. Advantages in using recirculating aquaculture systems (RAS) include:

1. Reduced water requirements
2. Low land requirements
3. Control of water temperature
4. Control of water quality
5. Independence from the adverse weather conditions
6. Year round production
7. Ability to use existing buildings
8. High yields per gallon of water
9. Improved feed conversion
10. Reduced reproduction

Recirculating systems also have some disadvantages:

1. High initial investment (for commercial size)
2. Complexity
3. Chronic sub-lethal effects of ammonia and carbon dioxide
4. Lack of successes for lenders (for commercial size)
5. Inefficiencies in filtration

Still, RAS units offer a great value as an educational lab. A potential aquaculturist has the ability to measure and control most of the variables that make up the environment of the recirculation system. This makes the recirculation system a good tool in teaching knowledge and skills for aquaculture (Figure 15-1).

Recirculation systems are ideal for learning aquaculture. Novices can learn to control and maintain the system to provide a favorable environment for the culture of fish or other aquaculture plants and animals. In the process of doing this, individuals gain competencies in agriculture, biology, physics, chemistry, mathematics, and agriculture. A recirculating used in the classroom should have education as its first purpose and not commercial production.

SYSTEM DESIGN

In commercial production, RAS is most often used where sufficient water is not available to wash fish wastes out of the production tank. By recirculating water through a water treatment system that removes ammonia and

FIGURE 15-1 A recirculating aquaculture system for classroom hands-on teaching. (Note how light from the outside is controlled.)

other waste products, the same effect as a flow-through configuration is achieved. A key to successful recirculation production systems is the use of cost-effective water treatment systems.

All RAS use processes to remove solid wastes, oxidize ammonia and **nitrite-nitrogen**, and aerate and/or oxygenate the water (see Figure 15-2). The basic system components include:

Biological filter
Solids filter
Tanks
Aeration
Buffering systems
Disinfection
Heaters/chillers
Lighting

A brief description of the individual water treatment processes follows.

Waste Solids

The major components of feeds used in aquaculture production consist of protein, carbohydrates, fat, ash, and water. The portion of feed not used by the fish is excreted as an organic waste (fecal solids). Bacteria break down these fecal solids, along with uneaten feed, in the system, consuming oxygen and generating **ammonia-nitrogen**. To minimize their impact on water quality, waste solids need to be removed from the system as quickly as possible. Waste solids can be classified into four categories:

1. Settleable
2. Suspended
3. Floatable
4. Dissolved solids

FIGURE 15-2 Required unit processes and typical components used in recirculating aquaculture production systems.

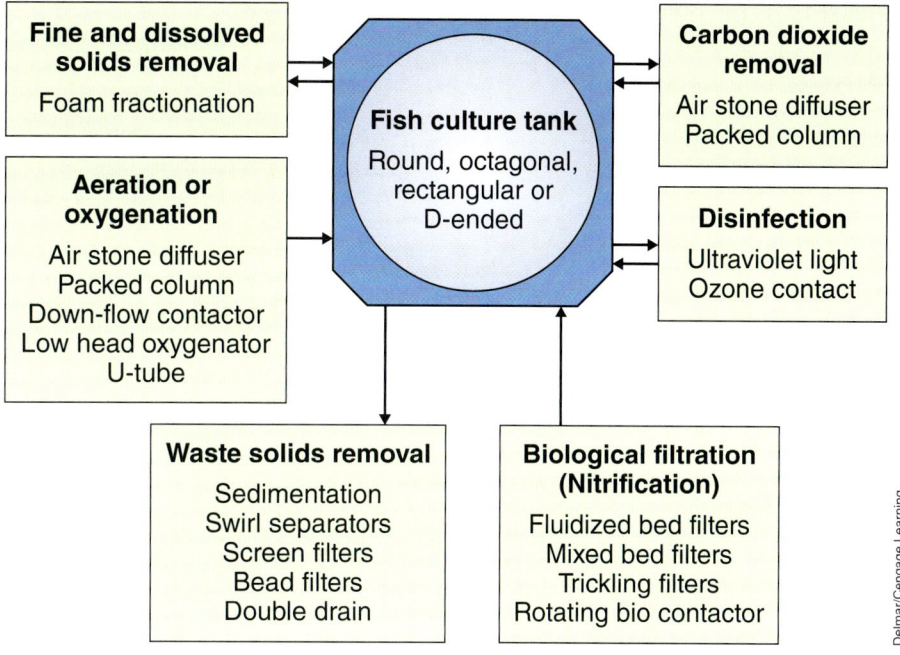

In recirculation systems, the first two are of primary concern, while the other two can become problems in systems with very little water exchange.

Settleable Solids

Settleable solids are generally the easiest of the four categories of waste solids to deal with. They should be removed from the water in the tank as rapidly as possible. Settleable solids settle out of water within one hour under still conditions. Settleable solids can either be allowed to settle within round culture tanks (where they accumulate on the bottom, in the center), or they can be kept in suspension with continuous agitation and removed with a properly designed sedimentation tank (clarifier) or filter. The sedimentation process can be enhanced through the addition of steeply inclined tubes (tube settlers) within the sedimentation tank to reduce flow turbulence and promote uniform flow distribution.

Suspended Solids

From a fish producer's point of view, the difference between suspended solids and settleable solids is a practical one. Suspended solids will not settle out of the water column under still conditions within one hour and would not be expected to be removed by conventional settling. If not removed, suspended solids can significantly limit the amount of fish that can be grown in the system and can interfere with and irritate the gills of fish.

The most popular treatment method for removing suspended solids generally involves some form of mechanical filtration. The two types of mechanical filtration most commonly used are screen filtration (Figure 15-3) or granular media filtration (sand or pelleted media).

FIGURE 15-3 Examples of how three screen filters work to capture and remove suspended solids.

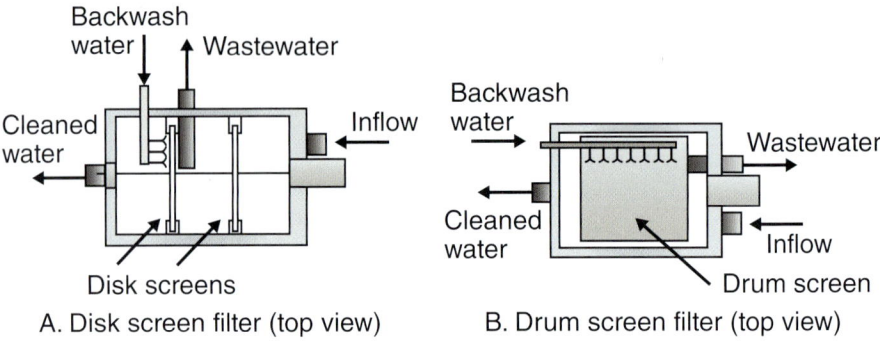
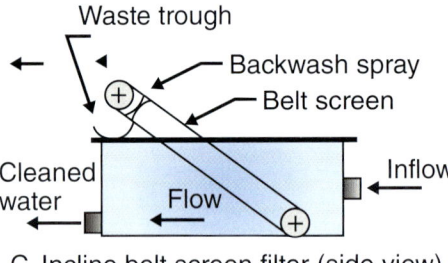

Fine suspended solids (less than 30 microns) contribute to more than 50 percent of the total suspended solids in a recirculation system. These solids increase the oxygen demand of the system and have been shown to cause gill irritation and damage in finfish. Additionally, dissolved organic solids (proteins) can contribute significantly to the oxygen demand of the total system.

Foam in the culture tank provides an opportunity to remove some proteins and very fine particulate that otherwise will remain in the water in spite of efforts to filter or settle it. Foam is the result of fine particles and proteins sticking to the outside of air bubbles and remaining above the surface, due to the water's surface tension. Foaming occurs when air is driven into the water, such as when water is being sprayed into the tank, near paddle-type aerators, or, most commonly, above air stones. A simple method of removing foam is to use a wet-dry vacuum cleaner when large amounts occur. A better method is to use a **foam fractionator** to constantly remove foam when it occurs. This dramatically improves water quality.

Fine and dissolved solids

Fine and dissolved solids cannot be easily removed by sedimentation or mechanical filtration technology. Foam fractionation (also referred to as protein skimming) is used to remove these solids in RAS. Foam fractionation, as used in aquaculture, introduces air bubbles at the bottom of a closed column of water that creates foam at the air/water interface. As the bubbles rise through the water column, fine suspended solid particles attach to the bubbles' surface, creating the foam at the top of the water column. The foam buildup is then channeled out of the fractionation unit

FIGURE 15-4 Diagram of how foam fractionation works. Particles attach to bubbles that rise through tube.

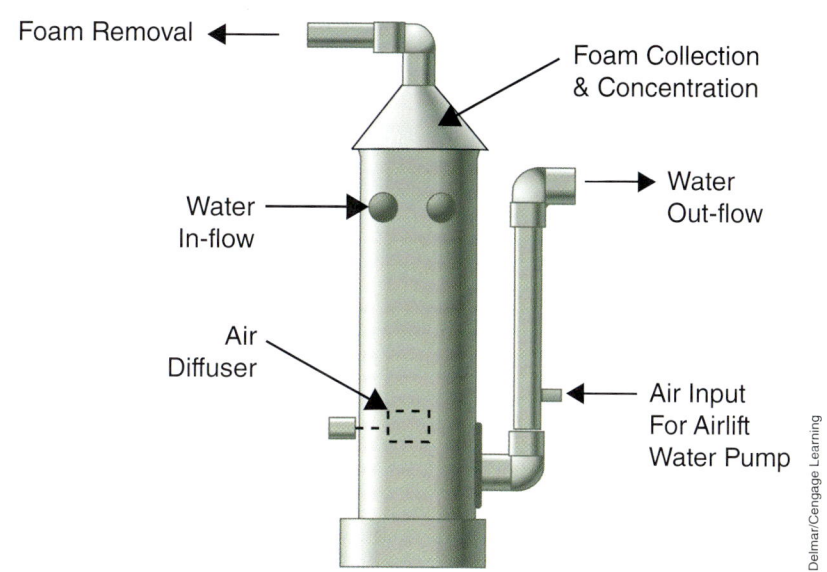

to a waste collection tank (Figure 15-4). The efficiency of a foam fractionation system is dependent on the properties of the water in the system (salt concentration, temperature, pH, and so on), but the system can significantly reduce water turbidity and oxygen demand in the culture tank.

Nitrogen

Total ammonia-nitrogen (TAN) consists of two chemical compounds:
- Unionized ammonia (NH_3)
- Ionized ammonia (NH_4^+)

TAN is the by-product of protein metabolism. It is excreted from the gills of fish as they assimilate feed and is produced when bacteria decompose organic waste solids within the aquaculture system. The un-ionized form of ammonia-nitrogen is extremely toxic to fish. The fraction of TAN in the un-ionized form is dependent upon the water's pH and temperature. At a pH of 7.0, most of the TAN is in the ionized form, whereas at a pH of 8.0 the majority is in the un-ionized form (Figure 15-5). Reduction in growth rates may be the most important sublethal effect. In general, the concentration of un-ionized ammonia-nitrogen in tanks should not exceed 0.05 mg/l.

Nitrification oxidizes ammonia and nitrite to nitrate. Nitrification depends on two bacteria: nitrosomonas and nitrobacter. These bacteria are the basis for biological filtration based on the following chemical reactions:

$$NH_3 + O_2 \xrightarrow{\text{nitrosomonas}} NO_2^- + O_2 \xrightarrow{\text{nitrobacter}} NO_3$$

These bacteria grow on the surface of the **biofilter substrate**, and, to some extent, on all production system components, including pipes, valves, tank walls, and so on. Although nitrite-nitrogen is not as toxic as

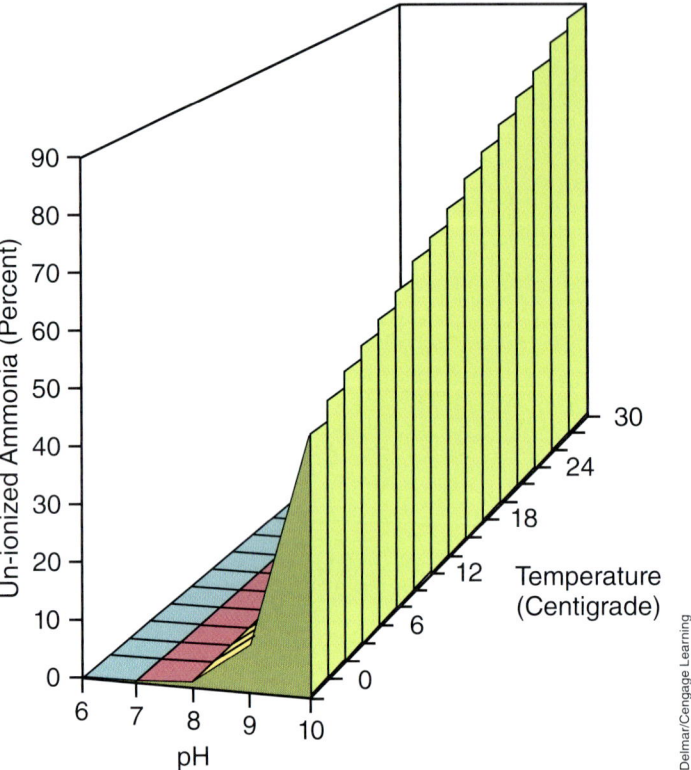

FIGURE 15-5 The relationship between temperature and pH on unionized ammonia. As temperature and pH increase so does unionized ammonia.

ammonia-nitrogen, it is harmful to aquatic species and must be removed from the system. Concentrations of nitrite-nitrogen should not exceed 0.5 mg/l for long periods of time. Nitrates are not generally of great concern to the aquaculturist. Some aquatic species can tolerate extremely high (greater than 100 mg/l) concentrations of nitrate-nitrogen in production systems.

Nitrate-nitrogen concentrations do not generally reach such high levels in recirculation systems, Nitrate-nitrogen is either flushed from a system during system maintenance operations (such as settled solids removal or filter backwashing) or **denitrification** occurs within a treatment system component, such as a settling tank. Denitrification is mainly due to the metabolism of nitrate-nitrogen by anaerobic bacteria producing nitrogen gas that is released to the atmosphere during **aeration** processes.

Nitrogen Control

Control of the concentration of un-ionized ammonia nitrogen (NH_3) in the culture tank is the primary objective of recirculation treatment system design. Ammonia-nitrogen must be removed from the culture tank at a rate equal to the rate of production to maintain a safe concentration. Technologies available for removing ammonia-nitrogen from the water include: air stripping, ion exchange, and biological filtration. Biological

FIGURE 15-6 Overview of a rotating biological contactor.

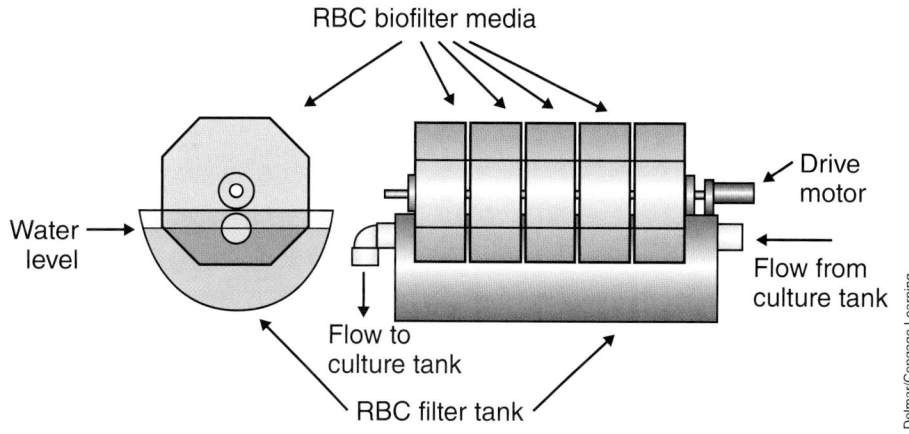

filtration is the most widely used. In biological filtration (biofiltration), a substrate with a large surface area is provided for nitrifying bacteria's attachment and growth. Gravel, sand, plastic beads, plastic rings, and plastic plates are commonly used biofiltration substrates. The substrate's configuration and its contact with wastewater define the water treatment characteristics of the biological filtration unit. The most common configurations for biological filters include rotating biological contactors (RBC), fixed film reactors, expandable media filters, and mixed bed reactors. In a recirculating system with a mature biofilter, nitrite-nitrogen concentrations should not exceed 10 mg/l for long periods of time and in most cases should remain below 1 mg/l.

Rotating biological contactors (RBC) have been used in the treatment of domestic wastewater for decades, and are now widely used as nitrifying filters in aquaculture applications (Figure 15-6). RBC technology is based on the rotation of a biofilter medium attached to a shaft partially submerged in water. Approximately 40 percent of the substrate is submerged in the recycle water. Nitrifying bacteria grow on the medium and rotate with the RBC, alternately contacting the nitrogen-rich water and the air. As the RBC rotates, it exchanges carbon dioxide (generated by the bacteria and fish) for oxygen from the air. The advantages of RBC technology are simplicity of operation, the ability to remove carbon dioxide and add dissolved oxygen, and a self-cleaning capacity. Major disadvantages are high capital cost and mechanical instability.

pH and Alkalinity

Fish metabolism and bacterial nitrification result in the formation of acids that lessen the buffering capacity of water and lower the pH. Most fish can tolerate a pH range of 5–10, however, a range of 6.5–8.5 is preferred for most aquaculture species. To replace lost alkalinity and sustain the buffering capacity of water, bicarbonate (HCO_3^{2-}), in the form of limestone, bicarbonate of soda, or other common sources is added. Often, biofilter media (oyster shell) or some other component of the system (concrete

TABLE 15-1 RECOMMENDED WATER QUALITY REQUIREMENTS OF RECIRCULATING SYSTEMS

Component	Recommended value or range
Temperature	Optimum range for species cultured: less than 5° F as a rapid change
Dissolved oxygen (DO)	60 percent or more of saturation, usually 5 ppm or more for warmwater fish and greater than 2 ppm in biofilter effluent
Carbon dioxide (CO_2)	< 20 ppm
pH	7.0 to 8.0
Total alkalinity	50 to 100 ppm or more as $CaCO_3$
Total hardness	50 to 100 ppm or more as $CaCO_3$
Un-ionized ammonia-N	< 0.05 ppm
Nitrite-N	< 0.5 ppm
Salt	0.02 to 0.2 %

tanks) serve as a source of carbonates. Depending on the species cultured, frequent monitoring of water hardness, alkalinity, and pH may be required. Table 15-1 lists suggested water quality parameters.

Temperature Control

System water temperature will depend upon the requirements of the species being cultured. For most warmwater species, a temperature range of 75° to 80°F is ideal. If the air temperature of the room in which the system is housed and the water source is within this range, then no additional heat source is required. However, if the room temperature is below the desired operational temperature, some type of controllable heater will be needed.

When culturing coldwater species, a water temperature of 70°F or less is desirable. Unless the room and the water can be maintained at 65°F or less, a chiller will be required. Water chillers are generally expensive and will significantly increase a system's cost.

Aeration

For best fish growth, recirculating systems should maintain adequate dissolved oxygen (DO) concentrations of at least 6 mg/L and keep carbon dioxide (CO_2) concentrations at less than 25 mg/L. Pumps and blowers move water through the system and run some of the filters. Agitators aerate (add O_2) and degas (remove CO_2) the water. **Oxygenation** of the water occurs through some type of diffuser and can be either non-pressurized or pressurized. The term aeration generally refers to the dissolution of oxygen from the atmosphere into water. The transfer of pure oxygen gas to water is referred to as "oxygenation."

Adding oxygen to a recirculating system by aerating only the water flowing into the culture tank will not usually supply an adequate amount of oxygen for fish production. The amount of oxygen that can be carried to the fish in this way is limited by the flow rate and the generally low concentration of oxygen in water. Therefore, most aeration in recirculating systems occurs in the culture tank. The most efficient aeration devices are those that move water into contact with the atmosphere (paddlewheels, propeller-aspirators, vertical-lift pumps). However, these methods usually create too much turbulence within a culture tank to be useful. The most common way to aerate in a recirculating tank system is known as "diffused aeration." Diffused aeration systems provide low pressure air from some type of blower to some form of diffuser near or on the bottom of a culture tank. These diffusers produce small air bubbles that rise through the water column and transfer oxygen from the bubble to the water.

The density of fish production with aeration alone is usually limited to 0.25 to 0.33 pounds of fish/gal of culture-tank volume. In greenhouse systems where algal blooms are common, oxygen is generated during the daylight hours, and culture densities of up to 0.50 pounds of fish/gal. can be achieved.

Oxygenation

Pure oxygen is used in recirculating systems when the intensity of production causes the rate of oxygen consumption to exceed the maximum feasible rate of oxygen transfer through aeration. Sources of oxygen gas include compressed oxygen cylinders, liquid oxygen (often referred to as LOX), and on-site oxygen generators. In most applications, the choice is between bulk liquid oxygen and an oxygen generator. The selection of the oxygen source will be a function of the cost of bulk liquid oxygen and the reliability of the electrical service needed for generating oxygen on-site. Adding gaseous oxygen directly into the culture tank through diffusers is not the most efficient way to add pure oxygen gas to water. At best, the efficiency of such systems is less than 40 percent. A number of specialized components have been developed for use in aquaculture applications including: down-flow bubble contactor, U-tube diffusers, low head oxygenation system, and pressurized packed columns.

Due to the intensive nature of recirculation systems, as well as the number of hours that the system will be unattended on evenings and weekends, use of a simple alarm system is strongly recommended. Such a system monitors the air delivery line pressure, water level in the fish culture tanks, water temperature in the overall system, and flow status in the water delivery system. In the case of low air pressure, low water level, high or low water temperature, or no recycled water flow, the telephone dialer will dial and deliver an alarm message to predetermined telephone numbers.

Disinfection

Diseases can spread quickly because of the density of fish in recirculating systems. Some chemicals used to treat diseases have a devastating effect on the nitrifying bacteria within the biofilter and culture system. Alternatives to traditional chemical or antibiotic treatments include the continuous disinfection of the recycled water with ozone or ultraviolet irradiation.

Start-Up of a New System

Activating a new biofilter (developing a healthy population of nitrifying bacteria capable of removing the ammonia and nitrite produced at normal feeding rates) requires a least one month. During this activation period, the normal stocking and feeding rates should be greatly reduced. Prior to stocking it is advantageous, but not absolutely necessary, to pre-activate the biofilters. Pre-activation is accomplished by seeding the filter(s) with nitrifying bacteria (available commercially) and providing a synthetic growth medium for a period of two weeks. The growth medium contains a source of ammonia nitrogen (10 to 20 mg/l), trace elements and a buffer. The buffer (sodium bicarbonate) should be added to maintain a pH of 7.5. After the activation period the nutrient solution is discarded. Many fish can die during this period of biofilter activation. Managers have a tendency to overfeed, which leads to the generation of more ammonia than the biofilter can initially handle.

Fish

Many species are suitable for recirculation aquaculture, but tilapia are very tough fish and will survive when other species may die. Other types of fish to used while learning how to operate the RAS include: minnows, sunfish, and ornamentals such as gold fish and guppies.

SYSTEM MANAGEMENT

Management of an RAS includes routine maintenance and day-to-day management, operations and observations.

Routine Maintenance

Very little routine maintenance needs to be performed on the fish tanks. The smooth-bottomed, circular fish tanks provide a self-cleaning flow that directs most of the uneaten feed and fish feces to the center drain. Screens on the tank drains should be removed and cleaned as necessary. The solids removal basin will require regular cleaning. Any screens should be removed and cleaned by spraying. New water should be added to fish tanks to replace the water lost during solids removal, foam fractionator discharges and evaporation. Once a week or as needed, the drain lines should be cleaned. All other system components should be maintained according to the manufacturers' specifications and guidelines.

Day-to-Day

Day-to-day management of a recirculation system can be satisfying, confusing, and depressing all at the same time. Management parameters and observations should be recorded on a computer or on paper collecting the parameters and observations as indicated in Figures 15-7 and 15-8.

Recirculating system monitoring

Date	System I.D.	Average Total NH_3	Alkalinity	Chloride	Average nitrite	Temp.	Average DO	CO_2	Observations/comments

Delmar/Cengage Learning

FIGURE 15-7 Suggested parameters for monitoring a recirculation aquaculture system (RAS).

FIGURE 15-8 Suggested form for tracking feed consumption in a RAS.

Feed consumption chart for recirculating system

Date	Tank 1	Tank 2	Tank 3	Tank 4	Tank 5	Observations/comments

Delmar/Cengage Learning

Also here are some do's and don'ts that help with the day-to-day management of RAS and help the beginner become comfortable with the system:

- Do monitor your water quality daily with several oxygen tests both before and ½ to 1 hour after feeding.
- Do keep records of feed and growth of fish.
- Do keep records of water quality to see trends.
- Do keep a small quantity of floating feed to see if your fish are active and feeding.
- Do keep records of mortalities and fish removed from the system.
- Do make increases in feed amounts gradually.
- Do change feed sizes as your fish grow.
- Do allow makeup water that is chlorinated to stand for several days with an air stone in it before using in the system.
- Do be alert to changes in appetite or general behavior of the fish.
- Do have makeup water standing by for emergencies.
- Do handle fish gently and as little as possible.
- Do use some type of netting or screen over the tanks to prevent fish from jumping out of the tanks.
- Do make feed charts for others to follow.
- Do remember that decreases in feed will upset the biofilter as much as increases.
- Do remember that this is a learning process, and failures sometimes result in more gain than successes.
- Do shield your tanks and biofilter from excessive light because of algae and light will inhibit bacterial metabolism.
- Do have a plan of action for power failures.
- Do generate as much help and interest from others as possible.
- Don't get excited if a fish or two die.
- Don't get in a hurry to stock your fish before the biofilter is active.
- Don't expect clear water in your system; good quality water in a recirculation system will eventually look like good onion soup, without the cheese and croutons.
- Don't get upset if your fish don't grow perfectly evenly.
- Don't make rapid changes in any water quality parameter except oxygen.
- Don't feed any moldy feed.

Water Quality Monitoring

Water quality monitoring begins with observing the fish because their actions will often reveal many problems. Write down observations of day-to-day fish behavior along with daily recording of the test data on the system (See Figures 15-7 and 15-8). A systematic monitoring of both the water quality and fish behavior will pay dividends at a later time.

The monitoring equipment used is a matter of preference and budget. A kit that will allow you to test oxygen, nitrite, carbon dioxide, alkalinity, ammonia, and chlorides is essential. Oxygen requires several tests per day, so an oxygen meter is a good investment. An Imhoff cone is useful to determine the solids in a system and a good thermometer.

When correcting a parameter, managers do so gradually, as a rapid change, if made too quickly, may be more harmful in the long run. The key is to make the changes gradually, giving both the biofilter and the cultured species a chance to adapt to the new conditions. The possible exception to this rule is low dissolved oxygen (DO). If this is very low, it should be raised as quickly as possible.

Sunlight

Sunlight is not a water quality parameter, but it can affect the water quality. If the system is located near a window or in a greenhouse, some provision must be made to reduce the light because it will cause excessive algae growth, which in turn may cause large variations in pH, DO, CO_2, nitrate toxicity, ammonia toxicity, and biofilter fouling. Excess algae also may lead to off flavors in the fish at the time of harvest.

Stress and Disease Control

The key to fish management is stress management. Fish can be stressed by changes in temperature and water quality, by handling (including seining and hauling), by nutritional deficiencies, and by exposure to parasites and diseases. Stress increases the susceptibility of fish to disease, which can lead to catastrophic fish losses if not detected and treated quickly. To reduce stress, fish must be handled gently, kept under proper water quality conditions, and protected from exposure to poor water quality and diseases. Even sound and light can stress fish. Unexpected sounds or sudden flashes of light often trigger an escape response in fish. In a tank, this escape response may send fish into the side of the tank, causing injury. Fish are generally sensitive to light exposure, particularly if it is sudden or intense. For this reason, many recirculating systems have minimal lighting around the fish tanks.

Diseases

More than 100 known fish diseases do not seem to discriminate between species. Other diseases are very host specific. Organisms known to cause diseases and/or parasitize fish include viruses, bacteria, fungi, protozoa, crustaceans, flatworms, roundworms, and segmented worms. There are also non-infectious diseases. (Refer to Chapter 11, Health of Aquatic Animals.) Tables 15-2 and 15-3 provide some possible management options based on the observation of the fish, water quality and feed. The secret to managing the RAS is to be alert to changes in the system and respond quickly to those changes with correcting actions.

TABLE 15-2 OBSERVATION OF FISH AND POSSIBLE OPTIONS IN MANAGING A RECIRCULATING TANK SYSTEM[1]

Observation	Possible Cause	Possible Management
Excitable/ darting/ erratic swimming	Excess or intense sound/ lights Parasite High ammonia	Reduces sound level/ pad sides of tank/ reduce light intensity *Examine fish with symptoms Check ammonia concentration
Flashing/ whirling	Parasite	Examine fish with symptoms
Discolorations/ sores	Parasite/ bacteria	Examine fish with symptoms
Bloated/ eyes bulging out	Virus or bacteria Gas bubble disease	Examine fish with symptoms Check for super saturation and examine fish with symptoms
Lying at surface/ not swimming off	Parasite Low oxygen High ammonia or nitrite Bad feed High carbon dioxide	Examine fish with symptoms Check dissolved oxygen in tank Check ammonia and nitrite concentrations Check feed for discoloration/ clumping and check blood of fish
Crowding around water inflow/ aerators	Low oxygen Parasite/ disease High ammonia or nitrite Bad feed	Check dissolved oxygen in tank Examine fish with symptoms Check ammonia and nitrite concentrations Check feed for discoloration/ clumping and check blood of fish
Gulping at surface	Low oxygen Parasite/ disease High ammonia or nitrite High carbon dioxide Bad feed	Check dissolved oxygen in tank Examine fish with symptoms Check ammonia and nitrite concentrations Check carbon dioxide level Check feed for discoloration/ clumping and check blood of fish
Reducing feeding	Low oxygen Parasite/ disease High ammonia or nitrite Bad feed	Check dissolved oxygen in tank Examine fish with symptoms Check ammonia and nitrite concentrations Check feed for discoloration/ clumping and check blood of fish
Stopping feeding	Low oxygen Parasite/ disease High ammonia or nitrite	Check dissolved oxygen in tank Examine fish with symptoms Check ammonia and nitrite concentrations
Discolored blood Brown	High nitrite	Examine fish with symptoms; add 5 to 6 ppm chloride for each 1 ppm nitrite; purchase new feed and discard old feed
Clear (no blood)	Vitamin deficiency	Examine fish with symptoms; check for feed for discoloration/ clumping; purchase new feed and discard old feed
Broken back or "s" shaped backbone	Vitamin deficiency	Examine fish with symptom; purchase new feed and discard old feed

*Have fish examined by a qualified fish diagnostician.
[1]Source: SRAC Publication No. 452.

Off-flavor

Off-flavor in recirculating systems is a common and persistent problem. Many times, fish have to be moved into a clean system—one with clear, uncontaminated water—where they can be purged of off-flavor before being marketed.

TABLE 15-3 POSSIBLE MANAGEMENT CHANGES TO A RECIRCULATING SYSTEM BASED ON WATER QUALITY AND FEEDING OBSERVATIONS[1]

Observation	Possible Management Change
Low dissolved oxygen (< 5 ppm)	Increase aeration Stop feeding until corrected Watch for symptoms of parasite/ disease
High carbon dioxide (above 20 ppm)	Add air stripping column Increase aeration Watch for symptoms of new parasite/ disease
Low pH (< 6)	Add alkaline buffers (sodium bicarbonate, etc.) Reduce feeding rate Check ammonia and nitrite concentrations
High ammonia (above 0.05 ppm as unionized)	Exchange system water Reduce feeding rate Check biofilter, pH, alkalinity, hardness, and dissolved oxygen in the biofilter Watch for symptoms of new parasite/ bacteria
High nitrite (above 0.5p ppm)	Exchange system water Reduce feeding rate Add 5 to 6 ppm chloride per 1 ppm nitrite Check biofilter, pH, alkalinity, hardness, and dissolved oxygen in the biofilter Watch for symptoms of new parasite/ disease
Low alkalinity	Add alkaline buffers
Low hardness	Add calcium carbonate or calcium chloride
Discolored/ clumped feed	Purchase new feed and discard old feed Watch for symptoms of new parasite/ disease

[1]Source: SRAC Publication No. 452.

SUMMARY

Given enough of the right resources and management skills, most species of fish grown in ponds, floating net pens, or raceways could be reared in commercial scale recirculating systems. All aquaculture production systems must provide a suitable environment to promote fish growth. Critical environmental parameters include the concentrations of dissolved oxygen, un-ionized ammonia-nitrogen, nitrite-nitrogen, and carbon dioxide in the water of the culture system. Nitrate concentration, pH, and alkalinity levels within the system are also important. To produce fish in a cost effective manner, aquaculture production systems must maintain good water quality during periods of rapid fish growth. The by-products of fish metabolism include carbon dioxide, ammonia-nitrogen, and fecal solids. If uneaten feeds and metabolic byproducts are left within the culture system, they will generate additional carbon dioxide and ammonia-nitrogen, reduce the water's oxygen content, and have a direct detrimental impact on the health of the cultured fish. The operation and management of a recirculating aquaculture system requires the monitoring and controlling of these parameters to produce healthy fish.

STUDY/REVIEW

Success in any career requires knowledge. Test your knowledge of this chapter by answering these questions or solving these problems.

True or False
1. A biofilter removes the settleable solids from a recirculating aquaculture system.
2. Oxygenation is the injection of pure O_2 into the RAS.
3. Recirculating aquaculture systems are designed for raising only tilapia.
4. In a recirculating system, fish can thrive at a pH of 10.
5. Maintaining a recirculating aquaculture system a simple process.

Short Answer
1. List four advantages and four disadvantages of a recirculating aquaculture system.
2. Identify the four categories of waste solids.
3. Describe the foam fraction and its removal.
4. Name the two chemical compounds that contribute to total ammonia-nitrogen (TAN).
5. What is the purpose of the substrate in a biofilter?

Essay
1. What is the purpose of a biofilter, and how can you tell when it is working in the system?
2. Describe the conversion of NH_3 to NO_3.
3. What is foam fractionation?
4. The fish in a recirculating system are observed to be gulping at the surface. What are the possible causes of this and possible management changes to correct the problem?
5. The water pH is < 6. What management changes should be considered for a recirculating system?

KNOWLEDGE APPLIED

1. Using a goldfish in a bowl and either chemical and/or electronic monitoring, track the changes in water quality for a week without changing the water.
2. Collect the data (numbers and observations) and maintain the records (Figures 15-7 and 15-8) on a recirculating aquaculture system (RAS) for one month. Report on the management decisions made based on the data and observations.
3. Using the Internet develop a report on successful, commercial recirculation aquaculture systems. Include the type of aquatic product, annual production in dollar and pounds, location and market.
4. Define "stress" as it relates to fish. What is stress? What changes in physiology can it cause?
5. Develop a report on the nitrogen cycle and explain the connection of the nitrogen cycle to the raising of fish in a recirculating system.

LEARNING /TEACHING AIDS

Books

Costa-Pierce, B.A. Desbonnet, A. Edwards, P. and Baker, D. (2005) *Urban aquaculture*. Oxfordshire, UK: CABI Publishing.

Lekang, O-I. (2007). *Aquaculture engineering*. New York, NY: Wiley-Blackwell.

Timmons, M.B. and Ebeling J.M. (2007). *Recirculating Aquaculture*. Ithaca, NY: Cyuga Aqua Adventures, LLC.

Internet

Internet sites represent a vast resource of information. The URLs (uniform resource locator) for the World Wide Web sites can change. Using a search engine such as Google find more information by searching for these words or phrases: recirculating aquaculture system, aeration, biofilter, ammonia-nitrogen, denitrification, foam fractionator, nitrification, nitrite-nitrogen, oxygenation, settleable solids, and suspended solids.

Specific Web sites:

Advancing Indiana Aquaculture. (ND). Recirculating Aquaculture Systems: http://indianafishfarming.com/

ALEARN. (ND). Recirculating Aquaculture Systems: http://www.aces.edu/dept/fisheries/education/recirculatingaquaculture.php

Aquaculture and Aquaponics Resource Library: http://www.northernaquafarms.com/knowledgelibrary/index.html

For other specific Internet sites, refer to Appendix Tables A-11 and A-14

SUPPLIERS

Recirculating aquaculture systems are sold by a variety of suppliers. These provide the training and management skills to operate larger units or to guide the construction of a larger unit.

Aquatic Eco-Systems, Inc. (mini fish farm): http://www.aquaticeco.com/subcategories/2181/The-Mini-Fish-Farm

CropKing aquaculture unit: http://www.cropking.com/aqua

PondSolutions.com, Recirculating Systems: http://www.pondsolutions.com/recirculating-systems.htm

Also, complete plans are available for constructing inexpensive systems: http://www.aces.edu/dept/fisheries/education/documents/ConstructingSmRASForclassroom-SRAC4501.pdf

Concerns about growing populations, increased food scarcity, and the environment have led researchers, farmers, nonprofit organizations, governments, and industry representatives to work together to help find sustainable solutions to meet the world's growing demand for food, fuel,

CHAPTER 16

Sustainable Aquaculture and Aquaponics

and water. Agricultural practices need to be as sustainable as possible. This means getting more from natural resources with a lighter environmental footprint, and in an economical way. Perhaps some combination of aquaculture production with the production of plants from the water used to produce aquatic animals comes the closest to being a sustainable operation.

OBJECTIVES

After completing this chapter, the student should be able to:

- Discuss the concept of sustainability
- List 12 standards for sustainability
- Identify six areas of scientific research in aquaculture
- List three rules of risk management
- Identify three methods of managing production risks and marketing risks
- Describe the role of aquaculture in providing food in the future
- Identify a possible energy source that could come from aquaculture

- Explain the role aquaculture has on water pollution and the role water pollution plays on aquaculture
- Discuss reasons for developing recirculating systems
- Name three wastes from aquaculture
- Describe integrated pest management
- Discuss how an aquaculture operation can contribute to strong communities
- Explain how aquaponics serves as a model of sustainable food production

KEY TERMS

Anaerobic digestion
Aquaponics
Beneficials
Bio-intensive
Biosecurity
Effluent
Genomics
Market niches
Probiotics
Risk
Risk assessment
Science
Sustainable
Urbanization

ATTEMPTS TO DEFINE "SUSTAINABLE"

Aquaculture is a component of agriculture in general, and as such is part of the discussion of sustainability. Nowadays, every industry seems to promote "sustainability" in one way or another. Simply defined, **sustainable** means a method of harvesting or using a resource so that the resource is not depleted or permanently damaged

Nevertheless, sustainability means different things to different people. For example, some people believe that any food production system that depends on non-renewable resources, such as oil, is not sustainable. Other people argue that this is not an important aspect of sustainable food production because they believe that alternative energy sources will be found as oil becomes scarcer. The U.S. Congress defined sustainable in the 1990 Food, Agriculture, and Trade Act. According to their definition, there are seven key features of sustainable food production systems. Sustainable food production systems:

1. Will meet human needs for food now and far into the future
2. Integrate plant and animal production
3. Rely as much as possible on natural processes and cycles
4. Are designed specifically to fit the biological, social, and economic conditions of specific places (are site specific)
5. Provide a livable income for the farm family
6. Protect natural resources
7. Enhance the quality of life for farmers and for society as a whole

A search of the literature reveals many more ideas of "sustainable." Each idea contains elements of the others.

Standards are used to evaluate agricultural and aquaculture products. Standards describe the ideal. In order to evaluate the sustainability of an operation or make changes in an operation or a system, some standards are needed. These set the ideal for comparison and allow progress to be tracked.

STANDARDS OF SUSTAINABLE AQUACULTURE

Oftentimes, when the term "sustainable" is heard, people just seem to accept it as positive without really thinking about what is involved. To overcome any misunderstanding of sustainable aquaculture, the concept here is presented within the framework of 12 standards.

A sustainable system of aquaculture will meet the following 12 standards at some level:

1. Base direction and change on science
2. Honor market principles
3. Increase profitability and reduces risk
4. Satisfy human need for fiber and safe, nutritious food
5. Conserve and seek energy resources
6. Create and conserve healthy soil
7. Conserve and protect water resources
8. Recycle or manage waste products
9. Select animals and crops appropriate for environment and available resources
10. Manage pests with minimal environmental impact
11. Encourage strong communities
12. Promote social and environmental responsibility

An analysis a sustainable aquaculture focuses on these standards. The level to which each aquaculture operations applies these standards determines their sustainability. Each of the 12 standards will be described and related to aquaculture.

Standard 1: Base Direction and Change on Science

Science represents an organized body of knowledge derived from a practical or experimental approach to problems. Further, science deals only with rational beliefs that can be verified or disproved by observation or experiment. Progress in agriculture and aquaculture relies on scientific research using the scientific method (See Figure 16-1). Through research, production, processing, and storage, all improved.

The agricultural scientist can never be absolutely certain that the experiment has eliminated all of the variables that might influence the results. Experimental design and statistics are tools scientists use to deal with variables. Also, to be valid, experiments must be repeatable.

Why Science?

Science is necessarily objective and needs to be detached from emotion and prejudice. To keep its basic nature and succeed in dealing with contemporary problems, experimental science must remain objective and detached.

Ongoing Research and the Use of Science

To improve aquaculture production and to address sustainability, ongoing scientific research addresses many issues, for example:

- Genetic improvement (conserve, characterize, and use genetic resources, selective breeding for economically important traits,

FIGURE 16-1 The Scientific Method.

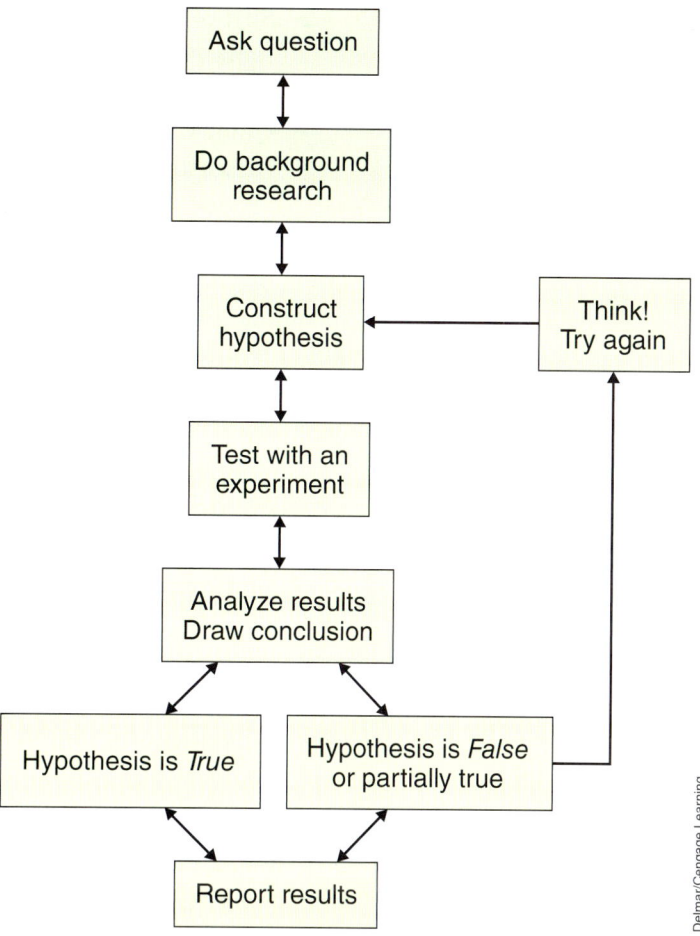

genomic resources, specific breeding aids, bioinformatics and statistical analysis tools)
- Aquatic animal health management (pathogen identification and disease diagnosis, vaccines and medicines, immunology and disease resistance, mechanisms of disease, epidemiology, microbial **genomics**, and aquatic animal health management)
- Reproduction and early development (control of reproduction, control of gender and fertility, gamete and zygote quality, gamete and embryo storage, cryopreservation (freezing), and use, early life stage development and survival)
- Growth, development, and nutrition (regulating feed intake, tissue growth and development, sustainable sources of nutrients, nutrient use and feed evaluation, interaction of gene regulation and nutrition, interactions affecting reproduction, effective **probiotics**, immune system enhancement [nutrients and immunostimulants])
- Aquaculture production systems (**biosecurity**, production intensity, integrated production systems, predator and fowl control, live aquatic animal handling, transport, and inventory, and culture of marine species in low-salinity water)

> Sustainability and environmental compatibility of aquaculture (aquaculture feeds, water use and reuse, effluent management control, social sustainability, and environmental sustainability)
> Quality, safety and variety of aquaculture products for consumers (tissue quality, interaction of genetics and nutrition, predicting product quality or defects, off-flavor delayed harvesting, off-flavor methodology, new uses for byproducts and processing)

Many private and public organizations conduct scientific research to aid the aquaculture industry, including producer conducted research (Figure 16-2). For details on specific research projects of the ARS at the USDA reports can be found on the aquaculture research site: http://www.ars.usda.gov/research/programs/programs.htm?np_code=106

Standard 2: Honor Market Principles

Sustainable aquaculture honors market principles: a market, marketing, a marketing plan, and profits from sales. Nothing happens until someone sells something. Clearly, nothing can be sold if a market for the goods or services does not exist. Marketing is the process of planning and executing the conception, pricing, promotion, and distribution of ideas, goods, and services to create exchanges that satisfy individual and organizational objectives.

The first step in marketing is a plan—a written statement to guide the marketing process. Writing requires considerable thought, time, energy, and information. A new product plan is prepared before a new product is produced. Marketing plans are usually dynamic. They can change during implementation. A good plan increases profits or the chance of profits for the business in sustainable aquaculture. In this book, Chapter 3 is entirely dedicated to the role of marketing in aquaculture.

FIGURE 16-2 Continued scientific research is essential to sustainable aquaculture.

Market Niches

Many of those pursuing sustainable aquaculture develop marketing niches—specialization in a specific, limited market sector. These producers develop marketing niches in order to take advantage of specific microclimates, regional demand, and their own special knowledge or skills. Marketing specialized products or services can be more difficult than marketing traditional products, and it can also be more risky. But market niches can be more profitable.

Standard 3: Increase Profitability and Reduces Risk

Aquaculture, like any agriculture venture, is risky. Often high risks are associated with profitability. Sustainable aquaculture increases profitability and reduces **risk**. The ability to minimize risk is a major factor in a successful sustainable venture. Profits represent the motivating force for going into business. Profits reward individuals their willingness to take the risk of operating a sustainable aquaculture/agriculture enterprise. Without risk, nothing sustainable can develop, and without a chance at creating profitability, no one will take risks (Figure 16-3).

Risk is manageable, whereas uncertainty has to be accepted. Those in sustainable aquaculture learn to manage risk and accept uncertainty. Managing risks requires excellent business skills, as described in Chapter 17.

Briefly, managing risk in aquaculture can include a number of options for any one risk. Often, several solutions are available. Using disease as an example, some producers might decide to implement an exclusionary approach, where the aim is to prevent the introduction of disease to their facility. Other producers may decide that post-infection treatment, where disease symptoms are treated after they appear. While both approaches can be effective, the better option will depend on how the producers respond the three guiding rules of risk management:

1. Don't risk more than can afford to be lost.
2. Don't risk a lot for a little.
3. Understand the likelihood and severity of possible losses.

FIGURE 16-3 Profitable, sustainable aquaculture operations understand and manage risks.

To effectively manage risks, the producer must know something about the likelihood and severity of loss from each possible risk and then decide on how best to manage each risk. This is called **risk assessment**. Options to manage risk in aquaculture include a wide range of techniques and approaches that can generally be grouped into two distinct categories: insurance and non-insurance options. Insurance options can be found through the USDA Risk Management Agency (RMA) on its Web site: http://www.rma.usda.gov/.

Non-insurance risks

Non-insurance risk management options include production risks and marketing risks. Management of production risks can include:
- Changing husbandry practices
- Careful observation of animal behavior
- Building redundancy (backup) into the operation
- Improving feed management
- Employing stringent biosecurity measures
- Minimizing the chance of disease introduction and cross contamination
- Knowing appropriate treatments

Management of marketing risks can include:
- Diversification to include other species and categories of production
- Creation of unique identity (brand)
- Development of a niche market or value-added products

While any number of options may exist to manage each potential risk, each option has associated costs and benefits that must weigh against the risk itself.

Standard 4: Satisfy Human Need for Fiber and Safe, Nutritious Foods

Food and shelter (fiber) are basic needs that humans seek to satisfy. When sufficient fiber and food are not supplied, people rush to blame various issues. Some blame the population; some blame the economy; some blame politics; some blame the environment.

Sustainable agriculture contributes to providing people with sufficient fiber (for protection from the elements) and safe, nutritious food. This contribution can be local, regional, national, or international. While adequate and continuous food and fiber production is essential to sustainability, the transportation and distribution systems are key elements to the success of sustainability.

The United Nations Food and Agriculture Organization (FAO) estimates that more than 40 percent of the seafood consumed worldwide is coming from aquaculture and that production must double to meet the expected demand by 2030. Momentum for marine-based systems is also growing. The future food needs represents a tremendous opportunity and challenge for aquaculture.

Urbanization

Worldwide, people are moving out of the country and into the cities. This is called **urbanization**. A continuous, massive amount of food is required to by the population. As people in the country move to cities worldwide, the number of cities with a population of 10 million will increase to 26. Several of these cities are in North America. Every city this size requires 6000 tons of food shipped in each day, or about 2.2 million tons per year. Many cities are made up of more than a million people, but less than 10 million. These cities also require the shipping of thousands of tons of food each day. Sustainable aquaculture operations will need to contribute to the human need for safe, nutritious foods.

Planning for sustainable agriculture must include ways of ensuring a continuous, healthy, and adequate supply of safe food for all, despite disruption from war or terrorism. This could require looking at local food systems to supply food, and keeping national aquaculture/agriculture alive and healthy so as to not become dependent on food production in other countries where production and trade policies can change with little notice or explanation.

Standard 5: Conserve and Seek Energy Resources

A sustainable system conserves energy resources and seeks new sources, recognizing that the energy resources of coal, petroleum, and natural gas are finite. Conservation can take the forms of using less resources overall, or using an alternative source that is renewable.

Energy conservation by using less energy uses often employs technology to increase the efficiency of some energy-consuming practice associated with agriculture: for example, pumping, transportation, storage, or harvesting. Clearly, another way to conserve is to cut back on any operations that use energy. Conservation by increasing efficiency or reducing operations continues, and organizations such as the U.S. Department of Energy, Energy Efficiency and Renewable Energy (http://www.eere.energy.gov/) provide information and resources to increase efficiency.

Energy Flow and Alternative Energy

Energy flow is enhanced through increased capture of solar energy and strategies to effectively use and store it. People are again thinking of the need for renewable/alternative resources, specifically wind, solar, bioenergy (biogas, butanol and ethanol), hydrogen (fuel cell), and nuclear. Most of these alternative renewable energy sources require more development and research to become renewable energy sources of the future for sustainable aquaculture and agriculture.

Sustainable aquaculture maintains a focus on reducing energy flow and the development and use of renewable and new energy sources. To succeed, this focus will likely need to be maintained by government programs and individual initiative. The combination of new technology, demand, research, and economics will lead to the incorporation of renewable energy sources in agricultural production.

Some of the current research related to aquaculture includes:
- Algal biofuels technology development
- Hydrogen production from algal systems
- Geothermal energy use
- Conservation of energy
- Solar energy development
- Biomass energy use

The future will likely see the development of clean energy technologies that strengthen the economy, protect the environment, and reduce dependence on foreign oil.

Standard 6: Create and Conserve Healthy Soil

A system of sustainable agriculture creates and conserves healthy soil. Erosion includes wind erosion and water erosion caused by rain and poor drainage, and it is accelerated by such human activities as forest removal, agriculture, grazing, construction, and mining. Whenever vegetation is removed and the ground is exposed to rainfall, soil erosion by water and wind may increase. On sloping land, the erosion rate increases. Erosion, widespread throughout the tropics, is a serious environmental and socio-economic problem.

Conservation methods include the general categories of tillage, vegetation, mulching, and cropping pattern. Well planned and managed vegetation cover can effectively control soil movement. Vegetation protects soil against erosion by reducing water movement and building soil structure. Mulching covers the soil with materials that reduce soil moisture evaporation and inhibit weed growth. Mulching also slows rainfall infiltration and protects the soil from direct impact from rain. Changes in a production system to help reduce soil movement include intercropping, alley farming, terracing, or use of grass strips.

Soil is more than ground-up rock fragments. Minuscule creatures are the life force of a healthy soil. These organisms help to purify water, air, detoxify pollutants, cycle crop nutrients, decompose plant residues, promote soil structure, and prevent disease outbreaks. Management of the land directly influences how well these organisms perform their important functions.

The food web does not just pertain to above ground collections of carnivores and herbivores. In fact, above ground herbivores are not responsible for most of the plant matter consumption in the world. In reality, decomposers in the soil do most of the work. Many organisms function in the soil food web, such as arthropods, bacteria (actinomycetes and the photosynthetic cyanobacteria), fungi, protozoa, nematodes, and earthworms. Some chemicals, production and tillage methods are detrimental to the organisms in the soil food web.

Critical information for conservation and soil health planning includes soil depth, soil type, drainage characteristics, slope of the land, chemical use, and tillage practices. The practices selected to control erosion and improve soil health should be based on a combination of principles. To be sustainable, the practices must be within the means of the producer and be perceived of high economic and social value or they will not be implemented or maintained.

A solid understanding of soil conservation and characteristics is essential for pond and raceway construction and management in aquaculture (See Chapter 13).

Standard 7: Conserves and Protects Water Resources

A system of sustainable aquaculture conserves and protects water resources. Freshwater from streams and rivers or from underground wells is the most fundamental of finite resources for sustainability. It has no substitutes, and it is expensive to transport. But freshwater sources are dwindling or becoming contaminated throughout the world. Chronic or acute water shortage is increasingly common in many countries with fast-growing populations, becoming a potential source of conflict (Figure 16-4).

FIGURE 16-4 Sustainable aquaculture conserves and protects the essential resource of water.

Water Use in Aquaculture

Farm ponds only account for about 3% of agricultural water use. Water is critical to commercial aquaculture production. There must be a watershed of sufficient size draining into a pond to provide enough water to fill it. The amount of water produced by a watershed depends on where it is. Climate, soil, and vegetation all influence how much water is available to fill a pond. For example, in the humid eastern states, only two to three acres of watershed would be needed for each acre-foot of pond water. For an average pond depth of 5 feet, this means 10 to 15 acres of watershed per acre of pond surface area are needed. In contrast, in the arid west, often 100 acres of watershed are needed per acre-foot. For levee ponds where there is no runoff water available from a watershed, another water source is needed, such as a well, spring, or reservoir. Usually, one well with a capacity of 2,000 to 3,000 gallons per minute will work to fill four 20-acre ponds. There must be enough water available to fill the pond in 10 days, or else there could be problems with growth of unwanted vegetation or water quality. One of the benefits of farm ponds is that they trap water and make it available for other uses, such as watering livestock or recreation, along with raising fish. However, open ponds lead to increased loss of water to the atmosphere through evaporation.

Conservation in Aquaculture

Water pumping restrictions are enacted in some areas because of the potential impact of aquaculture operations on lowering the underground water table. Farmers will reuse water as it becomes scarce and more expensive. In simple systems of water recycling, the main treatments may be aeration of water, disease control, or mechanical filtration to remove solids. For completely closed systems in which water is only used to initially fill the system and to replace losses due to evaporation, more sophisticated methods are being developed. Methods to improve sedimentation, mechanical or biological filtration, sterilization, oxygenation, aeration, degassing, cooling or heating, and pH control are being refined.

Water-recirculation technologies are continually being refined. Scientist research ways to maximize water quality in culture tanks by removing and concentrating waste. The water is treated for waste removal and recirculated through the culture tanks. High-intensity water reuse systems are proving economically viable for high-value species. Chapter 15 of this book discusses recirculating systems.

Loss of water from ponds through seepage into the ground is another area to consider for conservation. Seepage is like a bucket with lots of little holes in the bottom. Seepage requires constant addition of more water to keep the water level steady. In sandy soils, seepage is much higher (0.51 cm/day) than for ponds built in clay soils (0.05 cm/day).

Water Quality

Aquaculture requires water resources of high quality. Aquaculture itself, moreover, can impact on water quality. Water quality affects aquaculture, and aquaculture can affect water quality.

Aquaculture can contribute to water pollution. Ponds contain fish, fish wastes, excess feeds and fertilizers, as well as bacteria. In the majority of aquaculture ponds, waste water called **effluent** is discharged when fish are harvested. This effluent can be a pollutant if it decreases the water quality in natural waterways.

The water quality of effluent depends on stocking and feeding rates, use of aeration, water flow, other chemicals used to disinfect or control pests, and amount of water discharged. The timing of harvesting can vary between one and three years, depending on the management of the pond. This, too, will affect the water quality in the effluent. In addition, the size, flow, and water quality of the stream, lake, or river that the effluent is discharged into is critical. Rivers and streams have the natural capacity to use biological processes and recover from pollution, depending on time, distance, and the species living in them. Some streams are more able to absorb effluent without suffering damage than others.

Waste treatment from aquaculture ponds includes sedimentation to remove the large amount of particles in the water. Sedimentation alone results in low concentrations of pollutants in a high volume of water. If producers also use biological nutrient removal methods, such as varying the amount of time the effluent is in a settling pond, then solids are removed and biological oxygen demand (BOD) levels can be lowered.

Water Quality Affects Aquaculture Production

Water pollution can have a severe impact on fish production. Dissolved oxygen is probably the most important water quality factor that fish producers must understand. The primary source of dissolved oxygen in ponds is from microscopic plants called "phytoplankton" or "algae blooms." In the presence of sunlight, algae produce oxygen through photosynthesis. However, at night and on very cloudy days, phytoplankton only consume oxygen—as much as three to five times the amount of oxygen that the fish consume. Low oxygen is often the result of an imbalance in the amount of phytoplankton present in the pond. At times, low dissolved oxygen can be predicted before it occurs, but it may also develop suddenly and without warning. If dissolved oxygen drops too low, supplemental aeration will be needed.

Other water quality considerations for aquaculture can include temperature, carbon dioxide, total gas pressure, salinity, hardness, alkalinity, pH, ammonia, nitrate, iron, and hydrogen sulfide (See Table 16-1).

Chapter 12 of this book deals specifically with the water needed for aquaculture.

Standard 8: Recycle and Reduce Waste Products

Sustainable agriculture/aquaculture recycles and reduces wastes. The mindset about agricultural residues or wastes should be that of "a resource out of place," instead of simply "waste." This change in perspective permits the evaluation of waste from a positive standpoint. With appropriate techniques, wastes can be recycled to produce an important source of energy, a natural fertilizer, or a feed. The choice of recycling method will depend on the type of waste available and on the end use for the recycled waste. The mineral cycle is fostered by the cycling and recycling of wastes on-farm. Many examples of the past demonstrate that what was once thought of as a waste product often becomes a valuable by-product. For example, slaughterhouse and processing plant wastes that were once dumped became valuable feeds.

Aquacultural wastes are mainly undigested feed, manure, and the wastes from processing plants. The composition of manure depends upon fish-type digestibility, protein, fiber contents of rations, animal age,

TABLE 16-1 SOURCES OF WATER POLLUTION AND THEIR IMPACT ON FISH AND THE ENVIRONMENT

Type of Water Pollution	Effect on Fish and the Environment
Nutrients from yard waste, sewage, food processing, agriculture, livestock, and industry	Fish kills, low oxygen due to algae blooms, and reduced diversity of animals and plants
Heat from cities, power plants, industrial discharges, and the sun	Coldwater fish cannot survive, more disease occurs because of heat stress and lower oxygen levels
Toxic chemicals from landfills, agricultural and urban runoff, and industrial discharges	Reduced growth and survival of fish eggs; fish diseases and death of carnivores from chemicals being passed through the food chain
Disease organisms from raw or partially treated sewage and runoff from feedlots	Infection of fish and contamination of the environment
Sediments from agricultural fields, logged areas, feedlots, degraded stream banks, and road construction	Clogging of gills and filters, covering of spawning sites, reduced insects and plant growth

environment, and productivity. Nutrient content of manure is an important value in determining the recycling method. Five popular recycling methods for agricultural wastes are **anaerobic digestion**, refeeding, land application, composting, and incineration.

Anaerobic Decomposition

Biogas is produced when organic matter is degraded in the absence of oxygen—a process called anaerobic decomposition or anaerobic digestion. Animal wastes can be used to generate biogas, a mixture of methane and other gases formed from the decomposition of organic matter. Like other gas fuels, biogas can be used for cooking, lighting, and running small engines. Biogas production in other countries has been given major emphasis.

Standard 9: Select Animals and Crops Appropriate for Environment and Available Resources

Sustainable aquaculture producers look for aquatic animals and crops that are appropriate for the environment and available resources. This is especial true when considering some form of **aquaponics**—the combination of raising aquatic animals with the production of some plants. Aquatic animal effluent accumulates in water as a by-product of keeping animals in a closed system or tank. The effluent-rich water becomes high in plant nutrients, but this is correspondingly toxic to the aquatic animal. Plants are grown in a hydroponic system that enables them to use the nutrient-rich water. The plants take up the nutrients, reducing or eliminating the water's toxicity for the aquatic animal. Aquaponics systems are discussed at the end of this chapter.

Selection of the aquatic animals to be raised depends on many factors, including water resources markets, demand, and the abilities of the individuals involved in the production facility. For the selection of aquatic animals, Chapters 2, 6, 7, and 8 provide addition guidance.

Standard 10: Manage Pests with Minimal Environmental Impact

Sustainable agriculture manages pests with minimal environmental impact. Pest management is a challenge to all producers, but especially to those dedicated to sustainable, low-input practices. A wide array of techniques and control agents can effectively reduce or eliminate pest damage without sacrificing soil, water, or beneficial organisms.

Integrated Pest Management

Much of the philosophy of pest management in sustainable agriculture is expressed as some variation of Integrated Pest Management (IPM). Like many of the terms and practices in sustainable agriculture, IPM suggests different things to different people. Simply, IPM promotes minimized pesticide use, enhanced environmental stewardship, and sustainable systems.

Some individuals feel that traditional IPM still focuses too much on pesticide use. Perhaps a better term for the IPM in a sustainable system is "**bio-intensive** IPM." Bio-intensive IPM is a systems approach based on an understanding of pest ecology. Those who use bio-intensive IPM begin by accurately diagnosing the nature and source of pest problems, and then they rely on a range of preventive tactics and biological controls to keep pests within acceptable limits. Pesticides are used as a last resort, and with care, when other tactics fail.

Planning is key to using bio-intensive IPM. Planning occurs before production because many pest strategies require steps or inputs, such as beneficial organism habitat management, that must be considered well in advance.

Biological control is the use of living organisms such as parasites, predators, or pathogens to maintain pest populations below economically damaging levels. Biological controls can be natural or applied. Natural biological control is generally characteristic of biodiverse systems. It results when naturally occurring enemies maintain pests at a lower level than would occur without them.

Mechanical or physical controls use some physical component of the environment, such as temperature, humidity, or light as a deterrent to the pest.

Identification of the pest is a crucial step in any IPM program. The effectiveness of pest management measures depends on correct identification. Misidentification can actually be harmful and can cost time and money. Monitoring allows the grower to gather information about the crop, pests, and natural enemies. Monitoring requires a producer to systematically check for pests and **beneficials**, at regular intervals and at critical times. Pest management for different species can be found in Chapters 4, 6, 7, 8, and 11.

Standard 11: Encourage Strong Communities

Sustainable agriculture encourages strong communities. A strong community requires opportunities for young people, access to health care, living-wage jobs, and access to a quality education. This means that sustainable aquaculture must create and implement projects and industry that generate new wealth and jobs.

Lack of jobs (opportunities) and low-wage jobs are serious concerns. In general, jobs are viewed as a means to make communities vital and self-reliant and allow individuals and families to also be self-reliant in the context of community.

Sustainable aquaculture as a part of sustainable community development is based on finding the solutions by increasing the capacity of individuals and communities to work together to respond to the constant changes that are common.

The potential for aquaculture enterprises to support communities is great. The jobs and careers available to or needed by a community are described in Chapter 18. Of course, one big need is for the entrepreneurs to develop aquaculture enterprises.

10 QUESTIONS TO ANSWER FOR PEST MANAGEMENT DECISIONS

Good pest management decisions can be made only after answering these 10 questions:

1. What pests are present, and in what numbers and stages of development?
2. What conditions exist that may increase or decrease pest problems?
3. What natural enemies of the pests, such as parasites, predators, and diseases, are present that may play an important role in control?
4. What amount and type of damage are being caused or may soon be caused by pests?
5. What is the stage of development, condition, and value of the crop?
6. What is the potential for economical injury? How much damage is tolerable? Has the action threshold been reached?
7. What are the history and severity of previous infestations at the site? How were those infestations managed? What were the results?
8. What pest management options are available, and how do the advantages and disadvantages of each apply to the situation?
9. If alternatives are not available, is a pesticide treatment justified for the situation? If so, what is the material of choice?
10. If a pesticide is not justified, what approaches, if any, should be taken?

When pesticides are needed, the safest and most effective materials should be selected for use. The goals of IPM are to achieve the effective management of pests in the safest manner.

Standard 12: Promote Social and Environmental Responsibility

Standard 12 is a capstone standard because a sustainable system addressing the first eleven standards should automatically address standard 12. Standard 12 is also an initiator because colleges and universities launched sustainable agriculture programs in response to an emerging concern about natural resources, environmental, economic, and social dimensions of agriculture, including aquaculture.

New technologies spawn economic, social, and political changes, creating shifting relationships among these individuals and hindering a commitment to social and environmental responsibility. All of this is compounded by the shrinking of the world through modern communications and transportation.

Change is difficult. A diversity of strategies and approaches are necessary to create a more sustainable food system. Strategies and approaches will range from specific and concentrated efforts to alter specific policies or practices, to the longer-term tasks of reforming key institutions, rethinking economic priorities, and challenging widely held social values.

Conversion of agricultural land to urban uses is a particular concern, as rapid growth and escalating land values threaten existing operations and water supplies. Comprehensive new policies to protect sustainable aquaculture enterprises and regulate development are needed.

Policies and programs are needed to address labor, working toward socially just and safe employment that provides adequate wages, working conditions, health benefits, and chances for economic stability.

Food and agriculture represent the union of human rights, community, and environmental issues. The agriculture sector is by far the largest employer in the world, employing almost one-half of the world's workers. Every person relies on agricultural products for survival. It is also one of the most dangerous sectors. For food and agriculture companies, the global marketplace created increasingly complex supply chains, along with ever more challenging demands from the world's stakeholder communities.

Food and agriculture companies wanting to do business in a global economy are faced with an increasing assortment of international regulations governing the growing, producing, and selling of agricultural products. Likewise, the demands of non-government organizations (NGOs), shareholders, customers, governments, media, and other stakeholders have led to a dramatic growth in vendor codes of conduct and other external performance standards. In response, some food and agriculture companies are adopting international standards and certification schemes that address numerous environmental and social concerns. They are engaging with environmental, human rights, and development organizations to implement, monitor, and certify compliance with these largely voluntary codes of conduct. Further, these companies are increasing the transparency of their operations, policies, and practices to broaden the trust of the public and encourage fair, ethical business practices.

AQUAPONICS

Aquaponics, or the integration of hydroponics with aquaculture (Figure 16-5), is gaining increased attention as a bio-integrated food production system. Aquaponics serves as a model of sustainable food production by following certain principles: The waste products of one biological system serve as nutrients for a second biological system. The integration of fish and plants results in a polyculture that increases diversity and yields multiple products. Water is re-used through biological filtration and recirculation. Local food production provides access to healthy foods and enhances the local economy. In aquaponics, nutrient-rich effluent from fish tanks is used to provide plant nutrients in hydroponic production beds. This is good for the fish because plant roots and rhizobacteria remove nutrients from the water. These nutrients—generated from fish manure, algae, and decomposing fish feed—are contaminants that would otherwise build up to toxic levels

FIGURE 16-5 Aquaponics System.

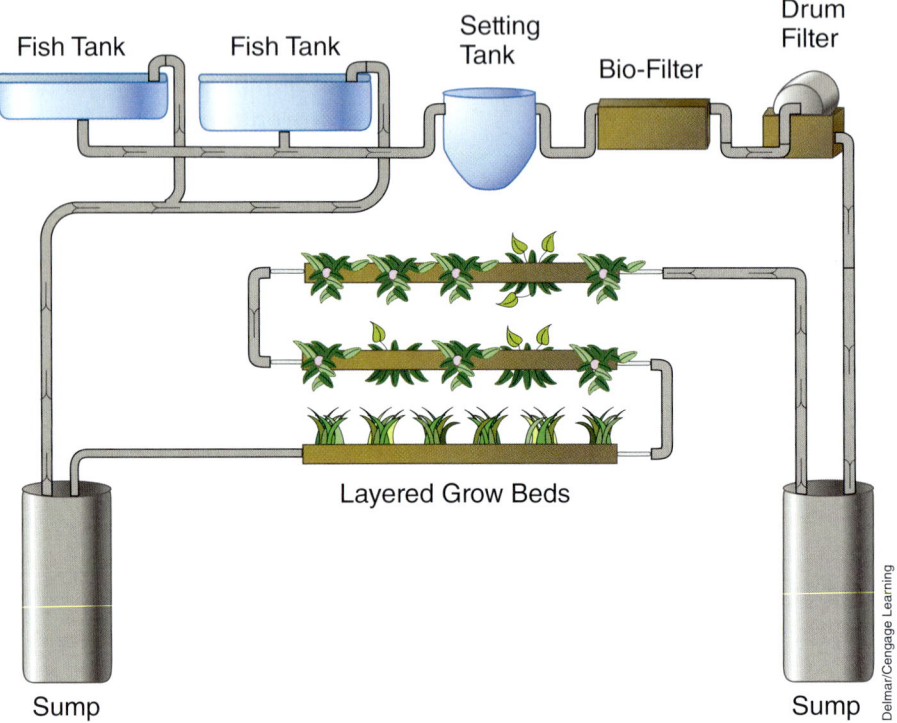

in the fish tanks, but instead they serve as liquid fertilizer to hydroponically grown plants. In turn, the hydroponic beds function as a biofilter—stripping off ammonia, nitrates, nitrites, and phosphorus—so the freshly cleansed water can then be recirculated back into the fish tanks. The nitrifying bacteria living in the gravel and in association with the plant roots play a critical role in nutrient cycling; without these microorganisms the whole system would stop functioning. (For more information, refer to Chapter 15, Recirculating Systems.)

Greenhouse growers and farmers are taking note of aquaponics for several reasons. Hydroponic growers view fish-manured irrigation water as a source of organic fertilizer that enables plants to grow well. Fish farmers view hydroponics as a biofiltration method to facilitate intensive recirculating aquaculture. Greenhouse growers view aquaponics as a way to introduce organic hydroponic produce into the marketplace because the only fertility input is fish feed and all of the nutrients pass through a biological process. Food-producing greenhouses—yielding two products from one production unit—are naturally appealing for niche marketing and green labeling. Aquaponics can enable the production of fresh vegetables and fish protein in arid regions and on water-limited farms, since it is a water re-use system. Aquaponics is a working model of sustainable food production, wherein plant and animal agriculture are integrated, and recycling of nutrients and water filtration are linked. In addition to commercial application, aquaponics has become a training aid on integrated bio-systems, popular with vocational agriculture programs and high school biology classes.

The technology associated with aquaponics is complex. It requires the ability to simultaneously manage the production and marketing of two different agricultural products. Until the 1980s, most attempts at integrated hydroponics and aquaculture had limited success. However, innovations since the 1980s have transformed aquaponics technology into a viable system of food production. Modern aquaponic systems can be highly successful, but they require intensive management, and they have special considerations.

The best plants to grow in an aquaponic setup are leafy greens and herbs. Because the main component of fish waste is nitrogen, and high-nitrogen fertilizer promotes lush foliage, leafy edibles such as chard and basil do well.

Fruiting plants such as tomatoes and cucumbers also seem to respond well. Root crops are not good candidates. Because the growing medium is typically coarse and heavy, potatoes and carrots would be misshapen and difficult to harvest.

SUSTAINABLE STANDARDS SCORECARD

Standards are meaningless if they are not tracked and evaluated. The following scorecard can be used to rate the sustainability level of each of the 12 Standards.

Standard	Score (1 to 5)
1. Base direction and change on science	
2. Honor market principles	
3. Increase profitability and reduces risk	
4. Satisfy human need for fiber and safe, nutritious food	
5. Conserve and seek energy resources	
6. Create and conserve healthy soil	
7. Conserve and protect water resources	
8. Recycle or manage waste products	
9. Select animals and crops appropriate for environment and available resources	
10. Manage pests with minimal environmental impact	
11. Encourage strong rural communities	
12. Promote social and environmental responsibility	
	Total

Merely to approach sustainability, each of the standards must receive a score of at least 3. A total score of 60 to 48 indicates a high level of sustainability. A score of less than 48 indicates the system needs more work to become sustainable. Of course, a low score of less than 3 for any standard should indicate that the operation is not meeting all 12 standards to an adequate level of sustainability and is therefore not sustainable.

SUMMARY

Reliance on food production systems that are not environmentally sound could lead to the destruction of the natural resources needed to produce food. If food production systems are not economically viable and risks are not managed, producers will go broke and their operations will cease to exist. If the systems are not socially sound, they will not meet people's needs.

One of the key factors in sustainable food production systems is to protect and conserve the natural resources on which food production depends. Water is a critical resource for all aquacultural production. It must be available in sufficient quantity, at a high enough quality, and at a cost that makes its use for aquacultural production possible.

Aquaponics is a working model of sustainable food production, wherein plant and animal agriculture are integrated, and recycling of nutrients and water filtration are linked.

The 12 standards provide a guidelines for developing sustainable operations and a means for evaluating sustainable operations.

STUDY/REVIEW

Success in any career requires knowledge. Test your knowledge of this chapter by answering these questions or solving these problems.

True or False

1. "Sustainable" means organic.
2. Risk a lot for a little in a sustainable aquaculture operation.
3. Market niches reduce risks.
4. Aquaponics is another word for hydroponics.
5. Fish manure is a waste product of aquaculture.

Short Answer

1. List 12 standards for sustainability.
2. Identify six areas of scientific research in aquaculture.
3. Identify three rules of risk management and three methods of managing production risks and marketing risks.
4. Briefly describe the role of aquaculture in providing food in the future.
5. How can an aquaculture operation can contribute to strong communities.
6. Name a possible energy source that could come from aquaculture.
7. Name three wastes from aquaculture.

Essay

1. Discuss the concept of sustainability.
2. Describe the technologies being developed due to high water prices, water scarcity, and high value aquaculture.
3. Describe integrated pest management.
4. Explain the role aquaculture has on water pollution and the role water pollution plays on aquaculture.
5. Explain how aquaponics serves as a model of sustainable food production.

KNOWLEDGE APPLIED

1. Search the Web to find four current projects where science is being used to improve aquaculture production.
2. Create a presentation on water use in the United States drawn from the U.S. Geological Survey at: http://water.usgs.gov/
3. Develop a report on some of the insurance options available to aquaculture producers at the USDA Risk Management Agency (http://www.rma.usda.gov/).

4. Visit an aquaculture facility (actual or online) and score its sustainability based on the scorecard at the end of this chapter.

5. Search the Web for a unique method of marketing a product from aquaculture, and then report on this marketing method.

6. Find plans for a simple aquaponics system and, as a team, construct it for a classroom demonstration.

LEARNING /TEACHING AIDS

Books

Barnabe, G., and Barnabe-Quet, R. (2002). *Ecology and management of coastal waters: The aquatic environment.* New York, NY: Springer Publishing.

Hutchinson, L. (2005). *Ecological Aquaculture: A sustainable solution.* Hampshire, England: Permanent Publications.

McClanahan, T. and Castilla, J. C. (2007). *Fisheries management : Progress toward sustainability.* Ames, IA: Blackwell Publishing Professional.

Pillay, T.V. R., and Kutty, M. N. (2005). *Aquaculture principles and practices.* Ames, IA: Blackwell Publishing Ltd.

Romanowski, N. (2007). *Sustainable freshwater aquaculture.* The complete guide from backyard to investor. Sydney, Australia: University of New South Wales Press.

World Bank, The. (2007). *Changing the face of the waters: The promise and challenge of sustainable aquaculture.* Washington, DC: The International Bank for Reconstruction and Development.

Internet

Internet sites represent a vast resource of information. The URLs (uniform resource locator) for the World Wide Web sites can change. Using a search engine such as Google find more information by searching for these words or phrases: sustainable aquaculture, aquaponics, bio-intensive, aquaculture effluent, market niches, risk management, and aquaculture integrated pest management.

For some specific Internet sites refer to Appendix Table A-11, A-12, and A-14.

Under the right conditions and with careful preparation, aquaculture can be profitable, both financially and emotionally. For someone poorly prepared and uninformed, aquaculture can be a disaster. Beginners should consider starting with small, simple systems. For example, much practical and relatively inexpensive experience can be gained by initially growing fish in a few floating cages or in an existing pond or with a small shellfish plot. As experience in production and marketing is gained, the business may expand into larger and more complex operations. Another option is to work with someone who successfully operates an aquabusiness.

CHAPTER 17

Aquaculture Business

OBJECTIVES

After completing this chapter, the student should be able to:

- Define terms related to aquacultural business management
- List reasons for keeping records
- Distinguish between basic kinds of records
- List guidelines for building and maintaining a good credit standing
- List factors that a lender looks for in a borrower
- List factors that a borrower looks for in a lender
- Identify indicators of good loan repayment ability
- List sources of credit for aquacultural enterprises

- List the essential components of all budgets
- Select budgeting principles from a list
- Define related management terms
- Describe functions in the management process
- Identify management considerations in planning an aquabusiness
- Explain important skills of managers
- Describe the importance of records and reports
- Explain important human relations skills

Understanding of this chapter will be enhanced if the following terms are known. Many are defined in the text, and others are defined in the glossary.

KEY TERMS

Accounting	Fixed costs
Accrual	Goals
Assets	Income
Balance sheet	Input costs
Break-even analysis	Interest
Capital	Net worth
Cash-basis	Profitability
Cash flow	Risks
Corporations	Shareholders
Economics	Partnerships
Enterprise budget	Sole proprietorship
Enterprises	Solvency
Equity	Strategic planning
Evaluation	Variable costs

COUNTING THE COST

No one should enter an aquaculture business without counting the costs. These costs can be personal or social considerations that may affect the success of the business venture, or they can be the actual costs of getting into and staying in business.

Personal/Social Costs

Some of the personal and social costs to consider when entering an aquaculture business come from the following checklist:

1. Are you willing to work long, hard, and irregular hours, for example, 16 hours a day, seven days a week?
2. Do you get along well and communicate effectively with people? (Small producers not only grow fish or shellfish, they must also promote and market themselves and their product.)
3. Are you comfortable with mathematical problem solving and mechanical troubleshooting?
4. Will you seek help when needed?
5. Do you personally have the technical expertise with fish or shellfish to manage the operation?
6. Can you afford to hire an experienced technician?
7. Do you know others in the business who will provide help or information?
8. Does your state have an aquaculture association you can join?
9. Are you willing to learn of current practices and new developments?
10. Are you familiar with legal issues of marketing your product?
11. Do you have the resources to construct and operate an approved facility if fish will be processed, for example, dressed and filleted?
12. Is the proposed culture site an unrestricted area, for example, not a right-of-way or wetland?
13. Is the prospective culture site located near the market and processing facilities?
14. Is the proposed site suitable for aquaculture?
15. Do you live close enough to the culture site to visit and monitor as needed and to ensure security?
16. Is an adequate supply of high-quality water available and suitable for aquaculture production?

17. Can you control water to, from, and within your system?
18. Can you effectively manage waste produced by your operation?
19. Will your neighbors and other user groups—for example, recreational and commercial fisheries—accept the aquaculture operation?
20. Have you discussed your planned operation with the appropriate local, state, or federal agencies?
21. Have you identified the permits required to construct and operate an aquaculture operation?
22. Have you determined what species you want to culture, and do you know its biology?
23. Have you explored the different production technologies available and identified one that satisfies your interests and resources?
24. Do you have the resources—financial, technical, and spatial—needed?
25. Are disease diagnostic services and dependable technical assistance readily available?
26. Do you have access to a dependable work force for physical labor?

Actual Dollar Costs

Anyone planning to enter an aquaculture business should make an economic evaluation of the investment. Making a realistic evaluation on paper improves the chances of success once money is committed and reduces the possibility of unpleasant surprises. Then, as the business matures, records are used to continually evaluate the economics of doing business.

Even before a single fish, crawfish, or clam is harvested, many input costs must be considered. These input costs include such items as:

Fry or fingerling (seed)	Processing and storage
Feed mixtures	Electricity or fuel power
Water chemicals	Antibiotics and sanitation
Structures	Labor
Equipment	Capital Water testing
Land Harvesting	Insurance
Collection	Record keeping
Aeration	Income taxes
Heater	Interest payments
Supplies	Depreciation
Transportation	Licenses

Table 17-1 lists the inputs for catfish production.

Enterprise Budgeting

Estimating the costs and returns for a particular activity is called "developing an enterprise budget." This procedure reflects the economic value of producing a specific output using a given set of inputs by following specific production practices. Profitability can be estimated by subtracting all the costs from the expected revenues.

TABLE 17-1 SELECTED NONDEPRECIABLE INPUTS USED IN PRODUCING CATFISH FOR FOOD, DELTA OF MISSISSIPPI

Item	Unit
Catfish floating feed (32% protein)	Ton
Fingerlings	Each
Diesel fuel	Gallon
Electricity	Kwh
Gasoline	Gallon
Earth moving	Cubic yard
Chemicals:	
Medicated feed (Romet-30)	Ton
Potassium permanganate	Pound
Copper sulfate	50-lb bag
Triple superphosphate	50-lb bag
Sodium chloride	50-lb bag
Test kits:	
Alkalinity	Each
Ammonia	Each
Carbon dioxide	Each
Chloride	Each
Nitrate/nitrite	Each
pH measurement	Each
Total hardness	Each

Two types of costs should be considered in developing enterprise budgets: variable and fixed. Variable costs are the expenses that vary based on production output, such as feed, fingerlings, and the like. Fixed costs are the expenses that do not change, regardless of whether production occurs, such as depreciation, interest on investment, insurance, taxes, and so on. Figure 17-1 shows that variable costs make up the largest portion of the total costs of fish production. In an examination of variable costs alone, feed comprises more than two-thirds of the costs, with fingerlings coming in a distant second, as shown in Figure 17-2.

Tables 17-2, 17-3, and 17-4 provide example enterprise budgets for catfish, trout, and crawfish.

Each producer faces different situations when trying to analyze the economic feasibility of fish production, so budget estimates in enterprise budgets should be used only as a starting point for planning.

MANAGING THE BUSINESS

Aquaculture is like any other agricultural business. It requires expert management. Management time is critical to the success of the business. Management involves:

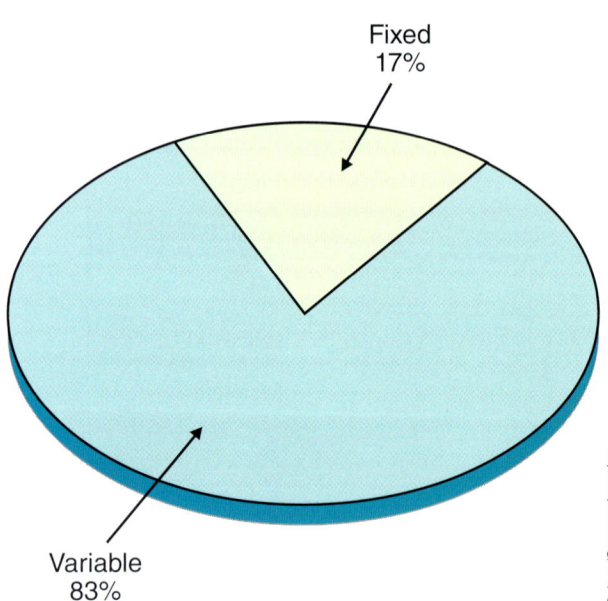

FIGURE 17-1 Variable costs make up the largest portion of the total cost of fish production.

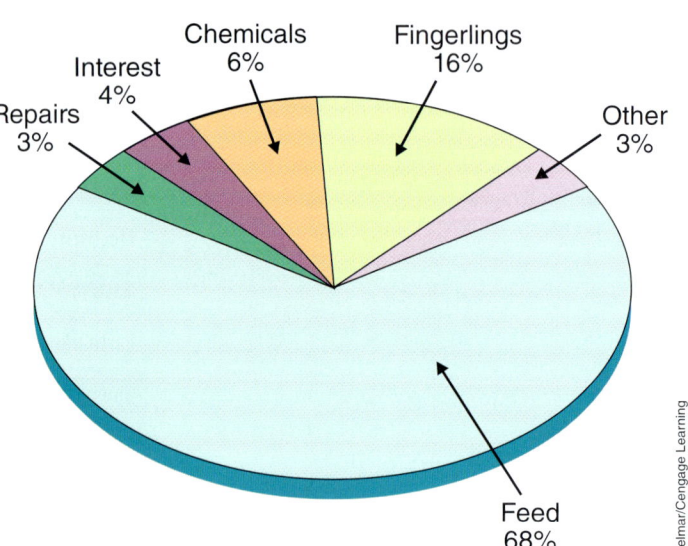

FIGURE 17-2 Feed comprises about two-thirds of the variable costs, with fingerlings coming in a distant second.

- Setting goals and objectives
- Recognizing and identifying problems
- Responding and acting when problems occur
- Seeking, compiling, and using relevant information
- Considering and analyzing alternative courses of action
- Making specific decisions
- Carrying out decisions or taking action
- Accepting responsibility for these decisions
- Evaluating the results of these decisions
- Developing training programs for family members and employees
- Directing and evaluating family members and employees
- Making buy and sell decisions
- Controlling financial operations
- Organizing the use of resources
- Establishing the timing of operations
- Monitoring operations and checking up on everything

Although the list overlaps, it could easily be expanded. The list does indicate a lot of things that managers do. That is why good full-time managers are crucial to most sizable businesses and why time must be set aside for management in any business, even if the main laborer is also the manager.

Management Functions

At the core of management concepts is a set of business goals and objectives that are developed and understood with clarity by the owner, by management, and by labor. Expectations about levels of annual earnings and

TABLE 17-2 CATFISH BUDGET FOR AN OPEN POND[1]

Item	Weight Each	Unit	Quantity	Price or Cost/Unit	Value or Cost
1. Gross Receipts Catfish	1.00	Lbs.	32,463.00	0.65	21,100.95
2. Variable Costs					
Fingerlings (4-in.)		Each	35,000.00	0.07	2,450.00
Floating Feed (32%)		Ton	34.68	250.00	8,670.70
Chemicals		Appl./acre	1.00	100.00	1,000.00
Harvest Labor		Hr.	32.00	4.00	128.00
Tractor (Fuel, Oil, Lube)		Hr.	82.00	2.25	184.50
Electricity		Kwh	10,800.00	0.07	756.00
Miscellaneous		Acre	10.00	5.00	50.00
Machine and Equipment (Repair)		Dol.			358.28
Interest on Operating Capital		Dol.	5,155.48	0.12	618.66
Total Variable Costs					14,216.14
3. Income above Variable Costs					6,884.81
4. Fixed Costs					
Interest on Building and Equipment		Dol.	9,424.50	0.12	1,130.94
Depreciation on Building and Equipment		Dol.			1,583.10
Other Fixed Charges on Building and Equipment		Dol.			133.30
Total Fixed Costs					2,947.34
5. Total of All Specified Expenses					17,163.48
6. Net Returns above All Specified Expenses					3,937.47
Net Returns per Acre:	Above Specified Variable Expenses				688.48
	Above Specified Total Expenses				393.75
Break-even Price (Per Cwt. Sold):	To Cover Specified Variable Expenses				43.79
	To Cover Specified Total Expenses				52.87

[1] This is only a sample budget with the elements that should be considered. Costs vary with time and location. Assumptions for 10 acres include: Stocking in spring, custom harvest in fall; estimated annual costs/returns; using recommended management practices; 3,500 fish stocked per acre; 20 lb/1,000 beginning weight; 2 lb of feed/lb of gain; 200 days in growing season; 1 lb ending weight; 7.25% death loss/unharvested fish. Net returns are to land, existing pond, operator's labor, and management. These estimates should be used as guides for planning purposes only.

production, maintenance of farm buildings and grounds, trade-offs between capital appreciation and current earnings, long-term growth, and achievements must be established. While these goals and objectives are not always formalized in writing, they need to be reasoned and discussed.

Figure 17-3 (page 493) suggests five basic management functions or activities used to achieve the goals and objectives of a business.

Planning

Though all five of the basic functions are important, planning is crucial because a good plan entails all the other functions. Planning involves:
- Setting daily priorities and schedules. What should be included in today's "to do" list? Who should complete each priority activity?

TABLE 17-3 COSTS AND RETURNS FOR COMMERCIAL TROUT PRODUCTION[1]

	Total Expenses	Total Income
Returns		
17,600 lb fish @ $1.50		$26,400
Cash Costs		
Repairs	$ 1,216	
Repair labor 60 hours @ $5.00	300	
Eggs 66,000 @ $43/1,000	2,840	
Feed 22,000 lb @ $.25/lb	5,500	
Electricity 450 kwh, $186.00/month	2,232	
Labor (feeding, processing, marketing, etc.)		
744 hours @ $5.00/hr.	3,720	
Insurance and license	1,000	
Miscellaneous, 5% of cash costs	782	
Interest on cash costs @ 12%	745	
Total cash costs	18,335	
Return over cash costs		8,065
Fixed Costs		
Interest on average investment	2,139	
Depreciation on buildings and equipment	1,690	
Return to management		$4,236

[1]This is only a sample budget with the elements that should be considered. Costs vary with time and location. The budget assumes the average cash expenses outstanding December through October at 12%. (Source: Southern Regional Aquaculture Center, Budgets for Trout Production)

TABLE 17-4 ESTIMATED ANNUAL OPERATING COSTS ASSOCIATED WITH A 40-ACRE CRAWFISH POND IN SOUTHWESTERN LOUISIANA[1]

	Cost
Variable Costs-Items	
Forage	$1,641
Fuel/Well	1,834
Repairs and Maintenance	1,059
Labor ($5/hr.)	2,918
Herbicides	157
Sacks	146
Bait ($0.16/lb.)	5,835
Total Variable Costs	$13,590
Fixed Costs	
Depreciation	6,978
Interest (12%)	3,605
Total Fixed Cost	$10,583
Total Annual Cost	$24,173

[1]All items represent annual operating cost the dollar amount will vary with location and time.

FIGURE 17-3 Five basic management functions or activities used to achieve the goals and objectives of a business.

- Recognizing problem areas and looking for alternative solutions. Why did production drop last month? Did we have the protein level right in the food? Did feed quality change? Should a consultant look at the ponds?
- Making a financial plan and cash-flow statement for the year, knowing when and how much credit must be obtained, and where the cash will come from to meet the regular obligations.
- Looking at alternative marketing plans.
- Establishing the overall enterprises for the business.
- Developing the business. How fast should the business grow? Is new staff needed? What professional development is needed for each manager?

It is not enough to plan just once a year. On the contrary, planning is an ongoing process. Plans get revised when goals are not being reached. Most planning deserves undivided attention without interruptions. A well-spent hour with a banker, with the computer, or in discussion with a trusted neighbor or partner can save a lot of money, time, and energy.

Organizing

Organizing is establishing an internal structure of the roles and activities required to meet the farm's goals. The manager must define the positions to be filled, the duties, responsibilities, and authority attached to each, and coordinate efforts among people. Organizing includes:

- Deciding who reports to whom. This is often referred to as the chain of command.
- Determining the functions in each position (job design), including the degree of authority.
- Establishing the work routines and standard operating procedures for each production enterprise.

Staffing

Staffing is as crucial a management function to a small or part-time business as it is to a much larger one. Often, the need to figure out how to get all the jobs done on time is even more critical. No business should try to operate without the possibility of hiring assistance when needed. Assistance can range from hiring a teenager to work after school, to help with a few operations, to contracting with an accountant in order to prepare tax records.

Recruiting and hiring workers

Whether a business needs one full-time worker, two or three part-time helpers on a regular basis, or hourly help for seasonal work, maintaining a competent force is essential. Labor management starts with obtaining qualified workers who understand what is expected of them. Written job descriptions ensure that the manager has thought through what is expected. Terms of compensation and benefits must be established—another reason for having something in writing.

Training and evaluating workers

Managing an aquabusiness means that someone takes responsibility for assigning tasks and making sure that workers understand how to do their jobs and what is expected of them. Incentives for high levels of achievement help. Telling people when they did something well may be even more important that telling them when they did something wrong. Both are necessary. This is even more important when all the workers are family members.

Directing

Directing is closely related to staffing. The smaller the business, the more the two are interlocked. Delegation of authority is often one of the most difficult things for the manager of a small business to accomplish. All the workers need to know their responsibilities and have a sense of when they can make decisions and when the boss must be involved. The lines of authority become more crucial with more employees. Motivation is part of directing. Knowing what is going on and listening to employee concerns help build communication and confidence. Managers should foster a team spirit, with every worker feeling some responsibility for the success or failure of the operation. Finally, a manager who is open and understanding wins respect in close working relationships.

Controlling

Controlling is another key function. Control is part of business management that determines what new methods are needed to turn out positive results when an investment decision is proven to be less profitable than planned. Control requires keeping track of expenses and income. It forces a manager to monitor what is happening every day. One of the important activities that plays a part in controlling is monitoring the records and accounts of the operations. These records can always be kept by someone besides the manager, but the work must be done on a regular basis. The manager analyzes these records to gain an understanding of what is going on.

As with any business, good managers pay close attention to those factors that affect profits most. Table 17-5 summarizes the sensitivity of major price and production efficiency factors and their effect on profit potential. For example, for every 0.1 lb that the feed conversion rate can be lowered, the cost of production would decrease by $1.69 for every 100 lbs of catfish produced.

Fish production requires a great deal of money. The overall net worth and cash flow of the potential producer should be large enough to withstand both the start-up period and any unforeseen setbacks.

Producers compare rates of production and levels of performance or productivity against established goals or generally accepted standards. Control ensures that these comparisons are made systematically and discussed with the people directly involved. Problems in production arising from natural causes need to be recognized and allowed for in good management.

TABLE 17-5 SENSITIVITY OF PRICE AND PRODUCTION FACTORS AND THEIR IMPACT ON PROFITS FOR A 10-ACRE POND

Item	Unit Change	Dollars Per Cwt. Sold
Pond Construction	$100/acre	0.42
Death Loss	1%	0.41
Stocking Rate	500/acre	2.57
Feed Conversion	0.1 lb.	1.69
Feed Price	$25/ton	2.24
Interest Rate	1 point	0.44
Fingerlings	1 cent each	1.13

Likewise, aquculturalists monitor production processes and making changes as necessary. Adjusting when to treat, when to feed, and when to start and stop harvest are all results of control. Keeping track of the work routines and making sure that plans are accomplished (or revised) make the difference.

Organizing and operating an aquaculture business require a manager to make and carry out many decisions. Some decisions take time and study. Others cannot wait until tomorrow. Part of the satisfaction of operating a business is seeing a major change in enterprises work better than expected, solving a nutrition problem that was finally recognized, or knowing that a pond was harvested when conditions where best. Good management allows the other farm resources to be used effectively. Understanding the functions of management is one step in becoming a better manager.

PLANNING—THE SECRET OF BUSINESS SUCCESS

What separates a successful aquabusiness from an unsuccessful one? Numerous factors: Quality of the land, water managerial skill, and sufficient equity capital are all important. Yet, some businesses that seem to have these basics are less successful than other businesses that are not so well endowed.

An important attribute of good business management is to be able to step away from the immediate concerns to look toward the future. Strategic planning is analyzing the business and the environment in which it operates in order to create a broad plan for the future. For smaller aquabusinesses, the most effective planning may take place at the kitchen table. To establish an appropriate atmosphere for strategic planning requires setting aside time away from day-to-day problems and interruptions so that the key participants—owners, managers, family members—can reach a common understanding about what they want to do in the next three to five years and how they want to do it.

Managers need to take a broad overview of the economy and the industry to determine the major opportunities and threats. Tactical planning is concerned with day-to-day and week-to-week decisions, such as what and how much pesticide to use, when to harvest fish, or whether to overhaul the old tractor or buy a new one (Figure 17-4). The results of strategic planning could lead to new enterprises, major capital investments, or perhaps even an exit from the business. This broader focus over a longer time distinguishes strategic planning from tactical planning.

Why Plan Strategically?

Strategic planning permits more profits, in the long run, by:
- Establishing a clear direction for management and employees to follow
- Defining in measurable terms what is most important for the firm

FIGURE 17-4 Sometimes the decision must be to hold on to the old tractor instead of buying a new one.

- Anticipating problems and taking steps to eliminate them
- Allocating resources (labor, machinery and equipment, building and capital) more efficiently
- Establishing a basis for evaluating the performance of management and key employees
- Providing a management framework that can be used to facilitate quick response to changed conditions, unplanned events, and deviations from plans.

Steps in Strategic Planning

Strategic planning involves the first seven steps shown in Figure 17-5. An eighth step—implementation—is strategic management.

Step 1: Define the Mission

The mission statement defines the purposes of the organization and answers the question, "What business or businesses are we in?" Defining the aquabusiness's mission forces the owner-operator or manager to identify carefully the products, enterprises, and/or services toward which the organization's production is oriented. This statement answers the question, "What is our current situation?" For example:

- What markets are likely to produce the best opportunities?
- What type of agricultural commodities or services can we produce to take advantage of these opportunities?
- What, if any, other activities are we involved in, and what are the priorities of these activities?

Answering these questions will suggest goals that will help to clarify objectives in the next step. A mission statement is not necessarily a long document. In fact, it should contain fewer than 100 words, and two or

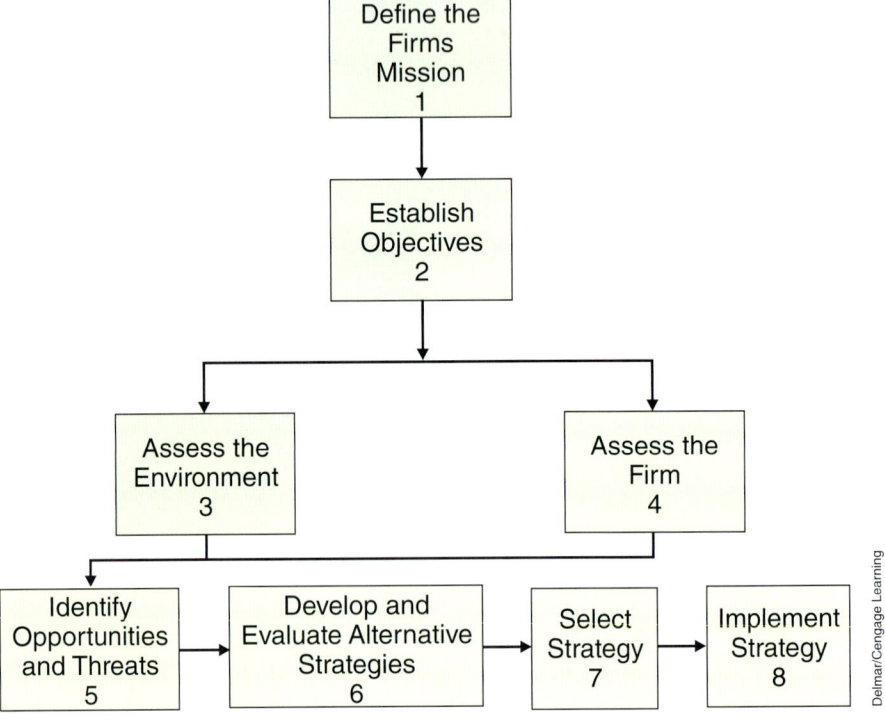

FIGURE 17-5 Strategic planning involves the first seven steps.

three sentences may be sufficient. Establishing strategic goals is the key element of the mission statement. The mission statement answers the question, "Why are we in business?"

Step 2: Establish Objectives

Goals, which are the more general, long-term desires of owner-operators or managers, clarify the aquabusiness's purpose. Objectives should translate the mission into concrete terms. Objectives should be measurable and straightforward statements, such as the following:

- Increase sales by 100 percent in the next five years
- Reduce labor costs by 25 percent in the next three years
- Increase production per acre of pond by 30 percent in the next five years
- Add four more raceways in the next two years

These objectives should be chosen in such a way that they contribute to reaching the goals identified in Step 1. Each objective is measurable and time-limited. This allows management to evaluate progress in implementing the plan.

Step 3: Assess the External Environment

Every agricultural firm faces uncertainties, threats, and opportunities that are beyond its control. Market forces may cause prices to plunge, either in the long run or short run. Overproduction, declining consumer demand, a strong dollar, high interest rates, changes in government policies, and regulation of labor and pesticides are external threats that can cut profits or make business more difficult. New market opportunities are created by demographic changes, changing consumer lifestyles, regional population growth, and technological breakthroughs.

The operator/manager must understand the economic, social, and technological forces that will affect the firm. Then reasonable expectations may be formulated about what will happen to product prices, interest rates, the rate of inflation, labor markets, and input prices over the next three to five years.

Step 4: Assess the Organization's Strengths and Weaknesses

The quality and quantity of resources within the control of the operator/manager is the first part of this assessment. What are the abilities and limitations of the operator/manager? What skills and abilities do the employees have? How modern and efficient is the physical plant? How large is the resource base? How much water is available? What is the cash position of the farm? These resources need to be compared against those of competitors.

Many farms have an unrealistic view of their own resources and operation because they do not compare themselves to others in the same business. The process of providing candid answers to these questions forces the operator/manager to recognize that every business is constrained in some way to its physical resources as well as its human skills and abilities.

Step 5: Identify Opportunities and Threats

This step combines the data gathered in Steps 3 and 4 to determine the threats and opportunities the business might encounter in the planning period. Difficulties in the external environment can present opportunities in another segment of agriculture. For example, concern about cholesterol in meat products created new markets for poultry and fish. Concern about carcinogens in the environment, including some pesticides, brought about new opportunities. Some firms have avoided problems by creatively turning an external threat into an opportunity.

Step 6: Develop and Evaluate Alternative Strategies

Steps 6 and 7 are at the heart of the strategic planning process. This is the point at which the business develops the alternative plans that describe the methods for achieving objectives and obtaining greater long-run profits.

In what ways can production agriculture firms gain a competitive advantage? The answer to this question depends to a great extent on whether or not the farmer is a price-taker. Farmers are traditionally price-takers. They take the price set by others who control the market.

Some types of strategies operator/managers use to gain a competitive advantage include:

1. Become more efficient. Increase profits by:
   - Reducing input use, holding product and price (quality) constant
   - Using more, or higher-quality, inputs increasing revenue more than costs.

2. Seek out alternative enterprises.
3. Exploit quality differences. Obtain price premiums for quality that more than offset the additional costs involved in producing higher quality commodities.
4. Integrate horizontally. Farm more units, or add enterprises, enlarge enterprises to gain more complete use of existing resources. Acquire additional resources. Spread fixed costs over more units of output.
5. Integrate vertically. Obtain more profit by moving higher or lower into the marketing and distribution channels—add storage or packing facilities, trucks to haul products, and direct marketing; acquire resources to produce inputs that formerly were purchased.
6. Reduce risks through diversification and hedging.
7. Identify new markets or narrow (niche) markets.

Organizations that have some degree of control in the market because of fewer competitors or because of the possibility for differentiation of products and services have the potential for additional strategies to gain a competitive advantage.

Possibly, no identified alternatives will permit a particular farm family to attain its objectives. Non-farming alternatives may need consideration. For example, to sell some farm assets, keeping part or all of the land and residence, and to seek off-farm employment. Non-farming alternatives should not be neglected in selecting alternatives.

Once alternatives are developed, Step 6 is only half completed. These alternative strategies must be evaluated. In practice, management may develop a long list of possible alternatives. These can usually be whittled down with reasoning and logic. Once the obvious losers are eliminated, pencil pushing is in order. No single or preferred method is used for evaluating alternatives, but some combination of the following may be used:

1. Budgeting alternatives, both profitability and cash flow
2. Break-even analysis
3. Projections of income, cash flow, and balance sheet statements
4. Computerized decision aids

Table 17-6 (page 502) shows a monthly cash flow budget for a trout farm. Income for the operation occurs only in November, yet expenses occur every month.

Step 7: Select a Strategy

From the analysis in Step 6, the firm selects a strategy—an alternative or a combination of alternatives—that will enable the operator/manager to achieve the desired objectives. Evaluating alternatives may show that the original objectives are not feasible. The operator/manager may have to move back to Step 2 and select new objectives or reformulate combinations of alternatives. Selection of a final strategy may involve trade-offs among goals. An alternative is seldom likely to be superior to all other alternatives for attainment of each of the goals of the operator/manager and his or her family. The process of strategic planning should be recognized more as an art than a science.

NO PLAN, NO MONEY

potential lender or investor wants to see a written business plan. The plan can be used as a guide in developing the business. Business plans can have different formats depending on the type and source of funding being sought or the general purpose. If funding is already in place, a plan can easily be used as a guide for strategy. Successful business plans contain certain elements.

Title page

The title page should minimally include four pieces of information: the name of the proposed project, the name of the business, the principals involved, and the address and phone number of the primary contact.

Table of contents

The table of contents should include the major topics of the body of the business plan and critical tables or figures that the investor should take particular notice.

Statement of purpose

This section is a brief mission statement of the aquaculture venture: what is to be done, why this project was chosen, and how it is to be accomplished.

Executive summary

This section presents the key elements of the business plan to prospective lenders/investors. The length of the executive summary should be under five pages to increase its likelihood of being read.

The business

The next section of the business plan provides details of the venture. The following points should be addressed:

- History
- Description
- Market
- Marketing
- Competition
- Operations
- Management
- Research and Development
- Personnel
- Loan Application and Effects
- Development Schedule
- Summary

Financial plan

This section includes sources and applications for capital, equipment list break-even analysis, pro forma balance sheet, pro forma income statement, pro forma cash-flow budget, historical financial statements, equity capitalization, debt capitalization, and supporting documents.

Still, those who fail to plan, plan to fail—especially without money.

Step 8: Implementation

Implementation is a crucial link in the strategic management chain. Management must periodically look back on the plan and determine how well the business is reaching its objectives. Assessing implementation will point to mid-course corrections. Assessment enables planners to understand the planning process. Perhaps objectives were set too optimistically,

TABLE 17-6 CASH FLOW AND MONTHLY LABOR REQUIREMENTS FOR TROUT[1]

Item	Total	Jan	Feb	Mar	April	May	Jue	July	Aug	Sept	Oct	Nov	Dec
Cash Receipts 17,600 lbs. @ $1.50	$26,400											$26,400	
Cash Expenses:													
Repairs	$1,216	$200				$200	$200	$200	$200	$100			$216
Eggs, eyes	2,840											$2,840	
Feed	5,500					400	600	700	900	1,100	1,200	600	
Electricity	2,232	186	186	186	186	186	186	186	186	186	186	186	186
Labor, all	3,720	60	200	200	250	200	200	200	200	200	150	1,800	60
Insurance and license	1,000							1,000					
Miscellaneous	782	67	65	65	65	65	65	65	65	65	65	65	65
Interest on cash expenses	625						625						
Monthly net	$8,485	(513)	(451)	(451)	(501)	(1,051)	(1,876)	(2,351)	(1,451)	(1,651)	(1,601)	$20,909	(527)
Labor hours	744	12	40	40	40	40	40	40	40	40	30	360	12

[1]Trout Farming in Washington, Cooperative Extension, Washington State University. The dollar amounts are not accurate, but these amounts do demonstrate the concept of a monthly cash-flow budget.

or perhaps critical threats or opportunities were not recognized. Recognizing and correcting the plan's weaknesses will improve strategic planning the next time the process is undertaken.

Strategic planning should not be viewed as a formidable task resulting in detailed plans. The plan should be written out, but a few pages will suffice. The process should include all the key players participating in the strategic management discussion. All individuals involved in managing the farm or ranch must understand where the business is going, how it plans to get there, and what problems or opportunities lie ahead.

SETTING GOALS FOR BUSINESS MANAGEMENT DECISIONS

Almost everyone is enthusiastic about goals. Most people like to discuss goals, and some boast of having them. Goals are definitively known and are important.

People who teach management also stress the importance of goals. Listen to almost any management guru to hear ideas like these:

- Identify your goals. Manage to reach them.
- Management is goal-directed.
- Take charge of your life and work. Set goals and attain them.
- Without goals, you cannot be a manager because you will not know what you want to achieve through your management decisions.

Almost everyone agrees on the importance of goals. The paradox of goals is this: Many people will publicly affirm that they have identified their goals and that goals are important. Most, though, cannot or will not record and communicate their desired outcomes in a goal statement that will guide their management decisions.

The communication, negotiation, and compromise required for goal identification yield additional important benefits. When goals are selected in a way that ensures that each person does work that he or she enjoys, motivation increases and management performance improves. Perceptions of reality are modified as participants gain a greater understanding of each other's roles, interests, and activities. Identifying goals has both immediate and long-term payoffs—the quality of daily management outcomes and focus of long-term decisions are improved.

Those who regularly set and write down goals report benefits like:

- Communication among family members improved
- Management decisions and work activities effectively focused on priority concerns
- Cash-flow management in the production unit and household improved, as impulse buying of production inputs and household items declined
- Borrowing, risk, and interest expense reduced
- Conflict reduced, and working relationships improved

> Expenses were kept under control, and profits increased
> Anxiety and concern over the present and future reduced
> A better balance between production activities and family life achieved

Goals and commitment—this combination, that cannot be beaten, ensures that the aquaculture business will grow, change, and remain profitable.

BUSINESS AND RISKY DECISIONS

Aquaculture is a high-stress industry. The management of the business is fraught with risk and uncertainty. Aquacultural managers must consider the risks associated with the ever-changing political, social, economic, and ecological environment in which they operate.

Types of Risks

Identifying the different events or sources of risk that affect the outcome of a decision is a crucial step in the decision-making process. The relative importance of the sources of agricultural risk differs among enterprises and changes over time. Risks in aquaculture include market, production, financial, obsolescence, casualty, legal, and human.

Market Risk

The variability and unpredictability of the prices that farmers receive for their products and what they pay for production costs are market risks. Fluctuating supply and demand conditions result in price variations.

Production Risk

This risk source is a result of the variability in production caused by such unpredictable factors as weather, disease, pests, genetic variations, and timing of practices. Examples include variations in yields, machinery breakdowns, and feed conversion efficiencies.

Financial Risk

Financing assets that the business controls creates risk (Figure 17-6). The increased use of borrowed capital leaves the operator vulnerable to not having enough cash to meet obligations or of not having adequate credit. Other examples of this source of risk include the possibility of losing the lease on the land and the ultimate disaster—bankruptcy.

Obsolescence Risk

The rapid development of new technology can make current production methods obsolete shortly after important investments have been made. The possibility of adopting new technologies too soon or too late is a risk farmers face.

FIGURE 17-6 Managers understand the financial risks of increasing productivity when variable inputs increase and fixed inputs remain flat.

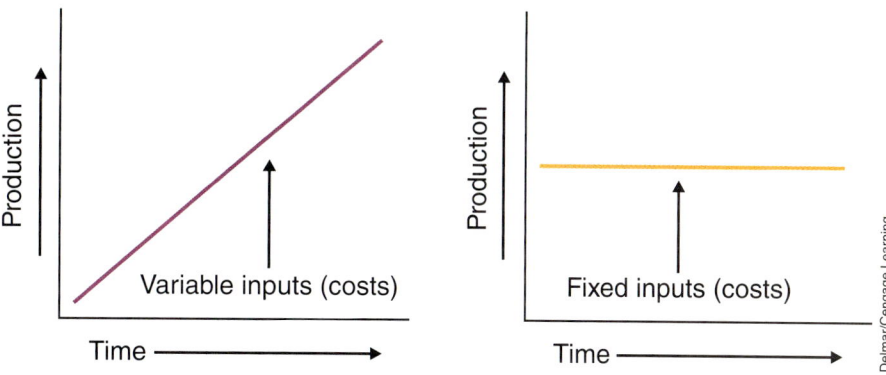

Casualty Loss Risk

This a traditional source of risk referring to the loss of assets as a result of such events as fire, wind, hail, flood, and theft.

Legal Risk

Governmental laws and regulations are a growing source of uncertainty for farmers. Changing social attitudes have resulted in laws and regulations governing environmental protection, water quality, food safety, and other farm-related matters. In addition, there is a risk of lawsuits resulting from accidents and other events.

Human Risk

The character, health, and behavior of individuals are unpredictable and contribute to the risk in farm management. The possibility of losing a key employee during a critical production period is one example of this type of risk. Dishonesty and undependability of business associates are other examples. Also, family needs and goals change, sometimes unpredictably.

Psychological studies suggest that business managers tend to overlook risk considerations as they make decisions. They do not deal with risk. When it comes to making decisions in today's risky agricultural climate, the wise manager must explicitly consider various sources of risk.

Managers respond to risk in different ways. Just as we classify people as being optimistic or pessimistic, conservative or liberal, we can also classify people according to their attitudes about taking risks—risk avoiders or risk takers.

The risk takers are the plungers, the more adventurous types who willingly make risky decisions. They are willing to accept greater risk in return for the small chance of a higher income.

Risk avoiders are the more conservative types who have a preference for less risky decisions. Risk avoiders are willing to sacrifice the small chance of higher income for less risk.

The framework for making risky decisions means that managers must choose among alternative actions, the outcomes of which depend on events beyond their control. The outcome of each combination of choices and events is known as a "payoff."

For some risky decisions, managers use intuitive thought processes. Many decisions are too complex and important to be handled by intuition alone. A more formal approach, such as a matrix or statistical comparison, provides the discipline to ensure that all available information has been used.

Risk analysis does not simplify decision making or eliminate the agony of making difficult choices. Risk analysis does not eliminate risk, but it can help the manager select the right risks to take in the often uncertain world of aquaculture business.

BUSINESS STRUCTURES

A business entity is the legal structure under which a farm or any business is organized and operated. Aquaculture owners can establish their businesses as sole proprietorships, partnerships, or corporations. Whether individuals inherit or receive a farm through gifts, they must decide on the type of business structure for the farm. The selected structure often changes as the farm grows or new individuals enter the business. Individuals who originally owned and operated their business as a sole proprietorship, for example, may choose to shift to a corporation, a partnership, or a multiple business organization.

The sole proprietorship is the most common form of business organization because most small businesses are owned and operated by a single individual. Sole proprietorships have a common-law origin and can be easily established and operated because the business structure is an extension of an individual's rights and responsibilities in property ownership and commercial transactions. Partnerships also have a common-law origin, and thus they have many of the characteristics of a sole proprietorship.

By contrast, incorporation has a statutory origin, which means that state laws prescribe a corporation's structure, procedures, and conditions of organization and operations. Incorporating a farm business requires a series of legal steps. Corporate activities are closely regulated.

Attracting Capital

The traditional sources of capital for small farms are the equity provided by family members, reinvestment of retained earnings, lease agreements, and loans. Capital sources are the same regardless of the organization's structure. The sole proprietorship may be the most limited in terms of capital acquisition because only one family is involved in the operations. Multiple ownership through a partnership or corporation allows the combining of funds from more than one family, which results in a larger business.

Table 17-7 indicates the capital needed to start a small trout operation. Depending on location, this can amount to an even higher amount.

TABLE 17-7 ESTABLISHMENT COSTS FOR A TROUT FARM PRODUCING LESS THAN 100,000 POUNDS[1]

Category	Units	Price	Quantity	Value
Site preparation[2]				$187.50
Concrete floor[3]	Yd.^{3}	$54.00	9.07	489.78
Concrete walls[3]	Yd.^{3}	$54.00	7.39	399.06
Reinforcing steel	Pair			232.50
Drainpipe	Pair			232.50
Screening	Pair			37.50
Tank forms	Pair			137.50
Snapties and wedges	Pair			250.00
Labor	Hourly	$5.00	197.50	987.50
Water intake assembly	Pair			812.50
Miscellaneous	Pair			625.00
Total Establishment Cost				$4,391.34

[1] The dollar amounts are not accurate and vary with location and time. These are only representative estimated costs for establishment of one pair of up to 10 pairs of tanks constructed in series. More than one series may be constructed. Expected annual production of 6,000 lbs per pair. (Source: Southern Regional Aquaculture Center, *Budgets for Trout Production*)
[2] Each tank based on dimensions of 35 ft long by 6 ft wide.
[3] Floor and walls at 6 in thick.

Federal Income Taxes

A sole proprietor's business pays no federal income tax. Instead, the taxable income of the business is included in the proprietor's personal income, and taxes are paid at the individual tax rates. Federal income taxes for a partnership are treated in a similar manner. The partnership files an information return showing the income and expenses, the names of the partners, and how the partnership earnings will be divided among the partners. The profits, losses, capital gains and losses, and tax credits are allocated to partners according to the terms of the partnership agreement. The partners pay taxes as individuals on their respective shares of partnership income.

Federal income tax savings may occur if a business incorporates and becomes subject to federal income taxation under Subchapter C of the Internal Revenue Code. Because a corporation is considered a separate taxpayer, the corporation can divide income among the corporation, owner-operator employees, and shareholders. The corporation pays individuals associated with the corporation for their contributions. Owner-employees receive a salary for their labor, and management and shareholders receive dividends for their capital investment. Residual income after all expenses are paid is taxed to the corporation at corporate income tax rates. Whether federal income taxes will be lower after incorporation depends upon the corporation's earnings level, the tax rates for individuals versus that for corporations, and the allocation of earnings.

Payroll Taxes

After incorporation, the sole proprietor or partner changes status from employer to employee. The business has at least one additional employee, if not more, which results in increased payroll taxes.

Social Security taxes are increased because the combined employee and employer rates under the corporate structure are higher than for self-employed individuals—partners or sole proprietors.

Stockholder-employees of corporations are also subject to Worker's Compensation charges on their salaries and are entitled to benefits under the Act. This is not true of sole proprietors or partners in a partnership. A stockholder-employee's salary may also be subject to the unemployment compensation tax.

Owners and operators of corporations and sole proprietors have to file personal income taxes through quarterly estimates or withholding rather than as a lump sum.

Structure Must Fit Objectives

The initial business organizational type for a small-scale family business is usually a sole proprietorship. When circumstances surrounding the operation suggest a partnership or corporation, an in-depth analysis needs to be made. An analysis of the organizational characteristics and the objectives of the family is perhaps the most important, but still the most neglected, phase of the process.

Usually, the decision does not need to be rushed. It is relatively easy and inexpensive to incorporate or form a partnership, but it may not be so easy and inexpensive to dissolve a corporation or partnership. Those thinking about changing business organizations should take enough time to weigh the advantages and disadvantages of each structure for their particular situation.

RECORDS IMPROVE PROFITABILITY

Managers need a complete and accurate farm records system in order to make informed management decisions that help maintain or improve business profitability. Records systems serve four functions:

1. As a service tool to assist in reporting to the Internal Revenue Service and other taxing entities, creditors, other asset owners, and to others who have a vested interest in the financial position of the business
2. As an indicator of progress
3. As a diagnostic tool for identifying strengths and weaknesses
4. As a planning tool

With the proper records, managers can determine the component cost of annual ownership and the annual operating cost per pound of product as shown in Tables 17-8 and 17-9 (page 510). This information helps the manager make management decisions by comparing these costs to

TABLE 17-8 ESTIMATED ANNUAL OWNERSHIP COST COMPONENTS EXPRESSED IN CENTS PER POUND OF CATFISH HARVESTED, DELTA OF MISSISSIPPI[1]

Item	Cents per Pound
Depreciation:	
Ponds	1.77
Water supply (well, pumps, motors, and outlet pipes)	0.23
Feeding (feeder with electronic scales and storage)	0.12
Disease, parasite, and weed control equipment (boat, motor, and trailer)	0.04
Miscellaneous equipment	2.65
Total Depreciation	4.81
Interest on Investment:	
Land	2.23
Pond construction	0.97
Water supply (wells, pumps, motors, and outlet pipes)	0.24
Feeding (feeder with electronic scales and storage)	0.10
Disease, parasite, and weed control equipment (boat, motor, and trailer)	0.01
Miscellaneous equipment	1.15
Total Interest on Investment	4.70
Taxes and Insurance	0.33
Total Annual Ownership Cost	9.85

[1]Source: Economic Analysis of Farm-Raised Catfish Production in Mississippi. Note: The amounts are not accurate and vary with location and time. The amounts demonstrate how the cost of ownership can be distributed across a pound of fish.

averages for the industry or historical information for the aquabusiness. Records can also help the manager plan and implement farm business arrangements and do estate and other transfer planning. Managers can use records to determine efficiencies and inefficiencies, measure progress of the business, and plan for the future.

Aquaculture business managers do not need to be accomplished accountants or experts on taxes and law. They do need to know how to keep the required records for their businesses; they must realize that all business decisions have income tax consequences; and they must be able to evaluate the accounting and legal professionals who serve their businesses.

Choosing a Records System

Records systems range from simple, hand-accounting systems using pencil and paper to sophisticated double-entry computer accounting systems. Some require a mix of hand and computer operations.

TABLE 17-9 ESTIMATED ANNUAL OPERATING COST COMPONENTS EXPRESSED IN CENTS PER POUND OF CATFISH HARVESTED, DELTA OF MISSISSIPPI[1]

Item	Cents per Pound
Repairs and Maintenance:	
Vegetative cover	0.23
Water supply (wells, pumps, motors, and outlet pipes)	0.13
Feeding (feeder with electronic scales and storage)	0.02
Disease, parasite, and weed control equipment (boat, motor, and trailer)	0.02
Miscellaneous equipment	1.31
Total Repairs and Maintenance	1.72
Fuel:	
Mowing	0.05
Feeding	0.31
Outboard motor	0.02
Electric floating paddlewheels	0.86
PTO-driven aerators and low-lift pump	0.45
Pumping	1.28
Transportation	0.50
Total Fuel	3.46
Chemicals	2.38
Telephone expenses	0.19
Test kits	0.04
Fingerlings	6.45
Feed (32% protein)	24.50
Labor:	
Operations management	2.74
Hired labor	4.85
Total labor	7.59
Harvesting and hauling	4.00
Liability insurance	0.25
Interest on operating capital	2.35
Interest on fish inventory	0.25
Total Annual Operating Cost	53.19

[1]Source: Economic Analysis of Farm-Raised Catfish Production in Mississippi.
Note: The amounts are not accurate and vary with location and time. The amounts only demonstrate how the operating costs can be distributed across a pound of fish.

A system should not only meet the accounting and planning needs of the operation, but it should also satisfy income tax, legal, and other outside reporting requirements. Programs should be selected with good detailed instructions for use.

Accounting Methods

Two types of accounting methods are used in farming—cash basis and accrual.

Cash-Basis Accounting

This method is used primarily for income tax reporting purposes in service industries. Generally in cash-basis accounting, income is recorded as income when it is received, and expenses are recorded as expenses when they are paid. Cash-basis accounting is simple and can provide some income tax advantages for businesses that are heavily dependent on inventory changes.

This method also has drawbacks. Cash-basis accounting can grossly distort the financial position, profitability measures, and operational results of the farm business. It is necessary to convert cash-basis accounting to accrual accounting for analysis and decision-making purposes.

Accrual Accounting

This method is required for tax purposes for most trading and manufacturing businesses. In accrual accounting, expenses are considered expenses when they are accrued (or committed), and income is counted as income when it is earned. This includes changes in inventories. The accrual accounting method does not depend on actual cash flow into the business. Expenses incurred are matched with related income to determine net income. This approach provides a better continuous picture of profitability. An assessment of cash flow is still needed to determine the financial feasibility of the business.

Basic Record Keeping

Record keeping need not be a complex managerial activity if some simple rules are followed. A well-designed farm record system makes the job easier as well as more efficient. (See Figure 17-7.)

FIGURE 17-7 Whether records are maintained by hand or with a computer, they are important for success.

Tips for Better Record Keeping

Six suggestions for better record keeping in an aquaculture business include:

1. Always record the gross or total amount. Never net it out (amount remaining after deductions).
2. Always go through all the steps for each transaction.
3. Run everything through a checking account.
4. Separate your business income and expenses from personal income and expenses.
5. Do periodic accuracy checks.
6. Staple your calculator tape to each page as you total your book so that you can refer back to it. Do not do your work twice—once is enough.

Items one and two fit together, as do items three and four.

Tax Records

The Internal Revenue Service requires a set of farm records to show all taxable income and expenses that are deductible. This can be done in many different formats. The manager or record keeper must maintain accounts to show the three different types of farm income: sale of "resale" (purchased) items, other ordinary income, and sale of capital items. Records must also be kept of the two types of expenses—ordinary expenses and capital expenses—along with some expenses that could be classified in either category. Included in the expense category is the annual depreciation record.

The record system chosen should support items on a tax return. The records must provide evidence of the types of income and expenses. This requires sales slips, invoices, receipts, deposit records, and canceled checks. Income and expenses should be clearly identified. Records of loans, debt repayment, and interest expenses must be kept as long as they have any income tax or legal ramifications.

Balance Sheet

The balance sheet is the first of the "big three" statements. (See Figure 17-8.) It summarizes three of the five accounts in a complete accounting system. The general accounting equation for the balance sheet is: assets equal debt plus equity. Phrased another way, assets minus debt equals equity.

The balance sheet is divided vertically into the left part, called assets (what the business owns), and the right part, called liabilities (what the business owes). (See Figure 17-9.) The total of the two parts must be equal. Two kinds of liabilities are included: (1) debt or outside capital and (2) equity (net worth) or inside capital. The debt represents claims lenders have on the assets, equity represents claims owners have on the assets.

FIGURE 17-8 The big three statements in an accounting system.

FORM 8A

BALANCE SHEET

NAME: Market   Depreciated cost   DATE:

CURRENT FARM ASSETS	Line no.	Value	CURRENT FARM LIABILITIES	Line no.	Amount
Cash, checking balance	14		Farm accounts payable and accrued expenses		
Prepaid expenses and supplies	15				
Growing crops	16				
Accounts receivable	17				
Hedging accounts	18				
	19				

Crops held for sale or feed	Line no.	Crop code	Quantity	Value					
	20				Judgements and liens				
	21				Estimated/accrued taxes:				
	22				Property				
	23				Income tax and social security				
	24				Accrued interest: Current				
	25				Intermediate				
	26				Long term				
	27				Subtotal accounts payable and accrued expenses			79	
	28				Current farm notes payable	Due date	Interest rate	Annual installment	Amount delinquent	Principal balance
	29									

Crops under govt. loan	Line no.	Crop code	Quantity						
	35								
	36								
	37								
	38				Total current farm liabilities				80	
	39				**INTERMEDIATE FARM LIABILITIES**					
	40				Description	Due date	Interest rate	Annual installment	Amount delinquent	Principal balance

Livestock held for sale	Line no.	Lvstk code	Quantity		
	45				
	46				
	47				
	48				
	49				
	50				
	51				

Total current farm assets		60			
INTERMEDIATE FARM ASSETS					
Breeding livestock	Number				
			Total intermediate farm liabilities	85	
			LONG TERM FARM LIABILITIES		
			Description	Due date	Interest rate	Annual installment	Amount delinquent	Principal balance
Farm machinery and equipment					
Total intermediate farm assets		65			
LONG TERM FARM ASSETS					
Farm real estate	Acres				
			Total long term farm liabilities	90	
			TOTAL FARM LIABILITIES		
FLB stock and co-op equity			**NONFARM LIABILITIES**		
Total long term farm assets		70	Nonfarm accounts payable and accrued expenses		
TOTAL FARM ASSETS					
NONFARM ASSETS					
Vehicles			Nonfarm notes payable	Due date	Interest rate	Annual installment	Amount delinquent	Principal balance
Household goods					
Cash value of life insurance					
Stocks and bonds					
			Total nonfarm liabilities	95	
Total nonfarm assets		75	**TOTAL LIABILITIES**		
TOTAL ASSETS			**NET WORTH**		

FIGURE 17-9 The balance sheet.

Horizontally, the balance sheet can be broken into three categories:
1. Current Assets. The first category, current assets, contains those assets that are in cash or are usually turned into cash during the course of the year. For tax purposes, they are assets that would be considered ordinary income if sold or ordinary expenses if purchased.
2. Intermediate Assets. The second category includes intermediate assets. They are not true current assets, but neither are they true long-term assets. They are assets used in the production of income and are generally viewed as non–real estate property, such as machinery and productive animals.
3. Long-Term Assets. The third asset category is composed of long-term assets. These generally include real estate property used for producing income. (See Figure 17-10.)

An asset's length of life is sometimes used to distinguish between the three asset types. For example, assets that last less than one year are current, those that last from one to 10 years are intermediate, and those that last more than 10 years are long-term. Some accountants use only two categories, current and long-term.

Farm Earnings Statement

The second of the "big three" statements focuses on current activity. It shows the income earned by the business before taxes and contains the other two of the five accounts in a complete accounting system. The general accounting equation is: sales minus cost of goods sold minus operating expenses, plus or minus inventory and capital adjustments equals income before taxes. Or, simply, revenue minus expenses equals income before taxes.

The earnings statement is divided into three sections:
1. Cash operating statement
2. Adjustments for inventory
3. Adjustments for capital items

FIGURE 17-10 Ponds represent a long-term asset.

The first section shows all cash income and cash expenses and produces a figure called "net cash farm income." The second section shows the inventory adjustment, which results in a figure called "adjusted net farm operating income." The inventory adjustment is the difference between the ending current assets and beginning current assets, adjusted for changes in accounts payable. The third section shows the capital account adjustment, which results in net farm earnings—or the return to unpaid labor, unpaid management, and equity capital. The capital adjustment is the difference between the intermediate and long-term assets at the end of the year and the intermediate and long-term assets at the beginning of the year.

The earnings statement ties together the information from the balance sheet with cash-basis income tax accounting data. The bottom line is an excellent measure of the profitability of the farm business.

Cash-Flow Statement

The most action-oriented of the "big three," the cash-flow statement shows how cash moves into and out of the business. (See Figure 17-11.) The general accounting equation is: inflows equal outflows. A complete cash-flow can also serve as a cash accuracy check. Many different formats for developing a cash-flow statement are available. One way is to divide the cash-flow into four sections:

1. Income, which is the marketing plan
2. Operating expenses, which is the production plan
3. Capital purchases, which is the investment plan
4. Principal, interest, and additional borrowing, which is the debt service plan.

This type of organization gives a better perspective on total cash-flow and aids in planning and control.

Three columns are necessary for each accounting period. Then, one set of these columns can be for each month or at least for each quarter. The first column would be called "projected," the second column would be called "actual," and the third column would be called "variance." In this fashion, the cash-flow statement can be used as a financial management control tool. In cash-flow planning for income, operating expenses, and investment, the business manager is asking, "How much am I going to sell or buy? At what unit price am I going to buy or sell? At what time am I going to buy or sell?" Debt-service information can be obtained from credit records and the balance sheet. A two- to three-year cash-flow history is useful. Then, the manager can find out how this year is going to differ from previous years. This helps make budgeting easier and more accurate.

The cash-flow statement is useful as an evaluation, control, and planning tool. But used by itself, it can relay false information because it only considers cash. For best results, the cash-flow statement should be used with the balance sheet and earnings statement. Used together, the "big three" provide a complete set of financial statements (Figure 17-12).

Cash-Flow Worksheet

FOR YEAR → 2000	JAN	FEB	MAR	APR	MAY	JUNE	JULY	TOTAL/YR
BALANCE ON HAND								
DESCRIPTION								
INCOME:								
CROP SALES								
FISH SALES								
CAPITAL ITEMS								
CUSTOM WORK								
INTEREST INCOME								
OTHER INCOME								
TOTAL INCOME								
EXPENSES:								
HIRED LABOR								
TAXES								
INSURANCE								
LEASE RENT								
LOAN PAYMENT								
CHEMICALS								
PESTICIDES								
FINGERLINGS								
FUEL, OIL, ETC.								
REPAIRS								
CUSTOM WORK								
FEED PURCHASES								
OTHER EXPENSES								
SUPPLIES								
OTHER EXPENSES								
TOTAL EXPENDITURES								
NET INCOME								
LIVING EXPENSES								
ENDING BALANCE								
AMOUNT TO BORROW								

FIGURE 17-11 A sample cash flow for six months used for projecting or tracking cash, income, and expenditures each month.

Other Key Accounts

Several other accounts feed into or supplement the five accounts in the "big three" financial statements. These include income accounts, expense accounts, capital item accounts, depreciation records, enterprise accounts, labor records, marketing records, feed records, experimental records, individual machine records, and family records.

FIGURE 17-12 Formulas for the big three.

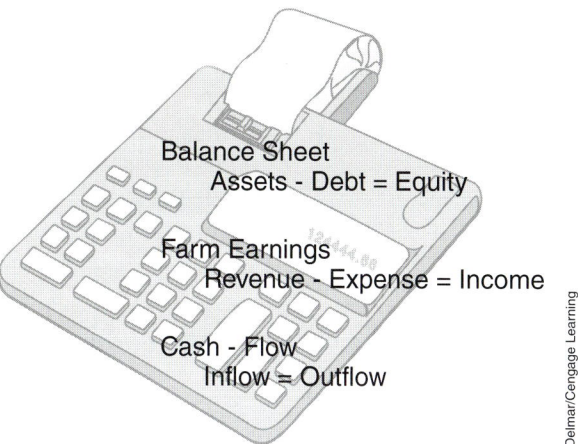

Checking for Accuracy

Single-entry cash-basis accounting can result in significant errors. It is best to balance the checkbook against the record book on a monthly basis. Then at the end of the year, the manager can make three accuracy checks:

1. Cash-flow
2. Profit/net worth
3. Liabilities

When these three accuracy checks balance, the business manager can proceed to file income tax and use the records to analyze and manage the business.

USING AN ACCOUNTING SYSTEM FOR ANALYSIS

Before decisions can be made or analyzed, the information necessary for the decisions must be available. The primary goal of any farm or ranch accounting system should be to provide business management analysis and control. The accounting system should be geared toward the farm or ranch manager. If the accounting system is not used, it is worthless. Many uses of the accounting system relate to individuals other than the manager, so the system must be able to provide financial information for them also. The accounting system supports major management functions by providing the information necessary for making decisions. The accounting system should supply three types of information:

➤ Scorekeeping, or evaluating performance (generally, a retrospective look available in the financial statements)
➤ Attention directing, to flag ongoing operating problems, inefficiencies, and opportunities (identified through analysis of the financial statements)
➤ Problem solving or analyzing the relative merits of alternative courses of action The accounting system provides information the farm manager needs for external reporting for tax and credit purposes. Also, accounting provides financial control of routine operations, business management analysis, and reporting to multiple owners.

Uses and Interpretations of the Statements

Just doing the scorekeeping—producing the financial statements—is not enough. It takes interpretation and analysis of the financial information to meet attention-directing and problem-solving needs.

Interpretation begins by evaluating net worth, a key measure of financial wealth. On a market-value basis, net worth shows what would be left if all assets were converted to cash and all liabilities paid. Next, the net income should be sufficient to meet withdrawals for family consumption. More cannot be taken out of the business than is earned. On a cost basis, the change in net worth from one year-end balance sheet to the next equals net income minus withdrawals. The next step is to use data from the financial statements for a systematic financial analysis of the operation.

Financial Analysis

The first step in financial analysis is to identify appropriate criteria that will facilitate a comprehensive analysis, and then measures for each criterion must be established. For each of the following five criteria (see Figure 17-13), one or more measures are suggested.

Liquidity

A concept describing a firm's ability to meet short-run obligations when due without disrupting the normal operation of the business. The ratio of current assets to current liabilities is a common measure.

Solvency

Solvency is a longer-run concept relating to capital structure and a firm's ability to pay all obligations if assets were liquidated. The focus is on total debt in relation to equity. It is a financial risk measure because the risk of not being able to repay borrowed capital and interest increases as the proportion of debt to net worth increases. Another equally useful measure is debt as a percentage of total assets.

Profitability

Relates to revenue less expenses, called "net farm income." But a dollar measure of net farm income is not sufficient because the size of the business is not considered. Furthermore, net farm income is typically a return

FIGURE 17-13 Criteria for financial analysis.

to unpaid labor, management, and capital, in contrast to other businesses, where it is a return only to capital. Return on assets and return on equity are two common measures of profitability. Net farm income is typically adjusted to get a return to capital expressed as a ratio to total assets.

Financial efficiency

A measure of the efficiency of a business is generating profit out of gross production. The secret of a business is to maximize the dollar value of profit out of each $1,000 value of farm production—a measure of gross production. Net farm income divided by value of farm production is one useful measure. Similarly, operating expenses, interest, and depreciation can individually be evaluated as a proportion of the value of farm production.

Repayment capacity

An assessment of the firm's ability to repay debt. Ability to repay capital debt and interest is a major concern. One measure is all interest plus principal payments on capital debts, expressed as a percentage of the value of farm production. A non-ratio method—capital debt repayment capacity—is calculated as net income plus depreciation less withdrawals.

How Are We Doing?

Financial measures and ratios can be interpreted three ways: (1) comparative analysis, (2) trend analysis, and (3) actual versus budgeted. Comparative analysis is a comparison of one operation's results with those of operations of comparable size and type. For example, if an operation's debt to asset ratio is 40 percent, how does this compare with the debt level of other successful operators? Trend analysis compares results in one year with results achieved in past years. A trend analysis shows strengths and weaknesses and helps focus attention on areas where further strengthening is needed. Comparison of actual performance with the cash-flow budget requires developing an operational plan for the year ahead and then comparing monthly or quarterly performance with projections. Management should focus on variances, or the differences between budgeted and actual performance.

COMPUTERS AND MANAGEMENT DECISIONS

Agribusiness has never been more dynamic and competitive than it is today. Each decision a manager makes, or fails to make, can significantly impact the business. In some cases, a decision can affect a single production cycle of one enterprise. In others, a decision can change the direction of an entire operation. In this fast-paced, high-risk climate, computers can play an important part in helping managers make crucial decisions about their farms. (See Figure 17-14.)

Computer programs help managers make a wide range of management decisions. Some programs are designed for strategic management, which is concerned with positioning the farm for success by matching the

FIGURE 17-14 Once a luxury, now computers are essential to successful business management.

business's long-range direction with resources, management capabilities, and the economic environment of the industry. Other programs address tactical management, which focuses on the day-to-day, season-to-season activities needed to carry out the long-range strategic plans.

The success of strategic or tactical decision making depends to a large degree on managers' access to relevant information and their ability to use that information effectively in making decisions. Today's computer programs can help gather important data, provide a framework for analyzing options, and perform calculations thoroughly and accurately in a fraction of the time it would take to do the same thing with pencil and paper.

Farmers' most important strategic management decisions deal with deciding the long-range direction of their farm businesses. Each must decide what enterprise, or combination of enterprises, offers the best long-term potential, how big the business should be, the type of financing needed, the amount of debt that can be handled, and how to ensure adequate profit from the business now and in the future. The most effective way to approach these and other strategic questions is to identify a wide range of options and then narrow the field to the most feasible plans. The decisions made must be consistent with both business and family goals, available resources, the management ability available, and the risk-bearing capacity of the business and the people involved.

As with any technology, the value of the new computerized management tools depends on how conscientiously and wisely they are used. For the system to reach its potential, the manager must be willing to record vital data on a regular basis and use the resulting information and analyses in making crucial business decisions.

OBTAINING CREDIT

Sound use of agricultural credit is a two-way street affecting both borrower and lender. The individual seeking credit must be prepared to demonstrate to the lending institution that the proposed financing is feasible.

In any borrower-lender relationship, the borrower supplies up-to-date financial and production records to provide an understanding of the business. Financial records include a balance sheet and an income statement, as well as historical and projected cash flows. If possible, three to five years of financial and production data are desirable. Many lenders today are asking for income tax returns for three years.

On the other hand, it is the lender's responsibility to analyze these documents in a logical and systematic manner. This results in a timely decision on the borrower's credit worthiness. Although good financial management is the primary responsibility of the borrower, both lender and borrower must use sound credit practices.

Selecting a Lender

Selecting a lender or lenders is a critical aspect of financial management. An owner/operator should shop for credit and investigate several sources before making a final decision. The borrower must be prepared to make judgments, as well as to be judged. Five guidelines to use in rating the quality of the credit service are:

1. Select a knowledgeable lender who understands aquaculture and agriculture today. Agriculture, undergoing rapid technological change, is beset with unique problems and opportunities. The lender must be able to demonstrate up-to-date knowledge of problems, trends, and modern aquaculture practices specific to the particular enterprise and geographic region. Lenders need a track record that shows an understanding of its customers and a genuine interest in and concern for the customer's welfare and financial progress.

2. Select a lender who has experience in agricultural credit and a commitment to agriculture. In some years, a depressed agricultural economy caused some lending institutions to exit agriculture. As a borrower, examine the lender's farm loan experience. Check its reputation by asking other aquaculturalists. Evaluate the lending institution's commitment to agriculture and service by looking at its track record during periods of adversity.

3. Choose a lender who is willing to discuss lending policies and terms and provide prompt action to credit requests. While investigating a source of credit and related services, compare the terms of credit with other available sources. Total credit charges are more important than interest rates alone. Credit expenses, such as up-front charges for farm loan application and closing fees, can add substantially to credit cost.

4. Choose a lender who has the capacity to meet anticipated credit needs. Agricultural businesses frequently need large sums of capital. This could create a roadblock in the credit process. Some institutions may have a statutory limit placed on the amount of credit they can extend to any one individual or business.

5. Select a lender who has a reputation for honesty and integrity. In our market-based economy, customer service is essential in doing business. Seek a lender familiar with and skilled in financial and production analysis. Periodic visits by the lender to an operation show sincerity and concern and enhance the lender's understanding of the business. Lenders should explain all services offered in practical and understandable terms.

A lender with a reputation for honesty will judge potential borrowers on the same basis. A strong borrower/lender relationship is one of mutual confidence. Maintaining confidentiality of information and objectively evaluating a situation—backing credit decisions with facts—are strong attributes to consider in selecting a lender.

Preparing for the Lender

Five tips help prepare for the credit request and negotiate a financial package with the lender:

1. As a borrower, you must provide current, accurate financial statements and supporting records. A current balance sheet with supporting schedules and inventories is essential. A record of earnings (usually an income statement) and a projected cash flow for your business are also needed. A good set of records showing production plans, short- and long-range goals, and procedures for implementation and evaluation enhance the chances of obtaining credit. Financial information that a particular bank wants may vary.

    A number of excellent farm record systems are available, many of which run on a microcomputer. Some will coordinate and reconcile income statements to balance sheets, ensuring consistent valuation and better tracking of financial progress.

2. Arrange credit in advance. Lenders do not like surprises. Do not inform the lender of a major decision after the fact. This can destroy trust and credibility and make future credit more difficult or impossible to obtain.

3. Allow your lender time to review your plans and make suggestions. Major purchase decisions are sometimes made on the basis of emotion rather than profitability. A lender can provide objectivity and counsel in reviewing your credit request. Explaining goals and plans builds confidence and trust, which strengthens the working relationship.

4. Keep your lender informed. Even the best of businesses face adversity that reduces the ability to repay. Inform your lender as soon as possible of changes in plans or unforeseen problems that will interfere with making loan payments. Communication is the key element in the initial request and throughout the credit process.

5. Maintain a high level of integrity. If you expect a lender to be honest and aboveboard at all times, then the same will be expected of you. Inaccurate information and failure to honor commitments will jeopardize the borrower-lender relationship.

Lender's Five Cs of Sound Credit
- Character
- Capacity
- Capital
- Collateral
- Conditions

FIGURE 17-15 The five Cs.

Managing Credit Use

Once you obtain credit, properly managing the credit becomes a major business challenge. Three basic financial statements—the balance sheet, income statement, and cash-flow statement—are tools used to monitor the financial strength of the business. When compiled and supported by accurate financial information, these tools can provide the support needed for many of the strategies you will devise and financial decisions you will face.

Any business—whether it is an agribusiness firm, farm, corporation, or small business—must meet certain criteria to be successful, particularly if credit is used. A successful business must exhibit strength in repayment ability and capacity, liquidity and solvency, and profitability and financial efficiency. Coincidentally, the lender's cornerstones of sound credit, the five Cs (Figure 17-15), encompass the same qualities:

1. Character (honesty, integrity, and management ability)
2. Capacity (repayment ability and profitability)
3. Capital (liquidity and solvency)
4. Collateral (minimizing risk to the lender)
5. Conditions (for granting and repaying the loan)

Both producer and lender can determine the financial status of the business with these criteria.

Any analysis of the use of credit is only as strong as the quality of financial and other information provided. Circumstances such as size and mix of enterprises, costs, values, commodity prices, collateral values, type of business entity, and time of year can all affect interpretation. Do not base final interpretation on any one factor, but rather on a balanced, comprehensive approach. Comprehensiveness is the number one factor in developing any valid analytical process.

HUMAN RESOURCES

Effective human resource management is an important part of running an aquabusiness, and it begins with planning. Using a plan requires that personnel be recruited and then managed effectively. Managing personnel (Figure 17-16) involves the major functions of work scheduling, training, motivation, evaluation, and discipline.

Personnel Planning

Effective personnel planning starts with a self-assessment by personnel managers. Their personal characteristics, attitudes, strengths and weaknesses, and supervisory skills directly affect the working relationships among employees and others in the farm business. Personnel needs depend on the work (tasks) to be done, the types of products grown, and the machinery and technology of each operation. An analysis of personnel needs should result in a statement of the kind and amount of work to be done, which, in turn, provides a basis for determining the number and types of workers needed. Matching current personnel—family and non-family—with tentative job descriptions is a critical step in developing job

FIGURE 17-16 Being a manager means working with groups of people—managing the human resources.

descriptions for new employees. Identifying mismatches between job descriptions and current responsibilities may help point up training needs, adjustments in job descriptions, shifts in responsibilities and, most important, tasks that cannot be adequately handled with existing personnel.

Hiring Employees

For a team of family and hired workers to function efficiently and effectively, one or more supervisors must carry out the following five personnel management functions: work scheduling, training, motivation, evaluation, and discipline.

Work Scheduling

Work planning and scheduling increase labor efficiency. Waiting for instructions, searching for a supervisor, duplicating the work of another employee, waiting for equipment to be available, doing maintenance work during critical periods of the production season, and wasting harvesting time because equipment was not ready for the season are examples of inefficiencies caused by poor work scheduling.

Work scheduling should be based on a list of tasks to be accomplished, the machinery and equipment needed for the tasks, the people available to do the tasks, and the time in which the work must be done. A task list identifies what needs to be done within the next period or periods of time. The work schedule accompanying the task list identifies the workers and equipment for the tasks. Providing instructions to workers about the tasks they are to do and when and where they are to do them is the final element of the work schedule. The instructions do not have to be given every day if employees are well trained and well supervised.

Training

Farm managers who hire workers with little aquaculture work experience must provide extensive training to new employees. The complexity of many farm tasks, the risk of injury to untrained workers and the labor inefficiencies that result from undirected, on-the-job stumbling make training essential.

Hiring experienced workers is sometimes considered an alternative to carefully planned and implemented training programs. In fact, all employees require training. Experienced employees may require considerable training to change poor work habits, inefficient practices, and lax attitudes toward safety that can endanger themselves and fellow workers. Some employers even prefer to hire inexperienced workers for some tasks because training can focus on the skills that are needed and not on retraining or changing old habits.

Motivation

Employees—family members included—do not change their behavior simply because someone tells them to do so. In fact, threats, bribery, and other types of manipulation may make little difference in an employee's work habits or attitude. The challenge for the farm manager is to balance workers' needs for job satisfaction with the farm's overall business goals. To do this, the farm manager must identify employees' most important unsatisfied needs and then determine the feasibility of satisfying those needs through work itself or conditions at the workplace.

A person working primarily to satisfy a need for social interaction may care little about labor productivity or sales. Can the person satisfy social needs at break times, before and after work, or through casual conversation during work? Or must the worker be disciplined for wasting time on the job?

Evaluation

A formal evaluation program lets employees know where they stand on a regular basis and includes guidelines for wage increases. The evaluation should tell employees how they are doing, identify areas where improvement occurred, and offer constructive suggestions for work improvement. Specific plans for training and job improvement should be discussed. Workers should also have the opportunity to make suggestions, raise questions, and air frustrations and complaints.

In addition to ongoing daily or other regular communications with workers, at least one formal evaluation meeting should be conducted with each employee every year. This meeting provides opportunities to review performance and progress during the past year and to establish performance goals for the coming year.

Compensation should be discussed during the evaluation meeting. Any changes in compensation should be consistent with the strengths and weaknesses discussed in the evaluation meeting. Merit increases should go only to those who have earned them, and employees should understand why they are or are not getting a raise.

Discipline

Workers function best when the rules are clear and they know the consequences of breaking them. Discipline problems can be minimized through careful employee recruitment and training, clear communication of work rules, and proper attention to human needs. When discipline is necessary, the supervisor should not sidestep the responsibility. Failure to provide discipline sends wrong and confusing messages to workers.

BUSINESS MANAGERS OF TOMORROW

The changing agricultural environment creates wide-reaching implications for managers of the future. A turbulent business environment arising from increased integration of the agricultural sector into the national and world economies, technological change, projected changes in government regulations, and expected changes in weather patterns will create new challenges for managers.

Other changes occurred in the agricultural sector that will affect the types of skills farm managers need. These include the movement toward fewer and larger commercial farms, a proliferation of part-time farmers near urban areas, increased vertical integration, and increased involvement of lending institutions in management and ownership. These changes affect the way operations are managed, the knowledge tomorrow's managers will need, and the forms their training will take.

Tomorrow's managers will be fewer in number, better educated, and more diverse than those of yesterday. They will use a broader set of managerial skills to meet the challenges of the turbulent business environment of the future. They will have access to new information and management skills that go beyond their formal education. This continuing educational process will take many forms, ranging from technology-based information transfer to intensive management development programs.

Tomorrow's Managers

The challenges of managing the aquaculture of tomorrow will be met by a wide and diverse group of managers. Farming the water is a business, and the successful farm operation will be managed as a business. The farms may be owned and operated by families, partnerships, or corporations, but the management will rely on business skills for success. Although experience is one method of developing these business skills, continuing education and training combined with farm experience is a faster, less risky means of developing sound business skills and practices.

Skills Managers Will Need

Tomorrow's managers will need an expanded set of managerial skills to succeed. Managerial skills in three areas—communication, business and economics, and technology—will be developed through formal university education, business experience, and continuing education.

Managing Innovation and Change

Changes in technology, information, and marketing systems occur at an increasing rate. As this rate of change continues, managers will be forced to adopt new practices and employ strategic thinking to survive. Changing consumer demand also will require increased innovation to fill existing or developing market niches.

Managing Risk

The growing exposure to global competition in production and financial markets, changing government policies related to trade, supply, and the environment, and changes in climate require new managerial skills in dealing with risk.

Tomorrow's farm and ranch managers will be forced to use the futures and options markets, contractual arrangements, and other risk-shifting tools to manage these risks.

Designing Effective Organizations

As the structure of farms and ranches evolves over the next several years, managers will confront many organizational challenges, including the need to develop organizations that use labor effectively and that can take advantage of new relationships with buyers, suppliers, and competitors.

Designing Information Systems

Increasing information and rapidly evolving information technologies create a distinctive set of challenges for farm and ranch managers. Practical information acquisition systems and computer decision support systems will continue to be developed to aid managers in this area.

Managing Human Resources

As the average size of commercial farms increases and the number of part-time farmers grows, human resource problems will become more important to farm and ranch managers. These problems may be increased by absentee land owners and managers. The challenges of dealing with more seasonal employees, reliance on specialized personnel, and expanded interactions with suppliers, buyers, and processors are likely to occupy more and more of the farm or ranch manager's time. Part-time farmers will face the need to balance farming demands with off-farm employment.

Strategic Planning

In addition to developing these five specific skills, managers will have to be strategic thinkers, capable of dealing with a turbulent environment by using the techniques and tools of strategic planning. These techniques will aid in managing technological innovations and in dealing with changing governmental policies, markets, weather, and business. Although based on theory and technical knowledge, these skills will best be developed through case-study learning experiences.

Global Marketplace

The managers of tomorrow will need to understand agricultural production and marketing in the global marketplace. (See Figure 17-17.) This implicitly includes an understanding of consumers and their evolving needs. As the firms that farm and ranch managers deal with as buyers, suppliers, and competitors become increasingly global, the managers of the production operation will need to understand their needs to better develop working relationships. Exciting opportunities may exist for cooperation between individual operations or among groups, particularly in satisfying consumer needs by filling niche markets. Managers will likely learn about changing consumer needs through nontraditional study, including internships, study abroad programs, and various conferences and institutes.

FIGURE 17-17 Managers of tomorrow will be successful in a complex world.

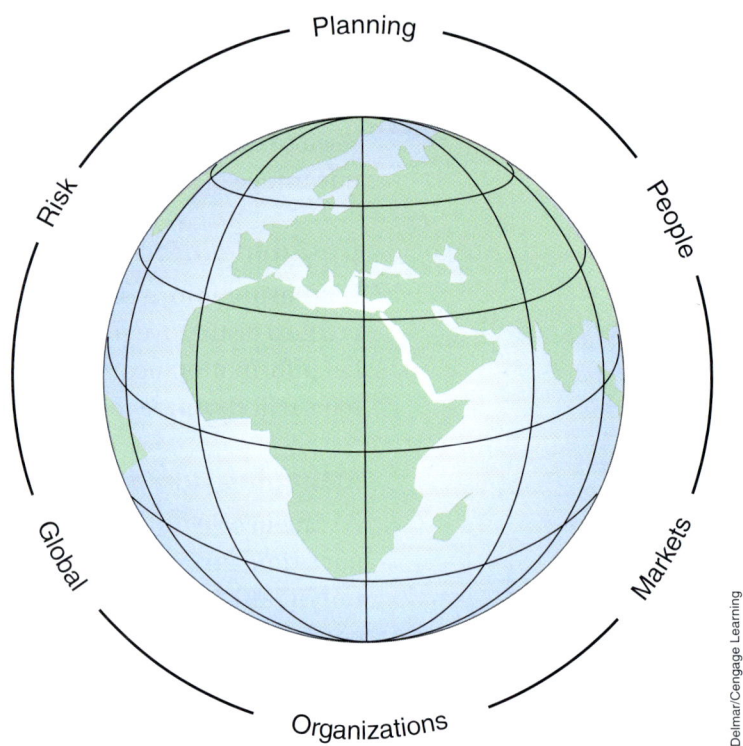

Computers and Management Tools

Managers of the future must also be computer literate. In addition to serving as a valuable information-management tool, the computer will become increasingly important to the manager as a decision aid and as a means of communicating with other producers, trade associations, private firms, and government agencies. With advances in cell phone and computer technology, the need for information literacy and technological savvy has never been greater. There are currently more than 3.4 billion cell phone users worldwide, making cell phones more common than computers. Cell phones are a modern phenomenon with a greater impact than the Internet, providing increased mobility, access to innumerable new technologies and capabilities.

Short courses and home study will likely provide important training, as will expanded use of computer simulations in traditional classroom settings. In this regard, computer technology and the Internet make new ways of learning available—online courses, video conferencing, webinars, and so on.

Agriculturists have always made decisions in a risky and uncertain environment. Increasing factors will affect that ambiguous environment. Computer programs help managers analyze the risks associated with their decisions. Almost all college graduates today are exposed to, understand, and accept computer technology, and they recognize that it can be used to sort vast quantities of information and aid in making decisions. As past generations of managers are followed by a computer-literate generation, demand will increase for continuing education on ways to use the latest information and technology for short- and long-term decision making.

Multiple Training Sources

To train tomorrow's managers, existing educational programs are likely to be modified—with a great deal of support from nontraditional sources. The following will be key sources of training for tomorrow's managers:

- Formal college education
- Management development programs
- Extension education
- Home study through new communication methods (online classes, conferences, and webinars)
- Professional and trade associations

The manager of the future will face an increasing need to understand the behavioral aspects of markets, employees, and competitors. Increased understanding of the liberal arts, as well as business concepts and practices, will be important. Effective networks among peers will also help. As technology changes, networking changes, due to the various evolving social networks on the Internet. The successful business manager will constantly learn and change with the times.

SUMMARY

No one should enter an aquaculture business without counting the personal, social, and financial costs. Management skills are necessary to operate the business successfully. These skills are gaining in importance as the economy changes and the workplace changes. Management involves the best use of human resources of the business and the best use of the financial resources of the business. Records are essential to good financial management.

Record systems can be simple or complex, hand-kept or computerized. Besides meeting tax requirements, they provide an effective part of planning and evaluating the business. When a business needs to borrow money, records support the need and ability to repay the debt.

Obtaining and using credit in an aquabusiness is a two-way street. Borrowers look for fair, understanding, and knowledgeable lenders. Lenders lend money to honest, knowledgeable borrowers who can demonstrate a plan and an ability to repay the loan. Both the borrower and lender manage the credit.

Business managers of the future face challenges. More will be required for success—more planning, more technology, more knowledge, and more human resources. Managers of the future will need to know how to learn and seek training and education from a variety of sources.

STUDY/REVIEW

Success in any career requires knowledge. Test your knowledge of this chapter by answering these questions or solving these problems.

True or False

1. People and businesses always write down their goals.
2. Most aquaculture businesses keep records at least for income tax purposes.
3. A balance sheet is a type of accounting method.
4. A planning document is one of the "big three" statements.
5. Cost of feed represents a fixed cost.

Short Answer

1. List six input costs for starting an aquaculture facility.
2. Name six variable costs associated with catfish production.
3. Give three examples of fixed costs associated with any aquaculture operation.
4. List five basic management functions used to achieve the goals and objectives of an aquabusiness.
5. List the eight steps of strategic planning.
6. List three benefits of setting business goals.
7. Identify eight general business risks.
8. What is the most common form of business organization?
9. Name two forms of multiple business ownership.
10. List the four functions for records.
11. Name two accounting methods.
12. List the three categories of a balance sheet.
13. What is the general accounting equation for a cash-flow statement?
14. Name three accuracy checks for a single-entry accounting system.
15. List five steps to financial analysis of an aquabusiness.
16. List the five Cs that are the cornerstone of sound credit.
17. Identify five functions of human resource management.
18. List six managerial skills needed by the manager of the future.

Essay

1. Indicate four personal and social costs associated with starting an aquacultural business.
2. Explain why enterprise budgeting is useful in establishing an aquaculture business.
3. Explain the difference between fixed cost and variable cost.
4. Define management.

5. Define the following: organizing, staffing, directing, and controlling.
6. Explain why motivation is an important part of management.
7. Explain the difference between tactical planning and strategic planning.
8. Define assets.
9. Define a farm earning statement.
10. A good accounting system supplies what three types of information?
11. Give three reasons why the computer helps make better management decisions.
12. Describe three guidelines for selecting a lender.
13. Give three tips for negotiating a loan.

KNOWLEDGE APPLIED

1. Invite a successful businessperson from the community—an entrepreneur—to class to describe start-up costs and planning for a new business. Discuss the variable costs and the fixed costs of the business.
2. Invite an accountant who specializes in farm accounts for aquaculture or agriculture to visit the class. Ask about the types of records provided to the farmer or aquabusiness. Find out how the records are generated. Discuss the value of the "big three" financial statements.
3. Determine the total cost to obtain the following production equipment for an aquaculture operation. Also, describe more completely the specifications of each piece of equipment. Use the table provided here.

Production Equipment

Item	Cost
Tractors (50 hp), units × cost/unit	
Feeder, units × cost/unit	
Feed bin (10 tons) with pad, units × cost/unit	
Trucks (1 ton), units × cost/unit	
Aluminum boat (14 ft)	
Boat motor (25 hp)	
Boat trailer	
Transport tank (125 gal) and equipment	
Fixed electric aerators, units × cost/unit	
Portable aerators, units × cost/unit	
Storage and service building (30 ft × 50 ft)	

(Continued)

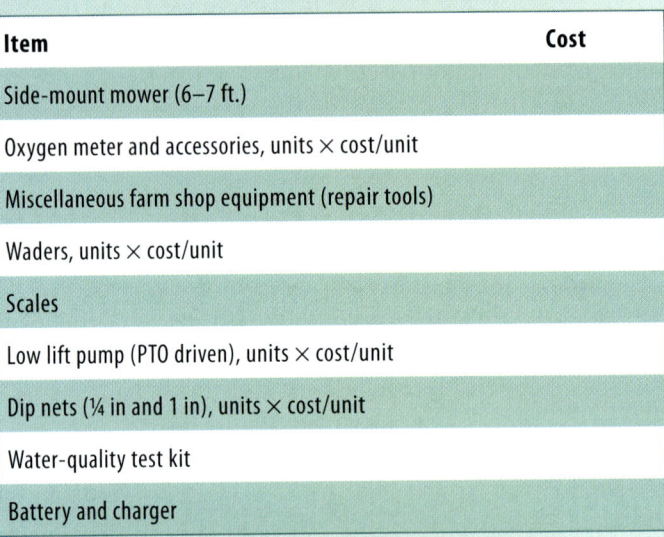

Item	Cost
Side-mount mower (6–7 ft.)	
Oxygen meter and accessories, units × cost/unit	
Miscellaneous farm shop equipment (repair tools)	
Waders, units × cost/unit	
Scales	
Low lift pump (PTO driven), units × cost/unit	
Dip nets (¼ in and 1 in), units × cost/unit	
Water-quality test kit	
Battery and charger	

4. Visit with a banker who specializes in agricultural or aquacultural loans. Find out what is required to obtain a loan and how the loans are managed by the lending institution. Discuss the roles of production knowledge and trust in the lending process.
5. Prepare an enterprise budget. Collect realistic values for the budget. Use mortality figures and feed-conversion rates normal for the species chosen. Show fixed costs and operating costs computed on an annual basis. Estimate the break-even and the payback. Use the table provided here.

Budget Analysis for the Annual Commercial Production of Fish

Item	Unit	Price or Cost/Unit	Quantity	Value or Cost per Pound	Your Cost
1. EXPECTED RECEIPTS					
Fingerlings					
Foodfish					
Other					
Total Value:					
2. OPERATING EXPENSES					
Fingerlings					
Feed					
Chemicals					
Fuel					
Pumping					
Aeration					

(Continued)

Item	Unit	Price or Cost/Unit	Quantity	Value or Cost per Pound	Your Cost
Feeding					
Transportation					
Repairs and Maintenance					
Ponds					
Water Supply					
Feeding Equipment					
Harvesting Equipment					
Miscellaneous Equipment					
Labor					
Manager					
Hired Full-Time					
Hired Part-Time					
Telephone					
Advertising					
Subtotal:					
Depreciation					
Pond Construction					
Water Supply					
Feeding Equipment					
Harvesting Equipment					
Miscellaneous Equipment					
Subtotal:					
3. INTEREST ON INVESTMENT					
Loan No. Purpose					
1	( )				
2	( )				
Subtotal:					

(Continued)

Item	Unit	Price or Cost/Unit	Quantity	Value or Cost per Pound	Your Cost
4. TAXES					
Property					
FICA					
State					
Federal					
Subtotal:					
5. INSURANCE					
Equipment					
Liability					
Life					
Subtotal:					
Total Costs:					

From your values, determine the following:

a. Income above Operating Expenses: _____

   Total Value Item 1 − Subtotal Item 2

b. Net Return to Land, Management, and Risk: _____

   Total Value Item 1 − Subtotal Items 2 + 3 + 4 + 5

c. Break-even Cash Price: _____

   Per Pound Value Item 1 − Per Pound Subtotal Item 2

d. Total Break-even Price: _____

   Per Pound Subtotals for Items 2 + 3 + 4 + 5

6. Visit with several local business owners and discuss the problems encountered in human relations and human resources in the workplace. Discuss the roles of leadership training and managing people.

7. Collect case studies of business problems, either financial or management. Suggest solutions for these problems. Match your skills against those of local business managers and owners.

8. Visit with a representative of a local small business association. Discuss the types of ownerships for a business—sole proprietorship, corporation, or partnership. Develop a list of advantages and disadvantages for each.

LEARNING/TEACHING AIDS

Books

Flynn, S. M. (2005). *Economics for dummies*. Hoboken, NJ: Wiley.

Nelson, B., Economy, P. and Blanchard, K. (2003) *Managing for dummies*. Hoboken, NJ: Wiley.

Romanowski, N. (2006) *Sustainable freshwater aquacultures: The complete guide from backyard to investor*. University of New South Wales Press.

Tiffany P. and Petersen, S.D. (2004). *Business plans for dummies*. Hoboken, NJ: Wiley.

Tyson, E. and Schell, J. (2008). *Small business for dummies*. Hoboken, NJ: Wiley.

World Bank. (2007). *Changing the face of the waters: The promise and challenge of sustainable aquaculture (Agriculture and Rural Development Series)*. World Bank Publications.

Internet

Internet sites represent a vast resource of information. The URLs (uniform resource locator) for the World Wide Web sites can change. Using a search engine such as Google, find more information by searching for these words or phrases: aquaculture business, cash-basis accounting, accrual accounting, enterprise budget, or strategic planning.

For some specific Internet sites refer to Appendix Tables A-11, A-12 and A-14.

One purpose of education and learning is to become employable and stay employable—to get and keep a job. People look for careers, and careers look for people. Two broad categories of career opportunities in aquaculture are working for someone else and working for yourself. Success in any career requires some

CHAPTER 18

Career Opportunities in Aquaculture

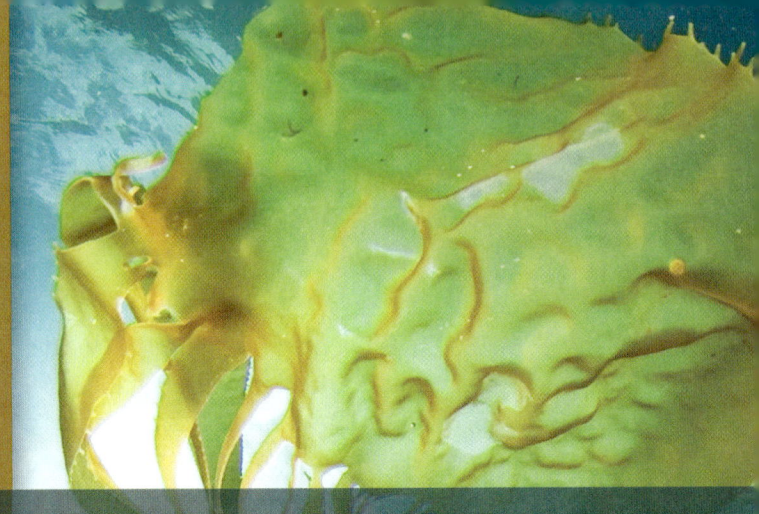

general skills and knowledge, as well as some very specific skills and knowledge unique to a chosen occupation in aquaculture.

OBJECTIVES

After completing this chapter, the student should be able to:

- List the basic skills and knowledge needed for successful employment and job advancement
- Describe the thinking skills needed for the workplace of today
- Identify the traits of an entrepreneur
- List six occupational areas of aquaculture
- Describe the general duties of the occupations in six areas of aquaculture
- Describe the education and experience needed to enter six areas of aquaculture
- List six general competencies needed in the workplace
- Describe five ways to identify potential jobs
- List eight guidelines for choosing a job
- List 10 guidelines for filling out an application form
- Describe a letter of inquiry or application
- List the elements of a resume or data sheet
- Describe ten reasons an interview may fail

- Identify the careers that require a science background
- Discuss what research studies indicate about basic skills and thinking skills for the workplace

Understanding of this chapter will be enhanced if the following terms are known. Many are defined in the text, and others are defined in the glossary.

KEY TERMS

Competencies
Creative thinking
Cultural diversity
Curricular
Data sheet
Demographic
Entrepreneur
Follow-up letter
Forecast
Interpersonal
Letter of application
Letter of inquiry
Resources
Résumé
Sociability
Systems
Visualization

GENERAL SKILLS AND KNOWLEDGE

Study after study indicates that potential employees seldom receive basic skills and knowledge. Without these basic skills and knowledge, the specific skills and knowledge for employment in aquaculture are of little value. The new workplace demands a better prepared individual than in the past. Finally, those individuals working for themselves must develop a trait called "entrepreneurship." This may also be a good trait for any employee.

Basic Skills

Success in the workplace requires that individuals possess skills in reading, writing, arithmetic, mathematics, listening, and speaking, at levels identified by employers nationwide.

Reading

An individual ready for the workplace of today and the future demonstrates reading (Figure 18-1) with the following **competencies**:

- Locates, understands, and interprets written information, including manuals, graphs, and schedules to perform job tasks
- Learns from text by determining the main idea or essential message
- Identifies relevant details, facts, and specifications
- Infers or locates the meaning of unknown or technical vocabulary
- Judges the accuracy, appropriateness, style, and plausibility of reports, proposals, or theories of other writers

Reading skills in aquaculture are necessary to keep up with new information and read directions for feeding or treating fish.

Writing

Individuals ready for the workplace of today and the future demonstrate writing abilities with the following competencies:

- Communicates thoughts, ideas, information, and messages
- Records information completely and accurately
- Composes and creates documents such as letters, directions, manuals, reports, proposals, graphs,

FIGURE 18-1 Reading skills are important to success.

and flow charts with the appropriate language, style, organization, and format
- Checks, edits, and revises for correct information, emphasis, form, grammar, spelling, and punctuation

In aquaculture, writing skills are necessary for such tasks as keeping pond or raceway records, describing disease conditions, or requesting a test.

Arithmetic and Mathematics

The workplace of today and the future requires individuals with competencies in arithmetic and mathematics. Arithmetic is computing with numbers by addition, subtraction, multiplication, and division. It is the application of mathematics. These important competencies are:
- Perform basic computations
- Use numerical concepts such as whole numbers, fractions, and percentages in practical situations
- Make reasonable estimates of arithmetic results without a calculator
- Use tables, graphs, diagrams, and charts to obtain or convey information
- Approach practical problems by choosing from a variety of mathematical techniques
- Use quantitative data to construct logical explanations of real-world situations
- Express mathematical ideas and concepts verbally and in writing
- Understand the role of chance in the occurrence and prediction of events.

Anyone not convinced of the value of arithmetic and mathematics to aquaculture should consider the skills required to figure pond volumes, treatment dosages, feed conversion ratios, and fish growth rates.

Listening

Individuals working today and in the future must demonstrate an ability to listen. This means to receive, attend to, and interpret verbal messages and other cues such as body language. Real listening means that the individual comprehends, learns, evaluates, appreciates, or supports the speaker.

Speaking

Finally, individuals successful in the workplace of today and tomorrow must demonstrate these speaking competencies:
- Organize ideas and communicate oral messages appropriate to listeners and situations
- Participate in conversation, discussion, and group presentations
- Use verbal language, body language, style, tone, and level of complexity appropriate for audience and occasion
- Speak clearly and communicate the message
- Understand and respond to listener feedback
- Ask questions when needed

Thinking Skills

Contrary to the old workplace, many research studies indicate that employers in the new workplace want workers who can think. (See Figure 18-2.) Employers search for individuals showing competencies in **creative thinking**, decision making, problem solving, mental **visualization**, knowing how to learn, and reasoning.

FIGURE 18-2 Thinking skills are important for successful employment.

Creative Thinking

Creative thinkers generate new ideas by making nonlinear or unusual connections or by changing or reshaping goals to imagine new possibilities. These individuals use imagination, freely combining ideas and information in new ways.

Decision Making

Individuals who use thinking skills to make decisions are able to specify goals and limitations to a problem. Next, they generate alternatives and consider the risks before choosing the best alternative.

Problem Solving

As silly as it may sound, the first step to problem solving is recognizing that a problem exists. After this, individuals with problem-solving skills identify possible reasons for the problem and then devise and begin a plan of action to resolve it. As the problem is being solved, problem solvers monitor the progress and fine-tune the plan. Being able to recognize a disease condition and look for solutions is a good example of problem solving in aquaculture.

Mental Visualization

This thinking skill is seeing things in the mind's eye by organizing and processing symbols, pictures, graphs, objects, or other information—for example, seeing a catfish pond from a diagram or a recirculating system's operation from a schematic.

Knowing How to Learn

Perhaps of all the thinking skills, this is most important because of rapid changes in technology. This skill is knowing how to learn techniques to apply and adjust existing and new knowledge and skills in familiar and changing situations. Knowing how to learn means awareness of personal learning styles and formal and informal learning strategies.

Reasoning

The individual who uses reasoning discovers the rule or principle connecting two or more objects and applies this to solving a problem. For example, chemistry teaches the theory of pH measurements, but the reasoning individual is able to use this information in understanding pH shifts in pond culture.

General Workplace Competencies

Besides the basic skills and the thinking skills, the workplace of today and tomorrow demands general competencies in the use of **resources, interpersonal skills**, information use, **systems**, and technology.

Resources

Resources of a business include time, money, materials, facilities, and people. Individuals in the workplace must know how to manage:
- Time, through goals, priorities, and schedules
- Money, with budgets and **forecasts**
- Material and facility resources, such as parts, equipment, space, and products
- Human resources, by determining knowledge, skills, and performance levels

Interpersonal

More than ever, people cannot act in a vacuum. Most people are members of a team (Figure 18-3), where they contribute to the group. They teach others in their workplace when new knowledge or skills are needed. More than ever, and at all levels, individuals must remember to serve customers and satisfy their expectations. Through teams, individuals frequently exercise leadership to communicate, justify, encourage, persuade, or motivate individuals or groups. As part of employment teams, individuals negotiate resources or interests to arrive at a decision. Finally, all interpersonal skills require individuals to work with and use **cultural diversity**.

Information

The information age is here. Individuals in the workplace must cope with and use information. Successful individuals will identify the need for information and evaluate the information as it relates to a specific job. With

FIGURE 18-3 Teams work together to solve problems and use resources efficiently.

FIGURE 18-4 Computers open access to more information than ever before.

the computer, individuals in the workplace must organize and process information in a systematic way. (See Figure 18-4.) Also, with all this information available, individuals must interpret and communicate information to others using oral, written, or graphic methods. For example, production data of raceways must be summarized. To manage production information, computer skills are the key.

Systems

No longer can any aspect of a business or industry be viewed as standing alone. Every activity is part of a system, and individuals now seek to understand systems, whether these are social, organizational, technological, or biological. With an understanding of the systems in a business, trends can be identified and predictions and diagnoses can be made. Individuals then modify the system to improve the product or service. Aquaculturalists view the pond as a system; production is another system.

Technology

Technology makes life easier only for those who know how to select it, use it, maintain it, and troubleshoot it. Technology is complicated. Successful individuals learn to apply appropriate new technology through all the basic skills, the thinking skills, and general workplace competencies.

Personal Qualities

After all the training in basic skills, thinking skills, and general workplace competencies, individuals still fail for lack of some personal qualities. These include responsibility, self-esteem, **sociability**, self-management, and integrity or honesty.

Responsible individuals work hard at tasks even when the task is unpleasant. Responsibility shows in high standards of attendance, punctuality, enthusiasm, vitality, and optimism in starting and finishing tasks.

Those possessing self-esteem believe in themselves and maintain a positive view of themselves. These individuals know their skills, abilities, and emotional capacity. They feel good about themselves.

Successful individuals demonstrate understanding, friendliness, adaptability, empathy, and politeness to other people. These skills are demonstrated in familiar and unfamiliar social situations. The best examples are sincere individuals who take an interest in what others say and do.

Along with self-esteem is self-management. Individuals successful in business accurately assess their own knowledge, skills, and abilities while setting well-defined and realistic personal goals. Once goals are set, those who manage themselves monitor their progress and motivate themselves through the achievement of goals. Self-management also implies a person who exhibits self-control and responds to feedback unemotionally and non-defensively.

Finally, to be successful in aquaculture, an employee or **entrepreneur** requires good old-fashioned honesty and integrity. Good ethics are still a part of good business.

INTANGIBLE SKILLS

More and more employers seek employees with intangible skills or soft skills, like balance in a person's life, communicating effectively, problem solving, decision making, resolving conflict, working with others, planning, conducting effective meetings, professional growth, ethics, community service, and volunteerism. These skills are outside of technical skills needed in a job or career. Yet they often are more important in determining an individual's success. Employers have been asking schools and colleges to teach these soft skills. Some schools and colleges have taught these skills, but not in formalized classroom settings with prepared lessons and assessments. Often, many of these skills are gained when the student takes an active role in a club or organization such as FFA or the National Postsecondary Agricultural (PAS) organization. Leadership roles in an organization are particularly effective in developing soft skills. Students who participate in contests such as the FFA Career Development Events (CDE) also seem to develop more of the soft skills important to success in the workplace.

Recognizing the need for these soft skills, many companies have developed formalized lessons, training, and assessment. For example the National FFA Organization has developed educational materials, called LifeKnowledge®, that provide training in four areas of the soft skills: personal, organizational, career, and community. LifeKnowledge® includes a set of 257 lesson plans structured to help FFA advisors and agricultural education teachers educate and prepare middle school and high school students to become proficient in 16 competency areas. In addition, another 100 lesson plans are available specifically for use at the collegiate level. The 16 competency areas and precept statements include:

Premier Leadership

A: Action

A1. Work independently and in groups to get things done
A2. Focus on results
A3. Plan effectively
A4. Identify and use resources
A5. Communicate effectively with others
A6. Take risks to get the job done
A7. Invest in others by enabling and empowering them
A8. Evaluate and reflect on actions taken and make appropriate modifications

B: Relationships

B1. Practice human relations skills including compassion, empathy, unselfishness, trustworthiness, reliability, and listening
B2. Interact and work with others
B3. Develop others
B4. Eliminate barriers in building relationships
B5. Participate effectively as a team member

C: Vision

C1. Contemplate the future
C2. Conceptualize ideas
C3. Demonstrate courage to take risks
C4. Adapt to opportunities and obstacles
C5. Persuade others to commit

D: Character

D1. Live with integrity
D2. Accurately assess my values
D3. Accept responsibility for personal actions
D4. Respect others
D5. Practice self-discipline
D6. Value service to others

E: Awareness

E1. Address issues important to the community
E2. Perform leadership tasks associated with citizenship
E3. Participate in activities that promote appreciation of diversity

F: Continuous Improvement

F1. Implement a leadership and personal growth plan
F2. Seek mentoring from others
F3. Use innovative problem-solving strategies
F4. Adapt to emerging technologies
F5. Acquire new knowledge

Personal Growth

G: Physical Growth

    G1. Practice healthy eating habits
    G2. Respect one's body
    G3. Participate in a fitness program
    G4. Set goals for long-term health

H: Social Growth

    H1. Acknowledge that differences exist among people
    H2. Present self appropriately in various settings
    H3. Develop and maintain relationships

I: Professional Growth

    I1. Plan and implement professional goals and priorities
    I2. Make clear decisions in my professional life
    I3. Demonstrate professional ethics
    I4. Balance personal and professional responsibilities
    I5. Demonstrate exemplary employability skills

J: Mental Growth

    J1. Think critically
    J2. Think creatively
    J3. Practice sound decision making
    J4. Solve problems
    J5. Commit to life-long learning
    J6. Persuade others
    J7. Practice sound study skills

K: Emotional Growth

    K1. Cope with life's trials
    K2. Live a compassionate and selfless life
    K3. Develop self-assurance and confidence
    K4. Embrace the emotional development process
    K5. Establish emotional well-being
    K6. Seek appropriate counsel
    K7. Practice healthy expressions of love

L: Spiritual Growth

    L1. Nurture a spiritual belief system
    L2. Respect and be sensitive to others' beliefs

Career Success

M: Communications

    M1. Demonstrate technical and business writing skills
    M2. Demonstrate professional job seeking skills

M3. Makes effective business presentations
M4. Communicate appropriately with coworkers and supervisors
M5. Operate effectively in the workplace

N: Decision Making

N1. Demonstrate the decision making process
N2. Demonstrates problem-solving skills
N3. Make ethical decisions
N4. Choose a career based on passion, abilities, and aptitudes

O: Flexibility and Adaptability

O1. Embraces emerging technology in the workplace
O2. Manages change
O3. Reacts with openness to feedback and professional growth opportunities
O4. Experiments and takes risks

P: Technical and Functional Skills in Agriculture and Natural Resources

Many high school agricultural programs and organizations such as 4-H, Block and Bridle, Alpha Zeta, PAS, FarmHouse, FFA, and Sigma Alpha endorse this training. Information about the training can be found on the LifeKnowledge Web site: www.CollegiateLifeKnowledge.org or on the FFA website: www.ffa.org/ageducators/lifeknowledge/. The common goal of this type of training is to help students succeed in their careers and to become society-ready graduates.

ENTREPRENEURSHIP

The most common view of an entrepreneur is one who takes risk and starts a new business. Though this may be true for some in aquaculture, some traits of entrepreneurship are desirable at many levels of employment in aquaculture. Within any organization, an entrepreneur may:

- Find a better or higher use for resources
- Apply technology in a new way
- Develop a new market for an existing product
- Use technology to develop a new approach to serving an existing market
- Develop a new idea that creates a new business or diversifies an existing business

Anyone can be an entrepreneur. The attitude of an entrepreneur includes:

- Risk taking with clear expectations of the odds
- Focusing on opportunities and not problems

FIGURE 18-5
Entrepreneurship requires mastery.

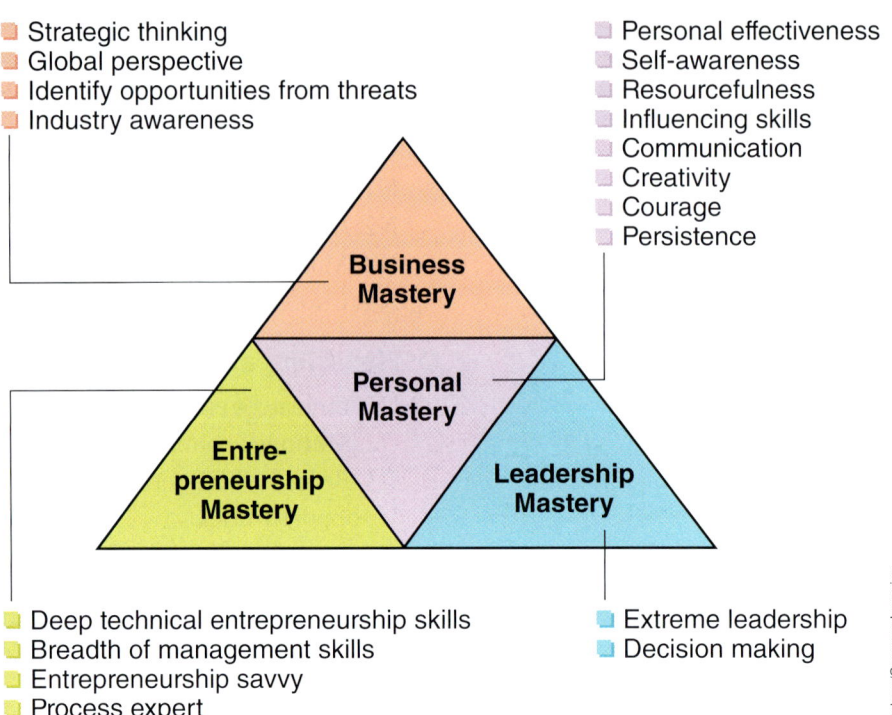

- Focusing on the customer
- Seeking constant improvement
- Emphasizing productivity over appearances
- Recognizing importance of example
- Keeping things simple
- Practicing open door and personal contact leadership
- Encouraging flexibility
- Being purposeful and communicating a vision

Entrepreneurs are ready for the unexpected, differences, new needs, change, demographic shifts, changes in perception, and new knowledge. Entrepreneurs are good employees and good employers. Entrepreneurs keep the aquaculture industry growing. (See Figure 18-5.)

JOBS IN AQUACULTURE

Some people consider the only jobs available in aquaculture to be those in the actual production or farm work. But the industry as a whole requires a large number of people to support the infrastructure of suppliers, producers, and marketers.

Specific jobs or employment opportunities in aquaculture can be grouped into general categories. These include supplies and services, training, production employment, marketing, inspection, and research and development. Each area requires some unique skills.

Supplies and Services

Occupations in the supplies and services area include those that support farm production and provide the inputs necessary for an operation to be productive. General areas of employment in supplies and services include financing, providing feed and other supplies, providing equipment, constructing facilities, and consulting.

Finance

Individuals involved in financing aquacultural operations provide money (capital) for the establishment and operation of facilities. They do this through loans and other types of financial assistance. Some examples of occupations in finance include:

- Banker (Figure 18-6)
- Loan officer
- Farm Credit System employee (http://www.farmcreditnetwork.com/).

FIGURE 18-6 Bankers help arrange credit for the growth of aquaculture and nutrition.

Besides a good knowledge of finance, a good knowledge of aquaculture is necessary. Typical duties include helping fill out loan application forms, evaluation of loan application forms, locating credit sources, arranging loans, financial advisement, and collecting debts. The responsibilities of aquaculture may occupy only part of the job, so other areas of agriculture could be part of the duties of an individual working in finance.

Feed and Supplies

Providing feed and supplies includes manufacturing, hauling, selling, storing, and purchasing feed, feed ingredients, and other supplies. Obviously, more and different variations of employment will be available in areas where more aquaculture is found. Some types of occupations include:

- Feed salesperson
- Feed mill purchasing agent
- Feed truck driver
- Feed mill operator
- Feed mill worker
- Feed mill scales operator
- Nutritionist

The knowledge and skills required for these jobs vary widely. For example, a feed salesperson (Figure 18-7) and a feed mill purchasing agent require a strong background in nutrition, but a nutritionist requires the greatest knowledge and training in nutrition. Nutritionists must understand the nutritional requirements of the species being fed and what nutrients each feedstuff provides. Truck drivers and feed mill workers need some mechanical abilities and need to know how to operate equipment safely. The skills of a feed mill operator combine some nutritional, mechanical, and engineering knowledge, plus a greater level of responsibility.

General duties of individuals providing feed and supplies include operating equipment, identifying feed ingredients, making careful measurements, product knowledge, following instructions, and consulting with farmers.

FIGURE 18-7 Selling feed requires an understanding of aquaculture and nutrition.

Equipment

Aquaculture requires some unique equipment and the development of new equipment. Equipment jobs include manufacturing, selling, hauling, and installing the equipment used in aquaculture facilities. Some examples are:

- Equipment engineer
- Manufacturing plant worker
- Equipment installer
- Equipment salesperson

Again, the exact skills and knowledge vary considerably depending upon the job. For example, the mechanical knowledge of an engineer is greater than that of the manufacturing plant worker, but the plant worker is probably better skilled in welding. All these jobs require some mechanical aptitude with skills and knowledge in welding, electricity, and hydraulics. Sales positions require product knowledge with strong communication and interpersonal skills.

Duties for individuals who choose jobs in areas providing equipment to aquaculture facilities include working in plants to assemble equipment and traveling to farms to sell, deliver, install, and repair equipment.

Construction

Constructing aquaculture facilities (Figure 18-8) involves designing, laying out designs, operating equipment, installing water facilities, and constructing buildings. Some examples of occupations include:

- Architect
- Carpenter
- Heavy equipment operator

FIGURE 18-8 To construct new ponds requires the services of surveyors, heavy equipment operators, well drillers, and welders.

- Electrician
- Well driller
- Surveyor

These occupations are needed for many types of industries. In areas where aquaculture is prominent, these occupations will be more specific. For example, in Mississippi, heavy equipment operators and surveyors develop skills for laying out and building catfish ponds. Electricians spend more time wiring pumps for aquaculture in Mississippi.

Typical duties for individuals in aquaculture construction include equipment operation, laying out facilities, installing pipe, making electrical connections, drawing and reading plans (blueprints), and working with all types of construction materials. Individuals interested in pursuing employment in construction areas need a mechanical aptitude.

Consulting

Consultants advise the aquaculturalist on how to establish, operate, and manage a business. A good consultant is worth good money. A bad consultant can cost lots of good money. Some examples of consultants include:

- General aquaculture consultant
- Extension aquaculture specialist (http://aquanic.org/jsa/federal_guide/stateextension.htm)
- Veterinarian
- Nutritionist

The knowledge and skills required vary with the type of consultant. Most require at least a bachelor's degree, but in many cases a master's or doctorate is preferred. Also, the best consultants combine education with practical experience. Consultants stay current with new developments and can sift through information to discover trends and make judgment calls.

Typical duties of consultants, depending on the type, include making on-farm visits, holding meetings, providing information, developing reports, making suggestions, providing close examinations of an operation, and providing advice on a wide range of subjects.

FIGURE 18-9 Good instructors are needed for classroom and laboratory training in aquaculture programs.

Training

With all the interest in aquaculture and with all the potential that it holds, employment opportunities exist in the area of training and education. Some typical occupations include:

- High school agriculture instructor (Figure 18-9)
- Postsecondary aquaculture instructor
- Fisheries and wildlife instructor
- Extension specialist

To become an instructor requires at least a bachelor's degree. Teaching in a high school requires certification. The best instructors also possess some good practical experience. Instructors and trainers must have excellent communication skills.

Duties of instructors include finding or preparing **curricular** materials for the classes, preparing lesson plans, using a variety of methods to transfer information, and evaluating the effectiveness of the information transfer. The duties of instructors last for full days and weeks at a time. Extension specialists provide training but on a short-term basis and generally to those already involved in aquaculture.

Production Employment

Employment in production involves all the occupations associated with reproducing, growing, and harvesting the aqua crop. These can be divided into management and worker categories.

Management

Management includes the planning, organizing, staffing, directing, and controlling of farm activities. Management requires responsibility, the level of which should be reflected in the wage and the necessary skills and knowledge. Management occupations include:

- Farm manager
- Site manager
- Hatchery manager

Managers possess a thorough knowledge of the production of the species being cultured. They know how to manage and supervise other people. Duties of managers can involve hiring and firing of workers, providing directions to workers, making decisions, meeting government regulations, selling the crop, buying feed and supplies, and keeping records. Managers need good communication skills to communicate with the owner and the workers.

Workers

Workers on an aquaculture facility can be very diverse. Workers can include unskilled as well as highly skilled and educated individuals. A few examples of some occupations include:

- Fishery technician
- Water technician
- Truck driver (Figure 18-10)
- Seine operator (Figure 18-11)
- Biologist
- Skilled laborer
- Unskilled laborer

The type of worker depends on the size and type of facility. Some facilities may be small enough that a worker could do several different jobs. Large facilities may have individuals for each job. The knowledge and skills vary. Technicians and biologists require knowledge and skills beyond the high school level and even to the bachelor's degree. Seine operators, truck drivers, and skilled and unskilled laborers receive short-term training and on-the-job training. Because the work is around ponds, raceways, and tanks, most of these jobs require good physical skills.

General duties of workers on an aquaculture facility include such things as checking water quality, moving fish, harvesting fish, feeding fish, hauling seines, and operating a variety of equipment. In general, these individuals work with water and with fish.

FIGURE 18-10 Truckers drivers use large trucks to move feed from the mill to the production facility.

FIGURE 18-11 Harvesting catfish requires someone who can drive a tractor and operate the reel and seine.

Marketing

Employment in marketing includes jobs in all activities that connect the product from the farm with the consumer. This involves jobs in processing and promoting.

Processing

Processing prepares the aquaculture crop for the consumer. Jobs are often in processing plants, many in assembly line settings (Figure 18-12). Some of the possible jobs include—
- Plant manager
- Supervisor
- Quality control specialist/technician
- Human resource manager
- Skilled laborer
- Unskilled laborer

Some positions in processing plants require performance of repetitive activities that can be learned quickly, whereas other positions require considerable experience. Some of the typical duties in processing include working with the live and dead fish, operating and cleaning equipment, cleaning the facilities, working around water, working in refrigerated areas, and inspecting fish and fish products.

Promoting

Promotion convinces the consumer to purchase the product. Promotion includes advertising, public relations, market analyses, demonstrations, and selling. Some examples of occupations in promotion are:
- Salesperson
- Advertising account representative
- Writer
- Photographer
- Association executive

FIGURE 18-12 Many jobs in processing plants are in assembly line settings.

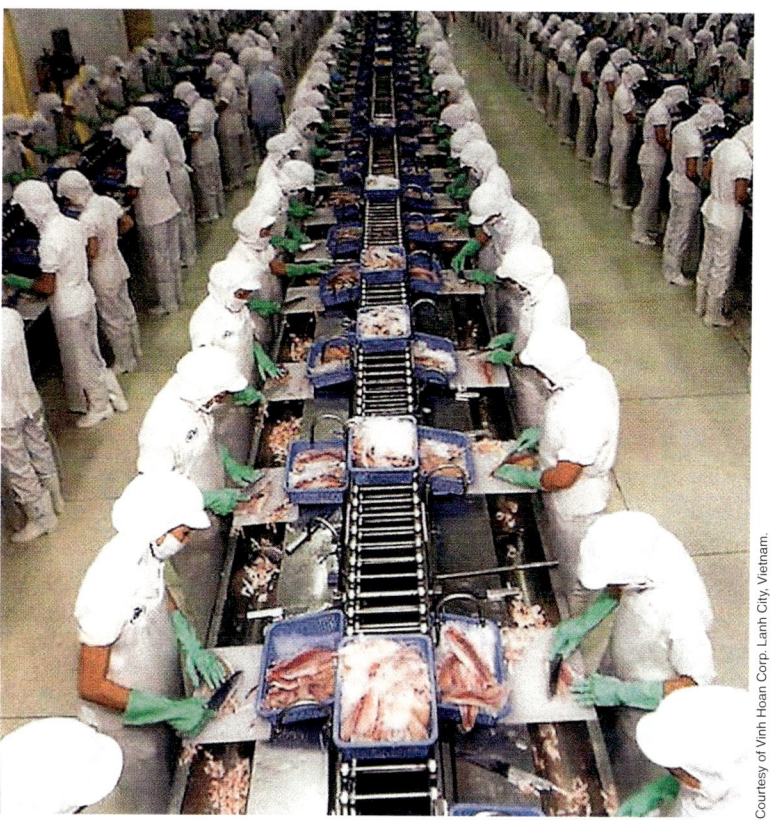

Occupations in promotion can take place in a city far removed from the area of production. Still, the work requires a practical knowledge of aquaculture. The ability to talk the language of aquaculture is important for success. Additionally, occupations in promotion require education in marketing, communication, and advertising. Duties may include writing articles, writing advertising, making sales presentations, placing orders, planning advertising campaigns, reading reports, and preparing reports—anything that promotes the consumption of the products of aquaculture.

Inspection

Every industry witnesses more regulation by government, often initiated by consumer advocacy groups. This creates inspection or monitoring jobs to issue permits and check for compliance with regulations in such areas as water quality, chemical use, and processing facility standards. Examples of occupations in inspection and monitoring include:

➤ Soil conservation technician (http://www.nrcs.usda.gov/about/employment.html)
➤ Fish and Wildlife Technician (http://www.fws.gov/jobs/dayinthe-life.html)
➤ Food inspector
➤ Water technician/tester

- Grader
- Quality control technician
- Laboratory technician

The primary focus of these occupations is on assuring quality products for the consumer while maintaining the environment. The work involves contact with farmers, processors, and others to enforce government regulations and laws. These occupations demand excellent human relations and communication skills.

Research and Development

Aquaculture progressed to the stage it is today because of a commitment to research and development. Occupations in research and development involve finding new and profitable ways to produce products in forms desired by consumers. Research may also identify new animals or plants for production. Some of the general occupations include:

- Research scientist/biologist
- Research technician (Figure 18-13)
- Laboratory assistant
- Unskilled assistant

Government agencies and private industries conduct research on new techniques and varieties. Depending upon the job, the duties include planning research, designing experiments or trials, selecting plants or animals for the trials, conducting the day-to-day steps of the experiment or trial, collecting data, laboratory analyses, analyzing data, reporting data, and writing final reports of the results. The work is often a mix of outdoor and indoor activities.

Depending on the job, the education required varies. The research scientist needs a doctorate in some related field—nutrition, fish biology, or water chemistry. Research technicians and laboratory assistants require some training beyond high school. Some of the training is often provided by the agency or industry that owns the laboratory.

SUPERVISED AGRICULTURAL EXPERIENCE

A Supervised Agricultural Experience (SAE) is designed to provide students the opportunity to gain experience in agricultural areas based on their interests. An SAE represents the actual, planned application of concepts and principles learned in agricultural education. Students experience and apply what is learned in the classroom to real-life situations. Students are supervised by agriculture teachers in cooperation with parents/guardians, employers, and other adults who assist them in the development and achievement of their educational goals. The purpose is to help students develop skills and abilities leading toward a career.

Planning and conducting an SAE for aquaculture could include areas of interest such as water technician, sales, fishery technician, hatchery technician, and instructor. Students should work with their instructors to:

FIGURE 18-13 Research keeps aquaculture moving forward, and jobs in research areas are available on many levels. Ecologist Rocky Smiley and technician Sarah Boone use a backpack electrofisher to assess fish communities in headwater streams and drainage ditches.

- Identify an appropriate SAE opportunity in the community
- Ensure that the SAE represents meaningful learning activities benefiting the student, the agriculture education program and the community
- Obtain classroom and individual instruction on SAE
- Adopt a suitable record keeping system
- Plan the SAE and acquire needed resources
- Coordinate release time and visits to SAE
- Sign a training agreement along with the employer, teacher and parent/guardian
- Report on and evaluate the SAE and records resulting from it

Additional help and ideas for planning and conducting an SAE can be found through the National FFA website (www.ffa.org).

EDUCATION AND EXPERIENCE

Requirements to begin working in aquaculture vary depending upon the level of work. One requirement common to all is practical work experience in aquaculture. Often, in order to gain this practical experience, the new employee begins at an entry-level job and then is advanced through the organization. The advancement depends on productivity, and on the skills and knowledge the employee brings to the job, skills, and knowledge gained on the job.

Entry-level educational requirements vary. The basic skills, thinking skills, and general workplace competencies discussed in this chapter are important. These skills should be obtained in high school and reinforced during additional training and schooling. More specialized education in aquaculture is offered in some high schools, community colleges, and universities.

Many high school programs in aquaculture provide the education necessary for lower-level positions. Often high school programs in aquaculture provide students with supervised work experience in some aspect of aquaculture. This is invaluable in getting a job and in helping individuals determine if they wish to pursue additional education. Some community colleges and other postsecondary schools provide specialized programs in aquaculture with practical experience as a part of the schooling. Programs at community colleges focus on entry-level technician jobs.

Universities and colleges offering bachelor's degrees, master's degrees, and doctorate programs provide highly specialized education in aquaculture. Depending on their location, they may specialize in warmwater culture, coldwater culture, genetics, nutrition, water chemistry, and other areas. A search of the Internet produces listings of high school, college, and university aquaculture programs; for example:

Sea Grant Program: http://www.oar.noaa.gov/programs/sgcolleges.html Campus Explorer (search "aquaculture"): http://www.campusexplorer.com, and

Arizona Aquaculture Education Programs: http://ag.arizona.edu/azaqua/azhighs.html.

IDENTIFYING A JOB

Finding that first job or switching job scan be difficult. Books, videos, and seminars are available on finding jobs. What follows are some suggestions. The Learning/Teaching Aids section at the end of this chapter contains more information. Sources for locating jobs include:

- Classified advertisements of newspapers
- Magazines or trade journals and publications
- Personal contacts
- Placement offices
- Employment or personnel office of company
- Public notices
- The Internet

Newspapers, magazines, trade journals, and publications can be good resources for locating a job. By reading the advertisements in these publications, the potential employee can determine the demand for his or her job skills. Also, the potential employee can compare his or her skills and training with those listed in the advertisements.

Another kind of classified advertisement is the "situation wanted" section of newspapers, magazines, and trade journals. Many people secure excellent jobs by advertising their skills in these sections. An employer may read about an individual's skills and realize he or she is the answer to the employer's needs.

Personal contacts are still a top source of jobs. Employers do not like to make mistakes. Some feel that if a trusted acquaintance makes a recommendation, this lessens the chances of making a mistake in hiring. Also, personal contacts may know of jobs opening up before they are publicly announced. This gives the potential employee more time to prepare and research the job. Personal contacts include friends, relatives, teachers, guidance counselors, and employees of the company.

Placement offices provide vocational counseling, give aptitude and interest tests, locate jobs, and arrange job interviews. Three types of placement offices are available: public, private, and school. These agencies work to match employers with prospective employees. Often too, an agency knows how to help a potential employee prepare and present himself or herself.

Public placement offices are supported by federal and state funds. Their services are free. Private placement offices charge for services they provide. This usually is a percentage of the beginning salary. Individuals using private placement services sign a contract before services are provided. High schools, trade schools, and colleges may maintain a placement service for their students. They also provide help for individuals to identify their aptitude or interest for a job and help in preparation for job interviews. Many companies support their own employment or personnel office. Individuals seeking employment can fill out application forms and leave **résumés** in case a job becomes available.

Finally, some companies seeking new employees may issue a public notice of some kind. This includes posters or fliers on bulletin boards around a community. Posters or flyers are sent to related businesses, which end up on their bulletin boards. Schools and colleges often receive public announcements of jobs.

Internet posting of jobs is another kind of a bulletin board. (See Figure 18-14.) Some companies and services maintain computerized databases of jobs. Interested individuals use the Internet to search for jobs that match their qualifications and desires. Often, résumés can be submitted over the Internet or at least faxed. This type of job listing opens the door wide to potential jobs—even those that are not local.

Career Clusters

Recently, a series of career clusters were developed to help students and instructor identify jobs and careers in 16 broad career areas. Career clusters provide a way for schools to organize instruction and student experiences around categories that encompass virtually all occupations—from entry through professional levels. The clusters provide information on the knowledge and skill required. One of the 16 clusters is Agriculture, Food and Natural Resources, which includes seven pathways:

➤ Food Products and Processing Systems
➤ Power, Structural and Technical Systems
➤ Plant Systems
➤ Natural Resource Systems
➤ Animal Systems

FIGURE 18-14 Advertisements for jobs posted on the Internet through AquaJobs.com.

- Environmental Service Systems
- Agribusiness Systems

Aquaculture occupations can be found in all the pathways. More information on the career clusters can be found at www.careerclusters.org/16clusters.htm or www.agrowknow.org. The additional pathway of Agricultural Biotechnology will likely be added to Agriculture, Food and Natural Resources in the future.

GETTING A JOB

Once job possibilities are identified, the work begins. Getting hired is difficult and requires preparation. Again, books, videos, and seminars teach how to get a job. A few tips follow.

Once you choose a job opening to pursue, do a little research on the company and the job before applying. Know these things about the job and the company:

- Name of the company
- Name of personnel manager
- Company address and phone number
- Position available
- Requirements for the position
- Geographic scope of the company (local, county, state, regional, national)
- Company's product(s)
- Recent company developments
- Responsibilities of the position
- Demand for the company's product(s)

Before you get too far along in the application process, be certain that the position is one that you want to pursue. Money is not everything in a job. Compare the characteristics of the occupation with those you possess by answering these questions:

- Does the job description fit your interests?
- Is this the level of occupation in which you wish to engage?
- Does this type of work appeal to your interests?
- Are the working conditions suitable to you?
- Will you be satisfied with the salaries and benefits offered?
- Can you advance in this occupation as rapidly as you would like?
- Does the future outlook satisfy you?
- Is there enough demand for this occupation that you should consider entering it?
- Do you have or can you get the education needed for the occupation?
- Can you get the finances needed to get into the occupation?
- Can you meet the health and physical requirements?
- Will you be able to meet the entry requirements?
- Are there any other reasons you might not be able to enter this occupation?
- Is the occupation available locally, or are you willing to move to a part of the country where it is available?

Application Forms

If the company requires an application form, remember that you are trying to sell yourself by the information given. Review the entire application form before you begin. Pay particular attention to any special instructions to print or write in your own handwriting. When answering ads that require potential employees to apply in person, be prepared to complete an application form on the spot. Take along a pen. Prepare a list of information you will need to complete the application form. The information may include your social security number; the addresses of schools you have attended; names, phone numbers, and addresses of previous employers and supervisors; names, phone numbers, and addresses of references. The following guidelines will provide you with some direction when completing application forms.

- Follow all instructions carefully and exactly.
- If you write by hand rather than type, write neatly and legibly. Handwritten answers should be printed unless otherwise directed.
- Application forms should be written in ink unless otherwise requested. If you make a mistake, mark through it with one neat line.
- Be honest and realistic.
- Give all the facts for each question.
- Keep answers brief.
- Fill in all the blanks. If the question does not pertain to you, write "not applicable" or "N/A." If there is no answer, write "none" or draw a short line through the blank.
- Many application forms ask what salary you expect. If you are not sure what is appropriate, write "negotiable," "open," or "scale" in the blank. Before applying, try to find out what the going rate for similar work is at other locations. Give a salary range rather than exact figure.
- Application forms submitted online require the same considerations as those filled out on site or filled out at home and taken to the employer.

Letters of Inquiry and Application

The purpose of a **letter of inquiry** is to obtain information about possible job vacancies. The purpose of a **letter of application** is to apply for a specific position that has been publicly advertised. Both letters indicate your interest in working for a particular company, acquaint employers with your qualifications, and encourage the employer to invite you for a job interview.

Letters of inquiry and application represent you. They should be accurate, informative, and attractive. Your written communications should present a strong, positive, professional image both as a job seeker and future employee. The following list should be used as a guide when writing letters of inquiry and application.

- Be short and specific (one or two pages). Use 8×11 in white typing paper, not personal or fancy paper.
- Be sure the letter is neatly typed and error free.

- Use an attractive form, free from smudges.
- Write to a specific person. Use "To Whom It May Concern" if answering a blind ad.
- Write logically organized paragraphs that are to the point.
- Write carefully constructed sentences free from spelling or grammatical errors.
- Be positive in tone.
- Express ideas in a clear, concise, direct manner.
- Avoid slang words and expressions.
- Avoid excessive use of the word "I."
- Avoid mentioning salary and fringe benefits.
- Write a first draft, then make revisions.
- Proofread final letter yourself, and also have someone else proofread it.
- Address and sign correctly. Type envelope addresses. This information should be included in a letter of inquiry:
  - Specify the reasons why you are interested in working for the company and ask if there are any positions available now or in the near future.
  - Express your interest in being considered a candidate for a position when one becomes available.
  - Because you are not applying for a particular position, you cannot relate your qualifications directly to job requirements. (You can explain how your personal qualifications and work experience would help meet the needs of the company.)
  - Mention and include your résumé.
  - State your willingness to meet with a company representative to discuss your background and qualifications. Include your address and phone number where you can be reached.
  - Address letters of inquiry to the "Personnel Manager" unless you know his or her name.

A letter of application should include:
- Indicate your source of the job lead.
- Specify the particular job you are applying for and the reason for your interest in the position and the company.
- Explain how your personal qualifications meet the needs of the employer.
- Explain how your work experience relates to job requirements.
- Mention and include your résumé.
- Request an interview and state your willingness. Include your address and phone number where you can be reached.

Résumé or Data Sheet

Some jobs require a résumé or data sheet. The following information should be considered when writing a résumé or data sheet:
- Name, address, phone number, and e-mail address
- Brief, specific statement of career objective

- Educational background—names of schools, dates, major field of study, degrees, or diplomas—listed in reverse chronological order
- Leadership activities, honors, and accomplishments
- Work experience listed in reverse chronological order
- Special technical skills and interests related to job
- References
- Limit to one page if possible
- Neatly typed and error free
- Logically organized
- Honest statement of qualifications and experiences

Employers look for a quick overview of who you are and how you fit into their business. On the first reading, an employer may only spend 10 to 15 seconds reading a resume. Be sure to present relevant information clearly and concisely in an eye-catching format. (See Figure 18-15.)

FIGURE 18-15 Resumes are neat, specific, logical, and one page long.

RESUME
Jane Jones

Current Address
PO Box 1238
Anywhere, ID 00000

Tel: 000/888-8888; e-mail address: jjones2@micron.com

Education
Local High School, Anywhere, ID: Graduated 2008.

College of Southern Idaho, Twin Falls, ID, 2008–2010: A.A.S., Fish Technology.

Career Objectives
Obtain satisfying job in the aquaculture industry that provides advancement opportunities during my career.

Activities and Honors
- Active member of 4-H Club for three years.
- Member FFA for four years and was elected president during senior year.
- Member Postsecondary Agricultural Student (PAS) organization 2008–2010.
- Advisor to local 4-H Club 2009 to present.

Employment and Work Experiences
January 2009 to Present: Fish R Us, Arco, ID; general help; sell aquarium supplies and ornamental fish, and maintain fish in 20 aquarium tanks.

July 2006 to December 2008: ABC Grocery, Anywhere, ID; restocked shelves; boxed groceries; worked into checker position.

References

Available on request.

The Interview

The next step in the job-hunting process is the interview. Though many dos and don'ts of an interview are available, perhaps the best advice comes from the interviewer's side of the desk. This list of items includes common reasons interviewers give for not being able to place applicants in a job:

- Poor attitude
- Unstable work record
- Bad references
- Lack of self-selling ability
- Lack of skill and experience
- Not really anxious to work
- Bad-mouthing former employers
- Too demanding (wanting too much money or to work only under certain conditions)
- Unable to be available for interviews
- Poor appearance
- Lack of manners and personal courtesy
- No attempt to establish rapport; not looking the interviewer in the eye

WOW! MY FIRST JOB INTERVIEW

You just got your first job interview, or maybe you just bombed two job interviews and you have a chance at another one. What should you do?

Check over this list of dos. Do:

1. Find out about the company before you interview—its products, who its customers are, and the like.
2. Be neat and well-groomed. Dress conservatively.
3. Be punctual—15 to 20 minutes early.
4. Have your résumé and examples of your work available for quick reference.
5. Have a pen and notepad to take notes.
6. Have a prepared list of questions regarding the job. These may be answered by the interviewer during the course of the interview.
7. When meeting the receptionist, smile, introduce yourself, state that you have an appointment, follow the receptionist's instructions, and wait patiently.
8. Greet the interviewer with a smile and by name.
9. If the interviewer offers his or her hand, shake it firmly.
10. Introduce yourself and state the purpose of your appointment.
11. Be seated only after the interviewer has asked you to do so.
12. Sit and stand erect.
13. Be polite and courteous.
14. Be sincere, enthusiastic, friendly, and honest.
15. Let the interviewer take the lead in the conversation.
16. Be alert. Sit slightly forward in the chair to give an alert appearance.
17. Be confident, look directly at the interviewer.
18. Make an effort to express yourself clearly and distinctly.
19. Speak correctly, use proper grammar, speak in clear, moderate tones.

- Being interested only in the salary and benefits of the job
- Lack of confidence; being evasive
- Poor grammar, use of slang
- Not having any direction or goals

Follow-Up Letters

Follow-up letters are sent immediately after an interview. The follow-up letter demonstrates your knowledge of business etiquette and protocol. Always send a follow-up letter regardless of whether you had a good interviewing experience and regardless of whether you are interested in the position. When employers do not receive follow-up letters from job candidates, they often assume that the candidate is not aware of the professional protocol they will need to demonstrate on the job.

The major purpose of a follow-up letter is to thank those individuals who participated in your interview. In addition, a follow-up letter reinforces your name, application, and qualifications to the employer and indicates whether you are still interested in the job position.

20. Take time to think about your answer. Choose your words carefully.
21. Answer questions completely, but give only essential facts.
22. Convey positive answers.
23. Speak positively of former employers and associates.
24. Watch for signs that the interview is over, such as the interviewer shuffling papers, moving chair around, and the like.
25. Thank the interviewer for his or her time.
26. Shake hands with the interviewer and leave promptly at the completion of the interview.
27. Write a follow-up letter to express your interest in the job and your appreciation for the opportunity to interview.

Some don'ts of interviewing include the following. Don't:

1. Take others with you to the interview—parents, friends, and so on.
2. Put your hat or coat on the interviewer's desk.
3. Use a limp or overpowering handshake.
4. Lean against a wall, chair, or desk.
5. Interrupt the interviewer.
6. Chew gum, smoke, eat candy, or the like.
7. Giggle, squirm in your chair, tap your fingers, swing a crossed leg, or the like.
8. Use slang or swear.
9. Talk too long.
10. Try to flatter the interviewer.
11. Give all yes or no answers.
12. Talk about personal problems.
13. Press for a decision on being hired. Employers have good and bad interviews, but they remember bad interviews longer.

SUMMARY

A primary goal of education and training is to become employable and stay employable—to get and keep a job or run a successful business. The world of work still requires people who can read, write, do math, and communicate. Rapidly advancing technology makes this even more apparent. Also, the modern workplace now looks for people who possess thinking skills. With a solid set of basic skills, future employees also need to relate to other people, to use information, and to understand the concept of systems and use technology. Old-fashioned ideas like responsibility, self-esteem, sociability, self-management, and integrity are not out of date either.

Aquaculture jobs range from those very closely tied to aquaculture to those that support aquaculture. In general, potential job areas include supplies and services, training, production, marketing, inspection, and research and development. Education and training for jobs in aquaculture vary from on-the-job training to high school to college degrees. After training and education, finding and getting the right job may still be a challenge. Several good resources exist for locating a job. Still, the best one is personal contact. Letters of inquiry, letters of application, a resume, and being prepared for the job interview help secure a job.

STUDY/REVIEW

Success in any career requires knowledge. Test your knowledge of this chapter by answering these questions or solving these problems.

True or False

1. Technology makes the basic skills of reading, writing, and math less necessary for today's workplace.
2. The only jobs in aquaculture are at trout and catfish facilities.
3. Jobs in aquaculture range from unskilled to highly skilled and highly educated.
4. Employers in the modern work force want employees who will take orders and suppress thinking skills.
5. Entrepreneurs are risk takers only involved in the development of a new aquabusiness.

Short Answer

6. List five basic skills necessary for employment.
7. List five thinking skills desired by today's employers.
8. List three personal qualities found in successful employees.
9. List five occupations in supplies and services for aquaculture.
10. An individual who wants to become an educator in aquaculture would look for jobs in what organizations?
11. Name two general areas in aquaculture that are apt to rely on unskilled labor.
12. Name the general occupation in aquaculture where the actual work can take place in a city far removed from the area of production.
13. Give three examples of occupations in the area of inspection.
14. If an individual wants to work in research and development in aquaculture, what occupations are available?
15. List five ways to find a job in aquaculture.
16. The following competency demonstrates an individual's writing ability:
    a. Identifies relevant details, facts, and specifications
    b. Checks for correct grammar and spelling
    c. Understands the role of change in the prediction of events
    d. Uses tables and graphs to obtain information
17. List three guidelines for filling out an application form.
18. What is the purpose of a follow-up letter?

Essay

19. Give an example of why basic skills would be important to each of the following occupations in aquaculture: loan officer, feed mill salesperson, equipment salesperson, and nutritionist.
20. Why are thinking skills important in any occupation in aquaculture?
21. Describe interpersonal skills.
22. Define an entrepreneur.
23. Describe how architects, carpenters, electricians, surveyors, and heavy equipment operators could be involved in occupations related to aquaculture.
24. Define a consultant for aquaculture and give examples of the duties of a consultant.
25. Briefly outline the process of getting a job in aquaculture.
26. Describe ten reasons an interview may fail.
27. What should a résumé contain and how long should it be?
28. What duties and responsibilities might be a part of working in a fish processing plant?

KNOWLEDGE APPLIED

1. Gather sample résumés and develop your own résumé or data sheet.
2. Use the Internet to find position announcements for jobs in aquaculture. Write a letter of job inquiry and a letter of job application for a selected job using this information.
3. Develop a list of questions frequently asked during an interview. Use the questions in role-playing job interviews and videotape the interviews.
4. Organize a field trip to a public or private placement office. Following the field trip, discuss the office's policies and how they affect job searchers and employers. Alternatively, invite a representative from a state employment agency to explain how employment agencies can help students gain employment.
5. Hold an aquaculture career field day. Invite individuals currently employed in aquaculture to present a panel discussion on career opportunities. For example, invite representatives from production, aquaculture, research, education, and government.
6. Search the Internet and select one career in aquaculture of interest and prepare a research paper on the career. The paper should identify the knowledge and skills required and the employment opportunities.
7. Collect pictures or photographs of people engaged in various careers in aquaculture and prepare a bulletin board collage.
8. Invite a resource person such as a business owner or personnel manager to discuss what he or she looks for in resumes, application letters and forms, and during interviews.
9. Invite a panel of local agribusiness people to discuss the importance of employee work habits, basic skills and attitudes, and how they affect the entire business.

LEARNING/TEACHING AIDS

Books

Business Council for Effective Literacy Bulletin. (1987). *Job-related basic skills: A guide for planners of employee programs*. New York: Business Council for Effective Literacy (BCEL).

Hull, D. *Career pathways: Education with a purpose*. (2005). Wavo, TX: CORD Communications.

Echaore-McDavid, S. and McDavid, R. (2010). *Career opportunities in agriculture, food, and natural resources*. New York, NY: Checkmark Books.

U.S. Department of Education, U.S. Department of Labor. (1988). *The bottom line: Basic skills in the workplace*. Washington, DC.

U.S. Department of Education. (1990). *Policy perspectives. Workplace Competencies: The Need to Improve Literacy and Employment Readiness*. Washington, DC. (See Workplace Basic Skills.com: (http://www.workplacebasicskills.com/)

U.S. Department of Education. (1991). *America 2000: An education strategy. Sourcebook*. Washington, DC.

U.S. Department of Labor. (1991). *What work requires of schools*. A SCANS Report for America 2000. Washington, DC. (See What Work Requires of Schools: http://wdr.doleta.gov/SCANS/whatwork/)

U.S. Department of Labor. (1992). *Learning a living: A blueprint for high performance*. A SCANS Report for America 2000. Washington, DC. (See Learning A Living: http://wdr.doleta.gov/SCANS/lal/)

U.S. Department of Labor, Bureau of Labor Statistics. (2010) *Occupational outlook handbook*. Washington, DC: http://www.bls.gov/oco/.

Web Sites

AgCareers.com: http//www.agcareers.com
AgriSeek: http://www.agriseek.com/
Aquaculture jobs.com: http://www.aquaculturejobs.com
Business.com: http://www.business.com/directory/agriculture/aquaculture/employment/weblistings.asp
JobMonkey: http://www.jobmonkey.com/aquaculturejobs/
Wikihow, How to get a job: http://www.wikihow.com/Get-a-Job
World Aquaculture Society Employment Service: http://darc.cms.udel.edu/wases/

Internet

Internet sites represent a vast resource of information. The URLs (uniform resource locator) for the World Wide Web sites can change. Using a engine such as Google, find more information by searching for these words or phrases: aquaculture careers, résumé, letter of appreciation, letter of inquiry, data sheet, basic skills or competencies (math, reading, English, writing, thinking), entrepreneurship, jobs in aquaculture (i.e., aquaculture consultant, extension, aquaculture specialist, veterinarian, nutritionist, instructor, hatchery manager, water technician, biologist, quality control specialist, laboratory technician, or research technician).

For some specific Internet sites refer to Appendix Tables A-11 and A-15.

Due to its location in a book and because of the implications of its name, an appendix is often ignored by the reader. But an appendix contains valuable information that can enhance a reader's understanding and learning. Moreover, the information in an appendix is quick and easy to find.

The information in this appendix includes a variety of useful conversions, conversion factors, measurement standards, common measures, mixing guides, contacts, sampling guides, a list of fish propagated by aquaculture, publications, training resources, and a key for pond water. Finally, this appendix contains some tables with Internet sites (URLs) that lead to many sources of data and information about aquaculture. By making full use of this appendix, the reader can understand, do, and learn more, and plan better.

TABLE A-1 MISCELLANEOUS CONVERSION FACTORS FOR AQUACULTURE USE	
1 acre-foot	= 43,560 cubic feet
1 acre-foot	= 325,850 gallons
1 acre-foot of water	= 2,718,144 pounds
1 cubic foot of water	= 62.4 pounds
1 gallon of water	= 8.34 pounds
1 gallon of water	= 3,785 grams
1 liter of water	= 1,000 grams
1 fluid ounce	= 29.57 grams
1 fluid ounce	= 1.043 ounces
1 grain per gallon	= 17.1 milligrams/liter
1 milliliter of water	= 1 gram
1 cubic meter of water	= 1 metric ton
1 quart of water	= 946 grams
1 teaspoon	= 4.9 milliliters
1 tablespoon	= 14.8 milliliters
1 cup	= 8 fluid ounces

(*Continued*)

TABLE A-1 MISCELLANEOUS CONVERSION FACTORS FOR AQUACULTURE USE (CONTINUED)

1 acre-foot/day of water	= 226.3 gallons/minute
1 acre-inch/day of water	= 18.9 gallons/minute
1 acre-inch/hour of water	= 452.6 gallons/minute
1 second foot of water	= 448.8 gallons/minute
1 cubic foot/second of water	= 448.8 gallons/minute
1 foot of water	= 0.43 pound/square inch
1 foot of water	= 0.88 inch of mercury (Hg)
1 horsepower	= 550 foot-pounds/second
1 horsepower	= 745.7 watts
1 kilowatt	= 1,000 watts
1 kilowatt	= 1.34 horsepower
1 hectare	= 10,000 square meters
1 hectare	= 2.47 acres
1 acre	= 4,048 square meters

TABLE A-2 CONVERSION FACTORS (C.F.): THE WEIGHT OF A CHEMICAL THAT MUST BE ADDED TO ONE UNIT VOLUME OF WATER TO GIVE ONE PART PER MILLION (PPM)

Amount of Chemical	To Equal
2.72 pounds per acre-foot	1 ppm
1,233 grams per acre-foot	1 ppm
0.0283 gram per cubic foot	1 ppm
0.0000624 pound per cubic foot	1 ppm
0.0038 gram per gallon	1 ppm
0.0584 grain per gallon	1 ppm
1 milligram per liter	1 ppm
0.001 gram per liter	1 ppm
8.34 pounds per million gallons of water	1 ppm
1 gram per cubic meter	1 ppm
1 milligram per kilogram	1 ppm
10 kilograms per hectare-meter	1 ppm

TABLE A-3 CONVERSION OF UNITS OF VOLUME

FROM	cm³	liter	m³	ft³	fl. oz.	fl. pt.	fl. qt.	gal.
				TO				
cm³	1	0.001	1×10^{-6}	3.53×10^{-5}	0.0338	0.00211	0.00106	2.64×10^{-4}
liter	1,000	1	0.001	0.0353	33.81	2.113	1.057	0.2642
m³	1×10^{6}	1,000	1	5.31	3.38×10^{4}	2,113	1,057	264.2
ft³	2.83×10^{4}	28.32	0.0283	1	957.5	59.84	29.92	7.481
fl. oz.	29.57	0.0296	2.96×10^{-5}	0.00104	1	0.0625	0.0313	0.0078
fl. pt.	473.2	0.4732	4.73×10^{-4}	0.0167	16	1	0.5000	0.1250
fl. qt.	946.4	0.9463	9.46×10^{-4}	0.0334	32	2	1	0.2500
gal.	3,785	3.785	0.0038	0.1337	128	8	4	1

cm³ = cubic centimeter = milliliter = ml; m³ = cubic meter; ft³ = cubic foot; fl oz = fluid ounce; fl pt = fluid pint; fl qt = fluid quart; gal = gallon.

TABLE A-4 CONVERSION FOR UNITS OF WEIGHT

FROM	gm	kg	gr	oz.	lb.
			TO		
gm	1	0.001	15.43	0.0353	0.0022
kg	1,000	1	1.54×10^{4}	35.27	2.205
gr	0.0648	6.48×10^{-5}	1	0.0023	1.43×10^{-4}
oz.	28.35	0.0284	437.5	1	0.0625
lb.	453.6	0.4536	7,000	16	1

gm = gram; kg = kilogram; gr = grain; oz = ounce; lb = pound.

TABLE A-5 CONVERSION FOR PARTS PER MILLION, PROPORTION, AND PERCENT

Parts per Million	Proportion	Percent
0.1	1:10,000,000	0.000010
0.25	1:4,000,000	0.000025
1.0	1:1,000,000	0.0001
2.0	1:500,000	0.0002
3.0	1:333,333	0.0003
4.0	1:250,000	0.0004
5.0	1:200,000	0.0005
8.4	1:119,047	0.00084
10.0	1:100,000	0.001

(Continued)

TABLE A-5 CONVERSION FOR PARTS PER MILLION, PROPORTION, AND PERCENT (CONTINUED)

Parts per Million	Proportion	Percent
15.0	1:66,667	0.0015
20.0	1:50,000	0.002
25.0	1:40,000	0.0025
50.0	1:20,000	0.005
100.0	1:10,000	0.01
150.0	1:6,667	0.015
167.0	1:6,000	0.0167
200.0	1:5,000	0.02
250.0	1:4,000	0.025
500.0	1:2,000	0.05
1,667.0	1:600	0.1667
5,000.0	1:200	0.5
6,667.0	1:150	0.667
30,000.0	1:33	3.0

TABLE A-6 PREPARATION OF STANDARD SOLUTIONS

Concentration of Standard Solution (ppm)	Amount of Concentrated Standard Solution (ml)	Volume Diluted to ml	Concentration of Final Solution (ppm)
1,000	1	1,000	1
1,000	5	1,000	5
1,000	10	1,000	10
500	2	1,000	1
500	10	1,000	5
500	20	1,000	10
250	4	1,000	1
250	20	1,000	5
250	40	1,000	10
100	1	100	1
100	5	100	5
100	10	100	10
50	2	100	1
50	10	100	5
50	20	100	10
10	10	100	1
10	50	100	5
1	1	10	0.1
1	5	10	0.5

TABLE A-7 SOLUBILITY OF OXYGEN IN PARTS PER MILLION (PPM) IN FRESHWATER AT VARIOUS TEMPERATURES AND AT A PRESSURE OF 760 MM HG (SEA LEVEL)

Temperature		Oxygen Concentration (ppm)	Temperature		Oxygen Concentration (ppm)
F°	C°		F°	C°	
32	0	14.6	69.8	21	9.0
33.8	1	14.2	71.6	22	8.8
35.6	2	13.8	73.4	23	8.7
37.4	3	13.5	75.2	24	8.5
39.2	4	13.1	77	25	8.4
41	5	12.8	78.8	26	8.2
42.8	6	12.5	80.6	27	8.1
44.6	7	12.2	82.4	28	7.9
46.4	8	11.9	84.2	29	7.8
48.2	9	11.6	86	30	7.6
50	10	11.3	87.8	31	7.5
51.8	11	11.1	89.6	32	7.4
53.6	12	10.8	91.4	33	7.3
55.4	13	10.6	93.2	34	7.2
57.2	14	10.4	95.0	35	7.1
59	15	10.2	96.8	36	7.0
60.8	16	10.0	98.6	37	6.8
62.6	17	9.7	100.4	38	6.7
64.4	18	9.5	102.2	39	6.6
66.2	19	9.4	104.0	40	6.5
68	20	9.2			

TABLE A-8 ALTITUDE CORRECTION FACTOR FOR THE SOLUBILITY OF OXYGEN IN FRESHWATER

Atmospheric Pressure (mm Hg)	or	Equivalent Altitude (ft.)	=	Correction Factor
775		540		1.02
760		0		1.00
745		542		.98
730		1,094		.96
714		1,688		.94
699		2,274		.92
684		2,864		.90
669		3,466		.88
654		4,082		.86
638		4,756		.84
623		5,403		.82
608		6,065		.80
593		6,744		.78
578		7,440		.76
562		8,204		.74
547		8,939		.72
532		9,694		.70
517		10,472		.68
502		11,273		.66

Example: Solubility of oxygen at sea level (760 mm Hg) at 20°C is 9.2 ppm. The solubility of oxygen at an altitude of 1,688 feet in 20°C water is 9.2 ppm x 0.94 (Correction Factor) = 8.65 ppm.

TABLE A-9 NUMBER OF WATER SAMPLES REQUIRED FROM PONDS TO ESTIMATE THE AVERAGES OF WATER QUALITY VARIABLES WITH 95 PERCENT CERTAINTY THAT ERRORS WILL NOT EXCEED THE SPECIFIED VALUES

Water Quality Variable	Number of Water Samples per Determination
Dissolved Oxygen	
± 0.5 ppm	6
± 1.0 ppm	2
pH	
± 0.5 unit	1
± 1.0 unit	1
Temperature	
± 0.5°C	2
± 1.0°C	1
Total Hardness	
± 1.0 ppm	1
Secchi disk (underwater) visibility	
± 5 cm	7
± 10 cm	2

TABLE A-10 FINFISH PROPAGATED IN AQUACULTURE

Scientific Family Name	Common Scientific Name	Scientific Name	Common Name
Acinpenseridae	Sturgeons	*Acipenser transmontanus* *Acipenser fulvescens* *Acipenser ruthensus* *Acipenser guldenstadti* *Acipenser nudiventris* *Acipenser stellatus* *Huso huso*	White sturgeon Lake sturgeon Sterlet Russian sturgeon Thorn sturgeon Starred sturgeon Beluga sturgeon
Amiidae	Bowfins	*Amia calva*	Bowfin
Anguillidae	Eels	*Anguilla anguilla* *Anguilla rostrata* *Anguilla japonica*	European eel American eel Japanese eel
Clupeidae	Herrings	*Dorsoma petenense* *Dorsoma cepedianum* *Alosa sapidissima*	Threadfin shad Gizzard shad American shad

(Continued)

TABLE A-10 FINFISH PROPAGATED IN AQUACULTURE (CONTINUED)

Scientific Family Name	Common Scientific Name	Scientific Name	Common Name
Salmonidae	Trout, Salmon, Chars	Salmo salar	Atlantic salmon
		Salmo clarki	Cutthroat trout
		Salmo trutta	Brown trout
		Salmo aguabonita	Golden trout
		Salvelinus fontinalis	Brook trout
		Salvelinus namaycush	Lake trout
		Salvelinus malma	Western brook char (or Dolly Varden)
		Oncorhynchus mykiss	Rainbow trout (many strains)
		Oncorhynchus nerka	Sockeye salmon
		Oncorhynchus gorbuscha	Pink salmon
		Oncorhynchus keta	Chum salmon
		Oncorhynchus tshawytscha	Chinook salmon
		Oncorhynchus kisutch	Coho salmon
		Oncorhynchus masu	Masu salmon
		Thymallus arcticus	Arctic grayling
		Coregonus spp.	Cisco and Whitefish
		Stenodus leucichthys	Sheerfish
Esocidae	Pikes	Esox lucius	Northern pike
		Esox masquinongy	Muskellunge
		Esox niger	Chain pickerel
Chanidae	Milkfishes	Chanos shanos	Milkfish
Cyprinidae	Minnows and Carps	Carassius auratus	Goldfish
		Cyprinus carpio	Common Carp
		Notemegonus crysoleucas	Golden shiner
		Notropis spp.	Shiners
		Pimephales spp.	Minnows
		Hypophthalmicthys molitrix	Silver carp
		Ctenopharyngodon idella	Grass carp
		Aristichthys nobilis	Bighead carp
		Abramis brama	Bream
		Tinca tinca	Tench
		Carassius carassius	Crucian carp
		Rutilus rutilus	Roach
		Cirrhinus molitorella	Mud carp
		Mylopharyngodon piceus	Black carp
		Labeo rohita	Rohu
		Catla catla	Catla or Bhakur
		Cirrhina mrigala	Mrigal or Naini
		Osteochilus hasselti	Nilem
		Puntius gonionotus	Puntius carp
Catastomidae	Suckers	Ictobius niger	Buffalofish
Ictaluridae	Catfishes	Ictalurus puctatus	Channel catfish
		Ictalurus furcatus	Blue catfish
		Ictalurus nebulosis	Brown bullhead
		Ictalurus catus	White catfish

(Continued)

TABLE A-10 FINFISH PROPAGATED IN AQUACULTURE (CONTINUED)

Scientific Family Name	Common Scientific Name	Scientific Name	Common Name
Claridae	Walking catfishes	Clarias batrachus Clarias macrocephalus Clarias fusais Clarias largera Clarias gariepinus	Walking catfish Walking catfish Walking catfish Walking catfish Sharp-toothed catfish
Siluridae	Freshwater sharks	Wallagonia attu Siluris glanis	Freshwater shark Sheatfish
Pangasidae	River catfishes	Pangasius sutchi Pangasius laurnaudi	River catfish River catfish
Atherinidae	Whitefishes	Odonthectes basilichthys Chirostoma spp.	Pejerrey Mexican whitefishes
Cyprinodontidae	Killifishes	Fundulus spp.	Topminnows and Killifishes
Poeciliidae	Livebearers	Gambusia spp.	Gambusias
Percichthyidae	Temperate basses	Morone saxatilis	Striped bass
Centrarchidae	Sunfishes	Lepomis spp. Micropterus spp. Pomoxix spp.	Sunfishes Basses Crappie
Percidae	Perches	Perca flavescens Stizostedion vitreum vitreum	Yellow perch Walleye
Carangidae	Jacks and Pompano	Trachinotus carolinus	Florida pompano
Sciaenidae	Drums	Aplodinotus grunniens Cynosion nebulosus Sciaenops ocellata	Freshwater drum Spotted seatrout Red drum
Mugilidae	Mullets	Mugil cephalus	Striped mullet
Cichlidae	Cichlids	Tilapia spp.	Tilapia
Scombridae	Tunas and Mackerels	Euthynnus spp. Thunnus albacares	Skipjacks Yellowfin tuna
Anoplopomatidae	Sablefishes	Anoplopoma fimbria	Sablefish
Bothidae	Lefteye flounders	Paralichthys lethostigma	Southern flounder
Pleuronectidae	Righteye flounders	Hippoglossus hippoglossus Hippoglossus stenolepis Pleuronichthys spp.	Atlantic halibut Pacific halibut Turbot
Ophicephalidae	Snakeheads	Ophicephalus spp.	Snakeheads
Osphronemidae	Gouramis	Osphronemus goramy Trichogaster pectoralis	Giant gourami Siamese gourami

TABLE A-11 WEB ADDRESSES FOR GOVERNMENT AGENCIES PROVIDING INFORMATION OR REGULATIONS TO AQUACULTURE

Web Site Description	URL
U.S. Department of Agriculture (USDA)	http://www.usda.gov/
Farm Service Agency (FSA)	http://www.fsa.usda.gov/pas/default.asp
Foreign Agricultural Service (FAS)	http://www.fas.usda.gov/
Risk Management Agency	http://www.rma.usda.gov/
Food Safety and Inspection Service (FSIS)	http://www.fsis.usda.gov/
Agricultural Marketing Service (AMS)	http://www.ams.usda.gov/AMSv1.0/
Animal Plant and Health Inspection Service (APHIS)	http://www.aphis.usda.gov/
Natural Resources Conservation Service (NRCS)	http://www.nrcs.usda.gov/
Agricultural Research Service (ARS)	http://www.ars.usda.gov/
National Agricultural Library (NAL)	http://www.nalusda.gov/
National Biological Information Infrastructure	http://www.nbii.gov/portal/community/Communities/Plants,_Animals_&_Other_Organisms/Fisheries_&_Aquatic_Resources/
U.S. Environmental Protection Agency-Agriculture-Aquaculture Operations	http://www.epa.gov/oecaagct/anaqupro.html
Guide to Federal Aquaculture Programs and Services	http://aquanic.org/jsa/federal_guide/index.htm
Alternative Farming Information Systems Information Center (Aquaculture)	http://www.nalusda.gov/afsic/afsaqua.htm
National Institute of Food and Agriculture (NIFA), formerly Cooperative State Research, Education, and Extension Service (CREES)	http://www.csrees.usda.gov/
Economic Research Service (ERS)	http://www.ers.usda.gov/
National Agricultural Statistics Service	http://www.nass.usda.gov/
U.S. Army Corps of Engineers (USACE)	http://www.usace.army.mil/
U.S. Department of Commerce	http://www.commerce.gov/
National Oceanic and Atmospheric Administration (NOAA)	http://www.noaa.gov/
National Marine Fisheries Service (NMFS)	http://www.nmfs.gov/
National Technical Information Service (NTIS)	http://www.ntis.gov/
National Environmental Satellite, Data and Information Service (NESDIS)	http://www.nesdis.noaa.gov/
National Sea Grant College Program	http://www.seagrantnews.org/
U.S. Department of Energy	http://www.energy.gov/

(Continued)

TABLE A-11 WEB ADDRESSES FOR GOVERNMENT AGENCIES PROVIDING INFORMATION OR REGULATIONS TO AQUACULTURE (CONTINUED)

Web Site Description	URL
National Biofuels Program	http://www.rpi.edu/dept/chem-eng/Biotech-Environ/Biomass/Biofuels.htm
U.S. Department of Health and Human Services (HHS)	http://www.hhs.gov/
Food and Drug Administration (FDA)	http://www.fda.gov/
Center for Veterinary Medicine (CVM)	http://www.fda.gov/AnimalVeterinary/default.htm
Center for Food Safety and Applied Nutrition	http://www.foodsafety.gov/
U.S. Department of the Interior	http://www.doi.gov/
Fish and Wildlife Service	http://www.fws.gov/
Geological Survey (USGS)	http://www.usgs.gov/
U.S. Agency for International Development (AID)	http://www.usaid.gov/
U.S. Environmental Protection Agency (EPA)	http://earth1.epa.gov/
National Science Foundation (NSF)	http://www.nsf.gov/
U.S. Small Business Administration (SBA)	http://www.sba.gov/
Tennessee Valley Authority (TVA)	http://www.tva.gov/
Food and Agriculture Organization of the United Nations (FAO)	http://www.fao.org/

TABLE A-12 WEB ADDRESSES FOR REGIONAL AQUACULTURE CENTERS (RAC)

Center	URL
Southern Regional Aquaculture Center (SRAC)	http://www.msstate.edu/dept/srac/index.html
North Central Regional Aquaculture Center (NCRAC)	http://www.ncrac.org/
Northeastern Regional Aquaculture Center (NRAC)	http://www.nrac.umd.edu/
Western Regional Aquaculture Center (WRAC)	http://www.fish.washington.edu/wrac/
Center for Tropical and Subtropical Aquaculture (CTSA)	http://www.ctsa.org/
Office of Aquaculture, USDA	http://www.usda.gov/wps/portal/!ut/p/_s.7_0_A/7_0_10B?navid=AQUACULTURE&parentnav=AGRICULTURE&navtype=RT
National Association of State Aquaculture Coordinators (NASAC)	http://www.nasda.org/cms/7195/8878/13164.aspx
AquaNIC—Aquaculture Network Information Center	http://aquanic.org/

TABLE A-13 AQUACULTURE PUBLICATIONS[1]
Advanced Aquarist's Online MagazineAquaculture
Aquaculture Asia
Aqua Culture Asia Pacific Magazine
Aquaculture Center
Aquaculture Compendium
Aquaculture Engineering
Aquaculture Europe Magazine
Asian Fisheries Society
Aquaculture Association of Canada
AquaFlow
Canadian Department of Fisheries and Oceans
Developments in Aquaculture and Fisheries Science
Elsevier Science Journals (Aquaculture and Aquaculture Engineering)
Federation of European Aquaculture Producers
Fish Farmer Magazine
Fish Farming International
Fish Farming News
Fisheries and Aquaculture in Europe Magazine
Fiskaren
Global Aquaculture Advocate
Hatchery International
Israeli Journal of Aquaculture
Journal of the World Aquaculture
Northern Aquaculture
Panorama Acuiclola (Mexico)
Panorama da AquicItura (Brazil)
Pond Dynamics/Aquaculture Collaborative Research Support Program publications (Funded by USAID)
Oceanus
Shrimp News International
Regional Commission for Fisheries
Turkish Journal of Fisheries and Aquatic Sciences
World Aquaculture Magazine
WorldFish Center Publications

[1]Source: AquaNIC (Aquaculture Network Information Center): http://aquanic.org/publications/ where all the publications are hyperlinked to their sources.

TABLE A-14 INTERNET RESOURCES FOR AQUACULTURE	
Topic	URL
American Fisheries Society	http://www.fisheries.org/
American Water Resources Association	http://www.awra.org/
Aquaculture Resources from the USDA	http://www.nal.usda.gov/afsic/afsaqua.htm
Aquaculture Health International	http://www.aquaculturehealth.com/
NetVet	http://netvet.wustl.edu/fish.htm#aquaculture
Aquaculture Network Information	http://www.aquanic.org/
National Shellfisheries Association	http://www.shellfish.org/
Aquaculture, Fish Farming News	http://www.thefishsite.com/
High School Aquaculture	http://darc.cms.udel.edu/hsaqlinks.html
European Aquaculture Society	http://www.easonline.org/
National Council for Agricultural Education	http://www.teamaged.org/council/
Seafood (aquaculture) Images	http://www.sea-ex.com/
Algae Image	Archivehttp://www.bgsu.edu/departments/biology/facilities/algae/html/Image_Archive.html
NOAA Photo Collection	http://www.photolib.noaa.gov/
National Fisheries Institute About Seafood	http://www.aboutseafood.com/
The National Renewable Energy Laboratory's (NREL)	http://www.nrel.gov/
Fish Jobs	http://fishjobs.com/
International Center for Aquaculture and Aquatic Environments	http://www.ag.auburn.edu/fish/international/
Thad Cochran National Warmwater Aquaculture Center	http://www.msstate.edu/dept/tcnwac/
United States Trout Farmers	http://www.ustfa.org/
California Aquaculture	http://aqua.ucdavis.edu/
World Aquaculture Society	http://www.was.org/
Joint Subcommittee on Aquaculture (resource page)	http://aquanic.org/jsa/
Fisheries Technology Associates, Inc:	http://www.ftai.com/links.htm

TABLE A-15 AQUACULTURE TRAINING AND EDUCATION AND GUIDE TO FEDERAL PROGRAMS AND SERVICES[1]	
Website	URL
Business.com Aquaculture Education and Training	http://www.business.com/directory/agriculture/aquaculture/education_and_training/
Guide to Federal Aquaculture Programs and Services	http://aquanic.org/jsa/federal_guide/index.htm
High School Aquaculture and Aquatic Science Programs	http://darc.cms.udel.edu/hsaqlinks.html
Arizona Aquaculture Education Programs	http://ag.arizona.edu/azaqua/azhighs.html
ALEARN: Aquaculture Education Programs Across the Country	http://www.aces.edu/dept/fisheries/education/programsinoperation.php

[1]Many universities offer specialized M.S. and Ph.D. degrees in aquaculture and have excellent research and experimental facilities. Land grant colleges and universities have a tradition of freshwater aquaculture (mainly biology) and good extension and continuing education services. The Sea Grant-funded universities and colleges deal mainly with brackish water and marine aquaculture.

TABLE A-16 DICHOTOMOUS KEY FOR POND WATER

1. Cells single or, if undergoing cell division, found in pairs 2
1. Cells numerous, multicellular; arranged in chains, filaments, or other multicellular organism 9
2(1) Cells or cultures of cells green in color 3
2(1) Cells or cultures of cells not green or a shade of green; generally yellow to brown in color 5
3(2) Cells motile by cilia .... ciliated protozoan
3(2) Cells nonmotile by flagellum 4
4(3) Cells with bright green (e.g., grass green) chloroplast, larger (10 to 15 microns in diameter), oval or flattened shape .... *Tetraselmis*
4(3) Cells a shade of green, generally small (2 to 5 microns in diameter), spherical or oval in shape .... *Nannochloropsis*
5(2) Cells with radial symmetry .... (centric diatoms) 6
5(2) Cells not radially symmetrical, with somewhat bilateral symmetry .... (pennate diatoms) 7
6(5) Cells large, spherical or oval, usually connected in chains of two or more cells, spines not present .... *Melosira*
6(5) Cells smaller; square, rectangular, or oval; single or in short chains, spines originating at the corner of each cell .... *Chaetoceros*
7(5) Cells tapering to a long, thin spine .... *Nitzschia closterium*
7(5) Cell endings rounded or in a point, but not tapering to a long spine 8
8(7) Cells bilaterally symmetrical in all views .... *Navicula* or other navicula-type diatom
8(7) Cells asymmetrical in some views .... *Achnanthes*
9(1) Cells photosynthetic, arranged in a filament, variable in color 10
9(1) Cells nonphotosynthetic, arranged into tissues and organs, generally clear or may be pigmented 12
10(1) Cells arranged as a branching filament, green in color .... *Cladophora*
10(1) Cells arranged in a nonbranching filament, brown in color 11
11(10) Cells generally in long chains of large, spherical cells, spines not present .... *Melosira*
11(10) Cells in short or long chains or smaller, square, or rectangular cells possessing spines that originate at the corner of each cell .... *Chaetoceros*
12(9) Body with segmented body parts, appearing shrimp-like, lacking cilia around the mouth .... copepod
12(9) Body smooth, non-segmented, appearing sac-like, with cilia around the mouth .... rotifer

TABLE A-17 TABLE OF FISH IN FDA REGULATORY FISH ENCYCLOPEDIA[1]

Common Name	Market Name	Scientific Name	Family
Albacore	Tuna	*Thunnus alalunga*	Scombridae (mackerels and tunas)
Arrowtooth Flounder	Flounder, Arrowtooth	*Reinhardtius stomias*	Pleuronectidae (righteye flounders)
Atlantic Cod	Cod	*Gadus morhua*	Gadidae (cods)
Atlantic Cutlassfish	Cutlassfish	*Trichiurus lepturus*	Trichiuridae (cutlassfishes)
Atlantic Salmon	Salmon, Atlantic	*Salmo salar*	Salmonidae (salmonids)
Atlantic Wolffish	Wolffish	*Anarhichas lupus*	Anarhichadidae (wolffishes)
Black Drum	Drum	*Pogonias cromis*	Sciaenidae (drums (croakers))
Black Pomfret	Pompano	*Parastromateus niger*	Carangidae (jacks and pompanos)
Blackback Flounder	Flounder or Sole	*Pseudopleuronectes americanus*	Pleuronectidae (righteye flounders)
Blacktip Shark	Shark	*Carcharhinus limbatus*	Carcharhinidae (requiem sharks)
Blue Catfish	Catfish	*Ictalurus furcatus*	Ictaluridae (North American freshwater catfishes)
Blue Crab	Crab, Blue	*Callinectes sapidus*	Portunidae (common edible crabs)
Blue Marlin	Marlin	*Makaira nigricans*	Xiphiidae (billfishes)
Brown Rockfish	Rockfish	*Sebastes auriculatus*	Scorpaenidae (scorpionfishes)
Bullseye Puffer	Puffer	*Sphoeroides annulatus*	Tetraodontidae (puffers)
California Scorpionfish	Scorpionfish	*Scorpaena guttata*	Scorpaenidae (scorpionfishes)
California Sheephead	Sheephead	*Semicossyphus pulcher*	Labridae (wrasses)
Canary Rockfish	Rockfish	*Sebastes pinniger*	Scorpaenidae (scorpionfishes)
Caribbean Red Snapper	Snapper	*Lutjanus purpureus*	Lutjanidae (snappers and fusiliers)
Channel Catfish	Catfish	*Ictalurus punctatus*	Ictaluridae (North American freshwater catfishes)
Chilipepper (Fish)	Rockfish	*Sebastes goodei*	Scorpaenidae (scorpionfishes)

(Continued)

TABLE A-17 TABLE OF FISH IN FDA REGULATORY FISH ENCYCLOPEDIA[1] (CONTINUED)

Common Name	Market Name	Scientific Name	Family
China Rockfish	Rockfish	*Sebastes nebulosus*	Scorpaenidae (scorpionfishes)
Chinook Salmon	Salmon, Chinook or King or Spring	*Oncorhynchus tshawytscha*	Salmonidae (salmonids)
Chub Mackerel	Mackerel, Chub	*Scomber japonicus*	Scombridae (mackerels and tunas)48
Coho Salmon	Salmon, Coho or Silver or Medium Red	*Oncorhynchus kisutch*	Salmonidae (salmonids)
Common Thresher Shark	Shark, Thresher	*Alopias vulpinus*	Alopiidae (thresher sharks)
Coney	Grouper	*Epinephelus fulva*	Serranidae (sea basses)
Cusk	Cusk	*Brosme brosme*	Gadidae (cods)
Dolphin	Mahi-mahi	*Coryphaena hippurus*	Coryphaenidae (dolphins)
Dover Sole	Sole	*Microstomus pacificus*	Pleuronectidae (righteye flounders)
English Sole	Sole	*Parophrys vetulus*	Pleuronectidae (righteye flounders)
Escolar	Escolar	*Lepidocybium flavobrunneum*	Gempylidae (snake mackerels)
European John Dory	Dory	*Zeus faber*	Zeidae (dories)
Golden Redfish	Perch, Ocean	*Sebastes norvegicus*	Scorpaenidae (scorpionfishes)
Gray Snapper	Snapper	*Lutjanus griseus*	Lutjanidae (snappers and fusiliers)
Gray Sole	Sole or Flounder	*Glyptocephalus cynoglossus*	Pleuronectidae (righteye flounders)
Great Barracuda	Barracuda	*Sphyraena barracuda*	Sphyraenidae (barracudas)
Haddock	Haddock	*Melanogrammus aeglefinus*	Gadidae (cods)
Kawakawa	Tuna	*Euthynnus affinis*	Scombridae (mackerels and tunas)
Lane Snapper	Snapper	*Lutjanus synagris*	Lutjanidae (snappers and fusiliers)
Lingcod	Lingcod	*Ophiodon elongatus*	Hexagrammidae (greenlings)
Milkfish	Milkfish	*Chanos chanos*	Chanidae (milkfish)
Mozambique Tilapia	Tilapia	*Oreochromis mossambicus*	Cichlidae (cichlids)

(Continued)

TABLE A-17 TABLE OF FISH IN FDA REGULATORY FISH ENCYCLOPEDIA[1] (CONTINUED)

Common Name	Market Name	Scientific Name	Family
Nile Tilapia	Tilapia	*Oreochromis niloticus*	Cichlidae (cichlids)
Northern Puffer	Puffer	*Sphoeroides maculatus*	Tetraodontidae (puffers)
Ocean Whitefish	Tilefish	*Caulolatilus princeps*	Malacanthidae (tilefishes)
Oilfish	Oilfish	*Ruvettus pretiosus*	Gempylidae (snake mackerels)
Orange Roughy	Roughy, Orange	*Hoplostethus atlanticus*	Trachichthyidae (roughies or slimeheads)
Pacific Barracuda	Barracuda	*Sphyraena argentea*	Sphyraenidae (barracudas)
Pacific Bonito	Bonito	*Sarda chiliensis*	Scombridae (mackerels and tunas)
Pacific Cod	Cod or Alaska Cod	*Gadus macrocephalus*	Gadidae (cods)
Pacific Crevalle Jack	Jack	*Caranx caninus*	Carangidae (jacks and pompanos)
Pacific Moonfish	Jack	*Selene peruviana*	Carangidae (jacks and pompanos)
Pacific Ocean Perch	Perch, Ocean	*Sebastes alutus*	Scorpaenidae (scorpionfishes)
Pacific Sierra	Mackerel, Spanish	*Scomberomorus sierra*	Scombridae (mackerels and tunas)
Pacific Snapper	Snapper	*Lutjanus peru*	Lutjanidae (snappers and fusiliers)
Patagonian Toothfish	Patagonian Toothfish	*Dissostichus eleginoides*	Nototheniidae (cod icefishes)
Petrale Sole	Sole or Flounder	*Eopsetta jordani*	Pleuronectidae (righteye flounders)
Pink Salmon	Salmon, Pink or Humpback	*Oncorhynchus gorbuscha*	Salmonidae (salmonids)
Pollock	Pollock	*Pollachius virens*	Gadidae (cods)
Quillback Rockfish	Rockfish	*Sebastes maliger*	Scorpaenidae (scorpionfishes)
Rainbow Trout	Trout, Rainbow or Steelhead	*Oncorhynchus mykiss*	Salmonidae (salmonids)
Red Drum	Drum or Redfish	*Sciaenops ocellatus*	Sciaenidae (drums (croakers)
Red Hawaiian Porgy	Porgy	*Chrysophrys auratus*	Family: Sparidae (porgies)
Red Snapper	Snapper	*Lutjanus campechanus*	Lutjanidae (snappers and fusiliers)

(Continued)

TABLE A-17 TABLE OF FISH IN FDA REGULATORY FISH ENCYCLOPEDIA[1] (CONTINUED)

Common Name	Market Name	Scientific Name	Family
Redstripe Rockfish	Rockfish	*Sebastes proriger*	Scorpaenidae (scorpionfishes)
Rex Sole	Sole or Flounder	*Glyptocephalus zachirus*	Pleuronectidae (righteye flounders)
Rougheye Rockfish	Rockfish	*Sebastes aleutianus*	Scorpaenidae (scorpionfishes)
Schoolmaster	Schoolmaster	*Lutjanus apodus*	Lutjanidae (snappers and fusiliers)
Sheepshead	Sheepshead	*Archosargus probatocephalus*	Sparidae (porgies)
Shortfin Mako Shark	Shark, Mako	*Isurus oxyrinchus*	Lamnidae (mackerel sharks)
Silk Snapper	Snapper	*Lutjanus vivanus*	Lutjanidae (snappers and fusiliers)
Silver Pomfret	Butterfish	*Pampus argenteus*	Stromateidae (butterfishes)
Silvergray Rockfish	Rockfish	*Sebastes brevispinis*	Scorpaenidae (scorpionfishes)
Skipjack Tuna	Tuna	*Katsuwonus pelamis*	Scombridae (mackerels and tunas)
Spinefoot	Spinefoot	*Siganus javus*	Siganidae (rabbitfishes)
Spotfin Croaker	Croaker or Corvina	*Roncador stearnsi*	Sciaenidae (drums croakers)
Starry Flounder	Flounder	*Platichthys stellatus*	Pleuronectidae (righteye flounders)
Striped Marlin	Marlin	*Tetrapturus audax*	Xiphiidae (billfishes)
Sunshine Bass	Bass	*Morone chrysops x saxatilis*	Percichthyidae (temperate perches)
Swordfish	Swordfish	*Xiphias gladius*	Xiphiidae (billfishes)
Tinfoil Barb	Barb	*Barbodes schwanefeldi*	Cyprinidae (carps)
Walleye Pollock	Pollock or Alaska Pollock	*Theragra chalcogramma*	Gadidae (cods)
White Hake	Hake	*Urophycis tenuis*	Gadidae (cods)
Widow Rockfish	Rockfish	*Sebastes entomelas*	Scorpaenidae (scorpionfishes)
Windowpane	Flounder	*Scophthalmus aquosus*	Bothidae (lefteye flounders)
Redlip Croaker	Croaker	*Larimichthys polyactis*	Sciaenidae (drums croakers)
Yelloweye Rockfish	Rockfish	*Sebastes ruberrimus*	Scorpaenidae (scorpionfishes)

(Continued)

TABLE A-17 TABLE OF FISH IN FDA REGULATORY FISH ENCYCLOPEDIA[1] (CONTINUED)			
Yellowfin Tuna	Tuna	*Thunnus albacares*	Scombridae (mackerels and tunas)
Yellowstripe Scad	Scad	*Selaroides leptolepis*	Carangidae (jacks and pompanos)
Yellowtail[181]	Amberjack or Yellowtail	*Seriola lalandei*	Carangidae (jacks and pompanos)
Yellowtail Flounder	Flounder	*Limanda ferruginea*	Pleuronectidae (righteye flounders)
Yellowtail Rockfish	Rockfish	*Sebastes flavidus*	Scorpaenidae (scorpionfishes)
Yellowtail Snapper	Snapper	*Ocyurus chrysurus*	Lutjanidae (snappers and fusiliers)

[1]Source: http://www.fda.gov/food/foodsafety/product-specificinformation/seafood/regulatoryfishencyclopediarfe/default.htm

GLOSSARY

Like a foreign language, terms unique to aquaculture can be baffling to the newcomer. An individual traveling to a foreign country to do business would be expected to know the language of the country. The same is true for the individual wanting to learn aquaculture. Successful individuals use the glossary and learn the language. Words not found in the glossary may be listed in the index and defined within the book.

A

Abdomen Belly; the ventral side of the fish surrounding the digestive and reproductive organs.

Abdominal Pertaining to the belly.

Ablation Removal of organs from the body by mechanical means.

Abnormalities Any deviation from a standard or from the expected.

Abrasion A spot scraped of skin, mucous membrane, or superficial epithelium.

Abscess A localized collection of necrotic (dead) debris and white blood cells surrounded by inflamed tissue.

Absorption The process by which water and dissolved substances pass into the cells.

Acclimated Gradually introduced to changes in water temperature and quality.

Acclimatization The adaptation of fishes to a new environment or habitat or to different climatic conditions.

Accounting A system of recording, classifying, and summarizing commercial transactions in terms of money.

Accrual Expenses are considered expenses when they are accrued (or committed) and income is counted as income when it is earned. This includes changes in inventories.

Acid A compound that yields hydrogen ions when dissolved in an ionizing solvent.

Acidosis Metabolic condition caused by low pH.

Acre-foot A water volume equivalent to that covering a surface area of one acre to a depth of one foot; equal to 325,850 gal. or 2,718,144 lbs. of water.

Activated sludge process A system in which organic waste is continually circulated in the presence of oxygen and digested by aerobic bacteria.

Acute Having a short and relatively severe course; for example, acute inflammation.

Acute catarrhal enteritis See Infectious pancreatic necrosis.

Acute toxicity Causing death or severe damage to an organism by poisoning during a brief exposure period, normally 96 hours or less. See Chronic.

Adaptation The process by which individuals, or parts of individuals, populations, or species change in form or function in order to better survive under given or changed environmental conditions. Also the result of this process.

Adductor Term used to describe muscle that draws toward the axis.

Adipose fin A small fleshy appendage located posterior to the main dorsal fin; present in Salmonidae and Ictaluridae.

Adipose tissue Tissue capable of storing large amounts of neutral fats.

Advertising The act or practice of attracting public notice to create an interest or induce to purchase.

Aerated lagoon A waste treatment pond in which the oxygen required for biological oxidation is supplied by mechanical aerators.

Aeration The mixing of air and water by wind action or by air forced through water; generally refers to a process by which oxygen is added to water.

Aerobes Organisms that can live and grow only where free oxygen is present.

Aerobic Referring to a process, for example, respiration; or organism, for example, a bacterium that requires oxygen.

Aggregate Soil made of a mixture of mineral particles.

Agitator Mechanism for stirring up and thus aerating water in hatching tanks and troughs.

Agriculture The art, science, and business of producing every kind of plant and animal useful to humans.

Air The gases surrounding the earth; consists of approximately 78 percent nitrogen, 21 percent oxygen, 0.9 percent argon, 0.03 percent carbon dioxide, and minute quantities of helium, krypton, neon, and xenon, plus water vapor.

Air bladder (swim bladder) An internal, inflatable gas bladder that enables a fish to regulate its buoyancy.

Air stripping Removal of dissolved gases from water to air by agitation of the water to increase the area of air-water contact.

Air stone A porous diffuser through which air is forced and subsequently releases tiny air bubbles into an aquarium; made of fused sand, fused glass spheres, plastics, or even lime wood.

Alevin A life stage of salmonid fish between hatching and feeding when the yolk sac is still present. Equivalent to sac fry in other fishes.

Alevins (sac fry) Fry that obtain nourishment from an attached yolk sac.

Algae a collective term that refers to several taxonomic groups; can be single-celled or multicellular; includes diatoms, seaweeds, and dinoflagellates

Algal bloom A high density or rapid increase in abundance of algae.

Algal toxicosis A poisoning resulting from the uptake or ingestion of toxins or toxin-producing algae; usually associated with blue-green algae or dinoflagellate blooms in fresh or marine water.

Alimentary tract The digestive tract, including all organs from the mouth to the anal opening.

Aliquot An equal part or sample of a larger quantity.

Alkaline Basic, pH greater than 7.

Alkalinity The power of a mineral solution to neutralize hydrogen ions; usually expressed as equivalents of calcium carbonate. Measure of pH buffering capacity.

Alkalosis Metabolic condition caused by a high pH.

Amino acid A building block for proteins; an organic acid containing one or more amino groups ($-NH_2$) and at least one carboxylic acid group ($-COOH$).

Ammonia The gas NH_3; highly soluble in water; toxic to fish in the un-ionized form, especially at low oxygen levels.

Ammonia nitrogen Also called "total ammonia." The summed weight of nitrogen in both the ionized (ammonium, NH_4^+) and molecular (NH_3) forms of dissolved ammonia (NH_4-N plus NH_3-N). Ammonia values are reported as N, the hydrogen being ignored in analyses.

Ammonium The ionized form of ammonia, NH_4^+.

Amphibian Any cold-blooded vertebrate of the class Amphibia, including frogs and salamanders; having an aquatic, gill-breathing tadpole stage and later developing lungs.

Anabolism Constructive metabolic processes in living organisms tissue building and growth.

Anadromous fish Fish that leave the sea and migrate up freshwater rivers to spawn.

Anaerobes Organisms that can live and grow where there is no free oxygen.

Anaerobic Referring to a process or organism not requiring oxygen.

Anaerobic digestion a series of processes in which microorganisms break down biodegradable material in the absence of oxygen; used for industrial or domestic purposes to manage waste and/or to release energy.

Anal Pertaining to the anus or vent.

Anal fin The fin on the ventral median line behind the anus.

Anal papilla A protuberance in front of the genital pore and behind the vent in certain groups of fishes.

Anatomy The structure of an organism or any of its parts.

Anchor ice Ice that forms from the bottom up in moving water.

Anemia A condition characterized by a deficiency of hemoglobin, packed cell volume, or erythrocytes (red blood cells). Anemias in fish include normocytic anemia, microcytic anemia, and macrocytic anemia.

Anesthetics Chemicals used to relax fish and facilitate the handling and spawning of fish. Commonly used agents include tricane methane sulfonate (MS-222), benzocain, quinaldine, and carbon dioxide.

Anions Negative ions.

Annulus A yearly mark formed on fish scales when rapid growth resumes after a period (usually in winter) of slow or no growth.

Anomalies Data values that deviate from the normal range but are accepted as accurate.

Anoxia Reduction of oxygen in the body to levels that can result in tissue damage.

Antennae Paired, lateral, moveable, jointed appendages on the head of crustaceans.

Anterior In front of or toward the head end.

Anthelmintic An agent that destroys or expels worm parasites.

Antibiotic A chemical produced by living organisms, usually molds or bacteria, capable of controlling or inhibiting the growth of other bacteria or microorganisms.

Antibody A specific protein produced by an organism in response to a foreign chemical (antigen) with which it reacts.

Antigen A large protein or complex sugar that stimulates the formation of an antibody. Generally, pathogens produce antigens, and the host protects itself by producing antibodies.

Antigenicity A measure of a substance's ability to cause the development of antibodies.

Antimicrobial Chemical that inhibits microorganisms.

Antinutrients Substances that occur naturally in plant or animal feedstuffs that adversely affect the performance of the animal.

Antioxidant A substance that chemically protects other compounds against oxidation, for example, ethoxyquin, BHT, BHA, and propyl gallate prevent oxidation and rancidity of fats.

Antiseptic A compound that kills or inhibits microorganisms, especially those infecting living tissues.

Antivitamin Substance chemically similar to a vitamin that can replace the vitamin or an essential compound but cannot perform its role.

Anus The external posterior opening of the alimentary tract; the vent.

Appendages Any part jointed to or diverging from the main body.

Aquaculture The art, science, and business of cultivating plants and animals in water.

Aquaponics The combination of raising aquatic animals with the production of some plants; fish wastes fertilize the plants and plants clean water for reuse in by the fish.

Aquatic Growing or living in or upon water.

Aquifer Underground supply of water.

Archive Historical records.

Artery A blood vessel carrying blood away from the heart.

Ascites The accumulation of serum-like fluid in the abdomen.

Ascorbic acid (vitamin C) A water-soluble vitamin important for the production of connective tissue; deficiencies cause spinal abnormalities and reduce wound-healing capabilities.

Asexual Reproduction without eggs and sperm.

Asphyxia Suffocation caused by too little oxygen or too much carbon dioxide in the blood.

Assembling Collecting aquaculture crops from different production sites to a central location so that the volume is large enough to efficiently use processing facilities.

Assets The property or resources owned and controlled by a business.

Assimilation The transformation of digested foods into an integral and homogenous part of the solids or fluids of the organism.

Asymptomatic carrier An individual that shows no signs of a disease but harbors and transmits it to others.

Atmosphere The envelope of gases surrounding the earth; also, pressure equal to air pressure at sea level, approximately 14.7 pounds per square inch (psi).

Atrophy A degeneration or diminution of a cell or body part due to disuse, defect, or nutritional deficiency.

Auditory Referring to the ear or to hearing.

Automatic feeder Feeder that is time controlled providing the correct amount of feed at set intervals of the day.

Autopsy A medical examination after death to ascertain the cause of death.

Available energy Energy available from nutrients after food is digested and absorbed.

Available oxygen Oxygen present in the water in excess of the amount required for minimum maintenance of a species and that can be used for metabolism and growth.

Avirulent Not capable of producing disease.

Avitaminosis (hypovitaminosis) A disease caused by deficiency of one or more vitamins in the diet.

Axilla The region just behind the pectoral fin base.

B

B.O.D. See Biochemical oxygen demand.

Bacteremia The presence of living bacteria in the blood with or without significant response by the host.

Bacteria See Bacterium.

Bacterial gill disease A disease usually associated with unfavorable environmental conditions followed by secondary invasion of opportunist bacteria. See Environmental gill disease.

Bacterial hemorrhagic septicemia A disease caused by many of the gram-negative rod-shaped bacteria, usually of the genera *Aeromonas* or *Pseudomonas*, that invade all tissues and blood of the fish. Other names include infectious dropsy, red pest, freshwater eel disease, redmouth disease, motile aeromonad septicemia (MAS).

Bacterial kidney disease An acute to chronic disease of salmonids caused by *Renibacterium salmoninarum*. Other names include corynebacterial kidney disease, Dee's disease, kidney disease.

Bacterin A vaccine prepared from bacteria and inactivated by heat or chemicals in a manner that does not alter the cell antigens.

Bacteriocidal Having the ability to kill bacteria.

Bacteriostat Chemical that stops the growth or multiplication of bacteria.

Bacteriostatic Having the ability to inhibit or retard the growth or reproduction of bacteria.

Bacterium (plural: bacteria) One of a large, widely distributed group of typically one-celled microorganisms, often parasitic or pathogenic.

Baffle Device, such as a screen, that interferes with water flow, thus stirring up and aerating the water.

Bait Any substance used to attract and catch fish on the end of a fishing hook, or inside a fish trap; traditionally, nightcrawlers, insects, and smaller fish used for this purpose

Balance sheet A statement of the assets owned and liabilities owed in dollars; it shows equity or net worth at a specific point in time.

Balanced diet (feed) A diet that provides adequate nutrients for normal growth and reproduction.

Bar marks Vertical color marks on fishes.

Barbel An elongated fleshy projection, usually of the lips.

Basal metabolic rate The oxygen consumed by a completely resting animal per unit weight and time.

Basal metabolism Minimum energy requirements to maintain vital body processes.

Bath A solution of therapeutic or prophylactic chemicals in which fish are immersed. See Dip; Short bath; Flush; Long bath; Constant-flow treatment.

Beer's law The light passing through a colored liquid decreases as the concentration of the substance dissolved in the liquid increases.

Benign Not endangering life or health.

Benthic Living on or in the bottom sediment of a pond.

Beneficials Any of a number of species like certain insects that perform valued services like pollination and pest control

Bermed Soil formed into a dike or ridge.

Best management practices For agricultural activities, land use practices or management styles that maintain land productivity and decrease pesticide and fertilizer expenditures.

Binders A substance promoting the cohesion of particles.

Bioassay Any test in which organisms are used to detect or measure the presence or effect of a chemical or condition.

Biochemical oxygen demand (BOD) The quantity of dissolved oxygen taken up by nonliving organic matter in the water.

Biocontrol Control of pests by hindering their biological status, through the introduction of a natural enemy or a pathogen into the environment.

Biofilter A device that uses living material to capture and biologically degrade pollutants; common uses include processing waste water in a recirculating system.

Bio-intensive An organic agricultural system that focuses on maximum yields from the minimum area of land, while simultaneously improving the soil; goal of method is long term sustainability on a closed system basis.

Biological control (biocontrol) Control of undesirable animals or plants by means of predators, parasites, pathogens, or genetic diseases (including sterilization).

Biological filtration Removal of ammonia from a recirculating system.

Biological oxidation Oxidation of organic matter by organisms in the presence of oxygen. Biologics A drug obtained from animal tissue or some other organic source.

Biomass Used to express total living weight.

Biosecurity A set of preventive measures designed to reduce the risk of transmission of infectious diseases, quarantined pests, invasive alien species, living modified organisms, and biological weapons.

Biotin (vitamin H) One of the water-soluble B-complex vitamins.

Bivalve Animals with two sides of a shell hinged together.

Black grub Black spots in the skin of fishes caused by metacercaria (larval stages) of the trematodes *Uvilifer ambloplitis*, *Cryptocotyle lingu*, and others. Black-spot disease is another name.

Black spot Usually refers to black cysts of intermediate stages of trematodes in fish. See Black grub.

Black-spot disease See Black grub.

Black-tail disease See Whirling disease.

Blank egg An unfertilized egg.

Blastopore Channel leading into a cavity in the egg where fertilization takes place and early cell division begins.

Blastula A hollow ball of cells, one of the early stages in embryological development.

Blood flagellates Flagellated whiplike tail protozoan parasites of the blood.

Bloom Used to describe flourishing algae in a pond.

Blue-sac disease A disease of sac fry characterized by opalescence and distention of the yolk sac with fluid, caused by previous partial asphyxia.

Blue slime Excessive mucus accumulation on fish, usually caused by skin irritation due to ectoparasites or malnutrition.

Blue-slime disease A skin condition associated with a deficiency of biotin in the diet.

Blue stone See Copper sulfate.

Boil A localized infection of skin and subcutaneous tissue developing into a solitary abscess that drains externally.

Borings Soil samples taken by drilling.

Brackish water A mixture of fresh and seawater; or water with total salt concentrations between 0.05 percent and 3.0 percent, or a salinity of 1 to 10 parts per thousand.

Branchiae (singular: branchia) Gills, the respiratory organs of fishes.

Branchiocranium The bony skeleton supporting the gill arches.

Branchiomycosis A fungal infection of the gills caused by *Branchiaomyces* spp. Other names include gill rot, European gill rot.

Branded Refers to unique products often identifying a single producer.

Breakdown Metabolic oxidation into component parts.

Break-even analysis Determining the point where income is equal to the total of the fixed costs and variable costs of doing business.

Broadcast To scatter feed or seed over a wide area.

Broodfish Fish held for spawning.

Broodstock Adult fish retained for spawning.

Bubble-Nest Builders Fish that create a nest suspended by a weave of tiny air bubbles used as a protective coating for the eggs and the newly hatched young.

Buccal cavity Mouth cavity.

Buccal incubation Incubation of eggs in the mouth; oral incubation.

Budget A formal plan that projects the use of assets for a future time; a schedule of expected returns or costs.

Buffer Any substance in a solution that tends to resist pH change by neutralizing any added acid or alkali. A chemical that by taking up or giving up hydrogen ions sustains pH within a narrow range.

Byssus A bunch of silky threads secreted by certain mollusks and serving as a means of attachment to an object.

C

CCVD Channel catfish virus disease.

Cfs Cubic feet per second.

Cage Encloses the fish in a container or basket that allows water to pass freely between the fish and the water source.

Caiman Any of several tropical American crocodilians of the genus Caiman.

Calcareous Composed of, containing, or like limestone or calcium carbonate.

Calcinosis The deposition of calcium salts in the tissues without detectable injury to the affected parts.

Calcium carbonate ($CaCO_3$) A relatively insoluble salt, the primary constituent of limestone and a common constituent of hard water.

Calcium cyanamide ($CaCN_2$) Used as a pond disinfectant, also known as lime nitrogen.

Calcium oxide See Lime.

Calorie The amount of heat required to raise the temperature of one gram of water one degree centigrade; a measure of energy.

Capital The amount of money that can be obtained through borrowing or selling of assets that is used to promote the production of other goods.

Carbohydrate Any of the various neutral compounds of carbon, hydrogen, and oxygen, such as sugars, starches, and celluloses, most of which can be used as an energy source by animals.

Carbon dioxide A colorless, odorless gas, CO_2, resulting from the oxidation of carbon-containing substances; highly soluble in water. Toxic to fish at high levels. Toxicity to fish increases at low levels of oxygen. May be used as an anesthetic.

Carbonate The CO_3^- ion, or any salt formed with it, such as calcium carbonate, $CaCO_3$.

Carcinogen Any agent or substance that produces cancer or accelerates the development of cancer.

Carnivore Animals feeding or preying on animals, eating only animal food.

Carotenoid A form of vitamin A responsible for coloration of flesh in fish.

Carrier An individual harboring a pathogen without indicating signs of the disease.

Carrier host (transport host) An animal in which the larval stage of a parasite will live but not develop.

Carrying capacity The population, number, or weight of a species that a given environment can support for a given time.

Cartilage A substance more flexible than bone but serving the same purpose.

Cash basis Income is recorded as income when it is received, and expenses are recorded as expenses when they are paid.

Cartilaginous Fish having a skeleton composed mainly of cartilage, a firm, elastic, whitish type of connective tissue.

Cash flow Actual money available in a business to pay salaries, expenses, dividends, purchase new equipment, and so on.

Cash-flow projection An estimate of monthly cash inflows and outflows over a period of time, usually one year.

Cash-flow summary A list or record of actual monthly cash levels for a business.

Cast To hurl or throw a fishing line with bait into the water

Catabolism The metabolic breakdown of materials with a resultant release of energy.
Catadromous Fish that leave freshwater and migrate to the sea to spawn.

Catalyst A substance that speeds up the rate of chemical reaction but is not itself used up in the reaction.

Cataract Partial or complete opacity of the crystalline lens of the eye or its capsule.

Cations Positive ions.

Caudal Pertaining to the posterior end; toward the tail.

Caudal fin The tail fin of fish.

Caudal peduncle The relatively thin posterior section of the body to which the caudal fin is attached; region between base of the caudal fin and base of the last ray of the anal fin.

Cecum (plural: ceca) A blind sac of the digestive tract, such as a pyloric cecum at the posterior end of the stomach.

Cellular Of, pertaining to, or like a cell or cells.

Chelator An organic substance that inactivates the metallic ions in a solution; for example, EDTA.

Chemical coagulation A process in which chemical coagulants are put into water causing suspended colloidal solids to flocculate and settle out.

Chemical oxygen demand (COD) A measure of the chemically oxidizable components in water, determined by the quantity of oxygen consumed.

Chemotherapy Cure or control of a disease by the use of chemicals (drugs).

Chinook salmon virus disease See Infectious hematopoietic necrosis.

Chitin A structural carbohydrate (sugar) found in the exoskeletons of crustaceans thatcontains some tightly bound noncarbohydrate material, including proteins and inorganic salts.

Chitinoverous Deriving nutrition from chitin, the support substance of the exoskeleton.

Chlorophyll Green pigment essential to photosynthesis in plants.

Chromatophores Colored pigment cells.

Chromosomes Structural units of heredity in the nuclei of cells.

Chronic Occurring or recurring over a long time.

Chronic inflammation Long-lasting inflammation.

Cichlids Freshwater fish of tropical America, Africa, and Asia similar to American sunfishes; some are food fishes; many small ones are popular in aquariums; fish of the family Cichlidae

Cilia Movable organelles (small organs) that project from some cells, used for locomotion of one-celled organisms or to create fluid currents past attached cells.

Ciliate protozoan One-celled animal bearing motile cilia.

Ciliated Having short, fine, hairlike growths that aid in movement, as found on many adult protozoans.

Circuli The more or less concentric growth marks in a fish scale.

Clarification Process of removing impurities or making clear.

Clean cropping Harvesting all fish at one time.

Clinical infection An infection or disease generating obvious symptoms and signs of pathology.

Cloaca The common cavity into which rectal, urinary, and genital ducts open. Common opening of intestine and reproductive system of male nematodes.

Closed-formula feed (proprietary feed) A diet for which the formula is known only to the manufacturer.

Clutch A hatch of eggs; the number of eggs produced or incubated at one time.

Cock Male trout.

Coelomic cavity The body cavity containing the internal organs.

Coelomic fluid Fluid inside the body cavity.

Coelozoic Living in a cavity, usually of the urinary tract or gall bladder.

Coenzyme A molecule that provides the transfer site for biochemical reactions catalyzed by an enzyme.

Coffee can Essential measuring device used by some fish culturalists in lieu of a graduated cylinder.

Coherent Sticking together, as with soil particles.

Coldwater disease See Peduncle disease or Fin rot disease.

Coldwater species Generally, fish that spawn in water temperatures below 55°F (12.8°C). The main cultured species are trout and salmon. See Coolwater species; Warmwater species.

Collateral Property, savings, stocks, and the like, deposited as security additional to one's personal or contractual obligations.

Colloid A substance so finely divided that it stays in suspension in water but does not pass through animal membranes.

Colorimetric Determining the quantity of a substance by the measurement of the intensity of light transmitted by a solution of the substance.

Columnaris disease An infection, usually of the skin and gills, by *Flexibacter columnaris*, a myxobacterium.

Commensal A plant or animal that lives in association with a host organism but is not injurious to it.

Communicable disease A disease that naturally is transmitted directly or indirectly from one individual to another.

Community Group of animal and plant populations living together in the same environment.

Compensation point That depth at which incident light penetration is just sufficient for plankton to photosynthetically produce enough oxygen to balance their respiration requirements.

Competencies Abilities or capabilities of employees.

Complete diet (complete feed) See Balanced diet.

Complete feed Feed that supplies 100 percent of the dietary requirements of the fish; used when there is little or no access to natural food.

Complicating disease An additional disease during the course of an already existing ailment.

Compressed Applied to fish, flattened from side to side, as in the case of a sunfish. See Depressed.

Conditioned response Behavior that is the result of experience or training.

Conductance The ability of a substance to allow the passage of electrical current.

Congenital disease A disease that is present at birth; may be infectious, nutritional, genetic, or developmental.

Congestion Unusual accumulation of blood in tissue; may be active (often called "hyperemia") or passive. Passive congestion is the result of abnormal venus return and is characterized by dark cyanotic blood.

Connective tissue Type of tissue that lies between groups of nerve, gland, and muscle cells and beneath the skin cells.

Constant-flow treatment Continuous automatic metering of a chemical to flowing water.

Contamination The presence of material or microorganisms making something impure or unclean.

Control (disease) Reduction of mortality or morbidity in a population, usually by use of drugs.

Control (experimental) Similar test specimens subjected to the same conditions as the experimental specimens except for the treatment variable under study.

Coolwater species Generally, fish that spawn in temperatures between 50° and 65°F (10.0° and 18.3°C). The main cultured coolwater species are northern pike, muskellunge, walleye, sauger, and yellow perch. See Coldwater species; Warmwater species.

Copepods Small, free-swimming, freshwater and marine crustaceans.

Copper sulfate (blue stone) Blue stone is copper sulfate pentahydrate ($CuSO_4 \cdot 5H_2O$); effective in the prevention and control of external protozoan parasites, fungal infections, and external bacterial diseases, highly toxic to fish.

Cornea Outer covering of the eye.

Corporation A body of people recognized by law as an individual person, having a name, rights, privileges, and liabilities distinct from the individual members.

Corynebacterial kidney disease See Bacterial kidney disease.

Costiasis An infection of the skin, fins, and gills by flagellated protozoans of the genus Costia.

Cranium The part of the skull enclosing the brain.

Creative thinking Ability to generate new ideas by making nonlinear or unusual connections or by changing or reshaping goals to imagine new possibilities; using imagination freely, combining ideas and information in new ways.

Crossbreeding The mating of unrelated strains of the same species to avoid inbreeding.

Crustacean Class in arthropod phylum containing crawfish, crabs, lobsters, shrimp, prawn, and others.

Cryogenically frozen Frozen at very low temperatures.

Cultural diversity Term used to describe the U.S. workplace as representing people from different backgrounds.

Culture The business of producing, propagating, transporting, possessing, and selling fish or shellfish raised in a private pond, raceway, or tank.

Cultured fish Farm-raised fish or shellfish.

Curricular Having to do with a course of study.

Cyanocobalamin (vitamin B_{12}) One of the water-soluble B-complex vitamins that is involved with folic acid in blood-cell production in fish; enhances growth in many animals.

Cyst Round, thick membrane with which some parasites are surrounded when in the resting state.

Cyst (host) A connective tissue capsule, liquid or semi-solid, produced around a parasite by the host.

Cyst (parasite) A noncellular capsule secreted by a parasite.

Cyst (protozoa) A resistant resting or reproductive stage of protozoa.

Cytoplasm The contents of a cell, exclusive of the nucleus.

D

Daily temperature unit (DTU) Equal to one degree Fahrenheit above freezing (32°F) for a 24-hour period.

Data sheet Similar to a résumé; contains pertinent information about potential employee.

Database Computerized collection of information that can be sorted and retrieved for various reports.

Decapod An animal with 10 legs, such as a crawfish.

Dechlorination Removal of the residual hypochlorite or chloramine from water to allow its use in fish culture. Charcoal is used frequently because it removes much of the hypochlorite and fluoride. Charcoal is inadequate for removing chloramine.

Dechlorinator Used to remove the residual hypochlorite or chloramine from water to allow its use in fish culture

Dee's disease See Bacterial kidney disease.

Deficiency A shortage of a substance necessary for health.

Deficiency disease A disease resulting from the lack of one or more essential constituents of the diet.

Deheading Removal of the head.

Demand The desire to possess a product combined with the ability to purchase.

Demand feeder Provides feed as animals desire.

Demography The study of vital and social statistics.

Denitrification A biochemical reaction in which nitrate (NO_3) is reduced to NO_2, N_2O, and nitrogen gas.

Density index The relationship of fish size to the water volume of a rearing unit; calculated by the formula: Density index = (weight of fish) / (fish length × volume of rearing unit).

Dentary bones The principal or anterior bones of the lower jaw or mandible. They usually bear teeth.

Depreciation The decrease in value resulting from the wear and tear of use, accident, destructive weather, poor management, and the obsolescence of equipment and processes.

Depressed Flattened in the vertical direction, as a flounder.

Depth of fish The greatest vertical dimension; usually measured just in front of the dorsal fin.

Depuration To rinse or wash free of impurities.

Dermal Pertaining to the skin.

Dermatomycosis Any fungus infection of the skin.

Detritus Debris from plants and animals.

Diagnosis The process of recognizing diseases by their characteristic signs.

Diarrhea Profuse discharge of fluid feces.

Diel Involving a 24-hour day that includes a day and the adjoining night.

Diet Food regularly provided and consumed.

Dietary fiber Non-digestible carbohydrate.

Diffusion The spreading out of molecules in a given space.

Digestion The breakdown of foods in the digestive tract to simple substances that may be absorbed by the body.

Diluent A substance used to dissolve and dilute another substance.

Dilution water Refers to the water used to dilute toxicants in aquatic toxicity studies.

Dip Brief immersion of fish into a concentrated solution of a treatment, usually for one minute or less.

Diplostomiasis An infection involving larvae of any species of the genus *Diplostomum*, Trematoda.

Discharge To release or to remove by unloading.

Disease Any departure from health; a particular destructive process in an organ or organism with a specific cause and symptoms.

Disease agent A physical, chemical, or biological factor that causes disease. Sometimes called "etiologic agent" and "pathogenic agent."

Disease prevention Steps taken to stop a disease outbreak before it occurs; may include environmental manipulation, immunization, administration of drugs, etc.

Disinfectant An agent that destroys infective agents.

Disinfection Destruction of pathogenic microorganisms or their toxins.

Dissolved oxygen (DO) The amount of elemental oxygen, O_2, in solution under existing atmospheric pressure and temperature.

Dissolved solids The residue of all dissolved materials when water is evaporated to dryness. See Salinity.

Distal The remote or extreme end of a structure.

Distributor An agent that sells merchandise.

Diurnal Relating to daylight; opposite of nocturnal.

Domestic Tame; bred and raised in captivity.

Dorsal Top side of the body, the back. Opposite of ventral.

Dorsal fin The fin on the back or dorsal side, in front of the adipose fin if the latter is present.

Dose A quantity of medication administered at one time.

Drainage May refer to methods of draining a pond or to surface water runoff.

Drawdown Process of lowering the water level in a pond, or completely draining a pond.

Dredges A large powerful scoop or suction apparatus for removing mud and gravel.

Dress To clean and eviscerate for marketing or consumption.

Dressed Killed and prepared for food market.

Drip treatment See Constant-flow treatment.

Dropsy See Edema.

Drug resistant A microorganism, usually a bacterium, that cannot be controlled (inhibited) or killed by a drug.

Drug sensitive A microorganism, usually a bacterium, that can be controlled (inhibited) or killed by use of a drug.

Dry feed A diet prepared from air-dried ingredients, formed into distinct particles and fed to fish.

Dysentery Disease characterized by passage of liquid feces containing blood and mucus; inflammation of the colon.

E

ERM See Enteric redmouth disease.

Economics Development and management of the material wealth of a government, community, or business.

Ecosystem A system of interrelated organisms and their physical-chemical environment.

Ectocommensal A commensal that lives on the surface of the host's body.

Ectoderm The outer layer of cells in an embryo that gives rise to various organs.

Ectoparasite Parasite that lives on the surface of the host.

Edema Excessive accumulation of fluid in tissue spaces.

Efficacy Ability to produce effects or intended results.

Effluent Water discharge from a rearing facility, treatment plant, or industry.

Egg The mature female germ cell, ovum.

Egg-Scatterers Disperse eggs haphazardly.

Egtved disease See Viral hemorrhagic septicemia.

Emaciation Wasting of the body.

Emarginate fin Fin with the margin containing a shallow notch, as in the caudal fin of the rock bass.

Emboli Abnormal materials carried by the bloodstream, such as blood clots, air bubbles, cancers or other tissue cells, fat, clumps of bacteria or foreign bodies, until they lodge in a blood vessel and obstruct it.

Embryo Developing organism before it is hatched or born.

Encyst To enclose or become enclosed in a cyst, capsule, or sac.

Endocrine A ductless gland or the hormone produced therein.

Endoparasite A parasite that lives in the host.

Endoskeleton The skeleton proper; the inner bony and cartilaginous framework of vertebrates.

Energy Capacity to do work.

Enrobing Coating fish products with flavors and various toppings such as bread crumbs.

Enteric redmouth disease (ERM) A disease, primarily of salmonids, characterized by general bacteremia. Caused by an enteric bacterium, *Yersinia ruckeri*. Synonym is Hagerman redmouth disease.

Enteritis Any inflammation of the intestinal tract.

Enterprise A specific process or activity that requires a certain amount of risk to make a profit.

Enterprise budget A look at the costs and risks involved with producing one commodity or making one product.

Entrepreneur One who starts and conducts a business assuming full control and risk. Environment All of the external conditions that affect growth and development of an organism.

Environmental gill disease Rapid growth of gill tissue (hyperplasia) caused by presence of a pollutant in the water that is a gill irritant. See Bacterial gill disease.

Enzootic A disease that is present in an animal population at all times but occurs in few individuals at any given time.

Enzyme A protein that catalyzes biochemical reactions in living organisms.

Epidermis The outer layer of the skin.

Epizootic A disease attacking many animals in a population at the same time widely diffused and rapidly spreading.

Epizootiology The study of epizootics the field of science dealing with relationships of various factors that determine the frequencies and distributions of diseases among animals.

Equity The value remaining in a business in excess of any liability or mortgage.

Equivalents per million (epm) A measure of ionized salts.

Eradication Removal of all recognizable units of an infecting agent from the environment.

Erosion The wearing away of land surface by wind and water. Erosion can occur naturally and by land use activities.

Esophagus The gullet; a muscular, membranous tube between the pharynx and the stomach.

Essential amino acids Those amino acids that must be supplied by the diet and cannot be synthesized within the body. In fish, these include: arginine, histidine, isoleucine, leucine, lysine, methionine, phenylalanine, threonine, tryptophan, and valine.

Essential fatty acids A fatty acid that must be supplied by the diet; for example, linoleic acid or linolenic acid.

Estate One's entire property and possessions.

Estuary The part of the mouth or lower course of a river in which the river's current meets the ocean's tide to create a mixing of fresh and salt water.

Etiologic agent See Disease agent.

Etiology The study of the causes of a disease, both direct and predisposing, and the mode of their operation; not synonymous with cause or pathogenesis of disease, but often used to mean pathogenesis.

European gill rot See Branchiomycosis.

Eutrophic Bodies of water that have excessive concentrations of plant nutrients causing excessive algal production and low transparency.

Evaluation Determining worth; appraisal.

Eviscerate To gut a fish; to remove the viscera.

Eviscerated Gutted; with internal organs removed.

Exclusive economic zone An area 200 nautical miles off the coast of a country reserved for exploring and exploiting, conserving, and managing the natural resources, whether living or nonliving.

Excretion The process of getting rid or throwing off metabolic waste products by an organism.

Exophthalmos Abnormal protrusion of the eyeball from the orbit.

Exoskeleton Hard outer shell that protects the body of an organism such as a crawfish, crab, or lobster.

Exotic A fish or animal that is not native to the state or locale.

Export To send merchandise or raw materials to other countries for sale or trade.

Extended aeration system A modification of the activated-sludge process in which the retention time is longer than in the conventional process.

Extensive culture Rearing of fish in ponds with low water exchange and at low densities; the fish utilize primarily natural foods.

Extensive production Raising of fish in low densities in ponds where the fish feed primarily on natural feeds.

Extruded Pushed through a die to give a certain shape; method of producing floating fish food.

Eyed stage Egg in which two black spots the developing eyes of the embryo can easily be seen.

F

Fps Feet per second.

F1 The first generation of a cross.

F2 The second filial generation obtained by random crossing of F1 (first filial) individuals.

Facultative Capable of living under varying conditions.

GLOSSARY

Farming The science or practice of agriculture; the business of operating a farm.

Fat An ester composed of fatty acid(s) and glycerol.

Fatty acid Organic acid present in lipids, varying in carbon content from 2 to 34 atoms (C_2 to C_{34}).

Fauna The animals inhabiting any region, taken collectively.

Fecundity Number of eggs in a female spawner.

Feed conversion ratio (FCR) The average number of pounds of feed needed to gain 1 pound in weight.

Feeding chart Guidelines provided by feed manufacturer.

Feeding level The amount of feed offered to fish over a unit time, usually given as percent of fish body weight per day.

Feeding ring Part of a cage that keeps feed from floating out of the cage.

Fertility Ability to produce viable offspring.

Fertilization (1) The union of sperm and egg; (2) addition of nutrients to a pond to stimulate natural food production.

Fillet Boneless sides of fish cut lengthwise away from the backbone.

Fin ray One of the cartilaginous rods that support the membranes of the fin.

Fin rot disease A chronic, necrotic disease of the fins caused by invasion of a myxobacterium into the fin tissue of an unhealthy fish.

Fines Small particles of feed.

Fingerling The stage in a fish's life between 1 in. (2.5 cm) and the length at 1 year of age.

Fish farm The property including private ponds, raceways, or tanks from which fish or shellfish are produced, propagated, transported, or sold.

Fish farmer Any person engaged in fish farming.

Fish farming The business of producing, propagating, transporting, possessing, and selling cultured fish or shellfish raised in a private pond, raceway, or tank.

Fistula An abnormal tubelike passage from an abscess or hollow organ to the skin.

Fixative A chemical agent chosen to penetrate tissues very soon after death and preserve the cellular components in an insoluble state as nearly lifelike as possible.

Fixed costs Costs that usually do not fluctuate with an increase or decrease in production.

Flagellum (plural: flagella) Whiplike locomotion organelle of single (usually free-living) cells.

Flashing Quick turning movements of fish, especially when fish are annoyed by external parasites, causing a momentary reflection of light from their sides and bellies. When flashing, fish often scrape themselves against objects to rid themselves of the parasites.

Flatland Area with not more than 3 percent slope.

Flow index The relationship of fish size to water inflow (flow rate) of a rearing unit; calculated by the formula: (fish weight) / (fish length × water inflow).

Flow rate The volume of water moving past a given point in a unit of time, usually expressed as cubic feet per second (cfs) or gallons per minute (gpm).

Flush A short bath in which the flow of water is not stopped, but a high concentration of chemical is added at the inlet and passed through the system as a pulse.

Foam fractionator A unit that introduces air bubbles at the bottom of a closed column of water creating foam at the air/water interface; as the bubbles rise through the water column, fine suspended solid particles attach to the bubbles' surface, creating the foam at the top of the water column; foam buildup channeled out of the unit to a waste collection tank.

Folic acid (folacin) A vitamin of the water-soluble B complex that is necessary for maturation of red blood cells and synthesis of nucleoproteins; deficiency results in anemia.

Follow-up letter Letter written immediately after a job interview.

Fomites Inanimate objects (brushes or dipnets) that may be contaminated with and transmit infectious organisms. See Vector.

Food chain Transfer of energy from one living thing to another in the form of food.

Food conversion A ratio of food intake to body weight gain; more generally, the total weight of all feed given to a lot of fish divided by the total weight gain of the fish lot. The units of weight and the time interval over which they are measured must be the same. The better the conversion, the lower the ratio.

Foot Drain end of holding trough.

Forage Food for animals taken by browsing or grazing, or the act of browsing or grazing to obtain food.

Forecast To calculate before hand.

Fork length The distance from the tip of the snout to the fork of the caudal fin.

Formalin Solution of approximately 37 percent by weight of formaldehyde gas in water. Effective in the control of external parasites and fungal infections on fish and eggs; also used as a tissue fixative.

Formulated feed A combination of ingredients that provides specific amounts of nutrients per weight of feed.

Fortification Addition of nutrients to feeds.

Free-living Not dependent on a host organism.

Freeboard Distance between pond surface and top of levees or dam generally between 1 and 2 ft. (30 to 61 cm).

Freshwater Water containing less than 0.05 percent total dissolved salts by weight.

Friable Easily crumbled or crushed into powder.

Fry The stage in a fish's life from the time it hatches until it reaches 1 in. (2.5 cm) in length.

Fungus Any of a group of primitive plants lacking chlorophyll, including molds, rusts, mildews, smuts, and mushrooms. Some kinds are parasitic on fishes. Reproduces by spores.

Fungus disease See Saprolegniasis.

Furuncle A localized infection of skin or subcutaneous tissue that develops a solitary abscess that may or may not drain externally.

Furunculosis A bacterial disease caused by Aeromonas salmonicida and characterized by the appearance of furuncles.

Fusiform Long and tapered toward the ends like a torpedo.

G

Gpm Gallons per minute.

Gall bladder The internal organ containing bile.

Gametes Sexual cells; eggs and sperm.

Gape The opening of the mouth.

Gas bladder See Air bladder.

Gas bubble disease Gas embolism in various organs and cavities of the fish, caused by supersaturation of gas (mainly nitrogen) in the blood.

Gastric Relating to the stomach.

Gastritis Inflammation of the stomach.

Gastroenteritis Inflammation of the mucosa of the stomach and intestines.

Gastrointestinal Referring to stomach and intestines, the digestive system.

Gastroliths Two small stones in the crawfish's stomach in which calcium carbonate is stored; used for hardening the shell after molting.

Gastropod Marine, freshwater, and land mollusks with one shell.

Gene The unit of inheritance. Genes are located at fixed loci in chromosomes and can exist in a series of alternative forms called "alleles."

Genetic dominant Character donated by one parent that masks in the progeny the recessive character derived from the other parent.

Genetics The science of heredity and variation.

Genital Pertaining to the reproductive organs.

Genital papilla Small nipple-like projection of tissue on male catfish.

Genomics Branch of genetics that studies organisms in terms of their full DNA sequences.

Genus A unit of scientific classification that includes one or several closely related species. The scientific name for each organism includes designations for genus and species.

Geographic distribution The geographic areas in which a condition or organism is known to occur.

Germinal disc The disc-like area of an egg yolk on which cell segmentation first appears.

Gill arch The U-shaped cartilage that supports the gill filaments.

Gill clefts (gill slits) Spaces between the gills connecting the pharyngeal cavity with the gill chamber.

Gill cover The flap-like cover of the gill and gill chamber; the operculum.

Gill disease See Bacterial gill disease and Environmental gill disease.

Gill filament The slender, delicate, fringe-like structure composing the gill.

Gill lamellae The subdivisions of a gill filament where most gas and some mineral exchanges occur between blood and the outside water.

Gill openings The external openings of the gill chambers, defined by the operculum.

Gill rakers A series of bony appendages, variously arranged along the anterior and often the posterior edges of the gill arches.

Gill rot See Branchiomycosis.

Gills The highly vascular, fleshy filaments used in aquatic respiration and excretion.

Glair Blue-like substance secreted by the female crawfish; used to attach laid eggs to her swimmerets.

Globulin One of a group of proteins insoluble in water, but soluble in dilute solutions of neutral salts.

Glycogen Animal starch, a carbohydrate storage product of animals.

Goals The end objectives or terminal points of a business.

Gonadotrophin Hormone produced by pituitary glands to stimulate sexual maturation.

Gonads The reproductive organs; testes or ovaries.

Grading Sorting of fish by size, usually by some mechanical device.

Gram-negative bacteria Bacteria that lose the purple stain of crystal violet and retain the counterstain, in the gram-staining process.

Gram-positive bacteria Bacteria that retain the purple stain of crystal violet in the gram-staining process.

Grants Money offered through the process of a proposal to an organization or a business for research.

Grazers Animals that eat continuously in well-defined bites.

Gregarine Name for a group of parasitic protozoans that live in insects, crustaceans, earthworms, and several other types of invertebrate animals.

Gross pathology Pathology apparent from the naked-eye appearance of tissues.

Groundwater The supply of water found beneath the Earth's surface.

Group immunity Immunity enjoyed by a susceptible individual by virtue of membership in a population with enough immune individuals to prevent a disease outbreak.

Grow-out Facilities that produce crops (fish) from the seed.

Gullet The esophagus.

Gyro infection An infection of any of the monogenetic trematodes or, more specifically, of *Gyrodactylus* species.

H

HACCP Stands for Hazard Analysis and Critical Control Point; a new method for ensuring the safety of food.

HRM See Enteric redmouth disease.

Habitat Those plants, animals, and physical components of the environment that constitute the natural food, physical-chemical conditions, and cover requirements of an organism.

Hagerman redmouth disease See Enteric redmouth disease.

Haptor Posterior attachment organ of monogenetic trematodes.

Hardness The power of water to neutralize soap, due to the presence of cations such as calcium and magnesium; usually expressed as parts per million equivalents of calcium carbonate. Refers to the calcium and magnesium ion concentration in water on a scale of very soft (0 to 20 ppm as CaCO3), soft (20 to 50 ppm), hard (50 to 500 ppm), and very hard (500+ ppm).

Harvesting Involves the gathering or capturing of the fish for marketing and processing. Aquaculture harvesting is typically topping or partial and total harvest.

Hatchery Produce the seed or young fish.

Hatchery constant A single value derived by combining the factors in the numerator of the feeding rate formula: Percent body weight fed daily = (3 × food conversion × daily length increase × 100) / length of fish. This value may be used to estimate feeding rates when water temperature, food conversion, and growth rate remain constant.

Hatchling Newly hatched alligator.

Head Inflow end of holding trough.

Headwaters Place of origin (where groundwater first surfaces) of a creek, stream, or river.

Heat capacity Characteristic of water making it resistant to temperature changes.

Heavy metals Metals such as cadmium (Cd), cobalt (Co), chromium (Cr), copper (Cu), lead (Pb), lithium (Li), manganese (Mn), mercury (Hg), nickel (Ni), zinc (Zn), or iron (Fe).

Heliciculture Culture of snails.

Hematocrit Percent of total blood volume that consists of cells; packed cell volume.

Hematoma A tumor-like enlargement in the tissue caused by blood escaping the vascular system.

Hematopoiesis The formation of blood or blood cells in the living body. The major hematopoietic tissue in fish is located in the anterior kidney.

Hematopoietic kidney The anterior portion of the kidney, involved in the production of blood cells.

Heme The iron-bearing constituent of hemoglobin.

Hemocoel The enclosed space around the organs of crustaceans that contain the animal's blood.

Hemoglobin The respiratory pigment of red blood cells that takes up oxygen at the gills or lungs and releases it at the tissues.

Hemorrhage An escape of blood from its vessels, through either intact or ruptured walls; bleeding.

Hen Female trout.

Hepatic Pertaining to the liver.

Hepatitis Inflammation of the liver.

Hepatoma A tumor with cells resembling those of liver; includes any tumor of the liver. Hepatoma is associated with mold toxins in feed eaten by cultured fishes. The toxin having the greatest affect on fishes is aflatoxin B_1, from *Aspergillus flavus*.

Herbivore Animals that subsist primarily on the available vegetation and decayed organic material in the environment.

Hermaphroditic Possesses gonads, testes, and ovaries for both sexes and can release both eggs and sperm.

Heterocercal Forked tail fin.

Heterotrophic bacteria Bacteria that oxidize organic material (carbohydrate, protein, fats) to CO_2, NH_4^-N, and water (H_2O) for their energy source.

Hide Raw or dressed pelt or skin of a large animal like an alligator.

Histology Microscopic study of cells, tissues, and organs.

Histopathology The study of microscopically visible changes in diseased tissues.

Homing Return of fish to their stream or lake of origin to spawn.

Homocercal Single-lobed tail fin.

Horizontal transmission Any transfer of a disease agent between individuals except for the special case of parent-to-progeny transfer via reproductive processes.

Hormone A chemical product of endocrine gland cells affecting organs that do not secrete it.

Host Animal on or in which a parasite lives.

Humectants Prevent bacterial growth.

Husbandry The occupation or business of farming.

Hyamine See Quaternary ammonium compounds.

Hybrid Crossbreeding fish of different varieties, races, or species.

Hybrid vigor Condition in which the offspring perform better than the parents. Sometimes called "heterosis."

Hybridization The crossing of different species.

Hydrate To combine with water.

Hydrogen ion concentration (activity) The cause of acidity in water. See pH.

Hydrogen sulfide An odorous, soluble gas, H_2S, resulting from anaerobic decomposition of sulfur-containing compounds, especially proteins.

Hydrological cycle Circular flow of water between the atmosphere and the Earth; precipitation, runoff, surface water, groundwater, evaporation, and transpiration.

Hydroponics The cultivation of land plants without soil, in a water solution.

Hyoid Bones in the floor of the mouth supporting the tongue.

Hyper A prefix denoting excessive, above normal, or situated above.

Hyperemia Increased blood resulting in distention of the blood vessels.

Hyperplasia Increased, abnormal tissue growth.

Hypersaline High salt content.

Hypo A prefix denoting deficiency, lack, below, beneath.

I

IHN See Infectious hematopoietic necrosis.

IPN See Infectious pancreatic necrosis.

Ich A protozoan disease caused by the ciliate *Ichthyophtherius multifilis*; sometimes called "white-spot disease."

Immune Unsusceptible to a disease.

Immunity Lack of susceptibility; resistance; an inherited or acquired status.

Immunization Process or procedure by which an individual is made resistant to disease, specifically infectious disease.

Import To receive merchandise or raw materials from other countries for sale or trade.

Impound To gather and enclose water for fish pond or irrigation.

Impoundment A dam, dike, floodgate, or other barrier confining a body of water.

Imprinting The imposition of a behavior pattern in a young animal by exposure to stimuli.

In berry Female crawfish carrying eggs.

In vitro Used in reference to tests or experiments conducted in an artificial environment, including cell or tissue culture.

In vivo Used in reference to tests or experiments conducted in or on intact, living organisms.

Inbred line A line produced by continued matings of brothers to sisters and progeny to parents over several generations.

Inbreeding Mating of closely related animals.

Incidence The number of new cases of a particular disease occurring within a specified period in a group of organisms.

Income Amount of money received periodically in return for goods, labor, or services.

Income statement Financial record that reflects the profitability of the business over a specified period of time; also known as a profit and loss statement or an operating statement.

Incubate To maintain at a favorable temperature and in other conditions promoting development of eggs until they hatch.

Incubation (disease) Period of time between the exposure of an individual to a pathogen and the appearance of the disease it causes.

Incubation (eggs) Period from fertilization of the egg until it hatches.

Incubator Device for artificial rearing of fertilized fish eggs and newly hatched fry.

Indigenous Refers to a species of fish, shellfish, or aquatic plant usually found in the public waters of the state.

Indispensable amino acid See Essential amino acids.

Inert gases Those gases in the atmosphere that are inert or nearly inert; nitrogen, argon, helium, xenon, krypton, and others. See Gas bubble disease.

Infection Contamination (external or internal) with a disease-causing organism or material, whether or not overt disease results.

Infection, focal A well-circumscribed or localized infection in or on a host.

Infection, secondary Infection of a host that already is infected by a different pathogen.

Infection, terminal An infection, often secondary, that leads to death of the host.

Infectious catarrhal enteritis See Infectious pancreatic necrosis.

Infectious disease A disease that can be transmitted between hosts.

Infectious pancreatic necrosis (IPN) A disease caused by an infectious pancreatic necrosis virus that presently has not been placed into a group. Sometimes called "infectious catarrhal enteritis," "chinook salmon virus disease," "Oregon sockeye salmon virus," and "Sacramento River chinook disease."

Inferior mouth Mouth on the under side of the head, opening downward.

Infiltration To filter through small gaps or passages.

Inflammation The reaction of the tissues to injury, characterized clinically by heat, swelling, redness, and pain.

Ingest To eat or take into the body.

Injection Method of introducing a drug or vaccine into the muscle or body cavity.

Inoculation The introduction of an organism into the tissues of a living organism or into a culture medium.

Inorganic Not characterized by life processes; not containing carbon.

Inputs Contribution of information, ideas, opinions, equipment, materials, and money to the successful operation of a business; the available data and other resources to solve a problem.

Input costs Money required to begin production.

Inspection Careful or critical examination of a product.

Instinct Inherited behavioral response.

Intensive culture Rearing of fish at densities greater than can be supported in the natural environment; utilizes high water flow or exchange rates and requires the feeding of formulated feeds.

Intensive production Raising of fish in densities higher than could be supported in the natural environment; requires feeding of formulated feeds.

Interest Payment for the use of money or credit.

Interpersonal Between people.

Interspinals Bones to which the rays of the fins are attached.

Intestine The lower part of the alimentary tract from the pyloric end of the stomach to the anus.

Intragravel water Water occupying interstitial spaces within gravel.

Intramuscular injection Administration of a substance into the muscles of an animal.

Intraperitoneal injection Administration of a substance into the body cavity (peritoneal cavity).

Inventory The value of goods or stock of a business.

Invertebrate Organism with a hard outer skeleton and lacking a spinal column.

Ion Electrically charged atom, radical, or molecule.

Ion exchange A process of exchanging certain cations or anions in water for sodium, hydrogen, or hydroxyl (OH^-) ions in a resinous material.

Isotonic No osmotic difference; one solution having the same osmotic pressure as another.

Isthmus The region just anterior to the breast of a fish where the gill membranes converge; the fleshy interspace between gill openings.

Iteroparous Spawning only several times in a lifetime.

K

Kidney One of the pair of glandular organs in the abdominal cavity that produces urine.

Kidney disease See Bacterial kidney disease.

Kilogram calorie The amount of heat required to raise the temperature of one kilogram of water one degree centigrade, also called "kilocalorie (kcal)," or "large calorie" a measure of energy.

Kype Upward-curving hook of lower jaw that occurs at spawning.

L

LDV See Lymphocystis disease.

Larva (plural: larvae) An immature form that must undergo change of appearance or pass through a metamorphic stage to reach the adult state.

Lateral Side of the body.

Lateral band A horizontal pigmented band along the sides of a fish.

Lateral line A series of sensory pores, sensitive to low-frequency vibrations, located laterally along both sides of the body.

Least-cost Method of feed formulation where the formula varies, within limits, as ingredient prices change.

Length May refer to the total length, fork length, or standard length.

Lesion Any visible alteration in the normal structure of organs, tissues, or cells.

Lethargy A state of sluggishness, inaction, indifference, or dullness.

Letter of application Sent with resume or data sheet when applying for a job.

Letter of inquiry Sent to potential employer requesting possibility of employment.

Leucocyte A white blood corpuscle.

Levee Earth dike used to enclose water.

Liabilities Just or legal responsibilities.

Lime (calcium oxide, quicklime, burnt lime) CaO; used as a disinfectant for fish-holding facilities. Produces heat and extreme alkaline conditions.

Line A cord used or made for angling (fishing): One end attaches to the reel the other end attaches to a hook or lure.

Line breeding Mating individuals so that their descendants will be kept closely related to an ancestor that is regarded as unusually desirable.

Linolenic acid An 18-carbon fatty acid with two double bonds; certain members of the series are essential for health, growth, and survival of some, if not most, fishes.

Lipid Any of a group of organic compounds consisting of the fats and other substances of similar properties. They are insoluble in water but soluble in fat solvents and alcohol.

Liquidity A business's ability to meet short-run obligations when due without disrupting the normal operation of the business.

Live bearers Fish that retain the eggs inside the body and give birth to live, free-swimming young.

Live-car Net attached to harvesting seine and used to crowd, grade, and hold fish in the pond; also called a "sock."

Live-haulers Individuals or business who specialize in trucking live fish.

Liver A large, reddish brown, glandular organ in vertebrates, located in the abdominal cavity; it functions in the secretion of bile and in essential metabolic processes.

Logarithm The exponent to which a base (usually 10) must be raised to produce a given number.

Long bath A type of bath frequently used in ponds. Low concentrations of chemicals are applied and allowed to disperse by natural processes.

Lymphocystis disease A virus disease of the skin and fins affecting many freshwater and marine fishes of the world; caused by the lymphocystis virus of the Iridovirus group.

M

MAS See motile aeromonas septicemia.

Macrominerals Minerals in the body in large quantities.

Macrophyte Vascular plants with true roots, stems, and leaves.

Malignant Progressive growth of certain tumors that may spread to distant sites or invade surrounding tissue and kill the host.

Malnutrition Faulty or inadequate nutrition.

Mandible Lower jaw.

Mantle Soft covering over the organs of a mollusk.

Mariculture Raising of organisms in the ocean.

Marine Of the sea or ocean.

Maritime Pertaining to the sea, its navigation and commerce.

Market Buyer of product; place to sell a product.

Marketing The process of getting the product from the producer to the consumer. It is the final step in food production.

Market niches A specific product with specific features aimed at satisfying specific market needs.

Mass selection Selection of individuals from a general population for use as parents in the next generation.

Matching funds Money provided on a pre-arranged split.

Mating system Any of a number of schemes by which individuals are assorted in pairs leading to sexual reproduction.

Maxilla or maxillary The hindmost bone of the upper jaw.

Mean The arithmetic average of a series of observations.

Mechanical damage Extensive connective tissue proliferation, leading to impaired growth and reproductive processes, caused by parasites migrating through tissue.

Median A value in a series below and above, which there are an equal number of values.

Melanophore A black pigment cell; large numbers of these give fish a dark color.

Menadione A fat-soluble vitamin; a form of vitamin K.

Meristic characters Body parts that can be counted, such as scales, gill rakers, and vertebrae; useful in species identifications.

Metabolic rate The amount of oxygen used for total metabolism per unit of time per unit of body weight.

Metabolism Vital processes involved in the release of body energy, the building and repair of body tissue, and the excretion of waste materials; combination of anabolism and catabolism.

Metabolites A product of the biochemical process of living organisms; a product of metabolism.

Metamorphosis Change from one form to another, as from a tadpole to a frog.

Methylene blue A quinoneimine dye effective against external protozoans and superficial bacterial infections.

Microbe Microorganism, such as a virus, bacterium, fungus, or protozoan.

Microencapsulation The coating of small feed particles with a substance that is insoluble in water but digestible by the enzymes in the digestive tract of the fish.

Microminerals Minerals in the body in small quantities.

Microohms Measure of electrical resistance; one-thousandth of an ohm.

Micropyle Opening in egg that allows entrance of the sperm.

Microscopic Small enough to be invisible or obscure except when observed through a microscope.

Microsporidean A small spore-producing organism.

Migration Movement of fish populations.

Milt Secretion that contains sperm produced by a male fish.

Mitigation To make less severe or intense.

Mitosis The process by which the nucleus is divided into two daughter nuclei with equivalent chromosome complements.

Molting For crustaceans, the shedding of the exoskeleton; molting occurs at intervals during a crustaceans's life and allows for expansion in size.

Monoculture Raising a single species in a pond or enclosure.

Monthly temperature unit (MTU) Equal to one degree Fahrenheit above freezing (32°F), based on the average monthly water temperature (30 days).

Morbid Caused by disease; unhealthy, diseased.

Morbidity The condition of being diseased.

Morbidity rate The proportion of individuals with a specific disease during a given time in a population.

Moribund Obviously progressing toward death; nearly dead.

Morphology The science of the form and structure of animals and plants.

Mortality Death, particularly death from disease or on a large scale.

Mortality rate The number of deaths per unit of population during a specified period. May be called "death rate," "crude mortality rate," or "fatality rate."

Motile aeromonas septicemia (MAS) An acute to chronic infectious disease caused by any motile bacteria belonging to the genus *Aeromonas*, primarily *Aeromonas hydrophila* or *Aeromonas punctate*. Sometimes called "bacterial hemorrhagic septicemia" or "pike pest."

Mottled Blotched; color spots running together.

Mouth-brooders Fish that hold the eggs and offspring in the mouth of the parent for extended periods of time.

Mouth fungus See Columnaris disease.

Mucking (egg) The addition of an inert substance such as clay or starch to adhesive eggs to prevent them from sticking together during spawn taking, commonly used with esocid and walleye eggs.

Mucus (mucous) A viscid or slimy substance secreted by the mucus glands of fish.

Mud line Bottom of a pond seine that keeps fish from escaping from under the seine.

Multiple harvest (partial harvest) Harvesting strategy that requires numerous yearly seinings or trappings and complete draining of a pond each 6 to 8 years.

Mutation A sudden heritable variation in a gene or in a chromosome structure.

Mycology The study of fungi.

Mycosis Any disease caused by an infectious fungus.

Mycotoxin Poisons derived from fungus.

Myomere An embryonic muscular segment that later becomes a section of the side muscle of a fish.

Myotome Muscle segment.

Mysis Developmental stage in the life cycle of a shrimp or prawn.

Myxobacteriosis A disease caused by any member of the Myxobacterales group of bacteria, for example, peduncle disease, coldwater disease, fin rot disease, or columnaris disease.

N

Nares The openings of the nasal cavity

Natal Pertaining to birth, usually in the context of fish that return to their place of birth to spawn.

Native fish All fish documented to live, spawn, or reproduce in public waters of a state, and whose first documented occurrence in public waters was not the result of direct or indirect importation by people.

Natural foods Plants and animals normally found in a pond or other water source, sometimes their production is enhanced by fertilization.

Nauplius Larval form in many crustaceans, with three pairs of appendages and a single median eye.

Necropsy A medical examination of the body after death to ascertain the cause of death; an autopsy in humans.

Necrosis Dying of cells or tissues within the living body.

Nematoda A diverse phylum of roundworms, many of which are plant or animal parasites.

Nephrocalcinosis A condition of kidney (renal) insufficiency due to the precipitation of calcium phosphate ($CaPO_4$) in the tubules of the kidney.

Nerve fibers Anatomical structures carrying nerve impulses.

Net pens Large cages used for raising fish, for example, salmon.

Net worth Assets minus liabilities.

Net worth statement Financial condition of a business at a definite point in time; it lists all assets, values of assets, and liabilities of a business; also known as a balance sheet, financial statement, or statement of financial condition.

Niacin One of the water-soluble B-complex vitamins, essential for maintenance of the health of skin and other epithelial tissues in fishes.

Nicotinic acid See Niacin.

Nitrification A method through which ammonia is biologically oxidized to nitrite and then nitrate.

Nitrite The NO_2^- ion.

Nitrogen (N₂) An odorless, gaseous element that makes up 78 percent of the earth's atmosphere and is a constituent of all living tissue. It is almost inert in its gaseous form.

Nitrogenous wastes Simple nitrogen compounds produced by the metabolism of proteins, such as urea and uric acid.

Nodule Small knot, knob, or lump of tissue.

Nonindigenous Refers to a species of fish, shellfish, or aquatic plant not usually found in public waters of the state.

Noninfectious Refers to diseases that are not contagious and that usually cannot be cured by medications.

Nonpathogenic Refers to an organism that may infect but causes no disease.

Nonpoint source pollution Pollution from a diffuse source.

Nostril See Nares.

Noxious Harmful or undesirable.

Nursery Ponds or tanks for newly hatched fry.

Nutria A large South American aquatic rodent; also called "coypu."

Nutrient A chemical used for growth and maintenance of an organism.

Nutrition All of the processes in which an animal takes in and uses food.

Nutritional gill disease Gill hyperplasia caused by deficiency of pantothenic acid in the diet.

Nymph The larva of various insects, especially dragonfly and mayfly larvae.

O

Ocean ranching Type of aquaculture involving the release of juvenile aquatic animals into marine waters to grow on natural foods to harvestable size.

Offal Waste parts of a slaughtered animal.

Off-flavor Musty-tasting or muddy-tasting fish flesh.

Omnivore Eating both vegetable and animal food.

On-line Refers to connecting computers via telephone lines or wirless technology for the purpose of transferring information.

Open-formula feed A diet in which all the ingredients and their proportions are public (nonproprietary).

Operculum A bony flaplike protective gill covering.

Opportunistic Waiting for a combination of favorable circumstances.

Optic Referring to the eye.

Organic Related to or derived from living organisms; contains carbon.

Osmoregulation The process by which organisms maintain stable osmotic pressures in their blood, tissues, and cells in the face of differing chemical properties among tissues and cells, and between the organism and the external environments.

Osmosis The diffusion of liquid that takes place through a semipermeable membrane between solutions starting at different osmotic pressures, and that tends to equalize those pressures. Water always will move toward the more concentrated solution, regardless of the substances dissolved, until the concentration of dissolved particles is equalized, regardless of electric charge.

Osmotic pressure The pressure needed to prevent water from flowing into a more concentrated solution from a less concentrated one across a semipermeable membrane.

Outfall Wastewater at its point of effluence or its entry into a river or other body of water.

Outliers Data values that lie outside the normal range and may be false (unacceptable) or true (acceptable anomaly).

Ovarian Having to do with ovaries, the female egg producing glands.

Ovarian fluid Fluid surrounding eggs inside the female's body.

Ovaries The female reproductive organs producing eggs or ova.

Overflow pipe Vertical pipe placed in a tank so that the top is at desired water height; water above this height drains from the tank.

Overt disease A disease, not necessarily infectious, that is apparent or obvious by gross inspection; a disease exhibiting clinical signs.

Oviduct The tube that carries eggs from the ovary to the exterior.

Oviparous Producing eggs that are fertilized, develop, and hatch outside the female body.

Ovoviviparous Producing eggs, usually with much yolk, that are fertilized internally. Little or no nourishment is furnished by the mother during development; hatching may occur before or after expulsion.

Ovulate Process of producing mature eggs (ova) capable of being fertilized.

Ovum (plural: ova) Egg cell or single egg.

Oxidation Combination with oxygen; removal of electrons to increase positive charge.

Oxygenation Process of providing or combining or treating with oxygen.

Oxygen transfer efficiency A measure of the percent of the total oxygen used that a device is able to put into solution.

Oxytetracycline (terramycin) One of the tetracycline antibiotics produced by *Streptomyces rimosus* and effective against a wide variety of bacteria pathogenic to fishes.

P

Paddlewheel Type of aeration device.

Pancreas The organ that functions as both an endocrine gland secreting insulin and an exocrine gland secreting digestive enzymes.

Pantothenic acid One of the essential water-soluble B-complex vitamins.

Para-aminobenzoic acid (PABA) A vitamin-like substance thought to be essential in the diet for maintenance of health of certain fishes.

Parasite An organism that lives in or on another organism (the host) and that depends on the host for its food, has a higher reproductive potential than the host, and may harm the host when present in large numbers.

Parasite, obligate An organism that cannot lead an independent, nonparasitic existence.

Parasiticide Antiparasite chemical (added to water) or drug (fed or injected).

Parasitology The study of parasites.

Parr A life stage of salmonid fishes that extends from the time feeding begins until the fish become sufficiently pigmented to obliterate the parr marks, usually ending during the first year.

Parr mark One of the vertical color bars found on young salmonids and certain other fishes.

Parts per billion (ppb) A concentration at which 1 unit is contained in a total of 1 billion units. Equivalent to 1 microgram per kilogram (1 mcg/ kg).

Parts per million (ppm) A concentration at which 1 unit is contained in a total of 1 million units. Equivalent to 1 milligram per kilogram (1 ml/kg) or 1 microliter per liter (1 ml/liter).

Parts per thousand (ppt) A concentration at which 1 unit is contained in a total of 1,000 units. Equivalent to 1 gram per kilogram (1 g/kg) or 1 milliliter per liter (1 ml/ liter). Normally, this term is used to specify salinity.

Partnership A form of business organization with multiple owners.

Pathogen Disease-causing organism.

Pathology The study of diseases and the structural and functional changes produced by them.

Payback Number of years it takes to recover the initial investment.

Pectoral fins The anterior and ventrally located fins whose principle function is locomotor maneuvering.

Peduncle disease A chronic, necrotic disease of the fins, primarily the caudal fin, caused by invasion of a myxobacterium (commonly *Cytophaga psychrophilia*) into fin and caudal peduncle tissue of an unhealthy fish. Other names include fin rot disease or coldwater disease.

Peeling plant Crawfish processing plant.

Pelvic fins Paired fins corresponding to the posterior limbs of the higher vertebrates (sometimes called "ventral fins"), located below or behind the pectoral fins.

Peptide bond Chemical connection between amino acids when forming proteins.

Percolation The process of a liquid passing through a filter.

Peritoneum The membrane lining the abdominal cavity.

Perivitelline fluid Fluid lying between the yolk and outer shell (chorion) of an egg.

Perivitelline space Area between yolk and chorion of an egg where embryo expansion occurs.

Permeability Rate of penetration by liquids.

Peroxide Containing oxygen.

Pesticide Broad name for chemicals that control or kill insects, fungi, parasites, and other pests.

Petechia A minute rounded spot of hemorrhage on a surface, usually less than one millimeter in diameter.

pH An expression of the acid-base relationship designated as the logarithm of the reciprocal of the hydrogen-ion activity; the value of 7.0 expresses neutral solutions; values below 7.0 represent increasing acidity; those above 7.0 represent increasingly basic solutions.

Pharynx The cavity between the mouth and esophagus.

Phenotype Appearance of an individual as contrasted with its genetic makeup or genotype. Also used to designate a group of individuals with similar appearance but not necessarily identical genotypes.

Photoperiod The number of daylight hours in 24 hours; exposure to light best suited to the growth and maturation of an organism.

Photosynthesis The formation of carbohydrates from carbon dioxide and water that takes place in the chlorophyll-containing tissues of plants exposed to light. Oxygen is produced as a by-product.

Phycocolloid Gelatin-like substance obtained from seaweed.

Physiology Study of body functions.

Phytin A form of phosphorus in seeds in which the availability of phosphorus to the animal is low.

Phytoplankton Minute plants suspended in water with little or no capability for controlling their position in the water mass; frequently referred to as algae.

Pig trough See Von Bayer trough.

Pigment The coloring matter in the cells of plants and animals.

Pigmentation Disposition of coloring matter in an organ or tissue.

Pituitary Small endocrine organ located near the brain.

Plankton Microscopic plants and animals.

Planktonic Describes the total of passively floating, drifting, or somewhat motile organisms occurring in a body of water, primarily comprising microscopic algae and protozoa.

Planting of fish The act of releasing fish from a hatchery into a specific lake or river. Sometimes called "distribution" or "stocking."

Plasma The fluid fraction of the blood, as distinguished from corpuscles. Plasma contains dissolved salts and proteins.

Plasticity Capacity of soil to be bent without breaking and to remain bent after force is removed.

Poikilothermic Having a body temperature that fluctuates with that of the environment.

Point source pollution Pollution from any single identifiable source.

Polliwog Another name for tadpole, the aquatic larva of frogs and toads, having internal gills and a tail.

Pollutant A term referring to a wide range of toxic chemicals and organic materials introduced into waterways from industrial plants and sewage wastes.

Pollution The addition of any substance not normally found in or occurring in a material or ecosystem.

Polyculture Raising two or more species in the same pond or enclosure.

Polysaccharide Class of carbohydrates of high molecular weight formed by the union of three or more monosaccharide molecules (sugar). Examples include starch, cellulose, dextrin, and glycogen.

Polytrophic Eating a wide variety of material, both plant and animal.

Pond run Ungraded by size or sex.

Population A coexisting and interbreeding group of individuals of the same species in a particular locality.

Population density The number of individuals of one population in a given area or volume.

Portal of entry The pathway by which pathogens or parasites enter the host.

Portal of exit The pathway by which pathogens or parasites leave or are shed by the host.

Posterior The back side.

Post-treatment Treatment of hatchery wastewater before it is discharged into the receiving water; pollution abatement.

Potassium permanganate $KMnO_4$, a strong oxidizing agent used as a disinfectant and to control external parasites.

Pox A disease sign in which eruptive lesions are observed primarily on the skin and mucous membranes.

Pox disease A common disease of fresh-water fishes, primarily minnows, characterized by small, flat epithelial growths and caused by a virus as yet unidentified. Sometimes called "carp pox" or "papilloma."

Precipitate To separate out from a solution.

Predator Animal that preys on, destroys, or eats other animals.

Premix Feed additive that contains vitamins and minerals.

Pretreatment Treatment of water before it enters the hatchery.

Primary producers Lowest link on the food chain.

Principal A capital sum as opposed to interest or profit.

Probiotics Cultures of beneficial microorganisms fed to production animals to improve digestion and improve health.

Processing All of the procedures to prepare a product of aquaculture for consumption.

Processor Business that takes the raw product to its final form.

Product pull Creating consumer demand.

Product push Providing product shelf space.

Production ponds Ponds used for the final growing stages.

Profit The money that remains after all fixed and variable costs are deducted from income.

Profitability Estimated by subtracting all costs from the expected revenues.

Progeny Offspring.

Progeny test A test of the value of an individual based on the performance of its offspring produced in some definite system of mating.

Prognosis Expected outcome of the disease.

Prolific Producing many young.

Promotion To encourage the growth or development of a business or industry.

Propagation Reproduction, raising, or breeding.

Prophylactic Activity or agent that prevents the occurrence of disease.

Protandrous Changing sex one or more times during life.

Protein Any of the numerous naturally occurring complex combinations of amino acids that contain the elements carbon, hydrogen, nitrogen, oxygen, and occasionally sulfur, phosphorus, or other elements.

Protozoa The phylum of mostly microscopic animals made up of a single cell or a group of more or less identical cells and living chiefly in water; includes many parasitic forms.

Proximate composition Analysis of protein, fat, and ash content.

Pseudobranch The remnant of the first gill arch that often does not have a respiratory function and is thought to be involved in hormone activation or secretion.

Pseudomonas septicemia A hemorrhagic, septicemic disease of fishes caused by infection of a member of the genus *Pseudomonas*. This is a stress-mediated disease that usually occurs as a generalized septicemia. See Bacterial hemorrhagic septicemia.

Purging Process of cleansing crawfish systems by not feeding and changing water often during the holding period.

Pus Yellowish-white liquid produced in certain infections.

Pyloric cecum See Cecum.

Pylorus Opening between the stomach and the start of the intestine in most vertebrates.

Pyridoxine (vitamin B_6) One of the water-soluble B-complex vitamins involved in fat metabolism but playing a more important role in protein metabolism; carnivorous fish have stringent requirements for this vitamin.

Q

Quality assurance The total integrated program for assuring reliability of monitoring and measurement data.

Quality assurance project plan A plan that details the monitoring objectives, program scope, methods, field and lab procedures, and the quality assurance and control activities necessary to meet the stated data quality objectives.

Quality control Routine application and procedures for obtaining prescribed standards of performance and for controlling the measurement process.

Quaternary ammonium compounds Several of the cationic surface-active agents and germicides, each with a quaternary ammonium structure; bactericidal, but will not kill external parasites of fish; used for controlling external bacterial pathogens and disinfecting hatching equipment.

R

Raceway An aquaculture rearing unit through which the water flows.

Radii of scale Lines on the proximal part of a scale, radiating from near center to the edge.

Ranching Raising an aquaculture crop under expansive (range) conditions, such as in the ocean.

Random mating Matings without consideration of definable characteristics of the broodfish; nonselective mating.

Ration A fixed allowance of food for a day or other unit of time.

Ray A supporting rod for a fin; two kinds: hard (spines) and soft rays.

Rearing unit Any facility in which fish are held during the rearing process, such as rectangular raceways, circular ponds, circulation raceways, and earth ponds.

Recessive Character possessed by one parent that is masked in the progeny by the corresponding alternative or dominant character derived from the other parent.

Reciprocal mating (crosses) Paired crosses in which both males and females of one parental line are mated with the other parental line.

Reconditioning treatment Treatment of water to allow its reuse for fish rearing.

Recruitment Production of young.

Rectum Most distal part of the intestine; repository for the feces.

Recycle/reuse The use of water more than one time for fish propagation. There may or may not be water treatment between uses, and different rearing units may be involved.

Red pest See Motile aeromonas disease.

Red sore disease See Vibriosis.

Redd Area of stream or lake bottom excavated by a female salmonid during spawning.

Redmouth disease An original name for bacterial hemorrhagic septicemia caused by an infection of *Aeromonas hydrophila* specifically. Synonyms: motile aeromonas disease; bacterial hemorrhagic septicemia.

Reel A winder consisting of a revolving spool with a handle, attached to a fishing rod; and used to keep the line from becoming tangled and for casting the line.

Regeneration Regrowing a lost body part, such as a claw.

Relift pump Moves water from one source to another.

Resistance The natural ability of an organism to withstand the effects of various physical, chemical, and biological agents that potentially are harmful to the organism.

Resources Available means or property; a supply that can be drawn on.

Respiration The process by which an animal or plant takes in oxygen from the air and gives off carbon dioxide and other products of oxidation.

Résumé A summary of an individual's employment record.

Retail sales Sales of fish in small quantities directly to the consumer.

Return The money available after production expenses are subtracted from total income. Riboflavin An essential water-soluble vitamin of the B-complex group (B_2).

Riparian area The vegetated area constituting a buffer zone between the stream bank and the beginning of land use.

Ripe Containing fully developed eggs; ready to spawn.

Risk A chance of encountering harm or loss.

Risk assessment Determination of quantitative (measurable) or qualitative (subjective) value of risk related to a actual situation and a recognized threat or hazard.

Rod Often made in sections, it carries a line used for angling and is used to support both the line and the reel and to cast the line.

Roe The eggs of fishes.

Rotational line-crossing System for maintaining broodstocks while preventing inbreeding.

Rotifers Many-celled, microscopic aquatic organisms having rings of cilia that in motion resemble revolving wheels.

Roundworm See Nematoda.

Runoff Rain that does not infiltrate the soil and so flows to ponds, streams, and depressions.

S

Sac fry A fish with an external yolk sac.

Safe concentration The maximum concentration of a material that produces no adverse sublethal or chronic effect.

Salinity Concentration of sodium, potassium, magnesium, calcium, bicarbonate, carbonate, sulfate, and halides (chloride, fluoride, bromide) in water. See Dissolved solids.

Salmonids Refers to trout and salmon.

Salt Compound resulting from an acid and a base.

Saltwater Water with a salinity of 30 to 35 parts per thousand.

Sample A part, piece, item, or observation taken or shown as representative of a total population.

Sample count A method of estimating fish population weight from individual weights of a small portion of the population.

Sanitizer A chemical that reduces microbial contamination on equipment.

Saprolegniasis An infection by fungi of the genus *Saprolegnia*, usually on the external surfaces of a fish body or on dead or dying fish eggs.

Saprophyte Organism that lives on dead or decaying organic matter.

Saturated Containing the maximum amount of solute capable of being dissolved under given conditions; thoroughly soaked with moisture (wet).

Saturation In solutions, the maximum amount of a substance that can be dissolved in a liquid without it being precipitated or released into the air.

Scale formula A conventional formula used in identifying fishes. "Scales 7 + 65 + 12," for example, indicates 7 scales above the lateral line, 65 along the lateral line, and 12 below it.

Scales Horny or bony plate-like outgrowth from skin of fish.

Scales above the lateral line Usually, the number of scales counted along an oblique row beginning with the first scale above the lateral line and running anteriorly to the base of the dorsal fin.

Scales below the lateral line The number of scales counted along a row beginning at the origin of the anal fin and running obliquely dorsally either forward or backward, to the lateral line. For certain species this count is made from the base of the pelvic fin.

Secchi disk A circular metal plate with the upper surface divided into four quadrants, two painted white and two painted black; lowered into the water on a graduated line, and the depth at which it disappears is noted as the limit of visibility.

Science Any systematic knowledge-base or prescriptive practice that is capable of resulting in a correct prediction, or reliably predictable type of outcome.

Second dorsal fin The posterior of two dorsal fins, usually the soft-rayed dorsal fin of spiny-rayed fishes.

Secondary Not primary or original.

Secondary invader An opportunist pathogen that obtains entrance to a host following breakdown of the first line of defense.

Sedentary Remaining in one place; attached or fixed.

Sediment Solids settling out to form bottom deposits.

Sedimentation pond (settling basin) A wastewater treatment facility in which solids settle out removing them from the hatchery effluent.

Seed Fertilized, matured ovule of a flowering plant, containing an embryo or rudimentary plant; any part of a plant that will reproduce, including tubers and bulbs; offspring or progeny.

Seed stock Larval fish or crustaceans, fry, small fingerlings.

Seeding Pumping plankton from a pond with bloom to a pond without bloom to promote plankton growth.

Seepage To flow out slowly through the pond bottom material.

Seine Harvesting net.

Selective breeding Selection of mates in a breeding program to produce offspring possessing certain defined characteristics.

Self-feeders Device that allows animals the choice of when to receive feed.

Semipermeable Permeable to different substances to different degrees.

Sensory receptors Organs or receptors that receive stimuli and convey these by the nerve fibers to the brain or spinal cord where they are interpreted.

Septicemia A clinical sign characterized by a severe bacteremic infection, generally involving the significant invasion of the bloodstream by microorganisms.

Serum The fluid portion of blood that remains after the blood is allowed to clot and the cells are removed.

Settleable solids That fraction of the suspended solids that will settle out of suspension under quiescent conditions.

Settling pond Area where the solids settle to the bottom and the top water is released into the environment or a stream.

Sex determination Genetic process leading to an organism being male or female.

Sexing Identifying males and females.

Shareholders Owners of shares of a company or a stockholder.

Shelf life Length of time a product maintains quality before being sold or used.

Shocking Act of mechanically agitating eggs, which ruptures the perivitelline membranes and turns infertile eggs white.

Short bath A type of bath most useful in facilities having a controllable rapid exchange of water. The water flow is stopped, and a relatively high concentration of chemical is thoroughly mixed in and retained for about one hour.

Side effect An effect of a chemical or treatment other than that intended.

Sign Any manifestation of disease, such as an aberration in structure, physiology, or behavior, as interpreted by an observer. (Note that the term "symptom" is only appropriate for human medicine because it includes the patient's feelings and sensations about the disease.)

Silo Deep cylindrical tanks that are similar in operation to horizontal raceways.

Silt Soil particles carried or deposited by moving water.

Single-pass system A system in which water is passed through fish-rearing units without being recycled and then discharged from the hatchery.

Sinuses Opening, hollow cavities.

Siphon Tubular structure for drawing in or expelling liquids.

Slope Incline from the level; slant.

Sludge The mixture of solids and water that is drawn off a settling chamber.

Slurry Thin, watery mixture of feed.

Smolt Juvenile salmonid at the time of physiological adaptation to life in the marine environment.

Snatch block Pulley.

Snout The portion of the head in front of the eyes. The snout is measured from its most anterior tip to the anterior margin of the eye socket.

Sociability The quality or character of being agreeable in company.

Sock Same as live-car.

Soft-egg disease Pathological softening of fish eggs during incubation, the etiological agent(s) being unknown but possibly a bacterium.

Soft fins Fins with soft rays only, designated as soft dorsal, and so on.

Soft rays Fin rays that are cross-striated or articulated, like a bamboo fishing pole.

Sole proprietorship Form of business organization in which one individual owns the business.

Solubility The degree to which a substance can be dissolved in a liquid; usually expressed as milligrams per liter or percent.

Solvency Having sufficient means to pay all debts.

Sp. and spp. Singular and plural abbreviations for species, respectively. The singular abbreviation is often used when identity of the genus of the organism is known but the exact species is not known.

Spat Spawn of the oyster; a young oyster.

Spawn The mass of eggs deposited in the water by fishes, amphibians, and other aquatic animals; to deposit eggs or sperm (milt).

Spawning (hatchery context) Act of obtaining eggs from female fish and sperm from male fish.

Spawning net Artificial nest, generally of Spanish moss or a synthetic material such as spandex, on which fish lay eggs.

Species The largest group of similar individuals that actually or potentially can successfully interbreed with one another but not with other such groups; a systematic unit including geographic races and varieties, and included in a genus.

Specific drug A drug that has therapeutic effect on one disease but not on others.

Spent Spawned out.

Spermatozoon A male reproductive cell, consisting usually of head, middle piece, and locomotory flagellum.

Spinal cord The cylindrical structure within the spinal canal, a part of the central nervous system.

Spines Unsegmented rays, commonly hard and pointed.

Spiny rays Stiff or non-cross-striated fin rays.

Spleen The site of red blood cell, thrombocyte, lymphocyte, and granulocyte production.

Sporadic disease A disease that occurs only occasionally and usually as a single case.

Spore Singe-cell reproductive unit capable of creating a new adult individual.

Sporophytes Form of plant that produces asexual spores to reproduce.

Stabilization pond A simple wastewater treatment facility in which organic matter is oxidized and stabilized converted to inert residue.

Standard environmental temperature (SET) The temperature at which all of the species' physiological systems operate optimally.

Standard length The distance from the most anterior portion of the body to the junction of the caudal peduncle and anal fin.

Standard metabolic rate The metabolic rate of poikilothermic animals under conditions of minimum activity, measured per unit time and body weight at a particular temperature. Close to basal metabolic rate, but animals rarely are at complete rest. See Basal metabolism.

Standing crop All of the fish in a pond or raceway.

Standing crop weight Total weight of all fish in a pond.

Stenohaline marine Fish unable to withstand a wide variation in water salinity.

Sterilant An agent that kills all microorganisms.

Sterilize To destroy all microorganisms and their spores in or about an object.

Stimuli Any factor or environmental change producing activity or response.

Stock Group of fish that share a common environment and gene pool.

Stocking Adding fish to a unit of water.

Stocker Fish 8 in. (20.3 cm) or over.

Stocking rate The number of fish per unit of water.

Stomach The expansion of the alimentary tract between the esophagus and the pyloric valve.

Strainers Fish that select food primarily by size rather than by type and strain water through gill rakers to remove food.

Strains Group of fish with presumed common ancestry.

Strategic planning Analyzing the business and the environment in which it operates to create a broad plan for the future.

Stratification Separation into distinct layers.

Stress A state manifested by a syndrome or bodily change caused by some force, condition, or circumstance a stressor in or on an organism or on one of its physiological or anatomical systems. Any condition that forces an organism to expend more energy to maintain stability.

Stressor Any stimulus, or succession of stimuli, that tends to disrupt the normal stability of an animal.

Stripping Manually releasing eggs and milt from broodfish.

Stun To render unconscious or incapable of action; performed on fish before entering processing plant.

Subacute Not lethal; between acute and chronic.

Substrata Subsoils.

Substrate An underlying substance on which something takes hold or takes root.

Suckers Fish that feed primarily on the bottom of their habitat sucking in mud, filtering and extracting digestible material.

Sulfadimethoxine sulfonamide Drug effective against certain bacterial pathogens of fishes.

Sulfaguanidine Sulfonamide drug used in combination with sulfamerazine to control certain bacterial pathogens of fishes.

Sulfamerazine Sulfonamide drug effective against certain bacterial pathogens of fish.

Sulfamethazine (sulmet) Sulfonamide drug effective against certain bacterial pathogens of fishes.

Sulfate Any salt of sulfuric acid; any salt containing the radical SO_4.

Sulfisoxasole Sulfonamide drug effective against certain bacterial pathogens of fishes.

Sulfonamides Antimicrobial compounds, for example, sulfamerazine or sulfamethazine.

Superior As applied to the mouth, opening in an upward direction.

Supersaturation Greater than normal solubility of a chemical oxygen and nitrogen, for example as a result of unusual temperatures or pressures.

Supplemental diet A diet used to augment available natural foods; used in extensive fish culture.

Supplier An individual or business that furnishes what is needed.

Surface runoff Any surface water from precipitation, snowmelt, or irrigation that runs off the land into any surface waterbody.

Susceptible Having little resistance to disease or to injurious agents.

Suspended solids Particles retained in suspension in the water column.

Sustainable A method of harvesting or using a resource so that the resource is not depleted or permanently damaged.

Swim bladder See Air bladder.

Swimmerets Appendages on the abdomen of a crustacean.

Swim-up Term used to describe fry when they begin active swimming in search of food.

Swim-up fry Fry that have lost their yolk sac and are ready for food.

Syndrome A group of signs that together characterize a disease.

Synthesize Assemble parts into a whole.

Systems Orderly combinations or arrangements of parts, elements, and the like, into a whole, especially such combinations according to some rational principle.

T

Tadpole Aquatic larva of frogs and toads, having internal gills and a tail.

Tackle Fishing gear.

Temper To allow fish to adjust to different water chemistry and temperature.

Temperature shock Physiological stress induced by sudden or rapid changes in temperature, defined by some as any change greater than three degrees per hour.

Tempering Gradually acclimating (accustoming) fish to changes in water chemistry and temperature.

Tender stage Period of early development, from a few hours after fertilization to the time pigmentation of the eyes becomes evident, during which the embryo is highly sensitive to shock. Also called "green-egg stage" or "sensitive stage."

Terramycin See Oxytetracycline.

Terrestrial Existing on land.

Territorial Behavior of an animal in defining and defending its specific area or territory.

Testes The male reproductive organs producing sperm cells and hormones.

Therapeutics Concerned with the treatment of disease; treatment strategies.

Thermal stress Stress caused by rapid temperature change or stress caused by extreme high or low temperature.

Thermocline Zone separating waters of varying densities.

Thiamine An essential water-soluble B-complex vitamin that maintains normal carbohydrate metabolism and essential for certain other metabolic processes.

Thiosulfate, sodium (sodium hyposulfite, hypo, antichlor) ($Na_2S_2O_3$) Used to remove chlorine from solution or as a titrant for determination of dissolved oxygen by the Winkler method.

Thorax Middle region of the body between the head and abdomen.

Tissue residue Quantity of a drug or other chemical remaining in body tissues after treatment or exposure is stopped.

Titrant The reagent or standard solution used in titration.

Titration A method of determining the strength (concentration) of a solution by adding known amounts of a reacting chemical until a color change is detected.

Titrimetric Analyses using a solution of known strength the titrant which is added to a known or specific volume of sample in the presence of an indicator. The indicator produces a color change indicating that the titration is complete.

Tocopherol Vitamin E, an essential vitamin that acts as a biological antioxidant.

Toggle A pin or rod inserted through a loop of haul line to attach it to the seine.

Tolerance Residue levels of a drug or chemical that are permitted by regulatory agencies in food eaten by humans.

Topical Local application of concentrated treatment directly onto a lesion.

Topography Surface features of a region; the lay of the land.

Topping Harvesting only those fish that have grown to marketable size.

Total dissolved solids (TDS) See Dissolved solids.

Total harvest Harvesting strategy that involves one-time seining or trapping and annual draining of a pond.

Total length The distance from the most anterior point to the most posterior tip of the fish tail.

Total solids All of the solids in the water, including dissolved, suspended, and settleable components.

Toxicity A relative measure of the ability of a chemical to be poisonous. Usually refers to the ability of a substance to kill or cause an adverse effect. High toxicity means that small amounts are capable of causing death or ill health.

Toxicology The study of the interactions between organisms and a toxicant.

Toxin A particular class of poisons, usually albuminous proteins of high molecular weight produced by animals or plants, to which the body may respond by the production of antitoxins.

Trademark The name or design officially registered and used by a merchant or manufacturer to identify goods and distinguish from those made by others.

Trammel seine A seine with two coarse outer nets that support a fine-mesh inner net that fish swim into and force through the coarse layers, trapping themselves as the fine-mesh net is forced in around them.

Transmission The transfer of a disease agent from one individual to another.

Transplanting The moving of shellfish from one growing area to another.

Trauma An injury caused by a mechanical or physical agent.

Treaty Formal agreement or compact duly concluded and ratified between two or more states or countries.

Trematoda The flukes from the subclass *Monogenea;* ectoparasitic in general, one host; from the subclass *Digenea;* endoparasitic in general, two hosts or more.

Tubercles Hornlike projections on the heads of breeding fathead minnows.

Tumor An abnormal mass of tissue, the growth of which exceeds and is uncoordinated with that of the tissues and persists in the same excessive manner after the disappearance of the stimuli that evoked the change.

Turbidity Presence of suspended or colloidal matter or planktonic organisms that reduces light penetration of water.

Turbulence Agitation of liquids by currents, jetting actions, winds, or stirring forces.

U

UDN See Ulcerative dermal necrosis.

Ubiquitous Existing everywhere at the same time.

Ulcer A break in the skin or mucous membrane with loss of surface tissue; disintegration and necrosis of epithelial tissue.

Ulcer disease An infectious disease of eastern brook trout caused by the bacterium *Hemophilus piscium*.

Ulcerative dermal necrosis (UDN) A disease of unknown etiology occurring in older fishes, usually during spawning, and primarily involving salmonids.

Under-stocking Periodically harvesting marketable fish from the rearing unit and stocking smaller fish in their place.

Unisex culture Raising only one sex usually male of a species.

United States Pharmacopeia (USP) An authoritative treatise on drugs, products used in medicine, formulas for mixtures, and chemical tests used for identity and purity of the above.

Urbanization Movement of people from rural to urban areas with population growth; increasing proportion of an entire population living in cities or suburbs of cities.

Urea One of the compounds in which nitrogen is excreted from fish in the urine. Most nitrogen is eliminated as ammonia through the gills.

Uremia The condition caused by faulty renal function and resulting in excessive nitrogenous compounds in the blood.

Urinary bladder The bladder attached to the kidneys; the kidneys drain into it.

Urinary ducts Tube for conveying urine.

Urogenital pore (urinary opening) External outlet for the urinary and genital ducts.

Uropod Last swimmeret on a crustacean that develops into a flipper.

V

VHS See Viral hemorrhagic septicemia.

Vaccine A preparation of nonvirulent disease organisms (dead or alive) that retains the capacity to stimulate production of antibodies against it. See Antigen.

Value-added Further processing to increase the value of a product.

Variable costs Costs that increase or decrease in relation to an increase or decrease in production.

Vector A living organism that carries an infectious agent from an infected individual to another, directly or indirectly.

Vein A tubular vessel that carries blood to the heart.

Vent The external posterior opening of the alimentary canal; the anus.

Ventral Underside of the body where the belly is located.

Ventral fins Pelvic fins.

Vertebrate Organism with an inner skeleton and a segmented spinal column.

Vertical transmission The parent-to-progeny transfer of disease agents via eggs or sperm.

Viable Alive.

Vibriosis An infectious disease caused by the bacterium Vibrio anguillarium. Also called "pike pest," "eel pest," or "red sore."

Viral hemorrhagic septicemia (VHS) A severe disease of trout caused by a virus of the Rhabdovirus group. Sometimes called "egtved disease," "infectious kidney swelling and liver degeneration (INUL)," or "trout pest."

Viremia The presence of virus in the blood stream.

Virulence The relative capacity of a pathogen to produce disease.

Virus Ultramicroscopic infective agent that is capable of multiplying in connection with living cells. Normally, viruses are many times smaller than bacteria.

Viscera The internal organs of the body, especially of the abdominal cavity.

Visualization Being able to see things in the mind's eye.

Vitamin An organic compound occurring in minute amounts in foods and essential for numerous metabolic reactions.

Vitamin D A radiated form of ergosterol that has not been proved essential for fish.

Vitamin K An essential, fat-soluble vitamin necessary for formation of prothrombin; deficiency causes reduced blood clotting.

Vitamin premix A mixture of crystalline vitamins or concentrates used to fortify a formulated feed.

Viviparous Bringing forth living young; the mother contributes food toward the development of the embryos.

Volumetric Measurement of substances by comparison of volume.

Vomer Bone of the anterior part of the roof of the mouth, commonly triangular and often with teeth.

Von Bayer trough A 12-in. (30.5 cm) V-shaped trough used to count eggs.

W

Warmwater species Generally, fish that spawn at temperatures exceeding 70°F (21.1°C). The chief cultured warmwater species are basses, sunfish, catfish, and minnows. See Coldwater species; Coolwater species.

Wastewater Water leaving a processing plant or production facility.

Water column Vertical pattern of water from the surface to the bottom of a water body.

Water hardening Process by which an egg absorbs water that accumulates in the perivitelline space.

Water hardness Measure of the total concentration of primarily calcium and magnesium expressed in milligrams per liter (mg/l) of equivalent calcium carbonate ($CaCO_3$).

Water quality As it relates to fish nutrition, involves dissolved mineral needs of fishes inhabiting that water (ionic strength).

Water table Level below which the ground is saturated with water.

Water treatment Primary: removal of a substantial amount of suspended matter, but little or no removal of colloidal and dissolved matter. Secondary: biological treatment methods, for example, by contact stabilization, extended aeration. Tertiary (advanced): removal of chemicals and dissolved solids.

Watershed The whole region from which a river receives its supply of water.

Weir A structure for measuring water flow.

Western gill disease See Nutritional gill disease.

Wetland Area that is covered with standing water or is saturated most of the year, and that supports mainly water-loving plants.

Whirling disease A disease of trout caused by *Myxosoma cerebalis*.

White grub An infestation of *Neodiplostomum multicellulata* in the livers of many freshwater fishes.

White spot disease A noninfectious malady of incubating eggs or on the yolk sac of alevins. The cause of the disease is thought to be mechanical damage. Also see Ich.

Wholesale sales Sales of fish in large quantities to buyer who then sells to distributor or retail market.

Withdrawal time Period of time that must pass after drug, chemical, or pesticide treatment before an animal can be eaten.

X

Xanthophyll Yellow to orange color; derivative of carotene. High levels in diet impart undesirable yellow color to light-fleshed fish.

Y

Yellow grub An infestation of *Clinostomum marginatum*.

Yolk The food part of an egg.

Yolk sac Source of nutrition for fish immediately after hatching.

Z

Zoeal Developmental stage in the life cycle of a shrimp or prawn; occurs between naplius and mysis.

Zooplankton Minute animals in water, chiefly rotifers and crustaceans, that depend upon water movement to carry them about, having only weak capabilities for movement; important prey for young fish.

Zoospores Motile spores of fungi.

Zygote Cell formed by the union of the male and female gametes the sperm and egg and the individual developing from this cell.

INDEX

NOTE: 'f' indicates a figure; 't' indicates a table

A

Abalone, 34f
   culture in California, 20
   management, 228–229
Abdomen, defined, 589
Ablation
   defined, 589
   shrimp eyestalk, 210
Abnormalities
   defined, 589
   health management, 311
Abrasion, defined, 589
Abscess, defined, 589
Absorption, 253, 589
Accessibility, 67
Acclimatization, defined, 589
Accounting
   aquaculture business methods, 487, 511
   defined, 589
Accrual
   aquaculture business, 487, 511
   defined, 589
Acetic acid, 323
Achlya genus, 326
Acid
   death points, 351, 351f
   defined, 589
   water requirements, 349
Acidosis
   defined, 589
   fish disease, 334
   health management, 311
Ackeley, H. A., 9
Acre-foot, defined, 589
Actinomycetes, and off-flavor catfish, 81
Activated sludge process, defined, 589

Actual versus budgeted analysis, financial analysis, 519
Acute, defined, 589
Acute catarrhal enteritis, defined, 589
Acute toxicity, defined, 589
Adaption, defined, 589
Adductor, defined, 589
Adipose fin, defined, 589
Adipose tissue, defined, 590
Adrenal glands, fish stress, 312, 313f, 314
Advertising, 64, 66
Aeration, 590
   and DO, 360, 360f
   RAS, 447, 454–455
   water requirements, 349, 360, 360f, 361
Aeration equipment, 415, 416f, 416–418, 417f, 418f
Aerobes, defined, 590
Aeromonas hydrophila, 327
Aeromonas salmonicida, 329
Agar, food preparation and pharmaceuticals, 244
Agencies
   government web addresses, 578t–579t
   water management regulation, 383
Aggregate, defined, 590
Agitator, defined, 590
Agriculture, defined, 3
Aguafeed tag, 271f
Ainsworth, S. H., 9
Air, defined, 590
Air bladder, defined, 590
Air stone, defined, 590
Air stripping, defined, 590
"Alarm reaction," fish, 313f
Alevin, defined, 590

Algae, 48, 246, 335, 590. *See also* Aquatic plants
   aquariums, 430, 435
   single-cell, 246
   types, 36f
Algae bloom, 474, 590
Algal toxicosis, defined, 590
Alimentary tract, defined, 590
Alkaline, 590
   death points, 351, 351f
   water requirements, 349
Alkalinity, 590
   RAS, 453–454
   water management, 373t
Alkalosis, 590
   fish disease, 333–334
   health management, 311
Alligators, 33–34, 234f
   culture method, 236–237
   deboned meat weights, 239t
   diseases, 238
   feeding, 238
   habitat, 234
   harvesting and marketing, 238–239
   seed stock and breeding, 234–236
   species, 233
   stocking rate, 237
Alternative energy, sustainability, 471–472
Altitude correction table, oxygen solubility, 574t
Alum, 334
American Fish Cultural Society, 10
American Fisheries Society, salmon report, 182
American Sportfishing Association, 176
Amino acids, 253, 261f, 262t, 590
Ammonia, 590
   TAN, 451
   water management, 373t
   water quality, 361, 363, 363f
Ammonia-nitrogen, 447, 448, 451–452, 590
Ammonium, defined, 590
Amoeba, plankton, 37t
Amphibians, 590
   frogs, 239
Anabolism, defined, 590
Anadromous fish, 590
   salmon, 108
Anaerobes, defined, 590
Anaerobic decomposition
   sustainability, 476
   water quality, 349, 361
Anaerobic digestion, 465, 591

Anal fin, defined, 591
Anal papilla, defined, 591
Anatomy, defined, 591
Anchor ice, defined, 591
Anchor parasites, 322–323, 323f
Anemia, 591
   in fish, 336–337
   iron deficiency, 268
Anesthetics, defined, 591
Angelfish, 188t, 189f, 191–192
Angulus genus, 323
Anions, 349, 350f, 591
   water requirements, 349
Annulus, defined, 591
Abnormalities, defined, 591
Anoxia, defined, 591
Antennae, 591
   crustaceans, 46
Anterior surface, aquatic animals, 40, 41f
Anthelmintic agent, defined, 591
Antibiotics
   bacteremia, 327
   in fish diets, 270
   health management, 311
Antibodies, 591
   fish infection barrier, 314, 315f
   health management, 311
Antigen, 591
   health management, 311
   protective antibodies, 314, 315f
Antigenicity, 343, 591
   health management, 311
Antimicrobial, defined, 591
Antinutrients, 591
   fish feed contamination, 271, 272t
Antioxidants, 591
   in fish diets, 270
Antiseptic, defined, 591
Antivitamin, defined, 591
Anus, defined, 591
Aponogetons, aquarium plants, 436, 437f
Appendages, 591
   crustaceans, 46
Application forms, 559
Aquaculture business
   business management, 489–490
   business structures, 506–508, 507t
   computer analysis, 519–520, 520f
   costs, 487–489, 489t
   credit, 520–523
   decision objectives, 503–504
   human resources, 523–525

management analysis, 517–519, 518f
management functions, 490–494, 493f
planning strategies, 496–501, 497f, 498f, 502t, 503
record keeping, 508–517, 509t, 510t, 512f, 513f, 516f
risks, 504–506, 505f
tomorrow's managers, 525–528, 528f
training resources, 529

Aquaculture
activities, 18–20
conversion factors, 569t–571t
defined, 591
demand, 22
environments, 4
FAO definition, 3
finfish propagated table, 575t–577t
future development, 20–22
history, 6–18
government regulations web addresses, 578t–579t
internet resources, 581t
management controls, 155–159
marketing and processing, 23
new species, 23
oxygen solubility tables, 573t–574t
publications list, 580t
research, 23–24
species cultured worldwide, 59t
standard solution preparations, 572t
techniques and technology, 23
top producing countries, 48t
traditional farming comparison, 4t
U.S. production all species 1983–2005, 6f
water quality sample table, 575t

Aquaponics, defined, 591
integrated production, 479–480, 480f
sustainability, 465, 476
technology, 481

Aquarium management
beginners, 439–440
checklist, 442
sample log, 440, 441f
tank cleaning, 439
tank feeding, 437–438
water quality, 441, 442f

Aquariums
established, 440f
establishing, 431
function of, 430
tank/supplies, 433, 434f, 435
tropical fish varieties, 431–433, 432f

Aquatic Animal Drug Approval Partnership (AADAP) program, 182
Aquatic animals. *See also* Crustaceans, Finfish, Fish, Mollusks
body systems, 41–43
characteristics, 35
crustacean anatomy, 45–47, 46f
finfish anatomy, 44f, 44–45
mollusk anatomy, 47–48, 48f
morphology, 41
physiology, 41–42
surfaces, 40, 41f
sustainability, 476
Aquatic insect control, water management, 377–378
Aquatic plants, 242–247
characteristics, 35
cultivation uses, 29
freshwater aquacrops, 245–248
land plant comparison, 48
marine aquacrops, 243–245
market, 247
nutrient acquisition, 305
nutrient uptake, 246
production considerations, 247–248
production worldwide 2006, 20
reproduction, 49, 246
species, 30t–31t
Aquatic structures, types of, 390
Aquatic vegetation control, water management, 378
Aquatic worms, whirling disease, 325
Aquifer, 591
Southern Idaho, 17, 17f
Arginine, amino acid, 261f
Arithmetic competencies, 538
Artemia, shrimp larvae feeding, 214–215
Artery, defined, 591
Artificial light periods, 94
Ascites, defined, 591
Ascorbic acid, 329, 591
Asexual reproduction, 43, 246, 592
Asian tapeworm
fish pathogen, 318
life cycle, 319f, 319–320
Asphyxia, defined, 592
Assembling, 64
Assessment, strategic planning, 498f, 498–499
Assets, 592
aquaculture business, 487, 514, 514f
Assimilation, 592
digestion, 42

Asymptomatic carrier, defined, 592
Atkins, G. C., 9
Atlantic pompano, 151
Atlantic salmon, 31t
Atlantic States, oyster hatcheries, 21
Atmosphere, defined, 592
Atrophy, defined, 592
Automatic feeder, 592
   catfish, 284, 285
   coolwater species, 294
   feed/feeding, 279
Autopsy, defined, 592
Avirulent, defined, 592
Axilla, defined, 592

B

Bacteremia, 327, 592
Bacteria, 326, 592
   diseases, 314, 317, 326–331
   health management, 311
   or virus, 330, 331
Bacterin, defined, 592
Baffle, defined, 592
Bait, 169, 171–172, 592
   market, 10, 19
Bait casting (spinning), 174
Baitfish
   management, 142–144
   types, 143f
   U.S. cultured, 21f
Balance sheet, 487, 512, 513f, 514, 592
Balanced diet, defined, 592
Bar grader, trout production, 116f
Bar marks, defined, 592
Barbel, defined, 593
Basel metabolic rate, defined, 593
Bass, feeding practices, 283, 297. *See also*
       Striped bass
Bass tapeworm, 318
Bath, 593
   disease treatment, 340
   health management, 311
Beer's law, 593
   colorimetric test, 375–376
   water requirements, 349
Beneficials, 593
   sustainability, 465
Benthic organisms, large fingerling feeding, 154
Benthic plants, 304, 593
Bermed, 593
   grow-out house, 236

Berners, Dame Juliana, 169
Bicarbonate (HCO_3^-), ionic component, 354
"Big three," 512, 512f, 514, 515
Bighead carp, 7t, 8, 31t
Binders, 593
   in fish diets, 269, 270
Bins, light control, 94f
Bioassay, defined, 593
Biochemical oxygen demand (BOD), 365, 373t, 593
Biocontrol, defined, 593
Biofilter, 593
   RAS substrate, 447, 451
Biofiltration, RAS, 452–453
Bio-intensive, 465, 477, 593
Biomass, defined, 593
Biosecurity, 465, 467, 593
Biotin, defined, 593
Bird predation, 157–158
Bivalve, defined, 593
   mollusks, 47
Black bass, spot fishing, 175
Black grub, 324, 325f, 593
Black spot, 593
Blank egg, defined, 593
Blastopore, defined, 593
Blastula, defined, 593
Bloom, phytoplankton, 29
Blower-type feeder, catfish, 284, 284f
Blue crab, 218
Blue gourami, 188t, 189f
Blue mussel, 226
Blue slime disease, 594
Blue-green algae, and off-flavor catfish, 81
Blue-sac disease, defined, 593
Boats, transport equipment, 423, 423f
Boil, defined, 594
Borger Color System, 367
Boring cage, aquatic structures, 391
Borings, 394, 594
Boron, 357
Bottom feeders, 41
Brackish water, 5, 594
Branchia, defined, 594
Branchiomycosis, defined, 594
Branded, 594
   products, 62
Breakdown of food (metabolic oxidation), 256
Break-even analysis, 487, 500, 501, 532, 534, 594

Brine shrimp, 438
Broadcast, defined, 594
"Broken back disease," 334, 335f
Bronze catfish, 188t, 189f
Broodfish, 594
   spawning transfer, 193
Broodstock, 6, 594
Brook trout, 9
Brown blood disease, 336
Brown spot disease, alligators, 238
Brush aerators, 418f
Bubble-nest builders, 594
   aquariums fish, 430, 432
Buccal cavity, defined, 594
Buccal incubation, defined, 594
Budget, defined, 594
Buffers, 349, 354, 594
Bulk feed tanks, catfish, 283, 284f
Bulldozers, 402
Bullfrogs, 33, 239f
Bulrushes, 49
Business plan, strategic planning, 501
Buyer
   characteristics, 61–62
   targeting, 60
Byssus, mussel secretion, 226, 594

C

Cabomba, aquarium plants, 436, 437f
Cage culture, 594
   advantages, 409
   disadvantages, 410
   tilapia feeding practices, 295, 296, 296t
Cages
   aquatic structures, 390, 408–409
   construction materials, 411–412
   construction/placement, 410–411, 411f, 412–415, 413f, 414f
Cahuila people, fish ponds, 9
Caiman, defined, 594
Calcareous shells, mollusks, 47, 594
Calcinosis, defined, 594
Calcium (Ca^{2+}), ionic component, 354
California Tray incubator, 111f
Calorie, defined, 594
Cannibalism, coolwater species, 294
Capital, 487, 506, 507t, 594
Carbohydrates, 259, 594
   fish diet, 279, 282
Carbon, organic matter breakdown, 365
Carbon dioxide (CO_2), 594
   RAS, 454
   water quality, 360–361, 361f, 373t
Carbonate (CO_3^{2-}), 354, 594
Career clusters, job search, 557–558
Career Development Events (CDE), 542
Career opportunities
   basic skills, 537–538
   education/experience, 555
   entrepreneurship, 545–546, 546f
   hiring tips, 558–562
   intangible skills, 542–545
   job search resources, 556–558, 557f
   supervised agricultural experience (SAE), 554–555
   suppliers/services, 546–554 547f, 549f, 550f, 551f, 552f, 553f
   thinking skills, 539f, 539–540
   workplace competencies, 540f, 540–542, 541f
Carnivores, 42, 254, 594
Carotenoids
   flesh pigmentation, 270
   ornamental fish food, 191
Carp, 6, 7t, 8, 31t, 32t, 137f
   amino acid requirements, 262t
   culture methods, 140
   dietary fat requirements, 261t
   diseases, 141
   feeding, 141
   first feeding, 298–299
   habitat, 138
   harvesting and yields, 141–142
   and koi, 190
   protein feed levels, 263t
   seed stock and breeding, 139–140
   types, 138f
Carrageen, aquatic plant polysaccharide, 29, 50, 244
Carrier, defined, 594
Carrier host, defined, 595
Carrion, alligator feeding, 238
Carrying capacity, defined, 595
Cartilage, defined, 595
Cash-basis, aquaculture business, 487, 511
Cash-flow, 595
   budget, 487, 500, 502t
   statement, 515
   worksheet, 516f
Cast fishing line, 171, 172
Casualty loss risk, aquaculture business, 505
Catabolism, defined, 595
Catalyst, defined, 595

Catfish, 10, 31t–33t, 97f. *See also* Channel catfish
  amino acid requirements, 262t
  bird predation, 157
  common feeding practices, 283–285, 286t, 287–288, 288f
  dietary fat requirements, 261t
  enterprise budget, 489, 489t, 491t
  ESC, 328
  feed conversion ratios, 287
  feed form/size, 283–284
  feeding practices, 287–288, 288f
  feeding rates, 285, 286t, 287
  fingerling composite length-weight chart, 106t
  first feeding, 298, 298f
  inspection, 79, 80
  and Mississippi, 10–15
  nutritional formula, 279, 280t, 281
  off-flavor, 80, 81–82
  ownership cost components, 509t
  ponds, 391
  production, 1980–2009, 15f
  protein feed levels, 263t
  species, 96, 97f
  U.S. cultured, 21f, 57
  weight-length relationship, 107f
  winter feeding, 288
Catfish feeders, 284f, 284–285
Catfish feeding guide, 286t
Catfish feedmill, 280
Catfish tapeworm, 318
Cations, 349, 350f, 595
Cattails, 49, 50f, 245
Caudal fin, defined, 595
Caudal peduncle, defined, 595
Cecum, defined, 595. *See also* Pyloric cecum
Cellular, health management, 311
Central America, marine shrimp farms, 21
Channel catfish, 96, 97f
  common feeding practices, 283–285, 286t, 287–288
  culture method, 104
  diseases, 105
  ESC, 328
  feeding, 105
  habitat, 96–98
  harvesting and yields, 106–107
  industry, 11
  life cycle, 99f
  processing and marketing, 107–108
  seed stock and breeding, 104
  spawning and egg production, 40t
  stocking rate, 104–105
  Vitamin C deficiency, 264f
Channel catfish virus disease (CCVD), 331–332, 332f
Chelator, 595
  EDTA, 214
Chemical additives
  wastewater treatment, 381, 383
  conversion factors, 570t
Chemical oxygen demand (COD), 365, 374t, 595
Chesapeake Bay, oysters, 18
Chicken/fish farming, 130
Chilling, processing, 73
Chilodenella cyprini, 239
Chilodenella hexasticha, 329, 329f
Chilodonelliasis, 329f, 329–330
China, 7f
  aquaculture, 6–8, 57, 58f, 59t
  aquatic plant production, 20
Chinese water chestnuts, 246–247
Chinook salmon, 182
  protein feed levels, 263t
Chitin, defined, 595
Chloride (Cl^-), ion, 354
  water quality, 364
Chlorine (Cl_2), water quality, 364
Chlorophyll, 48
  water quality test, 365
Chondrococcus columnaris, 327
Chronic, health management, 311
Cichlids, aquarium fish, 430, 432–433, 595
Cilia, defined, 595
Circular tanks, aquatic structures, 408, 408f
Circulatory system, 43
Circuli, defined, 595
Clams, 18, 34t, 224f
  anatomy, 48f
  management, 224–225
Clarification, defined, 595
Clay, pond construction, 394, 396
Clean cropping, defined, 595
Clear Springs Food Company, Idaho, 16
Cleidodiscus genus, 321
Clinostomum (yellow grub), 324
Cloaca, defined, 596
Closed systems, aquatic structures, 390
Closed-formula feed, defined, 596
Clown barb, 188t, 189f
Clutch, 596
  artificial, oysters, 221
  size, alligators, 235

Cock, defined, 596
Coelomic cavity, defined, 596
Coelozoic, defined, 596
Coenzymes, 596
   water-soluble vitamins as, 264
Coffee can, defined, 596
Coho salmon, 32t, 182–183, 183f
Coldwater aquaculture, 4
Coldwater hatcheries, fish culture, 294
Coldwater species
   defined, 596
   feed/feeding, 279
Collateral, defined, 596
College of Southern Idaho, 16, 17
Colloid, defined, 596
Color, water requirements, 367–368
Colorimetric, 349, 596
   water management test, 375
Columnaris disease, 317, 327–328, 596
Commensal, defined, 596
Commercial warmwater fishing, U.S., 10
Community, 596
   sustainability, 477
Comparative analysis, financial analysis, 519
Compensation point, defined, 596
Competencies, 596
   basic, 537–538
   LifeKnowledge®, 542–545
   workplace, 540f, 540–541, 541f
Compleat Angler (Walton), 169, 170f, 171f
   casting, 173
Compressed, defined, 596
Computers, 519–520, 520f
   future use, 528
   workplace competencies, 540–542, 541f
Conditioned response, defined, 596
Conductance, 596
   water, 352–353
   water requirements, 349
Congenital disease, defined, 596
Congestion, defined, 596
Connective tissue, 596
   formation and Vitamin C, 264
Construction, careers, 548–549, 549f
Consulting, careers, 549
Consumer, 60, 61
Consumption, 57
Contact feeding behavior, 256
Contamination, defined, 596
Control (disease), defined, 596

Control (experimental), defined, 597
Controlled light periods, 91, 93
Controlling, aquaculture business, 493f, 495–496
Conversion factors
   chemical additives, 570t
   miscellaneous, 569t–570t
   ppm, proportion, percent, 571t–572t
   units of volume, 571t
   units of weight, 571t
Coolwater species, 294, 597
   feed for 294
Coots, demand feeders, 285
Copepoda, 319, 319f, 597
Copper sulfate, 320, 321, 323, 326, 597
Cornea, defined, 597
Corporations, 487, 506, 507, 508, 597
Coste, M., 9
Costlasis (Costia), 323, 323f, 597
Cover, aquarium supplies, 433, 434f
Crabs, 34t, 218–219, 219f
Crappie, 147f
   management, 147–149
Crassiphiala (black spot), 324
   life cycle, 324, 325f
Crawfish (crayfish), 18, 34t, 35f, 202f
   anatomy, 46f
   culture method, 202–204
   diseases, 206
   enterprise budget, 489, 492t
   feeding, 205–206
   grading, 208
   habitat, 201
   harvesting and yields, 206–208
   marketing, 208
   operating cost components, 510t
   ready-to-eat, 84f
   stocking rate, 205
   U.S. cultured, 21f, 57
Creative thinking, 537, 539, 597
Credit
   managing use, 523, 523f
   obtaining, 520–522
Crocodile farming, 233
Crops
   processors, 552, 553f
   sustainability, 476
Crossbreeding, 162, 597
Crowding, 39
Crumbles, catfish feed form/size, 283
Crumbling, processing method, fish feed, 282, 283

Crustaceans. *See also* Aquatic animals
  anatomy, 45–47
  crawfish culture, 201
  external anatomy, 46f
  internal anatomy, 47f
  species, 34t
Cryogenically frozen, defined, 597
Cryptocorynes, aquarium plants, 436, 437f
Cultural diversity, 537, 540, 597
Culture, 3, 597
  species considerations, 37
Culture method
  alligators, 236–237
  crawfish, 202–204
  frogs, 241
  oysters, 222
  prawns, 210
  shrimp, 214–215
  catfish, 104
  hybrid striped bass, 134
  ornamental fish, 190
  salmon, 119–120
  tilapia, 125–128
  trout, 114
Cultured fish, defined, 597
Curricular material, 537, 550, 597
Cyanocobalamin, defined, 597
Cyclops, metazoa, 38t
Cyst, defined, 597
Cytophaga columnaris, 327
Cytoplasm, defined, 597

D

Dactylogyrus genus, 321
Daily feeding behavior, 256
Daily health records, fish disease, 338f, 338–339
Daily temperature unit (DTU), defined, 597
Dam, cross-section, 394f, 396
Data sheet, 537, 560–561, 597
Database, defined, 597
Decapods (ten legs), 45, 597
Dechlorinator, 430, 433, 434f, 597
Decision making, thinking skills, 539
Dee's disease, 597
Deep sea fishing, 174
Deficiency disease, defined, 597
Deheading, processing, 70, 72f, 598
Dehydrate, stress response, 314
Demand, consumer, 62, 598
Demand feeder, 598

  catfish, 284–285
  feed/feeding, 279
  trout, 289–290, 290f
Demography, 537, 598
Denitrification, 447, 598
Density index, rearing unit, 115, 598
Dentary bones, defined, 598
Department of Agriculture, Extension Service, 10
Depreciation, defined, 598
Depth of fish, defined, 598
Depuration, defined, 598
Dermal, defined, 598
Dermocystidium marinum (Dermo), oyster fungus, 222–223
Desmoids, algae, 36f, 48
Detritus, 598
  tilapia feeding, 129
Diagnosis, health management, 311
Diagnostic test, fish preparation, 339
Diapause, brine shrimp, 430, 438
Diarrhea, defined, 598
Diatoms, algae, 36f, 48
Diet, 598
  and energy, 257
Dietary fiber, defined, 598
Diffused aeration, RAS, 455
Diffuser aerators, aquatic equipment, 417–418, 418f
Diffusion, 598
  oxygen into blood, 42
Digestible energy (DE), 258
Digestion, 253, 255t, 598
Digestive system, aquatic animals, 42, 253, 255f, 255t
Diluent, defined, 598
Dip, 598
  health management, 311, 339
Dip net, aquarium supplies, 433, 434f
Diplostomiasis, defined, 598
Diquat, 328
Directing, aquaculture business, 493f, 494–495
Discharge, defined, 598
Discipline, human resources, 525
Discus fish, 188t, 193
Disease, 598
  in fish, 310, 312f
  management, 156. *See also* Disease treatment
  presence determination, 337–339
  prevention, 316, 598
  RAS monitoring control, 459, 460t
  resistance, 39, 314, 315f

Disease, parasite and weed control, 156t
Disease susceptibility
　anchor parasites, 322
　bacteremia, 327
　CCVD, 331
　Chilodonelliasis, 329
　Columnaris, 327
　Costia disease, 323
　ERM, 328
　ESC, 328
　fish grubs, 324
　fish lice, 323
　fungal infections, 326
　goldfish ulcer, 329
　Ich, 320
　infectious fish, 317, 318
　IHN, 332
　IPN, 333
　monogenetic flukes, 321
　tapeworms, 318
　trichodiniasis, 320
　whirling disease, 324, 325
Disease treatment
　acidosis, 334
　alkalosis, 334
　anchor parasites, 323
　anemia, 337
　bacteremia, 327
　brown blood disease, 336
　calculation formula, 341–342
　calculation, 341
　chilodonelliasis, 330
　columnaris, 328
　Costia disease, 323
　ERM, 328
　ESC, 329
　fish lice, 324
　fish poisoning, 336
　fungal infections, 326
　gas bubble disease, 337
　Ich, 321
　methods, 339–340
　monogenetic flukes, 322
　oxygen starvation, 333
　questions, 339
　trichodiniasis, 320
Diseases, in
　alligators, 238
　catfish, 105
　crawfish, 206
　hybrid striped bass, 135
　ornamental fish, 191

　oysters, 222–223
　prawns, 211
　salmon, 121–122
　shrimp, 217–218
　tilapia, 131
　trout, 117–118
Disinfection, 456, 598
Dissolved gases tests, water quality, 358–359, 359f
Dissolved oxygen (DO)
　defined, 598
　ledger, 358t
　RAS, 454
　water quality, 358, 374t
　water requirements, 349
Dissolved solids, 598
　RAS, 448, 449, 450–451
Distal, defined, 598
Distillers dried grains with solubles (DDGS), 265
Distribution networks, 62
Distributors, 60
Diurnal, defined, 598
Docosahexaenoic acid (DHA), essential fatty acid, 260
Domestic, defined, 598
Dorsal surface, aquatic animals, 40, 41f, 598
Drainage, 598
　ponds, 398, 398f, 402–403
Drawdown, defined, 598
Dredges, oyster harvesting, 223
Dress, defined, 599
Driftwood, aquarium supplies, 433, 434f
Drip treatment, defined, 599
Drug resistant, defined, 599
Drug sensitive, defined, 599
Dry feed, defined, 599
Dry fertilization, developed, 9
Duck/fish farming, 130
Duckweed, 49
Dysentery, defined, 599

E

Earth-bermed grow-out houses, alligators, 236
Economics, 487, 488, 526, 599
Ecosystem, defined, 599
Ectoderm, defined, 599
Ectoparasite, defined, 599
Ectothermic body temperature, 35
Edema, defined, 599

Education, web sites, 582t
Edwards, M. Miline, 9
Eels, 34
Effluent, 381, 599
　sustainability, 465, 474
　water requirements, 349
Egg-layers, aquarium fish, 432
Egg-scatterers, 430, 433, 599
Eggs, 8, 91, 599
　channel catfish, 99
　fertility, alligator, 235
　hatchery transfer, 101
　production, 40t
　threats to, 103
Egyptian aquaculture, 8
Eicosapentaeonic acid (EPA), essential fatty acid, 260
Electric paddlewheel units, aquatic equipment, 417, 417f
Electrolytes, 267
Electronic meters, water management tests, 376
Emaciation, defined, 599
Emarginate fin, defined, 599
Emboli, defined, 599
Embryo, defined, 599
Encyst, defined, 599
Endocrine glands, 91, 93t, 599
Endoparasite, defined, 599
Endoskeleton, defined, 599
Energy, 599
　fish diet, 279
　requirements and loses, 256–258, 258f
　sustainability, 471
England, aquaculture, 8
Enrobing, processing, 84, 599
Enteric redmouth (ERM) disease, 328, 599
Enteritis, defined, 599
Enterprise, 599
　aquaculture business, 487, 489t
　budget, 487, 488, 489, 491t, 492t, 599
Entrepreneur, 599
　careers, 537, 545–546, 546f
　personal qualities, 541, 542
Environmental diseases, fish, 317
Environmental gill disease, defined, 599
Environmental issues, 24
Environmental Protection Agency (EPA), 383
Environmental responsibility, 478–479
Environmental stimuli, hormone release, 91, 92f, 93–94
Enzootic, defined, 599

Enzymes, 264, 599
Epidermis, defined, 599
Epizootic, defined, 599
Epizootiology, defined, 600
Equipment
　aquatic structures, 390, 402, 415–423
　careers, 548
　direct market, 67
Equity, 487, 496, 506, 512, 515, 517f, 518, 519, 600
Equivalents per million (meq/l), 349, 600
Eradication, defined, 600
Escherichia coli 0157:H7, 77
Esophagus, defined, 600
Essential amino acids, 261
Essential fatty acids (EFA), 259, 260f, 260t, 279, 282, 600
Estate, defined, 600
Estimated feeding rates, catfish, 285, 286t
Estuary, defined, 600
Ethylene diamine tetraacetic acid (EDTA), chelator, 214
Etiology, defined, 600
Euglena, algae, 36f, 48
Europe, aquaculture, 8–9
Eutrophic bodies, defined, 600
Evaluation, 600
　aquaculture business, 487
　human resources, 525
　strategic planning, 498f, 499–500, 502t
Evaporation
　water management, 374t, 378
　water requirements, 367, 374t
Eviscerate, defined, 600
Evisceration, 71, 72f
Excavated pond, aquatic structures, 391, 393t
Exclusive economic zone, defined, 600
Excretion, defined, 600
Excretory system, aquatic animals, 42
Exophthalmos, defined, 600
Exoskeleton, defined, 600
Exotic, defined, 600
Export, defined, 600
Extruded, 600
　feed/feeding, 279
Extruded feed
　catfish feed form/size, 283–284
　diet composition, 282
Extrusion processing method, fish feed, 282
Eyed stage, 600
　trout eggs, 110, 110f

F

Facultative, defined, 600
Fairy shrimp, metazoa, 38t
Fan-Li, 6
Farm earning statement, 514–515
Farmers, aquaponics, 480
Farming, 600
  salmon, 120
Fat (lipids), 253, 259–260, 261t, 600
Fat soluble vitamins, 264–265
Fathead minnow, baitfish, 142, 143f
Fatty acid, defined, 600
Fauna, defined, 601
Feed
  calculation examples, 299–300
  catfish form/size, 283–284
  cost determination, 304, 304t
  fish as ingredients, market, 19
  major components, 448
  storage requirements, 300t, 300–301
  trout form/size, 289
Feed chart, defined, 601
Feed conversion ratio (FCR), 601
  calculation examples, 301–303
  catfish, 287
  description of, 301
  feed costs, 304, 304t
  feed/feeding, 279
Feed method, fish disease treatment, 340
Feed to food conversion, 254t
Feeding, of
  alligators, 238
  catfish, 105
  crawfish, 205–206
  frogs, 242
  hybrid striped bass, 135
  ornamental fish, 190
  oysters, 222
  prawns, 211
  salmon, 120
  shrimp, 216–217, 217t
  tilapia, 128–131
  trout, 117
Feeding behavior, 256
Feeding chart, trout production, 117
Feeding habits, aqua species, 39
Feeding rates
  catfish, 285, 287
  tilapia, 296, 296t
  trout, 291–293
Feeding ring, 391, 411, 601

Fee-fishing, 68, 69
  catfish marketing, 108
Fertility, defined, 601
Fertilization, 601
  water management tests, 376
  water requirements, 349
Fiber
  in fish diets, 269
  sustainability, 470
Fillets, 73, 83f, 601
Filter, aquarium supplies, 433, 434f
Filter feeders, clams, 35
Filtering systems, wastewater treatment, 381, 382
Filtration, mechanic methods, 449–450, 450f
Fin ray, defined, 601
Fin rot disease, defined, 601
Financial efficiency, financial analysis, 519
Financial risk, aquaculture business, 504, 505f
Finfish
  anatomy, 44–45
  aquaculture propagation table, 575t–577t
  baitfish, 142–144
  carp, 137–142
  channel catfish, 96–108
  common feeding practices, 283
  crappie, 147–149
  endocrine glands, 91, 92f
  external anatomy, 44f, 44
  genetic research, 159–162
  hybrid striped bass, 132–137
  internal anatomy, 44f
  management controls, 155–159
  marine species, 149–155
  red drum, 144–147
  salmon, 118–122
  seasonal breeding, 91
  spawning and egg production, 40t
  species, 31t–33t
  tilapia, 122–131
  trout, 108–118
Fingerlings, 601
  catfish farming, 14
Fish, 23. *See also* Aquatic animals, Finfish
  consumption in U.S., 60f
  cut forms, 83f
  diet considerations, 279
  diet ingredients, 278
  digestive system, 253, 255f, 255t
  disease types, 316
  DO impact, 359f
  FDA encyclopedia, 583t–587t

infection barriers, 314, 315f
infectious diseases, 317
noninfectious diseases, 317
resistance to disease, 314, 315f
sex determination, 95–96
species, 31t–33t
stress impact, 312–314, 313f, 315–316
transport equipment, 420–423, 422f
Fish and shellfish, per capita consumption by 2025, 22
Fish disease
  prevention, 319, 320
  signs/contributing factors, 319, 320
Fish farming, defined, 601
Fish Farming Experimental Stations, Arkansas, 10
Fish feed
  aquarium supplies, 433, 434f
  composition analysis, elements, 282
  mills, 280f
  preparation processes, 282
Fish grubs, 324
Fish Health Centers, 181
Fish lice, 323f, 323–324
Fish meals, fish feed, 292
Fish Technology Centers, 181
Fisheries Program (NFHS), 178–181
Fishing, history, 12–13
Fishing line, 171
Fish-out business, 69
Fistula, defined, 601
Five C's, 523, 523f
Fixative, defined, 601
Fixed costs, 487, 489, 601
Flagellum, defined, 601
Flashing, defined, 601
Flatland, defined, 601
Flesh analysis, water management tests, 376
Flexibacter columnaris, 327
Floatable solids, RAS, 448, 449
Floating heart, deep water plant, 245f
Floating pellets, catfish feed form/size, 283
Flooding, ponds, 398
Florida, ornamental fish culture, 189t, 197
Flow index, 391, 405–406, 601
Flow rate, defined, 601
Flukes, fish diseases, 321
Fluoride, ionic component, 354
Flush, 311, 601
  disease treatment, 340
Fly fishing (casting), 174
Foam fractionator, 447, 450–451, 601

Folic acid, 334, 601
Follow-up letter, 537, 563, 601
Fomites, defined, 601
Food
  market, 19
  sustainability, 470
Food, Agriculture, and Trade Act, 465
Food and Agriculture Organization (FAO), UN, 3
Food and Drug Administration (FDA)
  HACCP, 76
  regulatory fish encyclopedia, 583t–587t
Food chain, 35f, 39, 601
Food conversion, defined, 601–602
Food Safety and Inspection Service (FSIS), USDA, 80
Food-borne illness, prevention, 77
Foot, defined, 602
Forage, defined, 602
Forecasts, 537, 540, 602
Forel-Ule Color Scale, 367
Fork length, defined, 602
Formalin, 320, 321, 322, 323, 330, 602
Fortification, defined, 602
France, aquaculture, 9
Freeboard, 391, 399, 400, 401f, 602
Free-floating macrophytes, 49
Free-living, defined, 602
Freezing, processing, 73
Freshwater, defined, 602
Freshwater aquaculture, 4
Freshwater finfish, 31t–33t
Freshwater plant aquacrops, 245–248
Freshwater prawn, 209f
Frogs, 239
  life-cycle, 240f
  management, 240–242
Fry transfer, nursery ponds, 100–101
Fry, defined, 602
Fungal diseases, fish
Fungus
  fish diseases, 314, 317, 326, 326f
  health management, 311
Furunculosis, 329, 602
Furunde, defined, 602
Fusiform fish body, 41, 602
Future Farmers of America (FFA), 542

G

Gallbladder, 253, 255f, 255t, 602
Gametes, 602
  aquatic plants, 49

Gape, defined, 602
Garlick, Theodatus, 9
Gas bubble disease, 337, 602
Gases, water qualities, 349
Gastritis, defined, 602
Gastroenteritis, defined, 602
Gastrointestinal tract, 253, 255f
Gastroliths, defined, 602
Gastropods, 48, 602
Gehin, Antoine, 9
Genes, 95, 602
Genetic abnormalities, fish, *337*, 337
Genetic research, 159–162
Genetic selection, and breeding, 91
Genomics, 602, 602
   sustainability, 465, 467
Genus, defined, 602
Geographic distribution, defined, 602
Geosmin, and off-flavor catfish, 81
Germinal disc, defined, 602
Giant kelp, 48
Gill arch, defined, 602
Gill clefts (slits), defined, 602
Gill cover, defined, 603
Gill filament, defined, 603
Gill lamellae, defined, 603
Gill rakers, defined, 603
Gills, 45f, 603
   mollusk, 48f
   semipermeable membranes, 45
Glair, defined, 603
Global marketplace, 527, 528f
Globulin, defined, 603
Glycogen, defined, 603
Goals, 603
   aquaculture business, 487, 490–491, 493f
   management decisions, 503–504
Golden shiner, baitfish, 142, 143f, 312f
Goldfish, 142, 143f, 189t, 193–194
   ulcer disease, 329
"Goldfish furunculosis," 329
Gonadotrophin, defined, 603
Gonads, defined, 603
Government, programs/services web sites, 582t
Grade A inspection mark, 78, 79
Grade A inspection shield, 79f
Grade B inspection mark, 79
Grading, 64, 603
   crawfish, 208
   factors considered, 78–80
   hybrid striped bass, 136, 137f
   size, 73
   trout, 115–116
Grain, fish diet, 282
Grains per gallon, 349
Gram-negative bacteria, 603
Gram-positive bacteria, 603
Grants, defined, 603
Grass carp, 7t, 8, 32t
Gravel, aquarium supplies, 433, 434f
Grazers, 256, 603
Grease, water quality tests, 366
Green, Seth, 9, 10
Greenhouses, aquaponics, 480
Gregarine, defined, 603
Gross pathology, defined, 603
Ground water, 372, 603
Group immunity, defined, 603
Grow-out facilities, 19, 603
Gullet, defined, 603
Guppies, 188t, 194, 431, 432f
Gypsum, 334
Gyro infection, defined, 603
Gyrodactylus genus, 321

H

Habitat, defined, 603
Haptor, defined, 603
Hardness, 603
   water management, 374t
Harvesting, 19, 603
   equipment, 420
   pond techniques, 391, 395
Harvesting and yields
   catfish, 106–107, 108f
   crawfish, 206–208
   hybrid striped bass, 136
   ornamental fish, 191
   oysters, 223
   prawns, 211–212
   salmon, 122
   shrimp, 218
   tilapia, 131
   trout, 118
Hatcheries, 19, 603
   establishment of first, 9
   larval shrimp, 210
   maximum-production systems, 101
   national, 178t–180t
   saltwater shrimp, 213
   sport fish, 174
   troughs, 103f, 111f

trout incubator, 111f
water conditions, 101–102
Hatchery constant, defined, 603–604
Hatchling, 604
  alligators, 236
Hawaii
  aquaculture, 9
  Malaysian prawn culture, 20
Hazard Analysis and Critical Control Point (HACCP), 603
  advantages, 77
  manufacturing steps, 76–77
Head, defined, 604
Headwaters, defined, 604
Health management, fish, 311
Health records, fish, 338, *338*
Heat capacity, 604
  water requirements, 349, 352
Heater, aquarium supplies, 433, 434f
Heavy metals, 604
  flesh test, 376
  water requirements, 349
Heliciculture, snail raising, 226, 604
Hematocrit, defined, 604
Hematoma, defined, 604
Hematopoiesis, defined, 604
Heme, iron, 268, 604
Hemocoel, defined, 604
Hemoglobin, , 604
  brown blood disease, 336
  health management, 311
Hemorrhage, 604
  health management, 311
Hemorrhagic septicemia, 327
Hemorrhagic spots/ulcers, 317
Hen, defined, 604
Hepatitis, defined, 604
Hepatoma, defined, 604
Herbivores, 41, 42, 254, 604
Hermaphroditic animal, 604
  mollusks, 48
Heterocercal tail fins, 41
Heterotrophic bacteria, defined, 604
Hides, 604
  alligator, 239
Hiring, aquaculture business, 494, 524
Histology, defined, 604
Hobby fish, 29, 33
"Hole-in-the-head disease," 328–329
Homing, defined, 604
Homocercal tail fins, 41, 604
Honesty, personal qualities, 541, 542

Horizontal transmission, defined, 604
Hormones, 91, 93t, 604
  environmental stimuli to release, 91, 92f
  in fish diets, 270
  spawning induced by injection, 94–95
Host, defined, 604
Human health, Pfiesteria, 336
Human predation, 158
Human resources, 523–525, 527
Human risk, aquaculture business, 505
Humectants, 604
  feed/feeding, 279
  moist/semi-moist process, 282–283
Humphreys County, catfish, 14f
Hungingue, first hatchery established, 9
Husbandry, 8, 604
Hybrid striped bass, 132f
  culture method, 134
  diseases, 135
  feeding, 135
  habitat, 132
  harvesting and yields, 136
  processing and marketing, 137
  seed stock and breeding, 133
  species, 132
  stocking rate, 134–135
Hybrid, defined, 604
Hybrid vigor, 162, 604
Hybridization, 162, 604
Hydrate, 604
  health management, 311
Hydrogen sulfide (H_2S), 605
  water quality, 361–362
Hydrological cycle, 605
  pond culture, 201
Hydroponics, 242, 383, 605
  wastewater treatment, 381, 383
  water requirements, 349
Hyoid bones, defined, 605
Hyperemia, defined, 605
Hyperplasia, defined, 605
Hypersaline, 505,
  aquariums, 430

I

Ice fishing, 174
Ichthyophthiriasis (Ich), 320–321, 321f, 605
Ichthyophthiriasis multifiliis
  fish pathogen, 320, 320f, 321
  life cycle, 322f
Ichtyobodo genus, 323, 323f

Ionic components, tests for, 354–355
Immune, defined, 605
Immune system, health management, 311
Immunization, 605
   fish disease, 342
Implementation, strategic planning, 501, 503
Import, defined, 605
Impoundment, 605
   pond structure, 391, 392f, 393t
Imprinting, defined, 605
In berry, defined, 605
In vitro, defined, 605
In vivo, defined, 605
Inbred line, defined, 605
Inbreeding, 160, 605
Income, 487, 511, 514, 605
Income statement, 605
Income tax, 507
Incubate, defined, 605
Incubation
   artificial, alligators, 236
   disease, 605
   eggs, 8, 43, 605
Incubator, 605
   trout hatcheries systems, 111f
Indefinite treatment methods, fish disease, 340
Indigenous, defined, 605
Inert gas, defined, 605
Infection, defined, 605
Infectious diseases, 316, 606
Infectious hematopoletic necrosis (IHN), 332
Infectious pancreatic necrosis (IPN), 333, 606
Inferior mouth, defined, 606
Infiltration, 311, 606
   fish immunization, 342
Inflammation, 314, 606
Information management tools, 528
Information systems, design, 527
Information use skills, careers, 540
Ingest, defined, 606
Injection, defined, 606
Injections
   fish disease treatment, 340
   health management, 311
Innovation, managing, 526
Inoculation, 606
   fish immunization, 342
   health management, 311
Inorganic salts, 45, 606
Input costs, 487, 488, 489t, 606
Inputs, 606
   production, 60, 61

Insect control, water management, 377–378
Inspection, 553, 606
   fish-processing, 74–77
Instinct, defined, 606
Instructors, careers, 550, 550f
Integrated Pest Management (IPM), 476–477
Integrated systems, manuring, 130–131, 131f
Integrity, personal qualities, 541, 542
Intensive culture, defined, 606
Intensive production, defined, 606
Interest, 487, 512, 515, 518, 606
Intermediate buyer, 61, 62
Intermediate goods, 61
Internet, career opportunities, 557, 557f
Internet resources, 581t
Interoparous, defined, 606
Interpersonal, defined, 606
Interpersonal skills, careers, 537, 540
Interspinals, defined, 606
Intestine, 606, 253, 255f
Intragravel water, defined, 606
Intramuscular injection, defined, 606
Inventory, 606
   trout, 116–117
Invertebrate, defined, 606
Ion, defined, 349, 606
Ion exchange, defined, 606
Ionized ammonia (NH_4^+), RAS, 451
Ions, water requirements, 349
Irrigation
   water tests, 355
   wastewater treatment, 381, 382f
Isotonic, defined, 606
Isthmus, defined, 606

J

Jacobi, Stephan, 8
Japanese kelp (*Laminaria japonica*), 20
Job interview
   careers, 562–563
   do's/don'ts, 562, 563
Job placement, types of, 556
Jobbers, distributors, 65

K

Kelp (brown macrocysis), 245
Kidney, defined, 606
Kilogram calorie, defined, 606

Koi
  ornamental, 189t, 190, 194
  production plant, 195f
Kype, defined, 606

L

Lake trout, 32t
Largemouth bass, 32t, 35f
  spawning and egg production, 40t
Larvae, 18, 606
Larval fish
  common feeding practices, 283
  three groups, 296
Lateral band, defined, 607
Lateral line, 607
  finfish, 44f
  water vibrations, 43
Lateral, defined, 607
Leadership, soft skills, 542
Learning skills, thinking skills, 539
Least-cost formula
  catfish, 283
  feed/feeding, 279
Least-cost method, defined, 607
Legal risk, aquaculture business, 505
Lender
  preparing for, 522
  selecting, 521–522
Lernaea cyprinacea, 322, 323f
Lesion, defined, 607
Lethargy, 607
  health management, 311
  Trichodiniasis, 320
Letter of application, 537, 559, 560, 607
Letter of inquiry, 537, 559, 560, 607
Leucocyte, defined, 607
Levee, defined, 607
Levee-type pond
  aquatic structures, 391, 392f, 393t, 393, 397
  construction parameters, 399–401, 400f, 401f
  drains, 403, 403f
  self-loading pans, 402f, 402
Liabilities, defined, 607
LifeKnowledge®, 542–545
Light
  aquarium supplies, 433, 434f, 435
  RAS monitoring, 459
  water requirements, 367
Lime, defined, 607

Liming
  pH balance, 351
  water requirements, 349
  water test, 356
Line breeding, defined, 607
Linoleic acid, essential fatty acid, 260, 260f, 607
Lipids (fat), 253, 607
Liquidity, 518, 607
Listening competencies, 538
Live bearers, 607
  aquarium fish, 431, 432f
  ornamental fish, 188t, 191, 196
Live haulers, 61, 68, 70, 607
  minimum load, 107
Live-car, 391, 419f, 419–420, 607
Liver, 253, 255f, 255t, 607
Lobster, 34t, 219f, 219–220
Logarithm, defined, 607
Long bath, defined, 607
Lymphocystis disease, defined, 607

M

Macrocystis (kelp), 245
Macrominerals, 267, 268t, 607
Macrophytes, 49. *See also* Aquatic plants
Magnesium (Mg^{2+}), ionic component, 354
Maintenance, RAS, 456–457
Malaysia
  prawn culture, 20
  shrimp culture, 215
Malaysian prawn, 34t
Malignant, defined, 607
Malnutrition, defined, 607
Management
  careers, 550–551
  stress/disease prevention, 316
Mandible, defined, 607
Mantle, 607
  mollusks, 47, 48f
Manuring, food fish production, 130, 131f
Marginal macrophytes, 49
Mariculture, saltwater, 4, 5, 607
Marine, defined, 607
Marine finfish, 31t–33t
Marine plant aquacrops, 243–245. *See also* Seaweeds
Maritime, defined, 607
Market, 607
  niches, 465, 469, 607
  principles, sustainability, 468–469
  risk in aquaculture business, 504

Marketing, 57, 85, 607. *See also* Processing and marketing
  activities, 64–68
  careers, 552
  and competition, 57
  costs, 68
  crawfish, 208
  flowchart, 63f
  key plan elements, 60
  planning, 60–64
  selection, 66
Marketplace, global, 527, 528f
Markets, 19, 39
Masoten (Dylox), 322, 323, 324
Mass selection, defined, 608
Matching funds, defined, 608
Mathematic competencies, 538
Mating systems, defined, 608
Maxilla, defined, 608
Maya, Mesoamerica, irrigation systems, 9
McDonald hatching jar, 133f
Meader, Warren, 15
Meal, catfish feed form/size, 283
Mean, defined, 608
Meat consumption, U.S., 59f
Mechanical damage, defined, 608
Mechanical distribution system, trout (Salmonids), 290, 291f
Mechanical filtration, types, 449–450, 450f
Median, defined, 608
Medium processing, 19
Melanophore, defined, 608
Menadione, defined, 608
Meristic characters, defined, 608
Metabolic oxidation (breakdown of food), 256
Metabolic rate, 608
  and energy, 257
Metabolism, defined, 608
Metabolizable energy (ME), 258
Metamorphosis, 608
  oyster larvae, 222
Methane (CH_4), water quality, 364–365
Methionine, amino acid, 261f
Methylene blue, defined, 608
Micorominerals, 267
Microbe, defined, 608
Microencapsulation, 608
  feed/feeding, 279
  fish feed processing method, 282, 283
Microminerals, defined, 608
Microohms, defined, 608
Micropyle, defined, 608

Microsporidean, 608
  golden shiner disease, 144
Migration, defined, 608
Migratory Bird Treaty Act (MBTA), 157
Milligram per liter (mg/l), 349
Milt, 91
Minerals, 253, 265, 267–269, 268t
  supplements, 279
Minimal processing, 19
Mission, strategic planning, 497–498, 498f
Mississippi, catfish production, 11f, 11–15, 15f
Mitigation requirements, game fish, 181
Mitosis, defined, 608
Moist processing method, fish feed, 282–283
Mollies
  aquarium fish, 431, 432f
  ornamental fish, 189f, 195
Mollusks
  anatomy, 47–48, 48f
  species, 34t
Molting, 608
  crustaceans, 204
  softshell, 45
Monoculture, 18, 608
  tilapia males, 127
Monogenetic flukes, fish diseases, 321–322, 322f
Monospores, algae, 246
Monthly temperature unit (MTU), defined, 608
Morbidity, defined, 608
Morbidity rate, defined, 608
Morphology, defined, 608
Mortality, 608
  ERM, 328
  health management, 311
Mortality rate, defined, 608
Motile aeromonas septicemia (MAS), defined, 608
Motivation, human resources, 524–525
Mottled, defined, 609
Mouth-brooders, 609
  aquarium fish, 432
  ornamental fish, 187
  tilapia, 123, 124f
Mucking, defined, 609
Mucus, 609
  fish infection barrier, 314
  health management, 311
  stress impact, 315
Mud line, 609
  aquatic equipment, 391, 418

Mumford, NY, first public hatchery in U.S., 19
Muscular system, aquatic animals, 42
Muskellunge, coolwater species, 152, 294
Mussels, 34t, 226f
Mutation, defined, 609
Mycology, defined, 609
Mycosis, defined, 609
Mycotoxins, 609
  feed analysis, 282
  feed/feeding, 279
Myomere, defined, 609
Mysis stage, 609
  shrimp larvae, 210f, 214
Myxobacteriosis, defined, 609
Myxobolus cerebralis, 324

N

Natal river, salmon run, 182
National Aquaculture Development Plan (1996), 20
National Broodstock Program, 181
National Fish Hatchery System (NFHS), 177–178
  hatcheries, 178t–180t
  stocking, 181
National Institute of Food and Agriculture (NIFA), 383
National Marine Fisheries Service (NMFS), processing plant inspection, 75–76
National Pollution Discharge Elimination System (NPDES), 383
National Postsecondary Agricultural (PAS) organization, 542
National Research Council (NRC)
  amino acid requirements, 262t
  essential fatty acid requirements, 260t
  fish nutrition, 253
  vitamin recommendations, 266t
National Sea Grant program, 11, 555
  species development, 21–22
Native Americans, aquaculture, 9
Natural foods, 609
  coolwater species, 294
  feed/feeding, 279
Natural pond, aquatic structures, 391, 393t
Natural processes/environmental contamination, fish feed, 272t
Natural Resources Conservation Service (NRCS), 383, 396, 404
Nauplius stage, 609
  shrimp larvae, 190, 210f, 214

Necropsy, defined, 609
Necrosis, defined, 609
Nematoda, defined, 609
Nematode, metazoa, 38t
Nephrocalcinosis, defined, 609
Nerve fibers, defined, 609
Nervous system, 43
Net/pen culture, 20
Net pens, aquatic structures, 391, 609
Net worth, 487, 518, 609
New England, fish and game commissions, 19
Niacin, defined, 609
Nitrate (NO_3^-), ion, 354
  water management, 374t
Nitrate-nitrogen, RAS, 447, 448, 452
Nitrification, 447, 451, 609
Nitrite (NO_2^-), 364–365, 609
Nitrogen (N_2), 364–365, 374t, 609
  RAS, 451–453
Nitrogen cycle, 362f
"No blood disease," 334
Nodule, defined, 609
Nonindigeous species, defined, 609
Noninfectious diseases, 610
  fish, 316, 333–337
Nonpathogenic organism, defined, 610
Nonpoint source pollution, defined, 610
Nori, 20
Norris, T., 9
Northern pike, coolwater species, 150, 150f, 294
Noxious, defined, 610
Nursery, defined, 610
Nursery ponds, fry transfer, 100–101
Nutria, 610
  alligator feeding, 238
Nutrients, 610
  aquarium supplies, 435–436
  fish diet, 279
Nutrition, 610
  carbohydrates, 259
  energy loses, 258, 258f
  energy requirements, 256–258
  fat (lipids), 253, 259–260, 261t
  fish digestive system, 253–256
  minerals, 265–269, 268t
  protein, 253, 261–263, 263t
  stress/disease prevention, 316
  vitamins, 263–265, 266t
Nutritional deficiency, fish disease, 334, 335f
Nutritional gill disease, defined, 610

O

Objectives, strategic planning, 498, 498f
Obsolescence risk, aquaculture business, 504
Ocean ranching, 20, 610
Odor classifications, 368t
Odor, water requirements, 367
Offal, defined, 610
Offal collector, 71
Off-flavor seafood, 80, 81–82, 460, 610
　testing, 82
Office of Programs (OCIP), USDA, 80
Oils, water quality tests, 366
Omnivores, 42, 254, 610
On-line, defined, 610
Open-formula feed, defined, 610
Operculum, defined, 610
Opportunistic, defined, 610
Opportunities identification, strategic planning, 498f, 499
Optic, defined, 610
Organic, defined, 610
Organic matter
　breakdown products, 365
　surface water, 349
　water requirements, 349
Organizing, aquaculture business, 493f, 494
Ornamental fish, 29, 33
　culture methods, 190
　diseases, 191
　feeding, 190
　Florida culture, 189t, 197
　habitat, 187
　harvesting and yields, 191
　market, 19
　processing and marketing, 191
　seed stock and breeding, 187
　types, 188t, 189f
　U.S. production, 189t
Oscars, ornamental fish, 195–196
Osmoregulation, 610, 257
Osmosis, defined, 610
Osmotic pressure, defined, 610
Outfall, defined, 610
Ovarian fluid, defined, 610
Ovaries, defined, 610
Overflow pipe, defined, 610
Over-hydrate, stress response, 314
Overt disease, defined, 610
Oviduct, defined, 610
Oviparous, defined, 610
Ovulate, defined, 610
Ovum, defined, 610
Ownership changing, 65
Oxidation, 610
　fats, 259
Oxygen
　catfish requirements, 288
　dissolved, 349, 358–360, 359f, 374t
　solubility tables, 573t–574t
　testing equipment, 421, 421f
Oxygen starvation, 333
Oxygen transfer efficiency, defined, 611
Oxygenation, 611
　RAS, 447, 454–455
Oxytetracycline (terramycin), 327, 329, 611
Oysters, 17–18, 34t, 220f
　culture method, 222
　diseases, 222–223
　feeding, 222
　habitat, 221
　harvesting and yields, 223
　life cycle, 221f
　processing and marketing, 224
　seed stock and breeding, 221
　U.S. cultured, 21f

P

Pacific Northwest, oyster hatcheries, 21
Pacific salmon, 182–183
　amino acid requirements, 262t
Packaging, 65, 73–74
Paddlefish, 129
Paddlewheel, 611
　aerators, 416f, 416–418, 417f
　structures, 391, 416f, 416–417, 417f
Pancreas, defined, 611
Pantothenic acid, defined, 611
Para-aminobenzoic acid (PABA), defined, 611
Paramecium, plankton, 37t
Parasites, 611
　diseases, 314, 317, 318
　infection impact, 315–316
Parasiticide, defined, 611
Parasitology, defined, 611
Parr, defined, 611
Parr mark, defined, 611
Partnerships, 487, 506, 507, 611
Parts per billion (ppb), 349, 611
Parts per million (ppm), 349, 571t–572t, 611
Parts per thousand (ppt), 349, 352, 611

Pathogens, 310, 611
　types, 314
Pathology, defined, 611
Payback, defined, 611
Payroll taxes, 508
Pectoral fins, defined, 611
Peeling plant, defined, 611
Pellet binders, in fish diets, 270
Pelvic fins, defined, 611
Pen spawning, 101, 102f
Penduncle disease, defined, 611
Pens
　alligator grow-out, 237t
　　frogs, 241f
Peptide bonds, 261, 611
Percent, conversion factors, 571t–572t
Percolation, defined, 611
Percolation aquatic structures, 391, 397
　wastewater treatment, 381, 382
Peritoneum, defined, 611
Perivitelline fluid, defined, 611
Perivitelline space, defined, 611
Permeability, defined, 611
Peroxide danger, selenium protection, 268
Pest control, sustainability, 476, 478
Pesticides, 612
　in ponds, 399
　water quality tests, 365, 366
Petechia, defined, 611
Pets, market, 19
Pfiesteria, fish disease, 335
pH, 612
　RAS, 453–454
　water management, 374t
　water quality, 350–351, 351f
　water requirements, 349
Pharynx, 253, 255f
Phenotypes, 160, 612
Phenylalanine, amino acid, 261f
Phosphate (PO_4^{3-}), ionic component, 354
Phosphorus, 267
Photoperiods, 612
　spawning time, 91
Photosynthesis, 48, 304, 612
　chemical formulas, 49f
Phycocolloid, carrageen, 29, 612
Physiology, defined, 612
Phytin, seed phosphorus, 267, 612
Phytoplankton, 29, 35f, 474, 612
　aquariums, 430, 435
Pig/fish farming, 130, 131f
Pigment, defined, 612

Pigmentation, defined, 612
Pigments, in fish diets, 270
Pilot scale aquaculture, 22
Pinchon, Dom, 8
Pink salmon, 183
Pituitary, defined, 612
Plankton, 37f, 612
　coolwater species, 294
Planktonic organism, feed/feeding, 279
Planktonic plants, 304
Planning
　aquaculture business, 491, 493, 493f
　human resources, 523–524
Planting of fish, defined, 612
Plants, aquarium supplies, 436, 437f
Plasma, defined, 612
Plasticity, defined, 612
Platty, aquarium fish, 431, 432f
Poikilothermic, defined, 612
Point of source pollution, defined, 612
Poisoning, fish disease, 336
Polliwog, 241, 612
Pollutant, defined, 612
Pollution, 381, 612
　water sources, 475t
Polyculture, 7t, 39, 612
　tilapia, 127–128
Polysaccharide, 612
　carrageen, 29
Polytrophic, defined, 612
Pompano, 151f
Pond culture, shrimp, 212
Pond fertilization, water management tests, 376–377
Pond plankton metazoa, 38f
Pond run, 612
Pond spawning, 100
Pond water, dichotomous key, 582t
Ponds. *See also* Levee-type ponds
　aquatic structures, 390, 391
　as asset, 514f
　cage culture, 409f
　construction costs, 397
　design assistance, 404
　design/construction, 391, 393
　drainage, 398, 398, 402–404, 403f
　flooding, 398
　levee-type construction, 399–401
　orientation, 401
　other consideration, 395, 396
　pesticides, 399
　site preparation, 401

site selection, 397
size, 391–392, 399
soil characteristics, 394, 397
topography, 395
utility rights-of-ways, 398
water supplies, 393–394, 397
Pond-to-plate inspection program, 80
Population, defined, 612
Population density, defined, 612
Portal of entry, defined, 612
Portal of exit, defined, 612
Posterior surface, aquatic animals, 40, 41f
Posterior, defined, 612
Posthodiplostomum (white grub), 324
Post-treatment, defined, 612
Potassium (K^+), ionic component, 354
Potassium permanganate, 320, 321, 322, 323, 326, 328, 330, 613
Pox, defined, 613
Pox disease, defined, 613
Prawns, 34t, 209. *See also* Shrimp
  culture method, 210
  diseases, 211
  feeding, 211
  habitat, 209
  harvesting and yields, 211–212
  processing and marketing, 212
  seed stock and breeding, 210
Precipitate, defined, 613
Predators, 254, 613
Premixes, 279, 613
Pretreatment, defined, 613
Prevention
  acidosis, 334
  alkalosis, 334
  anchor parasites, 323
  anemia, 337
  Asian tapeworm, 319
  bacteremia, 327
  brown blood disease, 336
  CCVD, 332
  Chilodonelliasis, 330
  Columnaris, 328
  Costia disease, 323
  ERM, 328
  ESC, 329
  fish grubs, 324
  fish lice, 324
  fish nutritional deficiency, 335
  fish poisoning, 336
  fungal infections, 326
  gas bubble disease, 337

goldfish ulcer, 329
Ich, 321
IHN, 333
IPN, 333
monogenetic flukes, 322
oxygen starvation, 333
Trichodiniasis, 320
whirling disease, 325
Primary producers, defined, 613
Principal, defined, 613
Probiotics, 613
  sustainability, 465, 467
Problem solving, thinking skills, 539
"Processed Under Federal Inspection" (PUFI) mark, 76
Processing, 3, 19, 65, 82, 85, 613
  cuts, 83f
  enrobing, 84
  steps, 70–74
Processing and marketing
  catfish, 107–108
  hybrid striped bass, 137
  ornamental fish, 191
  oysters, 224
  prawns, 212
  salmon, 122
  shrimp, 218
  tilapia, 131
  trout, 118
Processing crops, careers, 552, 553f
Processors, buyer type, 60, 61, 68, 69
Product
  characteristics, 61
  forms, 73, 83f
  promotion, 66
Product pull, 62–63, 613
Product push, 63, 613
Production ponds, 106, 613
Production risk, aquaculture business, 504
Profit, 469, 469f, 613
Profitability, 66, 613
  aquaculture business, 487
  financial analysis, 518–519
  record keeping, 508–509, 509t, 510t
Progeny test, defined, 613
Prognosis, 613
  disease treatment, 339
  health management, 311
Prolific, defined, 613
Prolonged treatments, fish disease, 340
Promotion, 68, 613
  careers, 552–553

Propagation, defined, 613
Propeller-aspirator pump, aquatic equipment, 418
Prophylactic, defined, 613
Proportion, conversion factors, 571t–572t
Protandrous mollusks, 48, 613
Protective barriers
  effect of stress, 315
  fish, 314, 315f
Protein, 253, 261–263, 263t, 613
  alternative sources, 265
  fish diets, 279
  sources of, 281, 281t
Protozoa, 317, 320, 323, 323f, 324, 613
Proximate composition, 613
  feed analysis, 282
  feed/feeding, 279
Pseudobranch, defined, 613
Pseudomonas fluorescens, 327
Pseudomonas septicemia, defined, 613
Pump sprayer aerators, 416
Pumped water, water source, 372, 372f
Pumps, aquatic equipment, 415f, 415–416
Purging, defined, 613
Pus, defined, 613
Pylorus, defined, 613
Pyloric cecum, 255f, 255t
Pyridoxine, defined, 613

Q

Quality assurance (control), 68, 75, 80, 613
  procedures, 80–81
  procedures and functions, 75t
  project plan, 613–614
Quantity, direct/processing markets, 67
Quarternary ammonium compounds, defined, 614
Quick lime (Ca(OH)$_2$), 360

R

Raceways, 614
  aquatic structures, 390, 404–405, 405f
  bird predation netting, 157f
  catfish raising, 98f
  construction, 406–407
  trout raising, 114
Rainbow trout, 15, 32t, 109f
  amino acid requirements, 262t
  common feeding practices, 283, 288–289
  first feeding, 299
  IHN, 332
  life cycle, 113f
  water management, 378
  requirements, 367, 374t
Ranching, 614
  salmon, 119
Random mating, defined, 614
Ration, defined, 614
Ray, defined, 614
Reading competencies, 537, 539f
Rearing unit, defined, 614
Reasoning, thinking skills, 539
Receiving, processing, 70
Recessive, defined, 614
Reciprocal mating, defined, 614
Recirculating aquaculture system(RAS), 446
  advantages, 447
  disadvantages, 447
  fish varieties, 456
  start-up, 456
  system components, 448–456, 449f
  system design, 447–448, 448f
  system management, 456–461, 457f, 460t, 461t
Recirculating systems
  aquatic structures, 390
  monitoring log, 457f
Reconditioning treatment, defined, 614
Record keeping
  balance sheet, 512, 513f, 514
  cash-flow statement, 515
  farm earnings statement, 514–515
  methods, 511
  tax records, 512
  tips for, 512
Recreational fishing
  benefits, 177
  coho salmon, 182–183, 183f
  equipment, 169–172
  fish types, 174–175
  history, 169
  impact, 175–176
  methods, 172–174
  Pacific salmon, 182–183
  top 10 states for, 176
Recruitment, defined, 614
Rectum, defined, 614
Recycle, defined, 614
Recycling, sustainability, 475–476
Red algae, 244
Red drum, 32t, 144, 145f
  management, 145–147
Redd, defined, 614

Redmouth disease, defined, 614
Reed-mace, 50f
Reel, 169, 171, 614
Regeneration, 614
  crustacean limbs, 46
Regulation
  government web addresses, 578t–579t
  water management, 383
Regulatory Fish Encyclopedia (FDA), 583t–587t
Relift pump, aquatic equipment, 391, 422, 422f, 614
Remy, Joseph, 9
Repayment capacity, financial analysis, 519, 518f
Reproduction. *See also* Spawning
  aqua species, 39
  aquatic animals, 43
  aquatic plants, 49
  environmental stimulation, 91
  hormone control, 91
Research, needs, 23–24
Research and development, careers, 554, 554f
Resistance, defined, 614
Resources, 614
  careers, 537, 540
Respiration, 614
  chemical formulas, 49f
  purpose of, 304
Respiratory system, 42
Résumé, defined, 614
Résumés, careers, 537, 556, 560–561, 561f
Retail sales, 66, 70, 614
Return, defined, 614
Rice production, 242
Riparian area, defined, 614
Ripe, defined, 614
Risk, 614
  aquaculture business, 487, 504
  managing, 526
  sustainability, 465, 469–470
Risk assessment, 614–615
  sustainability, 465, 470
Risk avoiders, 505
Risk Management Agency (RMA), 470
Risk takers, 505
Rocks, aquarium supplies, 433, 434f
Rod, 170, 615
Roe, defined, 615
Roman aquaculture, 8
Rotating biological contactors (RBC), RAS, 453, 453f
Rotational line-crossing, 161, 161f, 615

Rotifers, 34, 38f, 615
Runoff, defined, 615

S

Sac fry, 615
  hatching trough, 103–104
Safe concentration, defined, 615
Sagittaria, aquarium plants, 436–437, 437f
Salinity, 5, 615
  and DO, 359f
  water management, 374t
  water requirements, 349, 352
Salmon, 29, 31t, 32t, 118f. *See also* Pacific salmon
  diseases, 121–122
  farming, 120
  feeding, 121
  habitat, 119
  harvesting and yields, 122
  life cycle, 121f
  processing and marketing, 122
  ranching, 119–120
  spawning and egg production, 40t
  species characteristics, 120t
  types, 119f
  U.S. cultured, 21f, 57
Salmonella enteritis, 77
Salmonids, 29, 108, 615. *See also* Salmon, Trout
  common feeding practices, 283, 288–289
  feed form/size, 289
  feeders, 289–290, 290f, 292f
  feeding rates, 291–293
  first feeding, 298, 298f, 299
  specialty feeds, 292, 293f
Salt, 336, 615
  disease treatment, 323
"Salt content," 349
Salts
  simple ions, 349
  water requirements, 349
Saltwater, defined, 615
Sample count, defined, 615
Sample, defined, 615
Sanitation, stress/disease prevention, 316
Sanitizer, defined, 615
Saprolegnia genus, 326
Saprolegniasis, defined, 615
Saprophyte, defined, 615
Saturated, defined, 615
Saturated fats, 259

Saturation, 615
  DO, 359–360
  water requirements, 349
Scale formula, defined, 615
Scales, 615
  fish infection barrier, 314
  stress impact, 315
Science, 615
  sustainability, 465, 466–468, 468f
Scientific method, 466, 467f
Sea Grant program, 11, 555
  species development, 21–22
Seasonal feeding behavior, 256
Seaweeds, 243–244
  culture, 245
  uses of, 244–245
Secchi disk, *357*, 374t, 375t, 615
Second dorsal fin, defined, 615
Secondary, defined, 615
Secondary invader, defined, 615
Secondary problem
  fish disease, 317
  health management, 311
Sedentary, defined, 615
Sedentary fish, metabolic rate, 257
Sediment, defined, 615
Sedimentation pond, defined, 615–616
Sediments, water management tests, 376
Seed (young), 18, 37, 616
Seed stock, defined, 616
Seed stock and breeding
  alligators, 234–236
  catfish, 98–104
  crawfish, 201
  hybrid striped bass, 133
  ornamental fish, 187
  oysters, 221
  prawns, 210
  salmon, 119
  shrimp, 213–214
  tilapia, 123–125
  trout, 110–113
Seeding, defined, 616
Seepage, 616
  water management, 374t, 378
Seines, 616
  aquatic equipment, 391, 395, 418, 418f
Seining, 136f
Selective breeding, 160–162, 616
  prawns, 210
Self-esteem, workplace competencies, 541, 542

Self-feeders, 17, 616
Self-loading pans, 402, 402f
Self-management, workplace competencies, 541, 542
Semi-moist processing method, fish feed, 282–283
Semipermeable, defined, 616
Semipermeable membranes, gills, 45
Sensory feeding behavior, 256
Sensory receptors, defined, 616
Sensory system, 43
Septicemia, defined, 616
Serum, defined, 616
Settleable solids, 616
  RAS, 447, 448–449
Settling pond, 616
  wastewater treatment, 381, 382
  water requirements, 349
Sex determination, defined, 616
Sexing, defined, 616
Shareholders, 616
  aquaculture business, 487, 507
Shelf life, 616
  and chilling, 73
Shelf-space incentives, 63
Shell erosion, lobsters, 220
Shocking, defined, 616
Short bath, defined, 616
Shrimp, 18, 34t, 212, 213f. *See also* Prawns
  culture method, 214–215
  diseases, 217–218
  feeding, 216–217, 217t
  gender in life cycle, 47
  habitat, 212–213
  harvesting and yields, 218
  life cycle, 210f
  processing and marketing, 218
  seed stock and breeding, 213–214
  species, 212
  stocking rate, 215–216
Side effect, defined, 616
Signs, 616
  ERM, 328
  IHN, 332
  acidosis, 334
  alkalosis, 334
  anchor parasites, 323
  anemia, 337
  Asian tapeworm, 319
  bacteremia, 327
  brown blood disease, 336
  CCVD, 331, 332f

Chilodonelliasis, 330
Columnaris, 327
Costia disease, 323
ESC, 329
fish grubs, 324
fish lice, 323
fish nutritional deficiency, 334
fish poisoning, 336
fungal infections, 326
gas bubble disease, 337
goldfish ulcer, 329
Ich, 320
IPN, 333
monogenetic flukes, 321
oxygen starvation, 333
Pfiesteria, 335
Trichodiniasis, 320
whirling disease, 325
Silica, ionic component, 354
Silos, 616
    aquatic structures, 390, 391, 407f, 407–408
Silt, defined, 616
Silver carp, 7t, 8
Single-cell algae, 246
Single-pass system, defined, 616
Sinking pellets
    catfish feed form/size, 283, 284
    tilapia, 295
Sinuses, defined, 616
Siphon, mollusks, 48, 48f, 616
Size grading, 73
Size/maturity, marketing considerations, 67
Skeletal system, aquatic animals, 42
Skin
    fish infection barrier, 314
    stress impact, 315
Skinning, processing, 72
Slope, defined, 616
Slopes, aquatic structures, 391, 395, 401–402, 402f
Sludge, defined, 616
Slurry, defined, 616
Smolt, defined, 616
Snails, 226, 227f
    management, 227–228
Snake River Canyon
    aquifer, 17f
    trout farming, 15
Snake River sturgeon, 152
Snake River Trout Company, 17
Snake River Valley, 40

Snake River, Idaho, sockeye salmon, 182
Snatch block, defined, 616
Snout, defined, 616
Sociability, 616
    workplace competencies, 537, 541
Social responsibility, sustainability, 478–479
Sock, 391, 419f, 419–420, 616
Sockeye salmon, 183
Sodium (Na^+), ionic component, 354
Sodium absorption ratio, water test, 355
Sodium chloride ($NaCl$), 336
    water quality, 352
Sodium percentage test, 355
Soft fins, defined, 616
Soft rays, defined, 616
Soft shell crawfish, 204
Soft skills, 542
Soft-egg disease, defined, 616
Soft-shelled crabs, 218
Soil
    conservation sustainability, 472–473
    pond characteristics, 394, 397
    water requirements, 366, 374t, 375t
Sole proprietorship, 617
    aquaculture business, 487, 506, 507
Solubility, 617
    water requirements, 349, 352
Solvency, 617
    aquaculture business, 487, 518
South Carolina, Malaysian prawn culture, 20
Southeastern Fish Cultural Laboratory, Alabama, 10
Soybean meal, fish diets, 265
Spat, 617
    young oyster, 221
Spawn, 8, 617
    channel catfish, 99
Spawning, 91, 40t, 617
    containers, 99 100f
    methods, 100
Spawning net, 617
Speaking competencies, 538
Specialty feeds, trout, 293, 293f
Species, 617
    market channels, 67
Spent, defined, 617
Spermatozoon, defined, 617
Spinal cord, defined, 617
Spiny rays, defined, 617
Spirostomum, plankton, 37t

Spirulina, 246
Spleen, defined, 617
Spores, defined, 617
 aquatic plants, 49
Sporophytes, 617
 mature algae, 246
Sport Fish Restoration and Boating Trust Fund, 177
Sport fishers, by gender and age, 176f
Sport fishing. *See also* Recreational fishing
 bait market, 19
 for release market, 19
Spray vaccination, 342, 343
Stabilization pond, defined, 617
Staffing, aquaculture business, 493f, 494, 523–524
Stand, aquarium supplies, 433, 434f
Standard environmental temperature (SET), defined, 617
Standard solutions preparations, 572t
Standing crop, 617
Standing crop, tilapia, 128
Starch to glucose, 259f
Steam pelleting, fish feed processing method, 282
Stenohaline marine fish, defined, 617
Stentor, plankton, 37t
Sterilant, defined, 617
Sterilize, defined, 617
Stew ponds, Middle Ages, 8
Still fishing, 172–173, 173f
Stimulants, in fish diets, 270–271
Stimuli, defined, 617
Stock, defined, 617
Stocker, defined, 617
Stocking, 617
 game fish, 181
Stocking rate
 alligators, 237
 catfish, 104–105
 crawfish, 205
 defined, 617
 hybrid striped bass, 134–135
 salmon, 121
 shrimp, 215–216
 tilapia, 128
 trout, 114–115
Stomach, defined, 617
Stone, Livingston, 9
Storage, 65, 74
Strainers, 256, 617
Strains, defined, 617

Strategic planning, 617
 aquaculture business, 487, 496–497, 497f
 future change, 527
 steps, 497–501, 498f, 502t, 503
Strategy selection, strategic planning, 500
Stratification, 617
 aquatic equipment, 391, 415
Stress, 618
 impact on protective barriers, 315–316
 prevention, 316
 RAS monitoring, 459
Stressors, 618
Stressors, fish, 312, 313t
Striped bass. *See also* Hybrid striped bass
 common feeding practices, 283, 298
 first feeding, 298
 U.S. production, 57
Striped mullet, 150
Stripping, defined, 618
Stun, defined, 618
Stunning, handling, 70
Sturgeon, 33t, 129, 151–153, 152f
Subacute, defined, 618
Substrate, 618
 RAS, 447
Suckermouth catfish, 192–193
Suckers, 256, 618
Sulfadimethoxine sulfonamide, 618
Sulfaguanidine, 618
Sulfamerazine, 328, 618
Sulfate (SO_4^{2-}), 354, 618
Sulfisoxasole, 618
Sulfonamides, 618
Sunfish, 153, 153f
Supersaturation, defined, 618
Supervised agricultural experience (SAE), 554–555
Supplemental diet, defined, 618
Suppliers, 60, 618
Surf fishing, 5f
Surface feeders, 41
Surface runoff, defined, 618
Surface spray pump, aquatic equipment, 415f, 415–416
Surface water, water source, 372
Susceptible, 618
 health management, 311, 314
Suspended solids, 618
 RAS, 447, 448–450
 water quality test, 356
Sustainable, defined, 465, 618
Sustainable aquaculture, 465

aquaponics, 465, 476, 479–481, 480f
scoreboard, 481t, 481–482
standards, 466–478
Swim bladder, defined, 618
Swimmerets, defined, 618
Swim-up, defined, 618
"Swim-up fry," 104
Switching costs, 62
Sword plants, aquarium plants, 436, 437f
Swordtail, 189f, 196, 431, 432f
Syndrome, defined, 618
Synthesize, defined, 618
Systems, 618
  careers, 540, 541

T

Tackle, defined, 618
"Tackle," 169
Tadpoles, 236, 240f, 241, 618
Tank culture, Tilapia, 297
Tanks, aquatic structures, 390, 406f, 408
Tapeworms, parasitic disease, 318, 318f, 319
Taverner, John, 8
Taxation, aquaculture business, 507–508
Teams, 540f
Technology, careers, 540, 541
Temper, defined, 618
Temperature
  and DO, 359f
  RAS, 454
  water management, 373t
  water requirements, 351–352, 375t
Temperature shock, defined, 618
Temperature stress, impact of, 316
Tempering, defined, 618
Tender stage, defined, 618
Terramycin. See Oxytetracycline
Terrestrial, defined, 619
Territorial nature, frogs, 241
Testes, defined, 619
Tetras, 188t, 189f, 196
Therapeutic agents, 319
Therapeutics, 619
  health management, 311
Thermal stress, 619
  water requirements, 349, 352
Thermocline, defined, 619
Thermometer, aquarium supplies, 433, 434f
"Theronts," 321
Thiamine, defined, 619
Thiosulfate, 619
Thorax, defined, 619

Threats identification, strategic planning, 498f, 499
Tilapia, 33t, 122f
  amino acid requirements, 262t
  cage culture, 295, 296, 296t
  common feeding practices, 283, 294–297, 296t
  cuts, 83f
  diseases, 131
  feeding rates, 296t, 296–297
  feeding, 128–130
  habitat, 123
  harvesting and yields, 131
  low on food chain, 39
  male monosex culture, 127
  mixed-sex culture, 125–126
  mouthbrooding, 123, 124f
  polyculture 127–128
  pond culture, 125
  processing and marketing, 131
  RAS, 456
  seed stock and breeding, 123–125
  species, 123
  stocking rate, 128
  tank culture, 297
  U.S. production, 57
Tingey, Jack, 15, 16
Tissue residue, defined, 619
Titrant, 375, 619
  water requirements, 349
Titration, defined, 619
Titrimetric analysis, 619
  water requirements test, 349, 375
Tocopherol, defined, 619
Toggle, defined, 619
Tolerance, defined, 619
Topical, defined, 619
Topography, defined, 619
  aquatic structures, 391, 395, 398
Topping, defined, 619
Total alkalinity test, water quality, 355–356
Total ammonia-nitrogen (TAN), RAS, 451
Total dissolved solids (TDS), 619
  water tests, 353–354, 375t
Total hardness test, water quality, 355–356
Total harvest, defined, 619
Total organic carbon test, 365
Total solids (TS), 619
  water tests, 353
Toxic metabolites, recirculating water, 114
Toxicity, defined, 619
Toxicology, defined, 619

Toxin, 619
  fish pathogen, 314
Toxins
  fish feed contamination, 271, 272t
  fish poisoning, 336
Trace elements tests, 357–358
Trace minerals, 269
Tractor power takeoffs (PTOs), 415
Tractors, aquatic equipment, 422, 422f
Trademark, defined, 619
Training
  human resources, 524
  sources of, 529
  web sites, 582
Trammel seine, defined, 619
Transmission, 619
  health management, 311, 316
Transplanting, defined, 619
Transport equipment, fish, 420–423, 422f
Transporting, 64
Traps, crawfish, 207f
Trauma, defined, 619
Treaty, defined, 619
Treatyse of Fysshynge wyth an Angle (Berners), 169, 172
Trematoda, defined, 619
Trend analysis, financial analysis, 519
Trichodina, 320, 320f
Trichodiniasis, 320
Trout, 9, 10, 29. *See also* Rainbow trout
  amino acid requirements, 262t
  cash-flow budget, 502t
  common feeding practices, 283, 288–289
  culture method, 114
  dietary fat requirements, 261t
  diseases, 117–118
  enterprise budget, 489, 492t
  feed form/size, 289
  feeders, 289–290, 290f, 292f
  feeding, 117
  feeding rates, 291
  first feeding, 298, 298f, 299
  grading, 115–116
  habitat, 110
  harvesting and yields, 118
  high on food chain, 39
  and Idaho, 15–17
  inspection, 78–79
  inventory, 116–117
  predator, 255
  processing and marketing, 118
  production 1933–2008, 16t
  protein feed levels, 263t
  spawning and egg production, 40t, 110
  specialty feeds, 292, 293f
  species, 31t–33t, 108, 109t, 109f
  stocking rate, 114–115
  U.S. cultured, 21f
Trucks, transport equipment, 423, 424f
Tubercles, defined, 619
Tumor, defined, 620
Turbidity, 620
  water quality test, 356, 357f
  water requirements, 349
Turbulence, defined, 620
2-methylisoborneal (MIB), and off-flavor catfish, 82

U

U.S. Fish and Wildlife Service, 10, 177, 180–182
  Fish Health Centers, 181
  NFHS hatcheries, 178t–180t
U.S. Army Corps of Engineers, 383, 398
U.S. Geological Survey, 397
U.S. Joint Subcommittee on Aquaculture (JSA), 20
Ulcer, defined, 620
Ulcer disease, in goldfish, 329, 620
Ulcerative dermal necrosis, defined, 620
Under-stocking, defined, 620
Un-ionized ammonia (NH_3), RAS, 451, 452, 452f
Unisex culture, defined, 620
United Nations Food and Agricultural Organization (FAO), 470
United States
  aquaculture development, 9–10
  catfish and Mississippi, 10–15
  ornamental fish production, 189t
  production all species 1983–2005, 6f
  seafood consumption, 57, 60f
  species cultured, 21f, 59t
  traditional fisheries, 20
  trout and Idaho, 15–17
United States Pharmacopeia (USP), 620
Univalve mollusks, 48
Urbanization, 620
  sustainability, 465, 471
Urea, 620
Uremia, 620
Urinary bladder, 620
Urinary ducts, 620

Urogenital pore, 620
Uropod, defined, 620
Utility-right-of-ways, ponds, 398

V

Vaccination methods, fish immunization, 342–343
Vaccines, 620
    ERM, 328
    fish immunization, 342
    health management, 311
    licensed fish, 343t
Valine, amino acid, 261f
Valisnerias, aquarium plants, 436, 437f
Value-added, 620
    processing, 19
    products, 67
Variable costs, 620
    business, 487, 489, 489t, 490f
Vector, defined, 620
Vegetation control, water management, 378
Vegetation excess, 159
Vein, defined, 620
Velvet cichlid, 195–196
Vent, defined, 620
Ventral fins, defined, 620
Ventral, defined, 620
Ventral surface, aquatic animals, 40, 41f
Vertebrate, defined, 620
Vertical pump, aquatic equipment, 415
Vertical transmission, defined, 620
Viable, defined, 620
Vibriosis, defined, 620
Viral diseases, fish, 317, 331–333
Viral hemorrhagic septicemia (VHS), 620
Viremia, defined, 620
Virulence, defined, 620
Virus, 621
    or bacteria, 330, 331
    fish pathogen, 314
    health management, 311
Viscera, defined, 621
Visceral cavity
    fish grubs, 324
    health management, 311
Visualization, 621
    careers, 537, 539
Vitamin C, 329
    fish disease, 334
    deficiency in channel catfish, 264f
Vitamin D, defined, 621
Vitamin K, defined, 621
Vitamin premix, defined, 621
Vitamins, 253, 263–265, 266t, 621
    feed/feeding, 279
Viviparous, defined, 621
Volume, conversion factors, 571t
Volumetric, defined, 621
Volumetric estimation fry number, 104
Volvox, algae, 36f, 48
Vomer, defined, 621
Von Bayer trough, defined, 621
Vorticella, plankton, 37t
Vrasski, V. P., 9

W

Wakame (*Undaria pinnatifa*), 20
Walleye, coolwater species, 153–154, 154f, 294
Walton, Izaak, 169
Warehousing, 74
Warmwater
    aquaculture, 4, 10
    hatchery fish culture, 294
    species, 621
Waste products, sustainability, 475–476
Waste solids, RAS, 448–459
Wastewater treatment, 381–383
Water
    aquaculture facility, 348, 350f
    basic qualities, 352
    calculating needs, 368–371, 369f
    estimating needs, 368
    in fish diets, 269
    management components, 372
    measurement units, 349
    quality measurements, 373
    sources of, 372–373
Water column, defined, 621
Water conditions
    clarification, 114
    hatching, 101–102
Water disposal, water management, 381–383
Water fleas, 38f
Water hardening, defined, 621
Water hardness, 621
    water requirements, 349
Water management
    calculating treatments, 378–380
    methods summary, 373t–375t
    other factors, 376–378
    quality measurements, 373
    regulations/agencies, 383

test methods, 375–376
water disposal, 381–383
Water pollution sources, 475t
Water quality, 349, 621
  aquariums, 441
  RAS, 454t, 458
  sample log, 442f
  samples required, 575t
  stress/disease prevention, 316
  sustainability, 474–475
Water resource conservation, sustainability, 473f, 473–474
Water soluble vitamins, 263–264
Water supply and levels, 156
Water table, defined, 621
Water temperature, 351–352, 375t
  and breeding, 94
  and feeding, 257f
Water tests, water qualities, 350–358, 353f, 357f
Water treatment, 621
  calculations formula, 379
  calculations, 378–380
Watercress, 31t, 245, 246
Watershed, 621
  aquatic structures, 391, 393
Watershed pond, 393–395, 395f
Weight, conversion factors, 571t
Weir, defined, 621
Welch, Terry, 23f
Wetlands, 621
Wetlands, levee site, 398
Whirling disease, 324, 621
White bass, spawning migrations, 133
White grub, defined, 621
White spot disease, defined, 621
White sucker, baitfish, 142, 143f
Whitefish, 154, 155f
Wholesale, 65, 70, 621
Wind speed, water management, 375t
Winter feeding, catfish, 288
Withdrawal time, defined, 621
Work schedule, human resources, 524
Workers, careers, 551, 551f, 552f
Workforce, aquaculture business, 494
Writing competencies, 537

X

Xanthophyll, defined, 621

Y

Yellow grub, defined, 621
Yellow perch, coolwater species, 33t, 154–155, 155f, 294
Yersinia ruckeri, 328
Yolk, defined, 621
Yolk sac, 621
  feed/feeding, 279
  first feeding, 298

Z

Zhujiang Delta, China, 7
Zinc, 268
Zoeal stage, defined, 622
  shrimp, 210f, 214
Zooplankton, 29, 34, 35f, 622
Zoospores, defined, 622
Zygote, 43, 622